INDUSTRIAL MINERALOGY

MATERIALS, PROCESSES, AND USES

Luke L.Y. Chang, B.S., Ph.D.

Professor of Geology
University of Maryland
College Park, Maryland

Prentice Hall
Upper Saddle River, New Jersey 07458

Library of Congress Cataloging-in-Publication Data

Chang, L. L. Y. (Luke L. Y.)
   Industrial mineralogy : materials, processes, and uses / Luke L.Y. Chang.
      p. cm.
   Includes bibliographical references and index.
   ISBN 0-13-917155-X
      1. Industrial minerals. 2. Mineralogy. I. Title.

TN799.5 .C48 2001
549—dc21

2001042438

Editor in Chief: Sheri L. Snavely
Senior Editor: Patrick Lynch
Editorial Assistant: Sean Hale
Marketing Manager: Christine Henry
Executive Managing Editor: Kathleen Schiaparelli
Assistant Managing Editor: Beth Sturla
Art Director: Jayne Conte
Cover Designer: Bruce Kenselaar
Manufacturing Manager: Trudy Pisciotti
Assistant Manufacturing Manager: Michael Bell
Production Supervision/Composition/Art Studio: Carlisle Communications, Ltd.
Art Editor: Shannon Sims
Managing Editor, Audio/Video Assets: Grace Hazeldine
Vice President of Production and Manufacturing: David W. Riccardi

Prentice Hall

© 2002 by Prentice-Hall, Inc.
Upper Saddle River, New Jersey 07458

10   9   8   7   6   5   4   3   2   1

ISBN 0-13-917155-X

Pearson Education Ltd., *London*
Pearson Education Australia Pty., Limited, *Sydney*
Pearson Education Singapore, Pte. Ltd.
Pearson Education North Asia Ltd., *Hong Kong*
Pearson Education Canada, Ltd., *Toronto*
Pearson Educacíon de Mexico, S.A. de C.V.
Pearson Education—Japan, *Tokyo*
Pearson Education Malaysia, Pte. Ltd.

Contents

Preface

The purpose of this book is twofold. First, it is written as a textbook designed to introduce science and engineering students to the fundamentals of industrial minerals and mineral-based materials. The approach I have used is to outline basic mineralogical properties, geological occurrence, distribution of deposits, industrial processes, and uses. By means of this outline, an industrial mineral is fully defined and can be utilized to its full value.

Second, for professionals in mineral industries, this book can be used as a reference in research and development. Forty-two industrial minerals and mineral groups have been selected for this book. Each has distinct and characteristic properties, and all are common and important industrial minerals. For each industrial mineral, a list of pertinent references is given.

In selecting the minerals discussed in this book, I have generally followed, with some exceptions, the traditional pattern established by previous publications, especially those by the American Institute of Mining, Metallurgical and Petroleum Engineers; the Society for Mining, Metallurgy, and Exploration; and Industrial Minerals Information Ltd. Ore minerals and industrial minerals with principal uses as chemical raw materials are excluded from the text.

In this book, I have first given a brief description of the crystal structure of each mineral. This is of foremost importance because crystal structure determines both the physical and chemical properties of solid compounds including industrial minerals. Next, for each industrial mineral, I have given a short overview of geological occurrence and distribution of deposits with an emphasis on major deposits in the United States, as well as in other mineral-producing countries. In this section, some classic publications are cited, because not only do they give comprehensive evaluations of the deposits discussed, but also because the scientific merit of these publications has withstood the test of time. Finally, I have included a discussion of industrial processes and ore beneficiation for each industrial mineral. It is well-known that industrial processes are dynamic. Modifications adopted by one plant may not be adequate in another plant, and procedures adopted at one time for one reason may have to be eliminated later for other reasons. Both common and generally accepted processes are described in this book. They may serve as guidelines for the application or design of industrial processes.

For industrial uses, efforts are made to give a full account of each industrial mineral. Certain aspects of the industrial minerals industry are not discussed in this book. These include supply systems, prices, transportation, taxes and tariffs, governmental regulations, and future trends. The interested reader can find discussions of these topics in many texts on mineral economics.

It is my sincere hope that this book can achieve its intended purposes: as an aid to teaching and research and development, and to make a contribution to the mineral industries.

Luke L. Y. Chang
College Park, Maryland

Acknowledgments

The author is grateful to the following publishers, and in some cases the individual authors, for their permission to reproduce the following copyrighted material: **(1)** Springer-Verlag New York, Inc., New York, NY for Fig. 99 from *Structural Chemistry of Silicates* (Liebau 1985); **(2)** Kluwer Academic Pub., Inc., Norwell, MA for Fig. 102a from *Zeolite Microporous Solids* (Newsan 1992); **(3)** McGraw-Hill, Inc., New York, NY for Fig. 9 from *Clay Mineralogy* (Grim 1968); **(4)** Academic Press, Inc., Orlando, FL for Figs. 54 and 56 from *Solid Lubricants and Self-Lubricating Solids* (Clauss 1972); **(5)** CRC Press, Inc., Boca Raton, FL for Figs. 102b and 107 from *Zeolite Catalysis: Principle and Application* (Bhatia 1989); **(6)** Economic Geology Publishing Co., Inc., Littleton, CO for Figs. 48 (Hewitt 1966); 49 (Beukes 1973); 60 (Riggs 1979); 61 (Notholt 1979); 67 (Worsley and Fuzesy 1979); and 82 (Kennedy 1950) from *Economic Geology;* **(7)** Elsevier Science Ltd., Oxford, UK for Fig. 6 from *Developments in Soil Science* (Valeton 1972); Fig. 30 from *Mineral Processing* (Pryor 1955); Fig. 46 from *Flotation: Theory, Reagents and Ore Testing* (Crozier 1992); Figs. 78 and 79 from *Palygorskite-Sepiolite: Occurrence, Genesis and Uses* (Weaver 1984); Fig. 17 (Christ et al. 1867), Fig. 19 (Rearden and Armstrong 1987), and Fig. 91 (Jones et al. 1977) from *Geochimica Cosmochimica Acta;* **(8)** Geological Society Publ. House, Bath, UK for Fig. 36 from *Jour. Geol. Soc. London* (Robertson 1975); **(9)** Industrial Minerals Information, Ltd., London, UK for Figs. 7, 32, 35, 40, and 59 from *Industrial Minerals, Geology and World Deposits* (Harben and Bates 1990); Fig. 12 (O'Driscoll 1988); Fig. 20 (Önal and Dogan 1988); Figs. 41 and 42 (Russell 1988); Fig. 46 (Burt et al. 1988); Fig. 69 (Fujii 1983); and Table 13 (Power 1985); Table 18 (McMichael 1990); and Table 20 (Fugii 1983, Sang 1983) from *Industrial Minerals;* **(10)** University of California Press, Berkeley, CA for Fig. 104 from *Calif. Univ. Publ. Geol. Ser.* (Hay 1963); **(11)** University of Chicago Press, Chicago, IL for Fig. 83 (Folk 1954); and Fig. 98 (Boettcher 1967) from *Jour. Geol.;* **(12)** John Wiley & Sons and Interscience, New York, NY for Figs. 101, 102d, 102e, 102f, 105, 106, and Table 25 from *Zeolite Molecular Sieve* (Breck 1974); and Fig. 28 (Thyler 1992); Fig. 29 (Hupp 1992); and Table 27 from *Kirk-Othmer Encyclopedia,* 4th ed.; **(13)** Marvel Dekker, Inc., New York, NY for Fig. 62 (Williams and Zeller 1987) from *Fertilizer Processing;* **(14)** Techna Sri, Casella Postale, Faenza, Italy for Fig. 88 (Schneider 1979, 1980) from *Ceramurgia Intern.;* **(15)** Prentice Hall, Inc., Upper Saddle River,

NJ for Fig. 84 from *The Growth of Single Crystals* (Laudise 1970); and Fig. 80 from *Mineralogy, Concept and Principles* (Zoltai and Stout 1984); **(16)** Mining Journal, Ltd., London, UK for Figs. 14 (Schiller 1985); and 15 (Svensson 1980) from *Mining Magazine;* **(17)** Australasian Institute of Mining and Metallurgy, Victoria, Australia for Fig. 72 (Towner 1986) from *Australia: A World Source of Ilmenite, Rutile, Monazite, and Zircon;* **(18)** Society for Mining, Metallurgy and Exploration, Littleton, CO for Fig. 5 from *Mining Engineering* (Sackett 1962); **(19)** American Institute of Mining, Metallurgical and Petroleum Engineers, New York, NY for Figs. 3 (Mann 1983); 16 and 18 (Kistler and Smith 1983); 31 (Appleyard 1983); 37 (Hancock 1983); 43 (Carr and Rooney 1983); 47 (Comstock 1963), 65 (Adams 1983), 70 (Cornish 1983), 94 (Roe and Olson 1983), 97 (Lynd and Lefond 1983); and 100 (Elevatorski and Roe 1983) from *Industrial Minerals and Rocks,* 5th edition; Fig. 8 from *Bauxite, 1984 Symposium* (Csillag et al. 1984); Table 1 (Jensen 1960) from *Industrial Minerals and Rocks*, 2nd edition; Table 2 (Hose 1963) from *Extractive Metallurgy of Aluminum*; and Tables 3 and 4 (Shaffer 1983); Table 6 (Mikami 1983); Table 7 (Kadey 1983); Table 8 (Rogers and Neal 1983); Table 9 (Reckling, Hoy, and LeFond 1983); Table 11 (Hancock 1983); Table 20 (Cornich 1983); and Table 26 (Breck 1983) from *Industrial Minerals and Rocks,* 5th edition; **(20)** American Ceramic Society, Westerville, OH for Fig. 34 (Gardenier 1988) from *Ceramic Bulletin;* and Fig. 45 (Smoke 1951) from *Jour. Amer. Ceramic Soc.;* **(21)** The Royal Society of Chemistry, Cambridge, UK for Fig. 85 (Kleinschmit 1980); and Table 21 (Gschneidner 1980) from *Speciality Inorganic Chemicals;* **(22)** Mineralogical Society of America, Washington, D.C. for Figs. 13 (Artiolo et al. 1993) and 24 (Sriramadas 1957) from *Amer. Mineral.,* and Figs. 81 (Heaney 1984), 87 (Ribbe 1982) and 93 (Evans and Guggenheim 1988) from *Reviews in Mineralogy;* **(23)** Clay Mineral Society, Boulder, CO for Fig. 39 (Murray and Lyon 1956) and Fig. 77 (Preisinger 1963) from *Clays and Clay Minerals;* **(24)** Geological Society of America, Boulder, CO for Figs. 26 and 27 (Levin 1950) from *Bull. Geol. Soc. Amer.;* **(25)** Soil Science Society of America, Madison, WI for Fig. 68 (Kapusta and Wendt 1963) from *Fertilizer, Technology and Usage;* **(26)** American Chemical Society, Washington, D.C. for Fig. 15 (Nies and Hulbert 1967) from *Jour. Chem. Eng. Data;* **(27)** Mineralogical Association of Canada, Montreal, Canada for Fig. 4 (Hill 1977) from *Canadian Mineralogist;* **(28)** Mineralogical Society, London, UK for Fig. 10 (Newman 1987) from *Chemistry of Clays and Clay Minerals;* and Fig. 38 (Brindley and Robinson 1947) from *Mineral. Magazine;* **(29)** E. Schweizerbart'sche Verlag, Stuttgart, Germany for Fig. 2 (Cameron and Papike 1978) from *Fortsche Mineral.;* and Fig. 102c (Bergerhoff et al. 1958) from *Neues Jahr. Mineral. Monatsheft;* **(30)** Wiley-VCH Publishers, Inc., Weinhem, Germany for Figs. 23, 33, 44, 58, 63, 64, 73, 74, 75, and 76 from *Ullmann's Encyclopedia of Inorganic Chemistry* (1985); Fig. 22 from *Industrial Inorganic Chemistry* (1989); and Tables 14, 16, and 18 from *Ullman's Encyclopedia of Inorganic Chemistry;* **(31)** Fig. 92 (Parkinson 1977) from *Chemical Engineering;* **(32)** American Association of Petroleum Geologists for Table 12 (Leighton and Pendextor 1962) from *AAPG Memoir;* **(33)** Addison-Wesley-Longman for Table 15 (Deer, Howie, and Zussman 1992) from *Introduction to Rock-Forming Minerals;* and **(34)** Springer-Verlag for Table 19 (Nathan 1984) from *Geochemistry of Phosphorites.*

Introduction

An industrial mineral is defined as a mineral that, based upon its distinct physical properties and chemical composition, has direct industrial applications. Industrial mineralogy is the science of examining and evaluating mineralogical properties, geological occurrence, distribution of deposits, industrial processes, and uses of industrial minerals.

Industrial minerals are distinguished from ore minerals, which have as their major function the production of metals. The characteristics of industrial minerals may be illustrated by the following examples:

- Diamond: because of its hardness, it is used as an abrasive.
- Graphite: because of its softness, it is used as a lubricant.
- Rutile: because of its color and high refractive index, it is used as a white pigment.
- Kyanite: because of its high melting point, it is used as a refractory material.

Rutile, for example, if used in the production of titanium, is considered to be an ore mineral, thus falling into the same category as galena and sphalerite. Therefore, in some cases, the division between industrial minerals and ore minerals is not always clear-cut.

Industrial minerals are both diversified and numerous. At one end of this wide spectrum of industrial minerals are silica sand and clay. These minerals are marketed in large tonnages with low unit prices and must be obtained close to their point of use. At the other end of the spectrum is the industrial diamond, which is marketed in grams or kilograms with a high unit price. Transportation costs do not deter the use of industrial diamonds. All other industrial minerals fall somewhere between these two extremes of unit and place value. Within these limits, industrial minerals are generally divided in the market into two categories. One category includes industrial minerals such as salt, potash, phosphates, and soda ash. These minerals are considered to be commodities and are sold on a price basis against industry product specifications. The other category includes functional or specialty industrial minerals that are designed with the end user in mind and must meet performance specifications. The second category includes, for example, micronized mica, ground calcium carbonate, water-washed kaolinite, soda-exchanged Fuller's earth, glass-grade feldspar, and acid-grade fluorite.

Mineral industries are grouped by their use of industrial minerals. These uses include abrasives, aggregates, cements, ceramics, drilling fluids, electronics, fertilizers

1

and soil conditioners, fillers, filters and absorbants, fluxes, foundry sands, glass, insulation, lubricants, pigments, plasters, and refractories. The chemical and metallurgical industries use minerals more or less as basic raw materials. Industrial minerals with major uses as raw materials, such as sulfur and bromine, are not selected for treatment in this book.

How many industrial minerals are there? The answer is many, and the number varies from inventory to inventory dependent upon the purpose of evaluation. In fact, every mineral has its function, but its applicability in industry is judged by, among other things, economic and technical factors. In this presentation, forty-two industrial minerals have been selected; of these each has the well-established functions shown below:

Abrasives:	Bauxite, garnet, industrial diamond
Aggregates:	Diatomite, dolomite, limestone, perlite, pumice, vermiculite
Cements:	Asbestos, dolomite, gypsum, limestone
Ceramics:	Barite, bauxite, bentonite, beryllium minerals, feldspar, kaolinite, lithium minerals, manganese minerals, pyrophyllite, rare earth minerals, silica, talc, wollastonite, zircon
Drilling fluids:	Barite, bentonite, sepiolite and attapulgite
Electronics:	Beryllium minerals, graphite, manganese minerals, rare earth minerals, silica
Fertilizers and soil conditioners:	Dolomite, limestone, nitrates, phosphates, potassium salts
Fillers, filters, and absorbants:	Diatomite, kaolinite, mica, sepiolite and attapulgite, silica, talc, wollastonite, zeolites
Fluxes:	Fluorite, limestone
Foundry sands:	Bentonite, chromite, graphite, olivine, silica, zircon
Glass:	Borax and borates, celestite, feldspar, fluorite, limestone, lithium minerals, silica, soda ash, zircon
Insulation:	Asbestos, mica, perlite, pumice
Lubricants:	Graphite, molybdenite
Pigments:	Barite, iron oxide pigments, titanium minerals
Plasters:	Gypsum
Refractories:	Bauxite, chromite, graphite, magnesite, pyrophyllite, silica, sillimanite, andalusite, kyanite, zircon

Among the industrial minerals listed, several require a note of explanation, because by definition they are not minerals. Bauxite is an assemblage of aluminum hydroxide minerals and is mainly consumed during the production of aluminum; nevertheless, it is the principal source of alumina, a prominent component in abrasives and refractories. Limestone and diatomite are sedimentary rocks consisting mainly of calcite and silica, respectively. Limestone has distinct industrial applications in addition to its use as a basic material for chemical industries, whereas diatomite is used as a filter aid and a functional filler because of its highly porous structure. Perlite and pumice are volcanic rocks, usually of rhyolitic composition. Their characteristic light weight promotes their use in the preparation of construction materials.

The "clay minerals" constitute a large group of minerals that have similar crystal structure, chemical composition, and morphological characteristics. The selection of bentonite (beidellite-montmorilonite), kaolin and kaolinite, and attapulgite and sepiolite is based upon their distinct industrial applications.

The basics of industry of industrial minerals are to mine the mineral, beneficiate the mineral to a concentrate, and market the final product. Steps should be taken beyond these well-established limits into the processing and manufacturing of advanced materials. These steps will not only generate higher profits for the industry, but also buffer the industry from market fluctuations.

Asbestos

Asbestos is the generic name given to a group of fibrous silicate minerals. Two groups of asbestos are generally recognized: the fibrous form of serpentine known as chrysotile and the amphibole asbestos, which includes crocidolite, amosite, anthophyllite, tremolite, and actinolite. In the second group, the last two have no commercial importance. Asbestos fibers have great tensile strength, resist heat and chemical attack, and are incombustible. They can be spun into yarn and made into textiles. Based upon these characteristic properties, asbestos minerals have industrial applications in thermal and electric insulation, concrete products, friction materials, and textiles.

Asbestos has been found to be a health hazard. Breathing air ladened with asbestos for a prolonged period of time may cause asbestosis, mesothelioma, or other health problems. The problem is complex, and not all asbestos minerals are equally harmful (Newhouse 1979; Clarke 1982; Benarde 1990).

About 95% of world asbestos production is chrysotile, 70% of which comes from the former USSR and Canada (Virta 1996).

MINERALOGICAL PROPERTIES

1. Chrysotile

Chrysotile is a mineral of the serpentine group, which has been divided on structural grounds into three types: (1) chrysotile with a fibrous structure, (2) lizardite with a planar structure, and (3) antigorite with an alternating wave structure (Wicks and Whittaker 1970). In the serpentine structure, there are two basic sheets, the tetrahedral sheet and the octahedral sheet. The former is composed of SiO_4 arranged in six-membered rings with a hexagonal symmetry, and the latter is composed of MgO_6 with an arrangement similar to that of brucite. The two sheets are linked together through the apical oxygens of the tetrahedral sheet, which replace two-thirds of the hydroxyl in the base of the octahedral sheet. The misfit between these two sheets is responsible for the development of the three different structures of serpentine (Fig. 1).

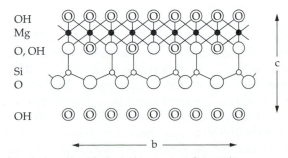

Figure 1   The linkage of tetrahedra and octahedra in the structure of serpentine.

Chrysotile is monoclinic with a = 5.34Å, b = 9.25Å, c = 14.65Å, and β = 93°16′. The elongation of the fibrous structure is along the a-axis, and the curvature of the fibrous structure may produce either spiral or concentric circular growth. The formation of such a curvature has been considered as a mechanism to relieve the misfit between the tetrahedral and the octahedral sheets (Yada 1971).

The composition of chrysotile is very close to its ideal formula, $Mg_6Si_4O_{10}(OH)_8$, but varies within the limits of substitution of Mg by Fe^{2+} or Fe^{3+}. It has been demonstrated by Wicks and Whittaker (1975) that there is no compositional break between the fibrous chrysotile and the planar lizardite, although, in general, the total Fe content of chrysotile is lower than that of lizardite (Whittaker and Wicks 1970).

Chrysotile is commonly gray, white, or yellow-green with a silky or pearly luster. Its specific gravity is 2.5 to 2.6 and hardness 2 1/2 to 3 1/2. Optically, chrysotile is biaxial negative, and 2V varies widely between 20° and 80°. Refractive indices have a range of 1.53 to 1.56 with a weak birefringence. In comparison with amphibole asbestos, chrysotile shows parallel extinction, lower refractive indices, and lower birefringence.

Phase-contrast optical microscopy is generally used for chrysotile analysis; however, chrysotile is often so finely crystallized that x-ray powder diffraction and electron microscopy may have to be applied to its identification. Refractive indices, birefringence, and sign of elongation of chrysotile fibers can be determined by the dispersion staining method (Uno 1992). In bulk samples, chrysotile asbestos may be quantitatively determined by combined Rietveld and RIR (Ref. Intensity Ratio) methods (Gualtieri and Artioli 1995) and by instrumental neutron activation analysis (Parekh et al. 1992). Oriented fibers with dimensions of not less than 0.2 mm long and not less than 0.02 mm thick may be used for single crystal x-ray diffraction (Wicks and Zussman 1975; Zussman 1979).

2. Crocidolite, Amosite, and Anthophyllite

The basic structural unit for all amphibole minerals is the double silicate chain (Whittaker 1979; Skinner et al. 1988). Successive SiO_4 tetrahedra share two and three oxygens, respectively, along the chain. This gives a formula of $(Si_8O_{22})_n$ for the infinite chain. These chains elongate along the crystallographic c-direction and repeat at intervals of approximately 5.3Å, which defines the c-dimension. Successive chains are inverted along the b-direction and stacked in opposite orientation along the a-direction as shown in Fig. 2. These chains are bonded together by actions in sites coordinated by six oxygens designated as C-sites (M_1, M_2, M_3, or M_4), by eight oxygens as B-sites (M_4), and by ten oxygens as A-sites. In addition to oxygen, there are hydroxyls present in the structure as shown in Fig. 2. For each unit formula $Si_8O_{22}(OH)_2$, there are two M_4, two M_1, two M_2, one M_3, and one A. The general formula for an amphibole is, then, $A_{0-1}B_2C_5Si_8O_{22}(OH)_2$. The most common cations that may enter each structural site are A: Na or K; B: Ca, Na, Mg or Fe^{2+}; and C: Mg, Fe^{2+}, Al, or Fe^{3+}. Aluminum is also present in the tetrahedral site, replacing silicon (Deer et al. 1992).

The relative position of adjacent chains is determined by the sizes of the cations at various sites. The resultant stacking sequence is such as to produce a unit cell of ei-

M4 M2 M1 M3 M1 M2 M4

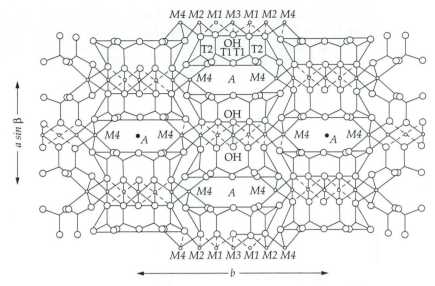

Figure 2   Idealized projection of an amphibole structure, after Cameron and Papike (1978).

ther a monoclinic or an orthorhombic symmetry. Crocidolite is the blue asbestiform variety of the soda amphibole riebeckite with a general composition of $Na_2(Fe^{2+},Mg,Fe^{3+})_5(Si_8O_{22})(OH)_2$. It is monoclinic with a = 9.75Å, b = 18.0Å, c = 5.3Å, and β = 103°. Crocidolite has a moderate hardness of 5.0 and specific gravity ranges from 3.2 to 3.4. Optically, it is biaxial negative with 2V varying between 90° and 50°. Refractive indices have a range of 1.70 to 1.72, and birefringence is moderate. Crocidolite has a silky or dull luster and soft to harsh texture. Its fibers are short to long with good flexibility.

Amosite is known as the brown asbestos. It consists chiefly of fibrous members of the cummingtonite-grunerite series with the general formula $(Fe^{2+},Mg)_7(Si_8O_{22})(OH)_2$. It has a vitreous or pearly luster and a coarse texture. Its fibers range from 2 to 11 inches in length with good flexibility. Amosite is monoclinic and has a = 9.6Å, b = 18.3Å, c = 5.3Å, and β = 101°50′. Its hardness is 5.5 to 6.0 and specific gravity 3.1 to 3.25. Optically, amosite is biaxial negative with 2V varying in a narrow range between 90° and 96°. Refractive indices have a range of 1.67 to 1.73, and birefringence is moderate.

Anthophyllite is an orthorhombic amphibole. Its chemical formula is identical to that of the cummingtonite-grunerite series, but anthophyllite is generally rich in magnesium. With an increase in Fe content from 0 to 60 mole%, cell dimensions of anthophyllite vary from 18.5 to 18.6Å in a, from 17.7 to 18.1Å in b, and from 5.27 to 5.32Å in c. Anthophyllite is white or grayish white to green or brownish green in color, and vitreous to pearly in luster. Its hardness ranges from 5.5 to 6.0, and specific gravity from 2.85 to 3.75. Its fibers are short with harsh texture and poor flexibility. Optically, anthophyllite is biaxial negative with 2V between 75° and 90°, and the Fe-rich anthophyllite is optically positive. Refractive indices have a range of 1.60 to 1.72 with a moderate birefringence.

Analytical methods used for the identification of chrysotile are also applied to amphibole asbestos. Amosite and crocidolite, although they are much alike in refractive indices, can be distinguished easily based upon the difference of dispersion color for the light vibrating parallel to the cross fiber axes (Uno 1992).

3. Chrysotile and Amphibole Asbestos

Mechanical Properties: The most important and most commonly quoted mechanical property of asbestos fibers is tensile strength. Results from measurements made on specimens of a standardized minimum length of 3 to 4 mm long (Hodgson 1979) are 31, 24, 25, and 35 ($\times 10^3$ kg/cm^2), respectively, for chrysotile, anthophyllite, amosite, and crocidolite. The Young's moduli for the same order of minerals are ($\times 10^6$ kg/cm^2) 1.65, 1.58, 1.58, and 1.9. Measurements of shear strength and shear modulus of the interfibrillar region have shown that increases in shear strength make the fiber more flaw sensitive and weaker, whereas increases in shear modulus tend to strengthen a fiber. Under the influence of heat in the temperature range up to 800°C, the tensile strength of chrysotile, amosite, and crocidolite fibers increases slightly before decreasing rapidly, as structural change and decomposition take place. The temperature range of strength increase for amosite and crocidolite is between 150° and 250°C, whereas for chrysotile it is up to 550°C. This marked increase is related to the increase in shear modulus. At temperatures above 500°C, the strength of asbestos fibers is only 10% of their original strength (Hodgson 1979).

Chrysotile usually appears as long thin fibers of uniform thickness. The flexibility varies from gently curved, flexible fibers to rigidly straight, inflexible fibers. This is an important property of chrysotile, but the cause of this variation is unknown. The spiral and concentric growths of chrysotile asbestos were observed by lattice-imaging techniques using high-resolution electron microscopy (Yada 1971). Based on the calculation of bending strain of maximally bent fiber segments, Germine and Puffer (1989) have made measurements of relative flexibility on amphibole fibers. The flexibility was found to increase as the diameter falls, and the variation is parallel with the variation in tensile strength with diameter. There is an inflection at about 3 μm, which appears to mark a transition from elastic to flexible behavior.

Electrical and Magnetic Properties: Volume resistivity is the only electrical property measured on asbestos. It is low for chrysotile (0.003 to 0.15 MΩ·cm) and crocidolite (0.2 to 0.5 MΩ·cm), varies between 2.5 and 7.5 MΩ·cm for anthophyllite, and has a range up to 500 MΩ·cm for amosite. The magnetic susceptibility of asbestos fibers is generally low. Chrysotile and anthophyllite have $\chi_g = 5.3 \times 10^{-8}$ and 14.3×10^{-8} at a magnetic field strength (H) of 7.98×10^5 A/m, respectively, whereas the Fe-rich amosite and crocidolite have 78.7×10^{-8} and 60.9×10^{-6} at identical H, respectively (Hodgson 1979).

Thermal Properties: Thermal analysis of chrysotile has been made by Ball and Taylor (1963), Brindley and Hayami (1965), and Hodgson (1979). The dehydroxylation of chrysotile takes place in the temperature range 600° to 780°C, and at 810°C, an exothermal peak indicates the formation of forsterite and quartz. A gradual weight loss between 100° and 600°C and a sharp loss above 600°C exist on the TGA curve. The total dehydration amounts to about 13% of H_2O.

Thermal decomposition of amphibole fibers occurs in three stages: (1) loss of physically combined water, (2) loss of chemically combined water, and (3) breakdown into pyroxene, iron oxides, and quartz in the temperature range 400° to 1050°C. On the DTA curve of crocidolite, an endothermal peak appears at 610°C for a dehydroxylation reaction in an inert atmosphere, which coincides with distinct steps on curves of weight loss and dynamic dehydration. The endothermal and exothermal peaks at 800° and 820°C correspond to the breakdown of amphibole and the formation of pyroxene, iron oxides, and quartz. In an oxidizing atmosphere, an oxidation process occurs at 400°C. The dehydration curve indicates the loss of water, which is not shown on the TGA curve. Addison et al. (1962) postulated a redox equation for the combined process of dehydrogenation and oxidation: $4Fe^{2+} + 4OH^- + O_2 \rightarrow 4Fe^{3+} + 4O^{2-} + 2H_2O$.

Asbestos fibers do not have a melting point; only the decomposition products melt. In the case of chrysotile and anthophyllite, the products consist of magnesium pyroxenes, which melt at temperatures above 1450°C. Crocidolite and amosite decompose to give iron pyroxenes, which generally melt at about 1000°C. The decomposition of asbestos fibers at high temperatures does not mean that their properties as heat insulators are diminished. The products of decomposition at a hot surface form a poorly conductive layer that provides protection to the succeeding layers of insulation.

Chemical Reactivity: Chrysotile fibers decompose rapidly in strong acids, whereas amphibole fibers show various degrees of resistance to reaction with acids. This distinction is a result of the structural differences between the two groups of asbestos fibers (Hodgson 1979). Hydrofluoric acid will completely decompose all asbestos, and concentrated acetic acid may cause the breakdown of chrysotile. All asbestos resist prolonged attack by alkalis, even strong ones such as, for example, 5M NaOH.

At temperatures below the thermal decomposition of chrysotile and amphiboles, asbestos fibers appear to be resistant to attack by molten metals and salts. Ball and Taylor (1963) have examined the hydrothermal reactions of chrysotile with oxides and silicates of the asbestos-cement assemblage. Results show that reactions between chrysotile and MgO, SiO_2, Al_2O_3, or Fe_2O_3 produce forsterite, talc, and brucite. Monticellite is formed in the reactions between chrysotile and tricalcium or dicalcium silicate.

Surface Properties: Chrysotile has a surface layer of hydroxyls connected to an adjacent inner layer of magnesium ions, which would be expected to exert a strong surface activity, either by partial solution or by attraction of ions of opposite charge (Pundsack 1955). The equilibrated pH in CO_2-free distilled water is 10.33. The surface charge in water is positive at a pH lower than 11.8 and rises to a maximum at a pH of 3 (Martinez and Zucker 1960). In contrast, the surface of amphibole asbestos has bands of silica that, being highly insoluble, would be expected to exert a weak surface charge solely by attraction of suitable ions. The surface of amphibole asbestos is like that of silica.

The surface area is the most important property of asbestos in its industrial applications. Each process requires asbestos fibers with a critical degree of fiberization, and the required degree of fiberization is achieved by milling raw asbestos. Raw amphibole asbestos has an absolute surface area of $6 - 20 \times 10^3$ cm^2/g, whereas, for chrysotile, the value may be up to 30×10^3 cm^2/g. The absolute surface area of typical asbestos fibers used in industrial processes is $30 - 90 \times 10^3$ cm^2/g.

GEOLOGICAL OCCURRENCE AND DISTRIBUTION OF DEPOSITS

1. Chrysotile Asbestos

Chrysotile asbestos usually occurs in veins in association with various serpentine and related minerals in serpentinized peridotites and dunites, and in serpentinized dolomitic marbles. The process of developing a chrysotile asbestos vein is a part of the serpentinization process, but not all serpentinization produces chrysotile asbestos deposits. In general there is a close relationship between the mineralogy of the chrysotile asbestos veins and the mineralogy of the host rock. Chrysotile asbestos may occur as a cross fiber where the fibers lie transverse to the vein, or as a slip fiber where the fiber lies in the plane of the vein (Gold 1967).

About 70% of world chrysotile production comes from the former USSR and Canada. The other major producers are South Africa, the United States, Brazil, Swaziland, Zimbabwe, China, and Australia. World production of asbestos has declined continuously in recent years. Canada's production has decreased from a high of 710,357 tons in 1988 to 515,341 tons in 1993 (U.S.B.M. Mineral Yearbook 1994).

Canada: Numerous asbestos deposits occur in a serpentine belt extending from the Baie Verte district of Newfoundland through the Thetford Mines, Black Lake, and Asbestos districts of Quebec, to Belvidere Mountain, Vermont (Lamarché and Riordon 1981). The belt is a typical ophiolitic complex of gabbroic and dioritic rocks, and of pyroxenite, peridotite, and dunite, serpentinized to various degrees with associated granite and talc-carbonate rocks. The serpentinized ultramafic rocks have undergone normal and reverse faulting and intense shearing. The chrysotile asbestos occurs mainly in the highly serpentinized peridotite and is found in numerous occurrences along the strike of the complex.

United States: Chrysotile asbestos is mined near Eden, Vermont, and in two localities in California, Copperopolis and New Idria. The Vermont asbestos deposit is the southern extension of the serpentine belt of Quebec down the Appalachian chain into northern Vermont (Rowbotham 1970). The host rock is a highly serpentinized peridotite and is emplaced in metamorphosed sedimentary and volcanic rocks. Slip fiber is the predominant variety mined but some cross fiber of good quality is also recovered.

The Pacific Asbestos deposit is located between Copperopolis and Sonora in the foothills of the Sierras, about 200 km east of San Francisco (Leney and Loeb 1972). The deposit occurs within a belt of peridotite and dunite that have been variously altered to serpentine. The New Idria–Coalinga chrysotile deposit is located in Fresno County, California (Munro and Reim 1962). The ultrabasic body has intruded into Jurassic sediments and is composed of highly sheared serpentine characterized by an extremely platy, slickenside nature. Boulders of massive serpentinized material occur in the loose platy serpentine. The asbestos is of short fiber similar to the Canadian variety.

The Former USSR: As in Canada and the United States, most of the chrysotile deposits in the former USSR are associated with ultrabasic rocks. Only a small percentage is found in dolomite (Petrov and Znamensky 1981). There are four major asbestos-producing districts: (1) the Bazhenovo deposits of the central and northern Urals, (2) the Dzhetygara deposits in the Southern Urals of western Kazakhstan, (3) the Kiembay deposits, also in the southern Urals, and (4) the Akdovurak deposits in Tuva, some 960 km west of the southwest end of Lake Baikal.

Brazil: The Cana Brava asbestos deposit is located about 190 km north of Brasilia and 95 km northeast of Uruacá in the State of Goiás. It contains chrysotile asbestos in compact veins in serpentine (Harben 1984; Harben and Kuzvart 1996).

South Africa and Swaziland: The Msauli mine, South Africa, and the Havelock mine of Swaziland are 12 km apart (Van Biljon 1964) and located about 23 and 25 km, respectively, south-southeast of Barberton in the eastern Transvaal. The asbestos deposits are sill-like bodies of serpentine intercalated in sedimentary sequence of Early Precambrian age. Chrysotile occurs as a stockwork of cross-fiber seams forming 3 to 4% of the rock.

Zimbabwe: The Shabani asbestos district is located some 160 km east of Bulawayo between Shabani and Mashaba (Laubscher 1968; Oldham 1968; Wilson 1968). The deposit occurs near the base of a lenticular sill that is composed of dunite at the base, overlain by peridotite, pyroxenite, and gabbro, and intrudes Early Precambrian gneisses. Strong shear zones are prevalent within the asbestos zone. The intrusion and the shearing have caused strong serpentinization of the ultrabasic rocks. The fiber occurs as a stockwork of cross-fiber veins with length up to 1 1/2 inches in places. About half of the world's textile asbestos fibers comes from this district.

Australia: The major chrysotile producer in Australia is the Woodsreef asbestos district (Butt 1981) near Barraba, north of Sydney, New South Wales. The deposit occurs in a layered peridotite-dunite complex that has intruded sediments of Mid-Paleozoic age. The fiber occurs mainly as a stockwork of cross-fiber seams and is generally short to medium in length.

China: Chrysotile has been mainly produced in the Lai-Yuan district of Hopeh and the Shihmien area of Szechwan (Virta and Mann 1994). Widespread mineralization of chrysotile asbestos was also found in southern Chaoyang, Liaoning (Li 1986). The asbestos veins intruded the host rocks, mainly as composite veins with lesser dendritic veins and stockwork. The host rocks are siliceous dolomite or siliceous calcareous dolomite.

2. Crocidolite and Amosite Asbestos

South Africa is the major producer of amphibole asbestos in the world. The Cape croci-dolite field is a 475 kilometer long belt stretching from the Orange River, northward past Griquatown and Kuraman to the border with Botswana (Coetzee et al. 1976). The rocks belong to the Transvaal Supergroup and are subdivided into the Ghaap, Postmasburg, and Olifanshoek groups. All the crocidolite occurs in the Ghaap group, mainly a dolomite containing a banded ironstone sequence (Hodgson 1979; Dreyer and Robinson 1981).

The Transvaal amosite field covers a 95 kilometer wide district from Chuniespoort to Kromellenboog. The Precambrian Transvaal Supergroup in this district is subdivided into four groups: (1) the Wolkbery Group of quartzite, (2) the Chuniespoort Group of carbonate, (3) the Malmani Dolomite, grading upward into banded ironstone, and (4) the Pretoria Group of quartzite and shale. Amosite occurs in the ironstone overly-ing the Malmani dolomite.

INDUSTRIAL PROCESSES AND USES

1. Milling and Beneficiation

Asbestos milling should recover as large a percentage as possible of long fiber, even at some sacrifice of the recovery of short fiber. Mill flow lines are designed to suit the char-acteristics of a given ore body, but the basic steps are similar for most milling processes. A generalization of the essential steps and equipment used in a Canadian chrysotile as-bestos mill, as described by Jenkins (1960), Mann (1983), and Virta and Mann (1994), is shown in Fig. 3. Primary crushing may be done in a surface plant when treating open pit ore, or underground in the case of an underground mine. Large jaw crushers are used in the majority of the Canadian plants, and gyratory crushers, which eliminate the need for feeders and bins, have gained in usage recently. Various types of rotary dryers, ver-tical tower dryers, and fluid-bed dryers are used in the drying process. The release and separation of fiber from gangue is accomplished by successive stages of crushing or comminution by impact. Finer fractions are generally screened out prior to air separa-tion. The concentrates undergo a series of cleaning operations designed to remove sand and dust. They are screened, and screened in some operations with suction. Trommels are used to further clean the fiber and screen it into standard-grade lengths.

In the grading mill the fiber is further separated into the approximate length brackets required for specific grades, and is subjected to several stages of screening us-ing various types of screens. An air filtering system must be used in the plant to mini-mize the dust content of the air. Virta and Mann (1994) described modifications of the primary rock line and fiber cleaning circuit, and for the primary and secondary crush-ing, drying, and ore concentration system in some Canadian asbestos mills.

African practice for the preparation of amphibole asbestos is generally very sim-ple, consisting, of a large degree, of hand sorting, crushing, and screening, as with the chrysotile in Canada. At the amphibole asbestos plant in Transvaal, milling practice is more elaborate because of the difficult cobbing due to the extreme toughness of the rock. The fiber is treated in different mill circuits variously equipped with jaw crushers, cone crushers, rolls, hammer mills, and trommels.

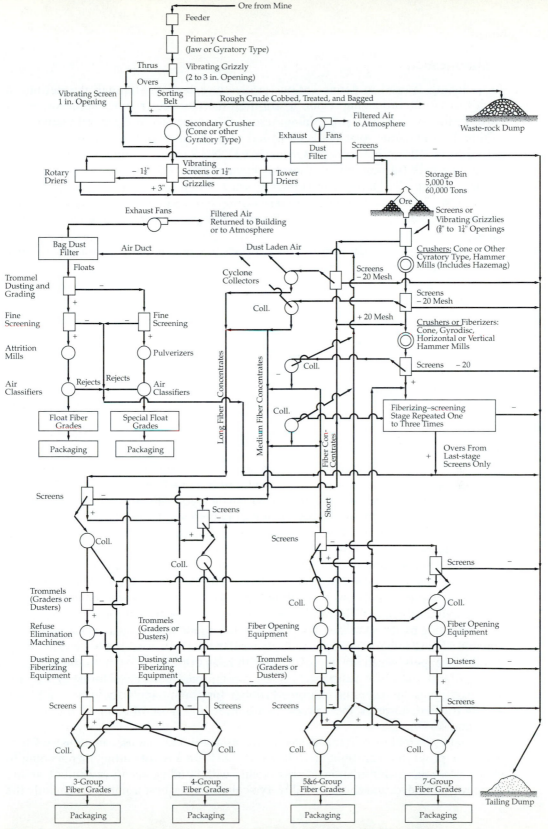

Figure 3 A flowsheet showing the milling process of a Canadian asbestos plant, after Mann (1983).

2. Specification

Fiber length is used to define various grades of asbestos. Although the definition varies from one producing region to another, the Quebec grading standards, as set forth by the Quebec Asbestos Mining Association, have received general recognition (Table 1). Chrysotile asbestos is classified into seven main groups, commonly referred to as follows: No. 1 crude (Group No. 1), No. 2 crude (Group No. 2), textile or shipping fibers (Group No. 3), asbestos cement fiber (Group No. 4), paper stock grades (Group No. 5), paper or shingle fibers (Group No. 6), and shorts and floats (Group No. 7). Two additional groups are Group No. 8, sands and gravels with a minimum 75 lbs. per cu ft; and Group No. 9, sands and gravels with more than 75 lbs. per cu ft.

Many users have developed different tests to evaluate the performance of fiber for market (Cossette and Delvaux 1979).

3. Industrial Uses

There are more than 3,000 uses for asbestos (Jenkins 1960; Michaels and Chissick 1979; Clarke 1982; Mann 1983; Griffiths 1986; Virta and Mann 1994). By far the largest quantities of asbestos are used in asbestos fiber composite materials for the reinforcement of inorganic (asbestos cement) and organic (PVC, rubber, plastics) bonding agents. The increase in tensile strength of asbestos with temperature up to 400°C is significant in some of its uses, and no substitutes have this characteristic (Derricott 1979).

Asbestos cement is widely used in building and in industry. Asbestos fibers, normally chrysotile, are added to hydrated Portland cement. This mixture, with 10 to 20 wt% asbestos, is compressed into flat or corrugated sheets or molded into boards, shingles, pipes, and other shapes. Pressure pipe is produced from high-quality Group 4 fiber, usually a blend of chrysotile and crocidolite, to ensure a good modulus of rupture. On the other hand, flexible asbestos-cement sheets may be made from a Group 6 fiber. Production of corrugated sheets generally includes some Group 5 fiber to improve adhesion of the wet sheets during the formation process. The slurry of asbestos cement, known as "sprayed asbestos," can also be produced by a spray process for fire-retardant layers as well as thermal and acoustic insulation. Glass-fiber-reinforced cement is made by adding approximately 5% of alkali-resistant glass fibers, which is relatively expensive, to hydrated cement. The resulting composite can be extruded, molded, or sprayed.

Molding materials made from phenol or melamine resins and asbestos, which are capable of hardening, have applications in insulation. The use of long fiber in the form of felts or papers, which are impregnated with resins, produces tough products of high strength and good heat resistance. Asbestos-reinforced polyvinylchloride (PVC) for floor coverings and asbestos-reinforced rubber for sealant applications have all been widely utilized; however, these products are more and more often being replaced by glass and carbon fibers.

Asbestos millboard is the most versatile asbestos material used in industry. Characteristic properties contributing to this versatility are ease of cutting or punching to shape, good thermal insulation, impregnability with bonding agents or cement, ability to be wet-moulded, and compressibility. Typical uses for asbestos millboard include the

Table 1 Asbestos grading by Quebec Asbestos Mining Association after Jenkins (1960). As of 1993 the grading standards have not been converted to the metric system (Virta and Mann 1994).

Group No. 1
  No. 1 Crude. Cross-fiber veins having 3/4-in. stable and longer.

Group No. 2
  No. 2 Crude. Cross-fiber veins having 3/8-in. staple up to 3/4-in. Run-of-mine Crude consists of unsorted crudes. Sundry Crudes consist of crudes other than above specified.

Group No. 3	Guaranteed Minimum Shipping Test			
(Commonly referred	1/2 in.,	4 Mesh,	10 Mesh,	Pan,
to as textile or shipping fibers)	oz	oz	oz	oz
3F	10.5	3.9	1.3	0.3
3K	7	7	1.5	0.5
3R	4	7	4	1
3T	2	8	4	2
3Z	1	9	4	2
Group No. 4				
(Commonly referred to as asbestos cement fiber)				
4A	0	8	6	2
4D	0	7	6	3
4H	0	5	8	3
4H	0	5	7	4
4K	0	4	9	3
4M	0	4	8	4
4R	0	3	9	4
4T	0	2	10	4
4Z	0	1.5	9.5	5
Group No. 5				
(Often referred to as paper stock grades)				
5D	0	0.5	10.5	5
5K	0	0	12	4
5M	0	0	11	5
5R	0	0	10	6
5Z	0	0	8.6	7.4
Group No. 6				
(Paper and shingle fibers)				
6D	0	0	7	9
6F	0	0	6	10
Group No. 7				
(Shorts and floats)				
7D	0	0	5	11
7F	0	0	4	12
7H	0	0	3	13
7K	0	0	2	14
7M	0	0	1	15
7R	0	0	0	16
7T	0	0	0	16
7RF and 7TF Floats	0	0	0	16
7W	0	0	0	16
Group No. 8 & 9				
(Sands and gravels)				
8S	0	0	0	16
		Minimum 50 lb per cu ft		
8T	0	0	0	16
		Minimum 75 lb per cu ft		
9T	0	0	0	16
		More than 75 lb per cu ft		

fabrication of rollers for transport of hot materials, formers for wire-wound electrical resistances, flange gaskets for joints in ducting and trunking used for gas transport, plugs and stoppers for molten metal containers, among other uses. Various blends of Groups 4, 5, and 6 fibers are used for this application.

Asbestos fibers have been widely used in the production of brake and clutch linings and pads because of their high frictional coefficient. Automobile brake linings are usually molded from a mixture of phenolic resin, fillers, and chrysotile of Groups 6 and 7 fibers.

Chrysotile fiber forms the basic raw material for almost all asbestos textiles. The length and flexibility of longer grades of chrysotile are such that spinning into yarn and cloth weaving are possible. They may be braced with an organic fiber and reinforced with incorporated wire or other yarn such as nylon, cotton, or polyester. Their major applications are in fire and heat protective clothing, blankets, curtains, and aprons, and for electrical insulation.

The uses of chrysotile asbestos as a filtration material range from the clarifying filtration of beverages to the removal of viruses and pyrogens from highly sensitive pharmacological solutions. The high specific surface area, positive zero-potential, and ion exchange capacity of chrysotile asbestos are a special combination and make it difficult to find an asbestos-free substance.

REFERENCES

Addison, C. C., W. E. Addison, G. H. Neal, and J. H. Sharp. 1962. Oxidation of crocidolite. *Jour. Chem. Soc.* (London) 1468–1471.

Ball, M. C., and H. F. W. Taylor. 1963. X-ray study of some reactions of chrysotile. *Jour. Appl. Chem.* (London) 13:145–150.

Benarde, M. A. 1990. "Adverse Health Effects of Asbestos." In *Asbestos: The Hazardous Fiber,* Ed. M. A. Benarde, 45–74. Boca Raton, Fl. CRC Press, Inc.

Brindley, G. W., and R. Hayami. 1965. Mechanism of formation of forsterite and enstatite from serpentine. *Mineral. Mag.* 35:189–195.

Butt, B. C. 1981. "Exploration Forecasts and Exploitation Realities at the Woodsreef Mine, New South Wales, Australia." In *Geology of Asbestos Deposits,* Ed. P. H. Riordon, 63–75. New York: American Institute of Mining, Metallurgical, and Petroleum Engineers.

Cameron, M., and J. J. Papike. 1978. Amphibole crystal chemistry: A review. *Fortsche. Minerals* 57:28–67.

Clarke, G. 1982. Asbestos—a versatile mineral under siege. *Ind. Minerals* (March): 21–29.

Coetzee, C. B., A. 1976. J. M. Brabers, S. J. Mulherbe, and W. J. van Biljon. 1976. "Asbestos." 5th ed. In *Mineral Resources of the Republic of South Africa,* Ed. C. B. Coetzee, 261–268. Dept. Mines/Geol. Sur., Handbook 7.

Cossette, M., and P. Delvaux. 1979. "Technical Evaluation of Chrysotile Asbestos Ore Bodies." In *Short Course in Mineralogical Techniques of Asbestos Determination,* 79–110. University of Laral, Quebec: Mineralogical Association of Canada.

Deer, W. A., R. A. Howie, and J. Zussman. 1992. *An Introduction to Rock-Forming Minerals.* 2d ed. Essex: Longman. 696p.

Derricott, R. 1979. "The Use of Asbestos and Asbestos-Free Substitutions in Buildings." Vol. 1, In *Asbestos—Properties, Applications, and Hazards,* Eds. L. Michaels and S. S. Chissick, 305–331. Chichester: John Wiley.

Dreyer, C. J. B., and F. A. Robinson. 1981. "Occurrence and Exploitation of Amphibole Asbestos in South Africa." In *Geology of Asbestos Deposits,* Ed. P. H. Riordon, 25–44. New York: American Institute of Mining, Metallurgical, and Petroleum Engineers.

Germine, M., and J. H. Puffer. 1989. Origin and development of flexibility in asbestiform fibres. *Mineral. Mag.* 53:327–335.

Gold, D. P. 1967. "Local Deformation Structures in Serpentinite." In *Ultramafic and Related Rocks,* Ed. P. J. Wylie, 200–202. New York: John Wiley & Sons.

Griffiths, J. 1986. Asbestos-Rising in east but setting in west. *Ind. Minerals* (April): 21–37.

Gualtieri, A., and Artioli, G. 1995. Quantitative determination of chrysotile asbestos in bulk materials by combined Rietveld and RIR methods. *Powder Diff.* 10: 269–277.

Harben, P. W. 1984. A profile of SA Minercão de Amianto—Brazil's asbestos producer. *Ind. Minerals* (March): 63–65.

Harben, P. W., and M. Kuzvart. 1996. *Industrial Minerals—A Global Geology.* Metal Bulletin PLC, London: Industrial Minerals Information Ltd., 462p.

Hodgson, A. A. 1979. "Chemistry and Physics of Asbestos." Vol. 1, In *Asbestos: Properties, Applications and Hazards,* Eds. L. Michaels and S. S. Chissick, 67–113. Chichester: John Wiley & Sons.

Jenkins, G. F. 1960. "Asbestos." In *Industrial Minerals and Rocks,* 3d ed., Ed. J. L. Gillson, 23–53. New York: American Institute of Mining, Metallurgical, and Petroleum Engineers.

Lamarché, R. Y., and P. H. Riordon. 1981. "Geology and Genesis of the Chrysotile Asbestos Deposits of Northern Appalachia." In *Geology of Asbestos Deposits,* Ed. P. H. Riordon, 11–23. New York: American Institute of Mining, Metallurgical, and Petroleum Engineers.

Laubscher, D. H. "The Origin and Occurrence of Chrysotile Asbestos in the Shabani and Mashaba Areas, Rhodesia." In *Symposium on Rhodesian Basement Complex* (Trans. Geol. Soc.) 71 (1968): 195–204.

Leney, G. W., and E. E. Loeb. 1972. Geology and mining operations at Pacific Asbestos Corporation. *Asbestos* 54 (4): 4–14.

Li, T. 1986. An investigation into the genesis of chrysotile asbestos deposits in southern Chaoyang, Liaoning Province. *Mineral Deposits* (in Chinese with English abstract). 5:75–86.

Mann, E. L. 1983. "Asbestos." 5th ed. In *Industrial Minerals and Rocks,* Ed. S. J. Lefond, 435–485. New York: Soc. Mining Engineers, AIME.

Martinez, E., and G. L. Zucker. 1960. Asbestos ore body minerals studies by zeta potential measurements. *Jour. Phys. Chem.* 64:924–926.

Michaels, L., and S. S. Chissick. 1979. *Asbestos: Properties, Applications, and Hazards.* Vol. 1. Chichester: John Wiley & Sons.

Munro, R. C., and K. M. Reim. 1962. Coalinga asbestos fiber—A newcomer to the asbestos industry. *Mining Engineering* 14:60–62.

Newhouse, M. 1979. "Epidemiology of Asbestos-Related Disease." Vol. 1, In *Asbestos: Properties, Applications, and Hazards,* Eds. L. Michaels and S. S. Chissick, 465–483. Chichester: John Wiley & Sons.

Oldham, J. W. 1968 "A Short Note on the Recent Geological Mapping of the Shabani Area." In *Symposium on Rhodesian Basement Complex* (Trans. Geol. Soc.) South Africa: 71:189–194.

Parekh, P. P., R. J. Janulis, J. S. Webber, and T. M. Semkow. 1992. Quantitation of asbestos in synthetic mixtures using instrumental neutron activation analysis. *Analytic Chem.* 64:320–325.

Petrov, V. P., and Y. S. Znamensky. 1981. "Asbestos Deposits of the USSR." In *Geology of Asbestos Deposits,* Ed. P. H. Riordon, 45–52. New York: American Institute of Mining, Metallurgical, and Petroleum Engineers.

Pundsack, F. L. 1955. The properties of asbestos. I. The colloidal and surface chemistry of chrysotile. *Jour. Phys. Chem.* 59:892–895.

Rowbotham, P. I. 1970. World asbestos industry. *Ind. Minerals.* (January):17–29.

Skinner, H. C. W., M. Ross, and C. Frondel. 1988. *Asbestos and Other Fibrous Materials.* New York and oxford: Oxford Press, 204p.

Uno, Y. 1992. Optical identification of asbestos by dispersion staining method. *Jour. Clay Sci. Soc.* (in Japanese with English abstract) Japan: 32 :42–52.

Van Biljon, W. J. 1964. "The Chrysotile Deposits of the Eastern Transvaal and Swaziland." Vol. 2, In *The Geology of Some Ore Deposits in Southern Africa,* Ed. S. H. Haughton, 625–669. South Africa: Geol. Soc.

Virta, R. L. 1996. "Asbestos." *U.S.G.S. Annual Review* (July): 7p.

Virta, R. L., and E. L. Mann. 1994. "Asbestos." 6th ed. In *Industrial Minerals and Rocks,* Ed. D. D. Carr, 97–124, Littleton, Col.: Soc. Mining, Metal. & Exploration.

Whittaker, E. J. W. 1979. "Mineralogy, Chemistry and Crystallography of Amphibole Asbestos." In *Mineralogical Techniques of Asbestos Determination,* Ed. R. L. Ledoux, 1–34. Quebec, Canada: Mineral. Assoc. Canada Short Course.

Whittaker, E. J. W., and F. J. Wicks. 1970. Chemical differences among the serpentine "polymorphs": A discussion. *Amer. Mineral.* 55:1025–1047.

Wicks, F. J., and E. J. W. Whittaker. 1975. A reappraisal of the structures of the serpentine minerals. *Can. Mineral.* 13:227–243.

Wicks, F. J., and J. Zussman. 1975. Microbeam x-ray diffraction patterns of the serpentine minerals. *Can. Mineral.* 13: 244–258.

Wilson, J. P. 1968. "The Mashuba Igneous Complex and its Subsequent Deformation." In *Symposium on Rhodesian Basement Complex* (Geol. Soc.) South Africa Annexure: 71 :175–188.

Yada, K. 1971. Study of microstructure of chrysotile asbestos by high resolution electron microscopy. *Acta Cryst.* A27:659–664.

Zussman, J. 1979. "The Mineralogy of Asbestos." Vol. 1, In *Asbestos: Properties, Applications, and Hazards,* Eds. L. Michaels and S. S. Chissick, 45–65. Chichester: John Wiley & Sons.

Barite

Barite ($BaSO_4$) is the most common barium mineral, and its abundance has facilitated the manufacture of barium chemicals. As an industrial mineral, barite is characterized by its specific gravity and refractive index. It is mainly used as a weighing material in well-drilling mud. Other uses are as a filler or extender in paint, plastic, and rubber products, as an oxidizer and decolorizer in glassmaking, and as a white pigment. Barium titanate and ferrite are important ceramics. The mineral is sometime referred to as "heavy spar" or "baryte."

Among a dozen major barite producers in the world, the top-ranked producers are China, the former USSR, Mexico, and the USA. World production of barite was estimated to be 5.5 M tons in 1992 (Searls 1992).

MINERALOGICAL PROPERTIES

Barite is orthorhombic with a=8.8842Å, b=5.4559Å, and c=7.1569Å (Chang et al. 1996). The structure consists of S and Ba polyhedra linked together by sharing oxygens as illustrated in Fig. 4. Both polyhedra show distortions. The interatomic distances for SO_4 range from 1.45 to 1.48Å (Colville and Staudhammer 1967; Hill 1977). The Ba atom is surrounded by a very irregular array of twelve oxygen atoms at distances from 2.8 to 3.3Å. Its environment is largely determined by the bonding requirements of the seven neighboring SO_4 groups. This large coordination site has set the limits for Ba substitution by common rock-forming cations that have, in general, smaller size. A complete series of solid solution forms between $BaSO_4$ and $SrSO_4$, but generally most barites have nearly pure $BaSO_4$ and contain remarkably little strontium. From synthetic studies and from Starke's compilation of 2,293 analyses of barite, Hanor (1968) concluded that $(Ba,Sr)SO_4$ is an inert phase for Ba-Sr exchange, and the inert behavior can account for the paucity of intermediate compositions in the barite-celestite series. Minor amounts of Ba being replaced by Pb and Ca are well documented, but rare. Soriano et al. (1991) used $d_{(210)}$ to determine strontium content in barite up to 12%, with a maximum absolute error of 0.5% $SrCO_3$, and $d_{(121)}$ has been used to estimate the composition in the $BaSO_4$-$SrSO_4$ series by Goldish (1989).

The solubility of barite is very low. Experimental data obtained demonstrate that in H_2O solution from 22° to 280°C and l to 1400 bars, and in 0.2 and 4 molal NaCl solutions from 100° to 250°C and 1 to 500 bars, the solubilities range between 10^{-6} and 10^{-8} molal (Blount 1977). Isobaric solubilities are maximum near 100°C in H_2O, and the maxima shift to higher temperature in solutions with increasing NaCl concentrations. Isothermal solubilities increase with rising NaCl concentration and pressure. For the reaction $BaSO_4(c) = Ba^{2+}(aq) + SO^{2-}_4(aq)$, logK, ΔG, ΔH, and ΔS at 298°K and l bar are 9.98, 13.61 kcal/mole, 6.3 kcal/mole, and −24.5 cal/mole–K, respectively. Barite has been prepared by combining dilute barium chloride and sodium sulfate solutions (Brower 1973; Blount 1974).

Most barite is white, but natural varieties in buff, gray, yellow, and red are also common, resulting from the presence of impurities, and blue barite is apparently a result of radiation from radium. The luster of barite is vitreous and its streak is white.

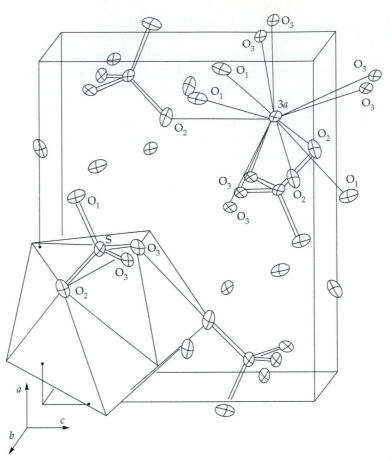

Figure 4 Crystal structure of barite showing the S and Ba polyhedra linkage in a unit cell. Thermal ellipsoids for all atoms represent 50% probability surface, after Hill (1977).

Crystals of barite are usually tabular, and also occur as globular concretions and as granular, fibrous, and earthy aggregates. Barite is brittle and shows perfect cleavage on {001}. Its hardness is 3 to 3 1/2 and its specific gravity about 4.5. The refractive indices of barite are $\alpha = 1.636\text{-}1.637$, $\beta = 1.637\text{-}1.639$, and $\gamma = 1.647\text{-}1.649$ (Burkhard 1978).

GEOLOGICAL OCCURRENCE AND DISTRIBUTION OF DEPOSITS

Deposits of barite are generally classified by their mode of occurrence into three major types: vein and cavity filling, residual, and bedded (Brobst 1958, 1970, 1984, 1994; Sackett 1962). The vein- and cavity-filling deposits occur mainly in fault and breccia zones at low temperatures and in association with dolomite, fluorite, and metalliferous ores. The barite deposits of the North Pennine ore fields in England consist of veins and replacements in limestone and have their origin in hydrothermal solutions derived from Pennine magma (Durham 1948). Fluorite is found in the inner zones and barite in the outer zones of low temperature. The barite vein deposits in the vicinity of Kutaisi in Georgia, the former USSR, varying from 0.3 to 1.0 m wide and up to 2,000 m long, perforate the Jurassic porphyry tuffs. The content of $BaSO_4$ varies between 81.6 and 98.6% (Tatarinov 1955). Vein- and cavity-filling deposits are widely scattered in the United States (Brobst 1983). Barite deposits in collapse and sink structures are common in the Central district of Missouri, and in the Appalachian states. Many deposits in the western states are associated with igneous rocks of the Tertiary age. In Mexico, large vein barite deposits occur in the Muzquiz, Coahuila, and Galeana districts. Most of the barite deposits in Morocco and Algeria are of the vein type (Brobst 1984, 1994; Griffiths 1995).

The barite of the residual deposits commonly occurs in clays derived from weathering of limestone (Brobst 1958, 1994). In southeast Missouri, extensive low-grade deposits of barite occur in residual clay derived from underlying dolomite. Many deposits are as large as 100 acres and 1.2 to 12 m in thickness. In central Missouri, high-grade barite deposits of both residual and vein fillings were found in carbonate host rocks with galena, sphalerite, and chalcopyrite (Leach 1980). Residual barite deposits also occur in the Cartersville district, Georgia, where iron oxide-coated barite is mined from yellow to brown clay, and in the Sweetwater district, Tennessee, where barite is mined from residual clay overlying bedrock with veins and shatter zones consisting of fluorite, pyrite, and barite.

The bedded deposits occur generally in shale and limestone, in which the barite is in association with chert, pyrite, iron oxides, carbonates, and clay minerals. The barite of this type, in contrast with the white barite from vein fillings and residual deposits, is usually dark gray to black and fine-grained. The "Barite belt" of Nevada covers an area across the central part of the state in a northeast-trending zone up to 95 m wide and 480 m long (Papike 1984). Barite beds generally interlayer with chert and minor amounts of limestone, and crop out as lensoid bodies. This led Poole (1988) to suggest a formation as mounds around hydrothermal veins. Large bedded barite deposits occur in the Ouachita Mountains of Arkansas. The deposits occur at the base of the Stanley shale of Mississippian age, overlying a thick sequence of bedded cherts. Pb and Zn mineralization is absent in the barite deposits (Mitchell 1984; Jewell and Stallard 1991). Bedded barite deposits of Middle Devonian age occur in Germany at Meggan in the Rhenish Schiefergebirge (Krebs 1981). Unlike the Nevada and Arkansas occurrences, there is a major accumulation of Pb-Zn sulfides at Meggen. Host sediments are black shales with some sandstone, similar to the shale of the Arkansas deposits, but chert is lacking at Meggen. Sizable deposits of barite in association with shale-hosted lead and zinc mineralization occur in the lower Paleozoic strata of the Selwyn basin in the Canadian Cordillera (Carne and Cathro 1982; Morin 1986).

Large bedded barite deposits occur in the lower Cambrian black shale series of south China in the Yangtze platform area and the adjacent Qinling geosyncline to the north. The former extends from Quangxi and Guizhou Provinces through Hunan and Jangxi to Anhui, whereas the latter has the range from Sichuan Province along the border between Hubei and Henan to Anhui (Chu 1989; Wang and Li 1991). The barite ore can be divided into four textural types: (1) massive, (2) laminated, (3) banded, and (4) nodular. There are carbonaceous mudstone interlayers in the barium ore beds, and organic carbon content may reach as high as 11 wt%.

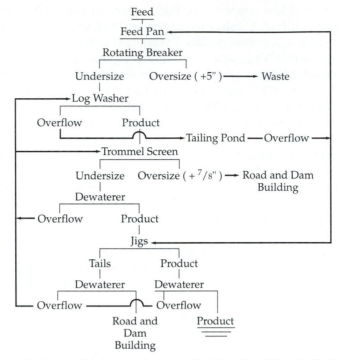

Figure 5   A flowsheet for the beneficiation of barite from residual deposits in Missouri, after Sackett (1962).

INDUSTRIAL PROCESSES AND USES

1. Milling and Beneficiation

Treatment of barite varies with the nature of the ore and the type of product to be made. For drilling mud, the only requirements for barite are to be fine grain and to have high specific gravity. Grinding may be the only treatment needed, and this can be done either wet or dry and in either ball mills or Raymond mills. When barite is used as a filler and extender, additional treatment such as bleaching is necessary in most cases to provide the required white color.

The milling process is relatively simple for barite from soft, residual deposits. Adhering clay is scraped off manually, using hatchets, or roughly picked and dried ore is rattled in a rocker on a screen and the undersize ore is hand-jigged. At larger plants, as illustrated in Fig. 5 for barite from Missouri, rotating breaker, log washer, and Trommel screen are used to separate the ore from clay- and gravel-sized gangue minerals (Sackett 1962). The residual barite from Georgia is usually hard and must be crushed to free the barite from the gangue minerals, and the milling process takes several stages of crushing, screening, jigging, and tabling. For barite coated with or in intimate association with iron oxides, it must go through magnetic separation to reduce the iron content to below 1%.

Hard, vein barite is usually pure enough for the market without beneficiation. The ore from some bedded deposits also requires no beneficiation to meet the specifications for drilling mud. Barite from either vein or bedded deposits commonly forms intergrowths with fluorite. Because of this, beneficiation by simple gravity separation is ineffective, and it becomes necessary to treat the barite with a flotation process (Eddy and Browning 1964). The bedded barite from Arkansas is so finely divided and so intimately mixed with gangue minerals that it must be milled by double crushing, jigging, and wet grinding in a ball mill to pass 325 mesh, followed by froth flotation (Brobst 1994). Barite floats readily with fatty-acid collectors and with the sulphate and sulphonate salts of the corresponding alcohols (Norman and Lindsay 1948). Sodium silicate is a sufficient depressant for iron-rich gangues, but calcium-bearing gangues requires metal salt-silicate treatment, and grade of concentrate is improved by adding dichromate in cleaning. Concentrates are filtered and dried in rotary kilns at temperatures high enough to destroy the reagents that might interfere with use in drilling mud.

The process for bleaching barite varies in different producing plants. Generally, the first step is to wet-ground the barite in a pebble mill. The thickened sludge is then treated in a ceramic-laid bleaching tank with hot, dilute sulfuric acid, agitated continuously for 8 to 12 hours, washed thoroughly, and dried.

2. Industrial Uses

By far the largest single use of barite is as a weighing agent in oil-well drilling mud. This use accounted for approximately 90% of total industrial consumption of barite in recent years. In the rotary drilling wells for oil and gas exploration, a fluid, generally known as drilling mud, circulates through the drilling string. Drilling mud has five essential functions: (1) to lift drill cuttings to the surface, (2) to control well pressure, (3) to maintain borehole stability, (4) to lubricate the drill string, and (5) to protect the producing zone (Lundie 1986). Because of its high specific gravity and chemical inertness, barite is added to drilling mud in order to provide the density needed to counteract dangerous formation pressure, and therefore to control well pressure. The deeper the well is drilled, the more barite is needed as a percentage of the total mud mix, because formation pressures rise sharply with increasing depth. In the past forty years, the average depth of oil wells in the United States has increased about 60% from 4,800 km to 8,000 km. The amount of barite required for 0.3m (1 ft) of drilling has increased almost eightfold, from 1 kg to 8.5 kg (Harben 1980). The major constituents of drilling fluids besides barite are bentonite, which is used as a viscosifier and a suspension agent; soda ash to precipitate dissolved calcium and magnesium from saline water; and caustic soda or lime to control pH and to reduce corrosion.

Current industrial specifications for barite for drilling mud, although under continual review, as set by the American Petroleum Institute (Bulletin RP 13A June 1985), are as follows: (1) specific gravity, 4.20, minimum; (2) soluble alkaline earth metals as calcium, 250 ppm, maximum; (3) wet screen analysis, residue on US sieve No. 200, 3.0%, maximum; and (4) particles less than 6 micrometers, 30%, maximum.

Barite acts as a principal source of barium oxide in the manufacture of glass. Barite is used in glass manufacturing to provide a high refractive index; to impart greater hardness and scratch resistance; to improve gloss, resistance to chemical attack, and flow properties to molten glass; and to act as a radiation filter in cathode-ray tubes (CRTs) (Griffiths 1984). Before barite can be used in glass production, it must first be converted to barium carbonate. This is accomplished by means of a reduction process which uses coal or other forms of carbon at 1000° to 1200°C in a rotary kiln (Griffiths 1984). The barium sulfide produced from this reduction process is then reacted with either carbon dioxide in solution or soda ash to produce barium carbonate. Specifications for barite in the manufacture of glass are set by various users, and range from a minimum of 98% $BaSO_4$ and a maximum of 1.5% SiO_2, 0.15% Al_2O_3, and 0.15% Fe_2O_3, with a mixture of sizes generally ranging from 40 to 140 mesh.

The barium carbonate produced from barite has many uses in ceramics: in ceramic tiles and bricks for better preservation of colors, in glaze mixes for better hardness and luster, and in enamels for better resistance to corrosion. Barium carbonate of high purity is the essential constituent in the manufacture of electroceramic barium titanate and magnetoceramic barium ferrites.

Ground barite is an important industrial filler and is used in paint, plastics, rubber, and paper (Brobst, 1994). The characteristic properties which contribute to this use are density, brightness, low abrasiveness, inertness, resistance to corrosion, and radiation absorption. The paint industry is by far the largest consumer of filler-grade barite, which accounts for more than 50% of total use, followed by rubber and plastic industries with an estimated 20 to 28% of the total use. Barite bleached with sulfuric acid is used as an extender in white lead paint because of its weight. The specifications of barite for filler and extender are set by various users, but most require a fine-grained product that is virtually all −325 mesh. Color is important to many users.

Two barium chemicals, Blanc fixe and lithopone, derived from barite, have well-established uses as filler, extender, and white pigment. Blanc fixe is a precipitated barium sulfate with characteristic properties such as insolubility in water and organic binder, and a high degree of whiteness and homogeneous granulation (Brobst 1983). Lithopone is a coprecipitate from solutions of barium sulfide and zinc sulfate and has a composition of approximately 70% $BaSO_4$ and 30% ZnS. It is used extensively in the manufacture of paints, both as a white pigment alone and mixed with other pigments. Lithopone is also used in the manufacture of high-graded rubber products such as automobile tires, and as a filler in linoleum and oilcloth. In recent years, Blanc fixe and lithopone have suffered from market competition from the use of titanium dioxide, calcium carbonate, and other pigments and fillers (Griffiths 1984).

In the construction industry, lump barite is used in concrete aggregates to weigh down pipelines buried in marshy areas and to shield nuclear reactors. A mixture of rubber and asphalt with about 10% barite has been used as a durable paving material for roads, runways, and parking lots.

Barite is used as an inert filler in friction materials including brake linings and clutch facings. Typically, brake linings require barite which is 99% less than 40 micrometers and has a minimum specific gravity of 4.3 (Griffiths 1992).

REFERENCES

Blount, C. W. 1974. Synthesis of barite, celestite, anglesite, therite, and strontianite from aqueous solutions. *Amer. Mineral.* 59:1209–1219.

Blount, C. W. 1977. Barite solubilities and thermodynamic quantities up to 300°C and 1400 bars. *Amer. Mineral.* 62:942–957.

Brobst, D. A. 1958. Barite resources of the United States. *U.S.G.S. Bull.,* 1072-B:67–130.

Brobst, D. A. 1970. Barite: world production, resources, and future prospects. *U.S.G.S. Bull.,* 1321:46p.

Brobst, D. A. 1983. "Barium Minerals." 5th ed. In *Industrial Minerals and Rocks,* Ed. S.J. Lefond, 485–501. New York: Soc. Mining Eng., AIME.

Brobst, D. A. 1984. The geological framework of barite resources. *Trans. Inst. Mining & Metall.* (London): Sect. A, Mining Industry, 93:123–130.

Brobst, D. A. 1994. "Barium Minerals." 6th ed. In *Industrial Minerals and Rocks,* Ed. D.D. Carr, 125–134. Littleton, Col.: Mining, Metallurgy, and Exploration, Inc.

Brower, E. 1973. Synthesis of barite, celestite and barium–strontium sulfate solid solution crystals. *Geochim. Cosmochim. Acta.* 37:155–158.

Burkhard, A. 1978. Barytre-Celestin und ihre Mischkristalle aus Schweizer Alpen und Jura. *Schweiz. Min. Petr., Mitt.* 58:1–96.

Carne, R. C., and R. J. Cathro. 1982. Sedimentary exhalation (sedex) zinc-lead-silver deposits, northern Canadian Cordillera. *Canadian Inst. Mining and Metall. Bull.* 75:66–78.

Chang, L. L. Y., R. A. Howie, and J. Zussman. 1996. "Rock-Forming Minerals." Vol. 5, In *B—Non-silicates: Sulphates, Carbonates, Phosphates, Halides.* Essex, England: Longman, 383p.

Chu, Y. 1989. Genetic types of barite deposits in China. *Mineral Deposits.* 8:91–96.

Colville, A. C., and K. Staudhammer. 1967. A refinement of the structure of barite. *Amer. Mineral.* 52:1877–1881.

Durham, K. C. 1948. "Geology of the Northern Pennine Orefield." 2d ed. Vol. 1, In *Tyne to Stainmore.* Great Britain: Mem. Geol. Survey, 128p.

Eddy, W. H., and J. S Browning. 1964. Selective flotation of a barite–fluorspar ore from Tennessee. *Rept. Invest.* RI–6419. U.S. Bureau of Mines, 8p.

Goldish, E. 1989. X-ray diffraction analysis of barium-strontium sulfate (barite–celestite) solid solutions. *Powder Diff.* 4:214–216.

Griffiths, J. 1984. Barytes: non-drilling applications. *Ind. Minerals.* (June):21–35.

Griffiths, J. 1992. Barytes fillers in recession—more of a weighting game. *Ind Minerals.* (February):39–49.

Griffiths, J. 1995. Barytes supply not at oil well. *Ind. Minerals.* (July):25–33.

Hanor, J. S. 1968. Frequency distribution of compositions in the barite–celestite series. *Amer. Mineral.* 53:1215–1222.

Harben, P. 1980. US market—Barytes pressured by drilling surge. *Ind. Minerals.* (March):74–77.

Hill, R. J. 1977. A further refinement of the barite structure. *Canadian Mineral* 15:522–526.

Jewell, P. W., and R. E. Stallard. 1991. Geochemistry and paleoceanographic setting of central Nevada bedded barites. *Jour. Geol.* 99:151–170.

Krebs, W. 1981. "The Geology of the Meggen Ore Deposit." In *Handbook of Stratabound and Stratiform Deposits,* Ed. K. H. Wolf, 509–549. Amsterdam: Elsevier.

Leach, D. L. 1980. Nature of mineralizing fluids in the barite deposits of central and southeast Missouri. *Econ. Geol.* 75:1168–1180.

Lundie, P. 1986. Standardization of minerals for drilling fluids. *Ind. Minerals.* (March):113–117.

Mitchell, A. W. 1984. "Barite in the Western Ouachita Mountains, Arkansas." In *A Guidebook to the Geology of the Central and Southern Ouachita Mountains, Arkansas,* Eds. C. G. Stone and B. R. Haley, 84-2: 124–131. Little Rock, Arkansas: Arkansas Geol. Comm. Guidebook.

Morin, J. A. 1986. "Mineral Deposits of the Northern Cordillera." Spe. Vol. 37, In *Proc. The Mineral Deposits of Northern Cordillera Sym.* Canadian Inst. Mining & Metall., 378p.

Norman, J., and B. S. Lindsey. 1948. Flotation of barite from Magnet Cove, Arkansas. *Mining Technology,* AIME, Tech. Publ., 1326.

Papike, K. G. 1984. Barite in Nevada: Nevada Bureau. *Mines and Geol., Bull.,* 98, Reno, Nevada.

Poole, F. G. 1988. "Stratiform Barite in Paleozoic Rocks of the Western United States. In *Proc., 7th IAGOD Symposium.* Stuttgart: E. Schweizerbartsche Verlagsbuch., 309–319.

Sackett, E. L. H. 1962. Barite, little-known industry that means "mud" to oil men. *Mining Eng.* 14:46–49.

Searls, J. P. 1992. Barite. *U.S.B.M. Minerals Yearbook,* 235–245.

Soriano, M. C. O., J. B. Ráfales, and J. G. Marfinez. 1991. X-ray diffraction analysis of strontium in barite. *Powder Diff.* 6:70–73.

Tatarinov, P. M. 1955. *Conditions of genesis of ore deposits and industrial minerals and rocks.* (in Russian). Moscow: Gostekhiz-dat., 1–280.

Wang, Z. C., and G. Li. 1991. Barite and witherite deposits in Lower Cambrian shales of South China: Stratigraphic Distribution and geochemical characterization. *Econ. Geol.* 86:354–363.

Bauxite

Bauxite is an earthy aggregate of aluminous minerals that is commonly used to represent the aluminum-rich members of the complete series of aluminous-ferruginous laterites. It is the basic ore of the aluminum industry. Over 90% of about 100 M tons mined annually is refined and smelted into aluminum. The remaining 10% is used to produce alumina for ceramics, refractories, abrasives, chemicals, and cements. The Bayer process, although modified through the years since it was first used in the 1930s, is the principal method used in the treatment of bauxite.

Bauxite deposits occur in every continent except Antarctica.

MINERALOGICAL PROPERTIES

Bauxite is composed principally of one or a combination of the following three aluminum hydrate minerals: (1) gibbsite ($Al(OH)_3$), (2) boehmite ($AlOOH$), and (3) diaspore ($AlOOH$). Generally, mixtures of gibbsite and boehmite are common, boehmite and diaspore less common, and gibbsite and diaspore rare. The mineralogical properties of bauxite as well as the chemical treatment used to process it are determined by the relative content of the three aluminum hydrate minerals. Each of them has characteristic mineralogical properties (Hose 1963; Wefers and Misra 1987).

Associated minerals include clay minerals (kaolinite, halloysite), quartz, iron oxide (hematite), and hydroxide (goethite), titanium minerals (rutile, anatase, leucoxene), iron sulfides (pyrite, marcasite), and carbonates (calcite, siderite, dolomite). The sulfides and carbonates are in minor amounts and of local occurrence only.

1. Aluminum Hydroxide Minerals

There are two distinct groups of aluminum hydrate minerals. The trihydrates ($Al(OH)_3$) include bayerite (α-form), nordstrandite (β-form), and gibbsite (γ-form), and the monohydrates ($AlOOH$) include diaspore (α-form) and boehmite (γ-form). In addition, gelatinous hydrates may consist of predominately amorphous form or gelatinous boehmite. All $Al(OH)_3$ forms have similar crystal structures, showing (OH) octahedra in which two-thirds of the centers are occupied by Al ions linked to layers with pseudohexagonal arrangement of Al ions. The various polymorphic forms are defined by the different stacking of the successive layers. The double layer in gibbsite as shown in Fig. 6a (Valeton 1972; Saalfeld and Wedde 1974) is located with respect to its neighbors so that (OH) of adjacent layers are situated directly opposite each other. The monoclinic and triclinic symmetries of gibbsite are the result of slight displacement of layers in the direction of a-axis and of both a- and b-axis, respectively. The (OH) octahedra are linked to one another through weak hydrogen bonds only.

The crystal structure of diaspore consists of layers of (OH), the sequence of which is that of a hexagonal close-packing. Aluminum ions occupy octahedrally coordinated sites between layers in such a way as to form strips of octahedra, the direction of which

27

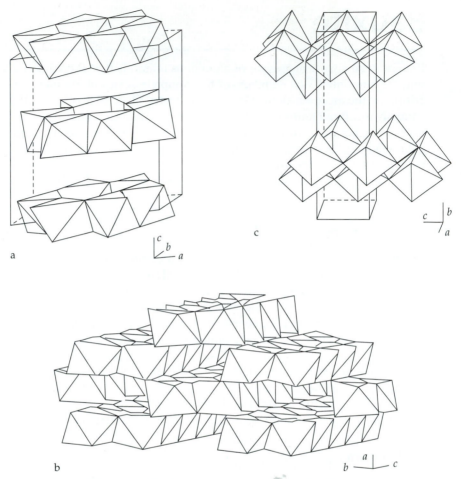

Figure 6   Crystal structures of (a) gibbsite, (b) diaspore, and (c) boehmite. Reprinted from Valeton, *Bauxite, Developments in Soil Science.* Copyright 1972, with permission from Elsevier Science.

defines the c-direction. The strips have the width of two octahedra and yield an or-thorhombic unit cell as shown in Fig. 6b. A neutron diffraction study of diaspore showed that the hydrogen ions adhere more closely to one of the two oxygen ions and are dis-placed from the straight line joining them (Busing and Levy 1958). The structure of boehmite is made up of double layers of (OH) octahedra with aluminum ions at their centers (Fig. 6c). The layers themselves are composed of chains of octahedra, the repeat distance of which defines the a-dimension. In the a-direction octahedra share edges, but the chains are linked laterally in the c-direction by sharing corners (Bezjak and Jelenic 1964; Christoph et al. 1979). The structure contrasts with that of diaspore in two princi-pal ways. First, in diaspore, every oxygen is hydrogen-bonded to one other oxygen; in boehmite, half of the oxygen is not involved in hydrogen-bonding at all, and each of the remaining is hydrogen-bonded to two oxygens. In diaspore the oxygens are in a close packed hexagonal layer; those within the double octahedral layers in boehmite are in a

Table 2 Properties of alumina hydrate minerals modified, after Hose (1963).

Mineral	Gibbsite	Boehmite	Diaspore
Chemical Formula	$Al_2O_3 \cdot 3H_2O$	$Al_2O_3 \cdot H_2O$	$Al_2O_3 \cdot H_2O$
$Al_2O_3\%$	65.4	85.0	85.0
$H_2O\%$	34.6	15.0	15.0
Crystallography			
Crystal System	Monoclinic	Orthorhombic	Orthorhombic
Unit Cell Parameters			
a, Å	8.64	3.69	4.40
b, Å	5.07	12.2	9.42
c, Å	9.72	2.86	2.85
beta	94°34′	90°	90°
Z	8	4	4
Refractive Index			
alpha	1.56 - 1.58	1.64 - 1.65	1.68 - 1.71
beta	1.56 - 1.58	1.65 - 1.66	1.71 - 1.72
gamma	1.58 - 1.60	1.65 - 1.67	1.73 - 1.75
Color	white, gray	white	colorless, yellow, brown
Cleavage	(001) perfect	(010) perfect	(010) perfect
Hardness (Moh's)	2½ - 3½	3¼ - 4	6½ - 7
Specific Gravity	2.4	3.1	3.3 - 3.5

cubic close-packed relationship. These differences in oxygen packing are consistent with the behavior of the two polymers of (AlOOH) on dehydration, in that diaspore yields α-Al_2O_3 (hexagonal), and boehmite yields γ-Al_2O_3 (cubic). Diaspore is the stable form of AlOOH, and boehmite is considered to be metastable, although it is the prevailing AlOOH that forms at temperatures below 300°C. Spontaneous nucleation of diaspore requires temperatures in excess of 300°C and pressures higher than 20 Mpa. Under equilibrium vapor pressure of water, gibbsite converts to AlOOH at about 100°C (Ervin and Osborn 1951; Neuhaus and Heide 1965).

A summary of the physical, crystallographical, and optical properties of gibbsite, boehmite, and diaspore is given in Table 2. Of the two polymors of AlOOH, diaspore has a greater hardness. This may be related to its tightly bonded and denser structure. The bond energy for the hydrogen bridge in diaspore is 28.7 Kj/mole, compared with 20.1 Kj/mole for boehmite, as determined in infrared studies (Glemser and Hartert 1956). Identification of the different hydroxides and oxides is best carried out by x-ray diffraction analysis.

Gibbsite is quite soluble in 10 to 20% solutions of sodium hydroxide at 142°C and a pressure below 10 atmospheres (Newsome et al. 1960). Under these conditions, boehmite is only slightly soluble. Boehmite, however, is quite soluble in 20 to 30% NaOH solutions at 225°C, but NaOH becomes reactive with hydrated aluminum silicates. Therefore boehmitic bauxite must contain low amounts of hydrated aluminum silicates if the production of aluminum trihydrate from them is to be economically feasible. Diaspore requires heating treatment at 600° to 700°C before it becomes soluble in sodium hydroxide solutions under the conditions at which boehmite is soluble.

2. Associated Minerals

Hematite (α-Fe_2O_3) and goethite (γ-FeOOH) are the most common iron minerals in bauxite (Hose 1963; Tardy and Nahon 1985). In both minerals, some of the iron may be replaced isomorphically by aluminum. This amount of aluminum is usually included in the chemical analysis of the bauxite, but is normally not extracted in the digest because these iron minerals are inert under the conditions of chemical treatments such as the Bayer process. Other iron minerals, including pyrite (FeS_2), siderite ($FeCO_3$), and magnetite (Fe_3O_4), may occur in some bauxite.

SiO_2 may occur as quartz, but most commonly it is present as a constituent in aluminosilicate minerals. Main aluminosilicate minerals are kaolinite, occurring in both well and poorly crystallized forms, and halloysite, which is commonly unrolled. Excess amounts of aluminosilicate cause loss of sodium hydroxide and extractable alumina because they react with sodium aluminate solution to form insoluble sodium aluminate silicates during digest in the Bayer process (Adamson et al. 1963).

Titanium dioxide, TiO_2, in the form of anatase is a common mineral in bauxite. Rutile and ilmenite are also present, but in minor quantities. The titanium oxides have no effect upon the digestion of bauxite.

3. Bauxite Mineralogy

Bauxite is an earthy aggregate of aluminum hydrate minerals occurring in a variety of colors that range from light gray, cream, pink or yellow, to dark brown and dark red. The color is largely determined by the type and particle size of the prevalent iron mineral. Highly dispersed goethite tends to be yellow and brown, whereas dark red tones usually are associated with coarser hematite. Bauxite has a cellular, porous, or fine-grained compact texture and conchoidal or uneven fractures. Oölitic and pisolitic textures are common, with rounded concretionary grains embedded in a clayey mass, and may extend throughout a deposit or series of deposits (Harder and Greig 1960; Hill and Ostojic 1984).

Geologically old diaspore bauxites are very hard and can reach a specific gravity of 3.6, whereas young bauxites with predominantly gibbsite are soft and have a specific gravity around 2.0 to 2.5. Chemical compositions of bauxite depend upon the type of aluminum hydrates it contains. Ranges of chemical compositions of bauxite from the world's major deposits are listed in Table 3.

There are different hypotheses used to explain the origin of bauxite formation, but it is agreed generally that most bauxite deposits are residual products resulting from highly intensified weathering of high alumina rocks by the solution and removal of constituents other than alumina such as soda, potash, lime, magensia, and silica. The favorable geological conditions for bauxite formation as described by Shaffer (1983) are (1) high rock permeability to permit desilication, (2) tropical climate with abundant rainfall and alternating wet and dry seasons to promote leaching, (3) low to moderate topographical relief to allow drainage and fluctuation in groundwater levels, and (4) a low rate of erosion and lengthy period of stability to permit the accumulation of weathered products. Kopeykin (1984) emphasized the temperature factor in lateritic bauxitization. Extensive deposits of bauxite developed in tropical zones with a rainwater temperature of 25°C and $P_{carbon\ dioxide} = 10^{-3.5}$ atm. A comparison of the chemical

Table 3 Range of major chemical constituents of bauxite, after Shaffer (1983).

	Al_2O_3, %	SiO_2, %	Fe_2O_3, %	TiO_2, %	Loss on Ignition, %
Australia					
Cape York	53–60	2–10	5–13	2.1–3.1	21–29
Gove	52	2–6	13–28	3.4	26.3
Darling Ranges	25	0.5–2.5	15–20		
Kimberly Region	47–50	2.5–3.5			
Brazil					
Minas Gerais	55–59	1.6–5.5	6.9–9.6		
Amazon	50–61	3.7–9.0	1.7–14	1.1–2.0	25–30
China	50–70	9–15	1–13	2	
Dominican Republic	46–49	1.6–5.2	19–21		
France	55–70	3–16	4–25	2–3.5	
Ghana					
Yenahin	41–63	0.2–3.1	1.2–30.9	1.5–5.3	20–29
Awaso	48–61	0.4–2.4	4–22	0.8–2.1	26–33
Kibi	32–60	0.3–2.9	6–45	2.0–6.2	13–30
Greece	35–65	0.4–3.0	7.5–30	1.3–3.2	
Guinea	40–65	0.5–5	2–30	3–5	22–32
Guyana	51–61	4–6	1–8	2–3	25–32
Haiti	46.8	3.4	21.9	2.8	24.1
Hungary	50–60	1–8	15–20	2–3	13–20
India	45–60	1–5	3–20	5–10	22–27
Indonesia	37–56	4–16	4–26		
Jamaica	49–51	0.7–1.6	19–21	2.5–2.7	25–27
Malaysia	38–60	1–13	3–21	1–2	
Sierra Leone	51–55	1.5–2	10–18	1.5	27–31
Suriname	50–60	2–6	2–15	2–3	29–31
United States					
Arkansas	45–57	5–24	2–12	1.6–2.4	22–28
Oregon & Washington	31–35	5–11	33–35	5–6	16–20
Southeastern states	51–56	12–15	1–5	1.5–3.5	22–30
USSR	26–52	2–32	1–45	1.4–3.2	
Yugoslavia	48–60	1–8	17–26	2.5–3.5	13–27
Romania	55	5	22	1–2	
Turkey	55–60	5–7	15–20	2–3	12–14

composition of parent rocks and associated bauxite is shown in Table 4. The relationship between carbonate rocks and the genesis of bauxite was calculated by Hidasi (1986) for some Hungarian bauxites. The weathering of 200 m of Triassic limestone, containing an average of 3.6% Al_2O_3, may produce 1 m of bauxite. The formation of residual bauxite deposits in Arkansas was described by Salter and Murray (1989) as having two stages: kaolinization of syenite followed by bauxitization.

In bauxite deposits, kaolinite and halloysite are by far the most common clay minerals. Bates (1962) observed a transition from halloysite through an intermediate stage of alumina gel to gibbsite. According to Wollast (1963), the solubility of amorphous aluminum hydrates exceeds the surrounding gibbsite values by a factor of 10^4. Therefore, clay minerals will normally form from the reaction of silica with amorphous aluminum

Table 4 Chemical compositions of parent rocks and associated bauxites after Shaffer (1983).

	Al_2O_3, %	SiO_2, %	Fe_2O_3 Plus FeO, %	TiO_2, %	Loss on Ignition, %
Arkansas					
Nepheline syenite	21.1	56.5	3.0	0.5	—
Bauxite	59.8	5.3	1.9	2.0	30.6
Australia					
Kaolinitic sandstone	5.2	91.1	0.5	0.3	1.8
Bauxite	50.0	9.6	9.2	2.2	26.6
Brazil					
Kaolinitic clay	37.9	30.4	14.5	1.5	15.8
Bauxite	51.3	3.0	17.5	1.1	27.1
Phonolite	20.9	53.1	5.3	0.4	1.7
Bauxite	55.1	5.5	9.6	1.8	27.7
Ghana					
Shale	18.2	61.0	4.7	1.0	7.5
Bauxite	60.5	1.4	9.8	2.2	25.6
Guinea					
Diabase	12.4	51.3	9.5	0.7	0.4
Bauxite	60.2	2.7	3.9	1.6	32.0
Guyana					
Epidiorite	18.9	49.1	6.4	0.9	0.4
Bauxite	56.4	3.1	11.6	0.8	27.8
Dolerite	17.3	52.0	11.2	0.5	—
Bauxite	59.0	9.8	1.1	0.7	28.1
India					
Basalt	16.8	47.4	14.9	1.7	2.2
Bauxite	56.4	1.7	7.1	8.9	25.4

hydrates. Garrels and Christ (1965) calculated the equilibrium of kaolinite and gibbsite at 25°C and 1 atmosphere. Results show that equilibrium is attained at a fixed value of the activity of dissolved silica of about 2 ppm. Below this value, kaolinite tends to dissolve incongruently to leave a residuum of gibbsite.

Instrumental analysis of bauxite has been described in some detail by Hose (1963) and Som and Bhattacharya (1995), including x-ray diffraction (XRD), differential thermal analysis, electron microscopy, and x-ray fluorescene (XRF). A basic program for quantitative interpretation of XRD data of bauxite has been developed by Bardossy et al. (1980) to identify phase assemblages and to calculate the percentage of crystalline components. Accurate mineralogical analysis of bauxite can be made by XRD with the Rietveld method (Perdikatsis 1992). A colorimetric test with malachite-green solution was used to differentiate bauxite and clay in exploration (Peachey et al. 1986). Using the external standard method, Schorin and Carias (1987) have successfully determined $Al(OH)_3$ as Al_2O_3 and hydrated ferric oxide as Fe_2O_3 in natural and beneficiated ferruginous bauxites by XRD. From routine XRD and XRF analyses, a stoichiometric method was presented by Bredell (1983) to estimate the available alumina in low-grade bauxite. De Weisse et al. (1978) applied neutron activation analysis as an analytical technique in the exploration of bauxites.

GEOLOGICAL OCCURRENCE AND DISTRIBUTION OF DEPOSITS

Based upon mode of occurrence and parent-rock petrography, bauxite deposits can be divided into two major groups: lateritic bauxites for deposits derived from the weathering of igneous or other rocks rich in aluminum silicates, and karst or limestone bauxites for deposits occurring as weathering residue of limestone and dolomite. Based on petrological and geochemical criteria, Dangic (1995) introduced a concept of karst bauxite facies.

Bauxite deposits occur in all geological ages from Precambrian to Recent. The formation of bauxite was particularly abundant in the Tertiary period with gibbsite as the principal aluminum mineral. Mesozoic bauxites, which are the second most common, are invariably derived from limestone, with boehmite as the dominant aluminum mineral. Paleozoic deposits are restricted and mainly contain diaspore and boehmite (Hill and Ostojic 1984).

World bauxite deposits can be found in provinces distributed in all continents except Antarctica, as shown in Fig. 7 (Patterson 1967, 1984; Harben and Bates 1990). In South America, the Guyana Shield Province has many economically important deposits in Guyana, Surinam, Venezuela, and Colombia. The Guyana Shield Province is flanked by the Brazilian Shield Province to the south and the Caribbean Province to the north, which has major deposits of karst bauxites in Jamaica and the Dominican Republic. South American and Caribbean bauxites generally are geologically young, gibbsitic ores. Boehmite is present in Jamaican bauxites. The large Guinea Shield Province of Africa extends from Guinea to Togo and branches into central Sahara, while the small Cameroon Province covers Cameroon and Zaire. The bauxites in these provinces are also gibbsitic. The Australian Province has deposits in Western Australia, Northern Territory, and Queensland. They are of Tertiary age, and predominantly gibbsitic ores. Other deposits, most of them small ones, include the Southern European Province, Russian Province, Chinese Province, Indian Province, and Arkansas Province. Bauxite deposits in Greece and France have boehmite as the predominant Al-mineral. Chinese bauxite is characterized by diaspore, and bauxite deposits in Arkansas are gibbsitic. The karst bauxite deposits in Spain have been described in detail by Molina (1991).

Bauxite deposits occur in a variety of shapes and sizes (Shaffer 1983). They are generally grouped into four main types:

1. Blanket deposits occurring at or near the surface in horizontal, tilted, or undulating sheets or groups of lenses. They are commonly found in tropical or semitropical regions, and tend to be Tertiary or younger. Horizontal blanket deposits may extend for kilometers or even hundred of kilometers with thickness ranging on average from 1.5 to 8.0 meters. Large blanket deposits occur in South America, Australia, and India.
2. Interlayered deposits occurring at definite stratigraphic horizons, interlayered with sediments or between sediments and igneous rocks, that resulted from the burial of Mesozoic or Cenozoic surface deposits. In some localities, the rocks enclosing the bauxite have been folded and faulted. Where the deformation was intense, partial transition from gibbsite and boehmite to diaspore and in some cases to corundum

Figure 7 Major bauxite provinces of the world, after Harben and Bates (1990).

was observed. Interlayered deposits are known in Guyana, Surinam, the former USSR, China, and southern Europe.

3. Pocket deposits occurring typically in irregular masses enclosed within limestone, clay, or igneous rocks. They commonly show sharp contacts with the enclosing rocks, while also occurring with gradational margins of kaolin or bauxitic rich clay. Bauxite deposits in Jamaica are representative examples.

4. Detrital deposits resulting from the accumulation of bauxite eroded from pre-existing bauxite deposits. They are exemplified by deposits in Arkansas (the United States).

INDUSTRIAL PROCESSES AND USES

1. Milling and Beneficiation

The industrial process used to beneficiate bauxite is relatively simple in comparison with other industrial minerals because most bauxite ore as mined is of an acceptable grade. The process consists of a series of standard procedures: crushing, washing, drying and grinding to remove clay and other impurities. The washing is generally performed in scrubbers equipped with vibrating screens, drying is done in rotary kilns, and grinding may be carried out in either a dry or wet condition. Heavy media separation has been used for the removal of iron minerals in lateritic bauxite. For siderite rich ores, the removal of $FeCO_3$ is accomplished by spiral concentrator and high intensity magnetic separation (Hunton and Dale 1973; Baumgardner 1990).

For bauxite characterized by specific mineral association, elaborate milling procedures must be used. This is exemplified by Yugoslavian and Hungarian bauxites which have excessive amounts of carbonates (calcite, dolomite, siderite) and silicates (clay minerals) (Csillag et al. 1984). Both ores are monohydratic bauxites, but the difference lies in their hardness: Hungarian bauxite is soft and Yugoslavian bauxite hard. Such a difference in hardness requires different procedures in selective comminution. Yugoslavian bauxite can be treated by either the dry or wet method, whereas only the wet method can be applied to Hungarian bauxite. Figure 8 (Csillag et al. 1984) shows a simplified treatment scheme with four different routes. Route 1 represents the process for treating the entire ore by magnetic separation. In Route 2, ores are divided by classification into a coarse fraction that is treated by magnetic separation, and a fine fraction that is treated by flotation. Route 3 is the direct removal of carbonates by flotation. All three routes are suitable for Yugoslavian bauxite, and all produce satisfactory results. In the case of Hungarian bauxite, Route 4 is the only adequate treatment because the fine fraction of Hungarian bauxite is almost carbonate free, and only the silicates need to be floated.

2. Processes and Uses

Bauxite is used in many industries including metals, refractories, chemicals, abrasives, and cements. For each use, there are well-defined specifications. Bauxite is a large volume and low unit price industrial material, therefore transportation charges are a major cost element of bauxite utilization. Currently, economic considerations favor processing bauxite into either alumina or aluminum ingot near the deposit and then shipping the product (Baumgardner 1990). Of the 100 M tons of bauxite produced annually (U.S.B.M. Mineral Yearbook 1993), about 40% comes from Australia. It is followed by Guinea (13.3%), Jamaica (10%), Brazil (8.8%), India (4.9%), the former USSR (3.8%), Surinam (3.2%), China (2.7%), and Guyana (2.0%).

Metallurgical Grade: Bauxite of metallurgical grade is used for the production of aluminum, and the treatment is carried out in two stages. Alumina is first produced by the Bayer process, and aluminum is then produced from alumina by the Hall-Heroult process. The important function of the Bayer process is to produce highly purified alumina (99% Al_2O_3 minimum) (Gerard and Stroup 1963; Gilzen 1976).

In the Bayer process, the dried, ground bauxite is digested in an autoclave using a solution of sodium hydroxide and sodium carbonate. A flocculent and lime are added to causticize the soda and act as a filter aid. This requires a concentration of 100 to 120 grams NaOH per liter and a pressure of 40 to 50 psi at 145°C to dissolve the contained gibbsite, and to obtain a stable supersaturated sodium aluminate liquor. Boehmitic, and particularly diasporic bauxites, require more drastic conditions. Digestion at 280 to 700 psi and 285°C, with 300 grams NaOH per liter, is the usual practice for European

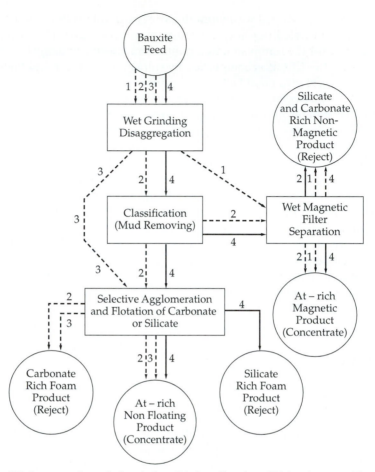

Figure 8 Simplified process scheme for bauxite beneficiation of bauxites of high carbonate and kaolinite contents, after Csillag et al. (1984).

boehmitic bauxites. In order to digest diasporic bauxite further, an increase in temperature and pressure is needed.

After digestion, temperature and pressure are reduced to atmospheric conditions and Al_2O_3 in solution and impurities in suspension in the caustic slurry produced are separated by sedimentation and filtration. The red mud consists mainly of iron oxides, titania, carbonated lime, and desilication product. The clarified (green) liquor is cooled to about 55°C, and a portion of the dissolved alumina is precipitated by introducing fine seeds of alumina trihydrate. The precipitated trihydrate is separated from the spent liquor, washed, and, for the most part, calcined to produce alumina. The spent liquor plus washings is evaporated and returned to process.

The solubility data show that the Al_2O_3 concentration in the process solution can be increased by increasing either the temperature, the NaOH concentration, or both. As a result, operating conditions in the Bayer process vary widely (Ginsberg and Wafers 1971). The digestibility of Chinese diasporic bauxite was found to be greatly improved by roasting. The extraction of Al_2O_3 from the roasted bauxite at 225°C can be equal to that from the original one at 260°C, other conditions remaining unchanged (Li et al. 1989).

Typical properties of metallurgical alumina are as follows:

1. particle size distribution (wt%) (in mesh): +100<5, +325>92, and −325<8;
2. bulk specific gravity: 0.95 to 1.00 (loose) and 1.05 to 1.10 (packed);
3. specific surface area: 50 to 80 m^2/gm;
4. moisture (to 300°): <1.0 wt%;
5. loss on ignition (300° to 1200°C): <1.0 wt%;
6. α-Al_2O_3 content: <20%; and
7. chemical analysis (wt%): Fe_2O_3<0.020, SiO_2<0.020, TiO_2<0.004, CaO<0.040, and Na_2O<0.050 (Andrews 1984).

Chemical Grade: Aluminum chemicals made from bauxite are aluminum sulfate, aluminum chloride, aluminum trihydrate, and sodium aluminate. The ratio of aluminum and iron is specified for chemical grade bauxite to be 23:1 or higher in terms of Al_2O_3/Fe_2O_3. Aluminum sulfate, which is made by digesting aluminum trihydrate taken from the Bayer process with sulfuric acid, is widely used in water treatment to coagulate suspended matter. Aluminum chloride is produced by treating heated bauxite with chlorine gas. It is used in refining mineral oils and in the manufacture of certain organic compounds, such as ethyl benzene and ethyl chloride. Sodium aluminate is produced in the Bayer process and is utilized in water purification, in the paper industry, and for treatment of titanium dioxide pigment (Shaffer 1983; Andrews 1984).

Aluminum trihydrate is produced in the Bayer process and is used as fillers in plastics, and to produce adsorbent and catalytic aluminas. As a filler, aluminum trihydrate acts as a fire retardant and smoke suppressant. Activated aluminas represent another group of important alumina chemicals that are produced by heating the hydroxide directly in rotary kilns in a temperature range from a low of 300° to a high of 1500°. Activated alumina is used for absorption of gas in water purification, and for selective absorption in the petroleum industry. It is used in many industrial catalytic processes, both as the catalyst and as a support for catalytically active components.

High-Alumina Cement Grade: High alumina cement is produced by fusing bauxite and limestone in rotary kilns or reverberatory hearth furnaces. The fused product is cooled and ground to cement fineness (Robson 1962). It has high resistance to chemical attack, especially from sulfides. Concrete made from refractory aggregates and high-alumina cement has particular applications in castable refractories. Typical specifications for bauxites used for high-alumina cements are:

Al_2O_3/SiO_2 = 10:1 (min), Al_2O_3/Fe_2O_3 = 20:1 (min), and Al_2O_3/TiO_2 = 16:1 (min).

Mineralogy and particle size are not critical for the cement grade.

Refractory Grade: Alumina refractories are produced from various aluminum-rich minerals with Al_2O_3 content ranging from 30 to 100% (Everts 1984). Apart from the lower end of this compositional series, which is based upon clays, and those based upon aluminum silicates, all refractories containing Al_2O_3 in excess of 45 to 55%

depend upon bauxite as raw material. In the process, bauxite is calcined in the temperature range of 1400° to 1800°C. The calcined bauxite is then mixed with binders and processed through a series of ceramic treatments, including pressing, extruding, casting, and shaping into a variety of forms for kiln firing (Gilzen 1976; Büchner et al. 1989).

Refractory alumina is manufactured by dry grinding calcined alumina to particle sizes smaller than 1 μm, and tabular aluminas are manufactured by grinding, shaping, and sintering calcined alumina. Thermal treatment at 1600° to 1800°C causes the oxide to re-crystallize into large tabular crystals of 0.2 to 0.3 mm. Tabular aluminas are used for the manufacture of refractory bricks, castables, electrical insulations, and high-grade ceramics.

Bauxite, to be acceptable for refractory use, must meet the following criteria: (1) high alumina content; (2) low iron oxide content, generally 2.5% maximum after cal-cination; (3) low titanium dioxide content, generally 4% maximum; (4) trace amounts of alkalis and alkaline earths; and (5) free silica content no greater than 10%.

Abrasive Grade: The two types of alumina abrasives produced are brown-fused alumina and white-fused alumina (Gilzen 1976; Power 1986). The former is directly made from bauxite which has a relatively high Al_2O_3 content and a typical specification of 82% Al_2O_3 (min), 8% SiO_2 (max), 8% Fe_2O_3 (max), and 4% TiO_2 (max) (Gilzen 1976; Power 1986). Alumina from the Bayer process is used for the latter and contains about 99% Al_2O_3 with approximately 0.5% Na_2O and small amounts of SiO_2 and Fe_2O_3.

Raw materials used in the production of brown-fused alumina abrasive are cal-cined bauxite, carbon, usually in the form of coke, and iron boring with a typical mass ratio of 80:15:5. They are mixed and fused in an electric-arc furnace. The iron and silica impurities in the bauxite are reduced by the coke and taken up by the iron boring in a ferrosilicon slag which separates out from the molten mass, and sinks to the bottom of the furnace. The addition of iron also ensures that the slag is magnetic for easy removal by electromagnetic separation. Three types of brown-fused alumina are currently used in the abrasives industry (McMichael 1989):

1. Regular brown-fused alumina is of a deep, dark brown color and contains 94 to 96% Al_2O_3, 2 to 3% TiO_2, 1 to 2% SiO_2, and small amounts of Fe_2O_3, CaO, ZrO_2, and MgO. Two sizes are commercially available: 300 microns produced by rapid cooling and 1,000 microns produced by slow cooling.
2. Semifriable fused alumina has a higher Al_2O_3 content of 96 to 98%, TiO_2 content ranges from 1.5 to 2.5%, and the SiO_2 content is below 1%. Its color is brown to red-dish brown, and it also has two sizes in use, 600 and 200 microns.
3. Microcrystalline fused alumina consists of fine alumina crystals for intrinsic strength. For rapid cooling, the melt is poured into small molds or onto pans to produce slabs.

The white-fused alumina abrasive is produced by direct fusion of alumina pro-duced by the Bayer process. Because the raw material is of high purity, only fusion and recrystallization take place. It contains from 99.5 to 99.9% Al_2O_3.

In addition to the brown- and white-fused alumina, gray monocrystalline alumina and zirconia modified fused alumina are manufactured for industrial uses. Monocrys-talline alumina is an abrasive of high purity, and it is produced from bauxite by single-stage fusion from sulfide melts. It contains in excess of 99% Al_2O_3 and about 0.5% TiO_2. Zirconia modified fused alumina is produced by fusion of baddeleyite or zircon sand with alumina. A special cooling effect is employed to give the grain a dendritic micro-crystalline intergrowth for higher grinding performance. Three varieties are available depending on the zirconia content: 10%, 25%, and 40%.

REFERENCES

Adamson, A. N., E. J. Bloore, and A. R. Carr. 1963. "Basic Principles of Bayer Process Design." Vol. 1, In *Extractive Metallurgy of Aluminum,* Ed. G. Gerard and P. T. Stroup, 23–56. New York: Interscience Publ.

Andrews, W. H. 1984. "Uses and Specifications of Bauxite." In *Bauxite,* Proc. 1984 Bauxite Symposium, Ed. L. Jacob, Jr., 49–66. New York: Soc. Mining Engineers, AIME.

Bardossy, G., L. Botyan, P. Gado, A. Griger, and J. Sasvari. 1980. Automated quantitative phase analysis of bauxites. *Amer. Mineral* 65:135–141.

Bates, T. F. 1962. Halloysite and gibbsite formation in Hawaii. *Clays and Clay Mineral* 9:315–328.

Baumgardner, L. H. 1990. "World Production and Economics of Alumina Chemicals." In *Alumina Chemicals, Science and Technology Handbook,* Ed. L. D. Hart, 7–11. Columbus, Ohio: Amer. Ceramic Soc.

Bezjak, A., and I. Jelenic. 1964. "The Crystal Structure of Boehmite and Bayerite." Vol. 2, In *Symposium sur les Bauxites, Oxides, et Hydroxides d'Aluminum,* Ed. M. Karulin, 105–111. Zagreb, Yugoslavia: Proc. 1st Int. Sym.

Bredell, J. H. 1983. Calculation of available alumina in bauxite during reconnaissance exploration. *Econ. Geol.* 78:319–325.

Büchner, W., R. Schlichs, G. Winter, and K. A. Brichel. 1989. *Industrial Inorganic Chemistry.* New York: VCH Publ., 614p.

Busing, W. R., and H. A. Levy. 1958. A single crystal neutron diffraction study of diaspore, A10(OH). *Acta Cryst.* 11:789–803.

Christoph, G. C., C. E. Corbato, D. A. Hoffmann, and R. T. Tettenhorst. 1979. The crystal structure of boehmite. *Clays and Clay Mineral* 27:81–86.

Csillag, Z., A. Csdordas-Toth, M. Ceh, and D. Ivankovic. 1984. "Role of Ore Dressing in Beneficiation of Monohydrate Bauxite." In *Bauxite,* 1984 Bauxite Symposium, Ed. L. Jacob, Jr., 704–725. New York: Soc. Mining Engineers, AIME.

Dangic, A. 1995. Karst bauxite facies: A new conception and related systematic. *Geol. Soc.* (Greece) 4:694–698.

De Weisse, G., U. Mannweiler, and L. Rabach. 1978. Rapid laboratory analysis by neutron activation: Experience in bauxite exploration. *Jour. Geochem. Expl.,* 9:93–102.

Ervin, G., and E. F. Osborn. 1951. The system alumina-water. *Jour. Geol,* 59:381–394.

Everts, J. A. 1984. "Calcined Bauxite." In *Bauxite,* Proc. 1984 Bauxite Symposium, Ed. L. Jacob, Jr., 84–96 New York: Soc. Mining Engineers, AIME.

Garrels, R. M., and G. L. Christ. 1965. *Solutions, Minerals and Equilibria.* New York: Harper and Row, 450p.

Gerard, G., and P. T. Stroup. 1963. "Extractive Metallurgy of Aluminum." Vol. 1, *Alumina,* New York: Interscience Publ., 355p.

Gilzen, W. H., ed. 1976. Alumina as a ceramic material. *Amer. Ceramic Soc.* (Columbus, Ohio) 253p.

Ginsberg, H., and K. Wafers. 1971. "Aluminum and Magnesium." Vol. 15, *Die Metallischen Rodrstaffe,* (Emke Verlag, Stuttgart).

Glemser, O., and E. Hartert. 1956. Hydrogen bonding in crystallized hydroxides. *Zeit. Anorg. Allg. Chem.,* 283:111–122.

Harben, P. W., and R. L. Bates. 1990. Industrial Minerals, Geology and World Deposits. *Metal Bulletin* (London) 312p.

Harder, E. C., and E. W. Greig. 1960. "Bauxite." 3d ed. In *Industrial Minerals and Rocks,* Ed. J.L. Gillson, 65–85. New York: Amer. Inst. Mining, Metallurg. and Petrol. Engineers.

Hidasi, J. 1986. Role of carbonate rocks in the genesis of bauxite. *Annales—Universitatis Scientiarum Budapestinensis,* Sectio Geologica, 26:179–186.

Hill, V. G., and S. Ostojic. 1984. "The Characteristics and Classification of Bauxites." In *Bauxite,* Proc. 1984 Bauxite Symposium, Ed. L. Jacob, Jr., 31–45. New York: Soc. Mining Engineers, AIME.

Hose, H. R. 1963. "Bauxite Mineralogy." Vol. 1, In *Extractive Metallurgy of Aluminum,* Ed. G. Gerard and P. T. Stroup, 3–20. New York: Interscience Publ.

Hunton, B. W., and N. W. Dale. 1973. Siderite removal by spiral concentrators and magnetic separators from Arkansas bauxites. *Light Metals* 2:843–860.

Kopeykin, V. A. 1984. The temperature factor in lateritic bauxitization. *Dokl. Acad. Sci.,* Earth Sci. Section 266:197–199.

Li, X. B., Z. Y. Yang, and Y. Z. Long. 1989. The effect of roasting on digestibility of diasporic bauxite. *Travaux du Comité International pur l'Etude des Bauxites,* de l'Alumine et de l'Aluminum (Zagreb) 19:379–386.

McMichael, B. 1989. Pyroprocessing techniques—The fiery world of industrial minerals. *Ind. Minerals* (August):54–69.

Molina, J. M. 1991. A review of karst bauxites and relted paleokarats in Spain. *Acta Gelogica Hungarica* 34:179–194.

Neuhaus, A., and H. Heide. 1965. Hydrothermal investigation in the system Al_2O_3-H_2O I. Phase boundaries and stability of boehmite, diapsore and corundum at pressures above 50 bars. *Ber. Deut. Keram. Ges.* 42:167–184.

Newsome, V. W., H. W. Heiser, A. S. Russell, and H. C. Stumpf. 1960. Alumina Properties. *Tech. Paper 10, Alcoa* (Pittsburgh, Penn.) 10p.

Patterson, S. H. 1967. Bauxite resources and potential aluminum resources of the world. *U.S.G.S. Bull.* 1228:176p.

Patterson, S. H. 1984. "Bauxite and Nonbauxite Aluminum Resources of the World, an Update." In *Bauxite,* Proc. 1984 Bauxite Symposium, Ed. L. Jacob, Jr., 3–20. New York: Soc. Mining Engineers, AIME.

Peachey, D., J. W. Aucott, J. L. Roberts, B. P. Vickers, and H. J. Bloodworth. 1986. Rapid colorimetric test to differentiate between bauxite-rich material and clay in exploration samples. *Appl. Geochem.* 1:527–529.

Perdikatsis, V. 1992. Quantitative mineralogical analysis of bauxites by x-ray diffraction with the Rietveld method. *Acta Geologica Hungarica* 35:447–457.

Power, T. 1985. Fused minerals—The high purity high performance oxides. *Ind. Minerals* (July):37–51.

Robson, T. D. 1962. *High Aluminum Cements and Concretes.* New York: John Wiley & Sons, 263p.

Saalfeld, H., and M. Wedde. 1974. Refinement of the crystal structure of gibbsite. *Zeit. Krist.* 139:129–135.

Salter, T. L., and H. H. Murray. 1989. Residual Bauxite genesis in Saline County, Arkansas. *Travaux du Comité International pour l'Etude des Bauxites,* de l'Alumine et de l'Aluminum (Zagreb) 19:213–223.

Schorin, H., and O. Carias. 1987. Analysis of natural and beneficiated ferruginous bauxites by both x-ray diffraction and x-ray fluorescence. *Chem. Geol.* 60:199–204.

Shaffer, J. W. 1983. "Bauxitic Raw Materials." 5th ed. In *Industrial Minerals and Rocks,* Ed S. J. Lefond, 503–527. New York: Soc. Mining Engineers, AIME.

Som, S. K., and A. K. Bhattacharya. 1995. Mineralogy of Panchpatmali bauxite deposits based on XRD, IR, DTA and SEM studies. *Jour. Geol. Soc.* (India) 45:427–437.

Tardy, Y., and D. Nahon. 1985. Geochemistry of laterites, stability of Al-goethite, Al-hematite, and Fe^{3+}-kaolinite in bauxites and ferricretes: An approach to the mechanism of concretion formation. *Amer. Jour. Sci.* 285:865–903.

Valeton, I. *Bauxite, Developments in Soil Science 1.* New York: Elsevier Publ., 226p., 1972.

Wefers, K., and C. Misra. 1987. Oxides and hydroxides of aluminum. *Tech. Paper 19, Alcoa* (Pittsburgh, Penn.) 92p.

Wollast, R. 1963. Aspect chimique du monde de formation desbauxites dans le Bas-Congo Confrontation des donnés thermodynamiques et experimentales. *Acad. Royal Sci. Outre-Mer. Bull. Belg.* (Sér) 7:392–412.

Bentonite

The term "Bentonite" is not an exact mineral name. It is generally defined as a clay which is composed predominantly of a smectite mineral, and whose physical properties are dictated by this clay mineral. In this definition, the mode of origin is not considered.

Bentonite has an extensive range of industrial applications in oil drilling, foundry sands, ore pelletizing, animal and poultry feed, ceramics, construction engineering, and others.

The term "bleaching earths" refers to clays that in their natural state (Fuller's earth), or after chemical or physical activation (activated clays), have the capacity to decolorize oil to ena-lingit to be of industrial value. Bleaching earths do not have a defined composition, but smectite is their main component.

MINERALOGICAL PROPERTIES

Smectite, the main component of bentonite, has a layered structure composed of two silica tetrahedral sheets and one central dioctahedral (Al) or trioctahedral (Mg) sheet as shown in Fig. 9 (Grim 1968). The layer has an unbalanced charge caused by substitution of silica by aluminum in the tetrahedral sheet, and substitution between magnesium and aluminum in the octahedral sheet. When the layers are stacked, the attraction holding them together is weak, and lowly charged cations and polar molecules can enter the interlayer, balance the charge, and cause lattice expansion. Dioctahedral smectites can be subdivided into the predominantly aluminum varieties, montmorillonite and beidellite, and the iron-rich variety, nontronite. The site of the negative charge on the layer is the basis used to distinguish between montmorillonite and beidellite. In montomorillonite, the ideal structural formula for which is $(Al_{3.15}Mg_{0.85})(Si_{8.00})O_{20}(OH)_4X_{0.85}nH_2O$, the charge arises from the divalent cations, usually Mg, in octahedral sites, whereas in beidellite, $(Al_{4.00})(Si_{7.15}Al_{0.85})O_{20}(OH)_4X_{0.85}nH_2O$, it arises from Al^{3+} in tetrahedral sites, with X representing the interlayer cation. Minerals with compositions intermediate between these end members are common. As a general rule, those minerals in which more than half of the charge originates in the octahedral sites are described as montmorillonites, and those in which tetrahedral charge predominates, as beidellite. The formula for the end member nontronite, as proposed by Brindley (1980), is $(Fe^{3+})_4(Si_{4-x}Al_x)_2O_{20}(OH)_4X^{1+}_xnH_2O$, which corresponds to a beidellite with all octahedral Al replaced by Fe^{3+}. Nontronite is generally used for dioctahedral smectite with octahedral $Fe^{3+} > 3$ per $O_{20}(OH)_4$. Trioctahedral smectites are also found in bentonite as saponites, which have magnesium predominant in the octahedral sites, and hectorite, a Li-rich variety (Newman and Brown 1987).

Members of the montmorillonite-beidellite series are by far the most common smectites in bentonite (Grim and Güven 1978; Elzea and Murray 1990). The smectite found in the well-known Wyoming bentonite has a composition close to that of the montmorillonite-beidellite boundary line, and, as analytical data have shown, this represents the composition of an appreciable number of bentonites. Magnesia content in

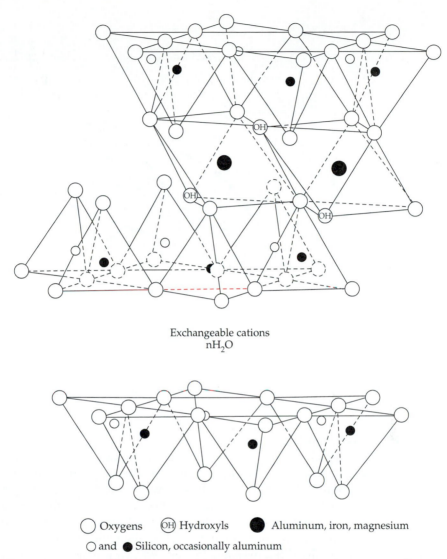

Exchangeable cations
nH₂O

○ Oxygens   ⊙ Hydroxyls   ● Aluminum, iron, magnesium

○ and ● Silicon, occasionally aluminum

Figure 9 Crystal structure of smectite. Reprinted from Grim, *Clay Mineralogy*, 1968, McGraw-Hill. Reproduced with permission of the McGraw-Hill Companies.

the series ranges from about 0.2 to 6.0%, whereas iron content (as Fe_2O_3) ranges from about 1.0 to 12.0%. The most common interlayer cations are Na and Ca.

The interlayer cations affect not only the extent of lattice expansion but also the absolute amount of water sorbed, the range of the sorption isotherm, and the acidity function of the water. Mooney et al. (1952) and Newman (1987) illustrated the variation of isotherm shape with interlayer cation in Wyoming bentonite, as given in Fig. 10, which also shows the basal spacing ($d_{(001)}$) at the regions where points of inflection occur on the isotherms. Na forms single-sheet complexes ($d_{(001)}$=12Å–12.5Å) at low humidities that expand to the two-sheet complexes ($d_{(001)}$=15Å) at p/p_o = 0.5–0.6. At high

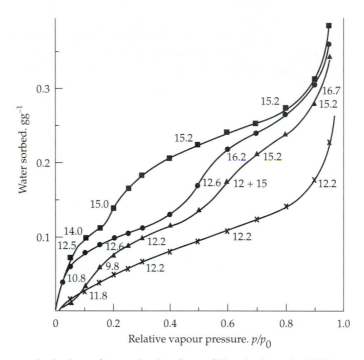

Figure 10   Water sorption isotherms for several cations forms of Wyoming bentonite. Solid square, solid circle, solid triangle, and cross represent, respectively, Ca, Ba, Na, and Cs. The numbers indicate $d_{(001)}$ in Å at various points on the isotherms. Published with the permission of the Mineralogical Society of Great Britain and Ireland.

humidities ($p/p_o > 0.95$), most montmorillonites and beidellites form three-sheet complexes ($d_{(001)} = 19Å$), and when immersed in water, swell macroscopically (Newman 1987). Cs is large enough to keep the 12Å interlayer spacing independent of water content. The reaction of divalent cations illustrates the effect of decreasing hydration energy, and the single-sheet complex becomes more stable through the sequence Mg → Ca → Sr → Ba. As shown in Fig. 10, the transition to the double-sheet complex of Ba occurs above $p/p_o = 0.5$, as with Na. It gives 19Å when immersed in water, but does not swell macroscopically (MacEwan and Wilson 1980; Newman 1987).

On heating, smectites lose H_2O and produce shrinkage at 100°–150°C. Up to 300°C, the lattice becomes non-expandable, and between 450°–700°C, the hydroxyl is lost. Dioctahedral smectites decompose at about 900°C, while the decomposition of trioctahedral smectites occurs at 700°C. Fusion temperature ranges from about 1000°C for iron-rich smectites to 1500°C for those low in iron (Grim 1968).

Quartz, feldspar, kaolinite, mica, illite, gypsum, calcite, and heavy minerals are minor components in bentonite. Most of these minerals are concentrated in the coarser fraction (>10μm), but mica and kaolinite occur in both fine and coarse fractions. Besides quartz, cristobalite and tridymite are important components in some bentonites, and mixed-layering of these two polymorphic forms is common. Zeolites are not often found in association with bentonite. Common heavy minerals are zircon, magnetite, apatite, magnesite, and pyroxene (Grim and Güven 1978).

Because bentonites contain minerals of wide and variable chemical composition, they vary greatly in physical properties (Elzea and Murray 1990). Bentonite from Wyoming is light yellow or greenish yellow in color, but from other localities it may be gray, brown, pink, or black. The appearance may be dull or powdery, but a fresh cut surface usually has a waxy luster. There is also a wide range of variation in morphological features. Grim and Güven (1978) described these features on the habit of individual crystallites and of their arrangements in the aggregates. Common ones are (1) lamellar aggregates, the Wyoming-type, which have distinct smectite lamellae; (2) mossy aggregates, the Cheto-type, which have a mossy appearance resulting from the curling of extremely thin layers of ribbons of crystallites; (3) reticulated aggregates, the Santa Rita-type, which consist of euhedral platelets forming a net-like arrangement; and (4) globular aggregates, the Otay-type, which consist of randomly arranged tiny globules. Smectite single crystallites occur in all types of aggregates.

Bentonite is exceedingly fine-grained and very plastic. The plasticity of a clay may be measured by the Water of Plasticity and the Atterberg values of Plastic Limit and Liquid Limit (Grim and Güven 1978). The Water of Plasticity of smectite, determined on a dry weight basis at 105°C, varies greatly from 83 to 250 wt%. The wide range is due to the variations in grain size, crystallinity, and the nature of the exchangeable cation. The value is higher for Na-smectite than for Ca-smectite. The Water of Plasticity of smectite has higher values than those of kaolinite (9 to 56 wt%) and illite (17 to 39 wt%). For Na-smectite and Ca-smectite, the Atterberg Plastic Limit and Liquid Limit are 75 and 500, and 90 and 160, respectively. Kaolinite's values are 35 to 45 for the Plastic Limit and 50 to 75 for the Liquid Limit. In comparison, smectites generally require more water than other clay minerals to develop plastic properties and have high drying shrinkage.

Some bentonites are highly absorbent, absorbing up to five times their weight or fifteen times their volume of water, with a consequent increase in volume and the formation of a gelatinous mass. Some non-swelling bentonites, when acid treated, acquire the ability to absorb and remove coloring material from oils, fats, and greases.

Mineralogical analysis of bentonite is made by a combination of x-ray diffraction, electron microscopy, and atomic absorption spectroscopy, for structural, morphological, and chemical characteristics, respectively.

GEOLOGICAL OCCURRENCE AND DISTRIBUTION OF DEPOSITS

By far, the most common mode of origin of bentonites is the in situ alteration of volcanic ash or tuff. The alteration process involves devitrification of the ash, hydration of the devitrified products, and crystallization of smectite. Geological evidence has demonstrated that during the alteration, there is a loss of alkali, downward migration of silica, and, in some instances, addition of magnesium (Grim and Güven 1978; Güven 1988). The environment in which the alteration takes place to produce bentonite varies widely, as indicated by an extensive range of associated sedimentary rocks. It includes (1) shallow marine (Wyoming, Texas, and Mississippi [the United States], England, and Germany); (2) freshwater (Canada, Czechoslovakia, New Zealand, and Yugoslavia); (3) estuaries and lagoons (Egypt and Pakistan); (4) desert pan (Australia); and (5) coal-forming (Australia and Canada) environments. In general, the alteration of ash or tuff to smectite is probably contemporaneous with the accumulation of igneous material. The devitrification process that generally takes place in water should not be considered to be the result of weathering processes. Hydrothermal alteration and deuteric alteration of igneous rocks are other processes that produce bentonite.

Bentonites are widely distributed in the United States, especially in formations of the Cretaceous and Tertiary ages. The largest producing deposit is in the Black Hills region of South Dakota, Wyoming, and Montana (Knechtel and Patterson 1962; Hosterman and Patterson 1992; Harben and Kuzvart 1996). The extent of the deposit has been well studied and extensive aerial surveying of the region has been done (Thorson 1997). There are many bentonite deposits in the Rocky Mountain states. The best known deposits are in the Cheto district of Arizona, and the Amargosa Valley of Nevada. Bentonite deposits are also found in Texas, Mississippi, and Alaska. Bentonite from the Black Hills region is a sodium bentonite whose quality ranks the highest in the world, and is commonly referred to as Wyoming bentonite. Bentonites from Arizona, Nevada, Texas, and Mississippi are calcium bentonite (Grim and Güven 1978; Güven 1988). The associated minerals in the Wyoming-type bentonites are quartz, cristobalite, feldspar, mica, and some zeolites, whereas the Arizona bentonite, known as the Cheto-type (Sloan and Gilbert 1966), contains kaolinite with a white to gray color. The Nevada bentonites are altered ash, but some of them are the result of hydrothermal action with rhyolitic volcanic rocks. Bentonites produced by hydrothermal alteration were also found in New Mexico. Saponite is the dominant smectite of some bentonites in the Amargosa Valley of Nevada (Papke 1969). Bentonite enriched with hectorite occurs near Hector, California (Ames et al. 1958; Odom 1992). The main Fuller's earth deposits in the United States are in the Paleocene Porters Creek formation distributed from Missouri through Illinois and Tennessee to Mississippi. The clay mineral assemblages of the formation consist of calcium-magnesium smectite, kaolinite, illite, and halloysite, and are in association with opal C-T and trace amounts of zeolite (Elzea and Murray 1994).

Bentonite deposits are found in many areas of Europe including England, Germany, Poland, Austria, Bulgaria, Cyprus, Greece, Hungary, Italy, Romania, Spain, Yugoslavia, and others (Elzea and Murray 1994; Harben and Kuzvart 1996). They are of the calcium smectite variety. In England, bentonites composed almost entirely of smectite are referred to as Fuller's earth, and are commonly found in formations of Cretaceous age in the London basin, and of Jurassic age in Somerset. In both locations, the bentonite is blue when fresh, and a yellow color when weathered. Fuller's earth is believed to be an alteration product of volcanic ash, but volcanic activity is absent in the region (Hallam and Selwood 1968; Cowperthwaite et al. l972). There are extensive deposits of calcium bentonite in Germany in the vicinity of Moosberg and Landshut in Bavaria. These occur in a marine section of marls and tuffaeous sands in the Upper Miocene molasse. The bentonite is light yellow to gray and has a waxy appearance (Siegi 1949). In Italy, large bentonite deposits are located on the islands of Ponza (DeAngeles and Novelli 1957) and Sardinia (Annedda 1956). The bentonites in both deposits are calcium bentonite of high purity. The Ponza bentonite is soft in the upper portion of the deposit, which has a 6 m thickness, and becomes hard in the lower portion with a color change from ivory to pale blue. The deposit is a product of combined deuteric and hydrothermal alterations of the rhyolite. The Sardinian bentonites are of two types: (1) hydrothermal altered trachytic tuffs, and (2) sedimentary bentonite derived from marine-altered ash. Hungarian bentonites were classified by Szeky-Fux (1957) into two groups on the basis of their origin: (1) bentonites produced by hydrothermal alteration of rhyolites and andesites as illustrated by deposits at Komloske, Mad, and Vegardo, and (2) bentonites produced by the alteration of rhyolitic-dacitic tuff which fell into the sea and was altered under marine conditions, as illustrated by deposits at Istenmezo, Nagyletent, and Band. All bentonites derived from rhyolite-andesite are the calcium variety, and many from the rhyolitic-dacitic tuff have incomplete alteration. In Greece, several of the volcanic islands have bentonite. The most extensively developed deposit, a Pliocene calcium bentonite, is on the island of Molos (Grim and Güven 1978). Bentonites of Cretaceous and Tertiary age are well developed in Romania and include the Banat and Transylvania regions (Cardew 1952). They are products of alteration of volcanic ash, and some bentonites occur in a lacustrine series of sediments. There are both calcium and sodium varieties. In Yugoslavia, bentonites of excellent quality are widely distributed (Lukas 1968). Nikolic and Obradoric (1963) reported a bentonite deposit from Blace in Serbia which occurs within a lignite coal seam.

Among the many bentonite deposits found in the former USSR, the most important ones are in Azerbaijan and Georgia. The Azerbaijan bentonites were formed mostly by the diagenetic alteration of ash and tuff in a marine environment, and the bentonite beds are interbedded with limestone and shale. The occurrence of both calcium and sodium varieties of bentonite has been reported (Seidov and Alizade 1970). In Georgia, extensive bentonite deposits are found in formations ranging in age from Jurassic to Tertiary. The best known deposits are of the so-called askanite from Askana. These bentonites resulted from alteration of pumice-like andesite-trachyte tuffs by either weathering (Belyankin and Petrov 1950) or hydrothermal process (Rateev 1967). Both sodium and calcium varieties were produced. In India, bentonite occurs in the Barmer region of Rajasthan (Siddiguie and Bahl 1963; Russell 1991) interbedded with calcareous sands and conglomerates. It is pure calcium bentonite. In Japan, well-developed bentonite deposits are found in Yamagata, Gumma, Niigata, and Nagano Perfectures, and in Hokkaido island (Iwao 1969). The deposits were mostly formed by alteration of ash, pumice, and tuff with contemporaneous deuteric or hydrothermal processes. Bentonite at the Kawasaki, Dobayama, and Kuroishi mines is hydrothermal in origin (Ito et al. 1999). China has extensive bentonite deposits, with two of the largest mines at Hei-Shin in Liaoning Province, and Linan in Zhejiang Province (Zhu 1985).

Bentonites are widely distributed in the Prairie Provinces of Canada (Byrne 1955; Bannatyne 1963; Harben and Kuzvart 1996). They are about the same geological age as the deposits in Montana and the Dakotas (the United States). The generally accepted mode of orgin for these bentonites is the alteration of dacitic and rhyolitic ash. The calcium variety is the dominant type, although some sodium bentonites are also known. An unusual occurrence of a natural hydrogen bentonite was found in the Pembina area of Manitoba. Oxidation of pyrite in the associated shales is believed to have produced an acid which reacted with the clay to form the acid bentonite. In Argentina, bentonites are best known from the Triassic formations in Mendoza and San Juan Provinces, and from the Paleocene and Eocene formations of Patagonia (Bordas 1943). The Triassic bentonites are altered tuffs interbedded with sands and comglomerates, but some of the beds contain detrital material in addition to the smectite, suggesting transportation and redeposition of the altered tuff. These are some of the few occurrences that could be classed as detrital bentonite.

Bentonite deposits are widely distributed in Algeria. Along the Chelif River near Mostaganem, bentonite beds occur with ashy silts of Miocene age. Ash structures are well preserved in the bentonite, suggesting an alteration origin. There is neither silicification of underlying beds nor any evidence of downward leaching. Some of the bentonite are sodium rich, although there are interlaminations with calcareous sediments (DeLapparent 1937). In the Oran area, bentonite deposits were formed from hydrothermal alteration of the rhyolite, by leaching out alkalies and silicon, and by adding magnesium (Sadran et al. 1955). In Morocco, there are many occurrences of bentonite in the Taourirt region (Frey et al. 1936). The geologic setting and the properties of bentonites are similar to those of Algeria.

INDUSTRIAL PROCESSES AND USES

1. Milling and Beneficiation

The milling process for bentonite is simple. Bentonite, as mined, commonly carries a considerable amount of moisture. It is a routine practice to air-dry the clay to reduce the moisture from about 30% down to about 7%. The dried clay is passed through a slugger-roll or other type of crusher, and fed by belt conveyor to rotary or oil-fired cylindrical dryers. Raymond roller millers are widely used to pulverize the dried bentonite, and cyclone dust collectors extract the −200 mesh product from the Raymond mill, depositing it in a bin for bagging. For processing Fuller's earth, a flowsheet illustrating the processes involved is shown in Fig. 11 (Siddiqui 1968).

Selective processing is used in various industries to produce special clay products. Extruders are used to pug the clay and mix it with additives that may improve its viscosity and dispersion properties. High-calcium bentonite, which is generally low-swelling, may be sodium-exchanged in solution with soda ash to enhance swelling characteristics. The drilling mud quality of bentonite may be greatly improved with additives such as high molecular weight polymers (Erdogan and Demirci 1996). To reduce iron content in high-iron bentonite, chemical treatment using sodium dithionite

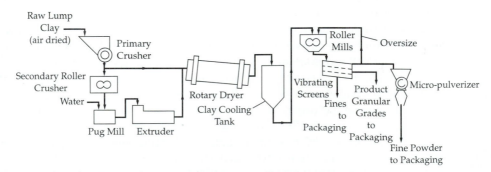

Figure 11   Flowsheet for the processing of Fuller's earth, after Siddiqui (1968).

and a mixture of sodium sulfide and oxalic acid was found to be satisfactory (Singh and Sharma 1994).

Acid-activated bentonite is generally produced by acidulation of high Ca-bentonite with mineral acids to enhance sorptive properties. The initial steps of treatment are identical to those for Fuller's earth, until the step in which acid is added and boiled by using steam for a period of one hour. After activation, the sludge is transferred to a continuous thickener system, where soluble salts and excess acid are washed off. The slurry is sent through a rotary filter to a rotary dryer, where the moisture is reduced to about 10 to 15%. In a Raymond mill or a fluid-energy grinding mill, the dried clay is pulverized to the desired size for packing.

Organic-clad bentonites are produced when sodium bentonites react with polar organic molecules (Grim 1962). The acidity of the reaction increases with the polarity of the organic molecule and may be sufficient to replace adsorbed water. In this form bentonite becomes water repellent or hydrophobic. In the case of highly polar molecules, layers several molecules thick can be adsorbed on the surface of bentonite. The bentonite can be completely clad with organic molecules that may interpose between the layers of the organic-clad bentonite, resulting in thixotropic thickening.

2. Industrial Uses

Bentonite is one of the most versatile minerals in its industrial uses as illustrated in Fig. 12 (O'Driscoll 1988). Major uses are in drilling fluid, foundry molding sand, bleaching, and iron ore pelletizing.

Drilling Muds: The function of the drilling fluid is to remove cuttings, control well pressure, maintain borehole zone stability, lubricate the drill string, and protect producing zone (Lundie 1986). Viscosities of the fluid greater than that of water are required to facilitate the removal of cuttings and should be markedly thixotropic, so that cuttings do not settle to the bottom of the drillhole. An impervious coating must be built up on the wall of the drillhole in order to impede the penetration of water from the drilling fluid into the formation, and must be thin in order not to retard the drilling operation.

Bentonite gives yields in excess of 100 barrels per ton, so that only about 5% clay is adequate to produce the desired viscosity. Bentonite is markedly thixotropic, has very high gel strength, and is outstanding for its low filter-cake permeability. With such a set of characteristic properties, bentonite can serve as both a viscosifier and a suspension agent in drilling fluid. The other property that determines its essential use in drilling fluid is its ability to produce an impervious clay layer on the wall of the drillhole. Na bentonites are extensively and widely used worldwide as ingredients in drilling fluids.

The critical specifications for bentonite that is to be used in drilling fluid are (1) suspension and filtrate properties, (2) amount of coarse material (wet screen analysis), and (3) moisture content. Several sets of testing procedures for these specifications are currently in use. Those recommended by the American Petroleum Institute (API STD 13a 5th edition 1969) are described below. In conducting the suspension test, 22.5 grams of bentonite are mixed with 350 cc of distilled water to prepare a suspension. The

Crude Bentonite	Activated with Acid (Activated Bleaching Earth)	Naturally Active (Na/Ca-Bentonite)	Alkaline Activated (Na-Exchanged Bentonite)	Organically Activated (Organophilic Bentonite)
Foodstuffs Industry	Refining, Decolourizing, Purifying, and Stabilizing of Vegetable and Animal Oils and Fats			
Sulphur Production	Refining, Decolourizing, Bitumen Extraction			
Forest and Water Conservation		Powder Fire-Extinguishing Agents/Binding Agents for Oil on Water		
Mineral-Oil Industry	Refining, Decolourizing, and Purifying of Mineral Oils, Fats, Waxes, Paraffin/Catalysts for Oil Cracking			Grease Thickening
Beverages and Sugar Industry		Fining of Wine, Must, and Juices/Beer Stabilization/Purifying of Saccharine Juice and Syrup		
Chemical Industry	Catalysts/Catalyst Carriers Insecticides and Fungicides/Fillers, Dehydrating Agents/Water and Waste-Water Purification/Absorbents for Radioactive Materials			
Paper Industry	Pigment and Colour Developer for Carbonless Copying Paper/Absorption of Impurities in White Water System			
Cleaning and Detergents	Regeneration or Organic Fluids for Dry Cleaning	Polishes and Dressings/Additives for Washing and Cleaning Agents and for Soap Production		
Pharmaceutical Industry		Starting Material for Healing Earths and Medicaments/Bases for Creams and Cosmetics		
Ore Production		Binding Agents for Ore Pelletizing		
Building Industry		Supporting Suspensions for Cut-Off Diaphragm Wall Constructions and Shield Tunnelling/Subsoil Sealing (e.g., Dumps)/Anti-Friction Agents for Pipejacking and Shaft Sinking/Additive for Soil Concrete, Concrete and Mortar		
Ceramics Industry		Plasticizing of Ceramic Compounds/Improvement of Strength/Fluxing Agents		
Horticulture, Agriculture, Animal Husbandry		Soil Improvement/Composting/Animal-Feed Pelletizing/Liquid-Manure Treatment/Cat Litter		
Drilling Industry		Borehole Scavenging for Saltwater	Thixotropic Suspensions for Borehole Scavenging	
Tar Exploitation			Emulsification and Thixotroping of Tar-Water Emulsions, Tar and Asphalt Coatings	
Paint and Varnish Industry				Thickening, Thixotroping, Stabilizing and Anti-Setting Agents for Paints, Varnishes, Coating Materials, Sealing Cements, Waxes, Adhesives
Foundries		Binding Agents for Special Moulding Sands	Binding Agents for Synthetic Moulding Sands, Core Sands	Binding Agents for Anhydrous Casting Sands/for Thickening Blackwashes

Figure 12 Major areas of applications for bentonites. © Industrial Minerals, July 1988.

viscosity is determined from the suspension and the yield point is calculated from readings at 300 and 600 rpm. The filtrate test is used to measure the volume of water lost from the prepared suspension in a filter press. The amount of coarse material (>200 mesh) in bentonite mud is determined in the wet screen analysis. It involves mixing 10 grams of bentonite and 350 cc of distilled water, stirring, aging, washing through a sieve, collecting coarse material from the sieve, drying, and weighing. The amount of the coarse material is calculated as a percentage of the original bentonite. For each of the properties, API has a recommended maximum value.

Foundry Molding Sands: The molding sands used in foundries are composed of sand and bentonite. These materials provide bonding strength and plasticity. Important foundry properties are (1) green compression strength, (2) dry compression strength, (3) hot strength, (4) flowability, (5) permeability, and (6) durability (Grim and Güven 1978). Both calcium and sodium bentonites are used as molding clays. Compared with sodium bentonite, calcium bentonite has a higher green compression strength, lower dry compression strength, lower hot strength, and better flowability. Both the American Foundrymen's Society (Foundry Sand Handbook 7th ed. 1963) and Steel Founder's Society of America (Designation BT-65 1965) have established procedures to measure compression strength and flowability.

Decolorization of Oils: Acid-activated bentonites and Fuller's earth are widely used to decolorize mineral, vegetable, and animal oils. They also serve to deodorize, dehydrate, and neutralize the oils. In the process, the oils may be filtered through a granular product of 10- to 60-mesh particles, or the oil may be placed in contact with finely ground clay of approximately −200 mesh; then the oil is separated from the clay by filtration (Grim and Güven 1978). In decolorization, bentonite and Fuller's earth are also used for the preparation and decolorization of other foodstuffs. Both may be added as a solid or as a suspension in water. Their effectiveness in the treatment of sugar beet juice (Erdogan et al. 1996), for example, matches the decolorization result achieved by diatomite and sepiolite, two decolorizing minerals that are discussed elsewhere. The clay, in order to be used for decolorizing oils and foodstuffs, must have low oil retention and good filtration characteristics. It must not change the nature of the oil or the foodstuff and give an objectionable taste to them.

The American Oil Chemists' Society has offered procedures for conducting bleaching tests for bentonite and Fuller's earth (Bleaching Tests, A.O.C.S. Official Method, Cc 8a 52, American Oil Chemists' Soc. 1958).

Iron-Ore Pelletization: Bentonites are used extensively in pelletizing iron ores, a process for which De Vaney (1956) received a patent (U.S. Patent No. 274372). For better blast furnace feed, finely pulverized ore concentrate is pelletized into small balls of 2.45 cm or more in diameter before furnace treatment. The amount of bentonite required for pelletization is about 0.5% of the ore. Because of its superior dry strength, sodium bentonite is preferred (Sastry and Fuerstenau 1971; Wakemen 1972).

Other Uses: In addition to the industrial uses of bentonite as listed in Fig. 12 (O'Driscoll 1988), bentonites have been examined in many studies to evaluate their possible application as backfill materials in nuclear waste repositories (Madsen 1998). Radionuclide transport through bentonite is possible only through diffusion in stagnant porewater. The apparent diffusion coefficients for radionuclides in compacted bentonite for 20 to 25% water content are very low, suggesting long breakthrough times after significant decay.

REFERENCES

Ames, L. L., L. B. Sand, and S. C. Goldich. 1958. A Contribution on the Hector, California, bentonite deposit. *Econ. Geol.* 53:22–37.

Annedda, V. 1956. Deposits of bentonite on Sadali and Vallanova Tulo Territories, Sardinia. *Resoconti Assoc. Min. Sard.* 60:5–9.

Bannatyne, B. B. 1963. Cretaceous bentonite deposits in Manitoba. *Manitoba Dept. Mines & Nat. Resources,* Mines Branch Publ. 62–5, 20p.

Belyankin, D. S., and V. P. Retrov. 1950. Petrographic composition and origin of Askana clays. *Bull. Acad. Sci.,* Ser. Geol. (USSR) 2:33–44.

Bordas, A. F. 1943. Contribution al conocimientos de las bentonites argentonas. *Rev. Geol. Mineral.* 14:1–60.

Brindley, G. W. 1980. "Order-Disorder in Clay Mineral Structures." In *Crystal Structures of Clay Minerals and Their X-ray Identification,* Ed. G. W. Brindley and G. Brown, Monograph 5, 146–196. London: Mineral. Soc.

Byrne, P. J. S. 1955. Bentonites in Alberta. *Res. Council Alberta,* Rept. 71, 20p.

Cardew, J. 1952. Bentonites in Rumania's chemical industry. *Chem. Age* (London) 67:862.

Cowperthwaite, A., F. I. Fitch, J. A. Miller, J. C. Mitchell, and R. H. S. Robertson. 1972. Sedimentation, petrogenesis and radioisotopic age of the Cretaceous Fuller's earth of southern England. *Clay Mineral.* 9:309–327.

DeAngeles, G. C., and G. Novelli. 1957. Bentonite from Ponza. *Geotectonia* 4:1–7.

DeLapparent, J. 1937. Origin of bentonite from North Africa. *C.R. Soc. Geol. Fr.* 10:126–128.

Elzea, J. M., and H. H. Murray. 1990. Variation in the mineralogical, chemical and physical properties of the Cretaceous clay spur bentonite in Wyoming and Montana (USA). *Appl. Clay Sci.* 5:229–248.

Elzea, J. M., and H. H. Murray. 1994. "Bentonite." 6th ed. In *Industrial Minerals and Rocks,* Ed. D. D. Carr, 233–246. Littleton, Col.: Soc. Mining, Metal., and Exploration.

Erdoan, B., and S. Demirci. 1996. Activation of some Turkish bentonites to improve their drilling fluid properties. *Appl. Clay Sci.* 10:401–410.

Erdogan, B., S. Demirci, and Y. Akay. 1996. Treatment of sugar beet juice with bentonite, sepiolite, diatomite with quartamin to remove color and turbidity. *Appl. Clay Sci.* 11:55–67.

Frey, R., B. Yovanovich, and J. Burghelle. 1962. 1936. Composition and probable genesis of some decolorizing clays of North Africa. *Serv. Min. and Carte Geol. Moroccco,* Mem. 38:65p.

Grim, R. E. 1962. *Applied Clay Mineralogy.* New York: McGraw-Hill, 422p.

Grim, R. E. 1968. *Clay Mineralogy.* 2d ed. New York: McGraw-Hill, 598p.

Grim, R. E., and N. Güven. 1978. *Bentonites: Developments in Sedimentology 24.* Amsterdam: Elsevier Publ., 256p.

Güven, N. 1988. "Smectites." Vol. 19, In *Reviews in Mineralogy,* Ed. S. W. Bailey, 497–559. Washington, D.C.: Min. Soc. Amer.

Hallam, A., and B. W. Selwood. 1968. Origin of the Fuller's earth in the Mesozoic of Southern England. *Nature* 220:1193–1195.

Harben, P. W., and M. Kuzvart. 1996. Industrial minerals—A global geology. *Industrial Minerals Information, Ltd., Metal Bulletin* (London) 462p.

Hosterman, J. W., and S. H. Patterson. 1992. Bentonite and Fuller's earth resources of the United States. *U.S.G.S. Prof. Paper* 1522:45p.

Ito, M., T. Ishii, H. Nakashima, and Y. Hirata. 1999. The study of genesis and formation condition of bentonite. (in Japanese with English abstract) *Jour. Clay Sci. Soc.* (Japan) 38:181–187.

Iwao, S. 1969. The clays of Japan. *Int. Clay Conf. Proc.* 209p.

Knechtel, M. M., and S. H. Patterson. 1962. Bentonite deposits of the Northern Black Hill District, Wyoming, Montana, and South Dakota. *U.S.G.S. Bull.* 1082:893–1029.

Lukas, E. 1968. State of exploration of Zaluska Goriea bentonites in comparison with other Yugoslavian bentonites. *Rud. Metal. Zb.* 3:277–286.

Lundie, P. 1986. Standardization of minerals for drilling fluids. *Ind. Minerals* (March):113–117.

MacEwan, D. N. C., and M. J. Wilson. 1980. "Interlayer and Intercalation Complexes of Clay Minerals." In *Crystal Structures of Clay Minerals and Their X-ray Identification,* Ed. G. W. Brindley and G. Brown, Monograph 5, 197–248. London: Min. Soc.

Madsen, F. J. 1998. Clay mineralogical investigations related to nuclear waste disposal. *Clay Minerals* 33:109–129.

Mooney, R. W., A. C. Keenab, and L. A. Wood. 1952. Adsorption of water by montmorillonite. *Jour. Amer. Chem. Soc.* 74:1367–1374.

Newman, A. C. D. 1987. "The Interaction of Water with Clay Mineral Surface." In *Chemistry of Clays and Clay Minerals,* Ed. A. C. D. Newman, Monograph 6, 237–274. London: Min. Soc.

Newman, A. C. D., and G. Brown. 1987. "The Chemical Constitution of Clays." In *Chemistry of Clays and Clay Minerals,* Ed. A. C. D. Newman, Monograph 6, 1–128. London: Min. Soc.

Nikolic, N., and J. Obradoric. 1963. Bentonite from Blace, Serbia. *Geol. An. Balken Poluostrva* 30:125–128.

Odom, I. E. 1992. Hectorit deposits in the McDermitt caldern of Nevada. Preprint No. 92–155. *Soc. Mining, Metal., and Exploration* (Littleton, Colo.) 12p.

O'Driscoll, M. 1988. Benonite, overcapacity in need of markets. *Ind. Minerals.* (July):43–67.

Papke, K. S. 1969. Montmorillonite deposits in Nevada. *Clays and Clay Minerals* 17:211–222.

Rateev, M. A. 1967. Sequence of hydrothermal alteration of volcanic rocks into bentonite clays of the Askana deposit in the Georgian SSR. *Dokl. Akad. Nauk SSSR* 175:675–678.

Russell, A. 1991. India—A treasure trove of minerals. Lifting the lid on export. *Ind. Minerals* (January):17–33.

Sadran, G., G. Millot, and M. Bonifas. 1955. The origin of bentonite deposits at Lalla Maghnia (Oran). *Ser. Carte Geol., Algeria, Bull.* 5:213–234.

Sastry, K. V. S., and D. W. Fuerstenau. 1971. A laboratory method for determining the balling behavior of taconite concentrates. *Trans. Soc. Mining Engineers, AIME,* 250:64–67.

Seidov, A. G., and K. A. Alizade. 1970. Mineralogy and origin of the bentonite clays of Azerbaidia. *Izd. Akad. Nauk Azerb. SSR, Baka* 190p.

Siddiguie, N. N., and D. P. Bahl. 1963. Geology of the bentonite deposits of the Barmer District, Rajasthan. *Mem. Gel. Surv.* (India) 96:36p.

Hasnuddin Siddiqui, M. K. 1968. *Bleaching Earth.* New York: Pergamon Press, 86p.

Siegi, W. 1949. Glassy stuff in the Molesse of Upper Bavaria and its application in bleaching earths. *Neues Jahr. Min., Monatsh., Abl.* A 77–82.

Singh, P. K., and V. P. Sharma. 1994. Beneficiation of low grade bentonite clay by chemical methods and their utilisation in oil industries. *Indian Jour. Engin. & Mat. Soc.* 1:284–288.

Sloan, R. L., and J. M. Gilbert. 1966. Electron-optical study of alteration in the Cheto clay deposit. *Clays and Clay Mineral.* 15:35–44.

Szeky-Fux, V. 1957. The hydrothermal genesis of bentonite on the basis of studies in Komloska. *Acta Geol. Acad. Sci. Hung.* 4:361–382.

Thorson, T. A. 1997. Aerial surveying of Wyoming bentonite. *Appl. Clay Sci.* 11:329–335.

Wakeman, J. S. 1972. Method for evaluation of data from the batch testing of green pellets. *Trans. Soc. Mining Soc., AIME,* 252:83–86.

Zhu, G. 1985. The bentonite industry in China. *Ind. Minerals* (January):53–59.

Beryllium Minerals

Of the more than forty beryllium minerals known, only two, beryl and bertrandite, have industrial importance for the production of beryllium and beryllia. The metal has outstanding thermal, mechanical, electrical, and nuclear properties. It is also an extremely valuable alloying component, especially with copper. The oxide is characterized by its unusual ability to conduct heat in combination with its electrical insulating properties. Both beryllium and beryllia are toxic. Precautions must be taken in the handling and processing of them to eliminate health hazards.

The United States, the former USSR, and Brazil are the world's leading producers of beryllium minerals, with a production of 6,810 tons in 1995. The United States accounts for about 70% of annual production (Harben and Kuzvart 1996).

MINERALOGICAL PROPERTIES

1. Beryl

Beryl, with an ideal composition of $Be_3Al_2(SiO_3)_6$, has the most elegant structure among all silicate minerals. Its structure is characterized by hexagonal rings of six SiO_4 tetrahedra arranged one above another in planes to form hollow columns parallel to 0001. These rings are linked together both laterally and vertically by BeO_4 tetrahedra and AlO_6 octahedra to produce a three-dimensional structure (Fig. 13) (Artioli et al. 1993). Beryl is hexagonal, with a = 9.208Å and c = 9.197Å.

Beryl usually contains some alkalis, and in certain varieties the total alkali content may reach an amount of approximately 5 to 7%. Lithium is the most common substitution of beryllium, whereas the larger alkalis enter the hollow columns of the beryl structure, frequently in the presence of H_2O. A classification of beryl on the basis of alkalis has been proposed, including (1) alkali-free, (2) sodipotassic, (3) sodic, (4) sodic lithian, and (5) lithian caesian beryls. The octahedral Al is often replaced by Fe^{2+} and Mg with minor amounts of Fe^{3+}, Cr^{3+}, V^{3+}, Mn^{2+}, and Ti^{4+}. Fe_2O_3 up to 6.22 wt% and MgO up to 2.51 wt% have been reported (Duchi et al. 1993). SiO_2 content in beryl is quite constant, varying in range from 63 to 66 wt% (Aurisicchio et al. 1988). The BeO content in beryl varies from 12 to 13.5%.

Beryl has many colors: colorless, white, bluish green, greenish yellow, yellow, blue, and rose. Its hardness and specific gravity are, respectively, 7.5 to 8 and 2.63 to 2.91. Crystals with good clarity suitable for gems are classed as aquamarine (bluish green), emerald (green), helidor (golden yellow), and morganite (pink), pigmented by Fe^{2+}, Cr^{3+}, Fe^{3+}, and Mn^{2+}, respectively. Optically, beryl is uniaxial negative with ε = 1.565 to 1.599 and ω = 1.569 to 1.610. Both refractive indices increase with an increase in the alkali content. The refractive index ω increases with an increase in specific gravity and decreases with an increase in BeO content (Cerny and Hawthorne 1976).

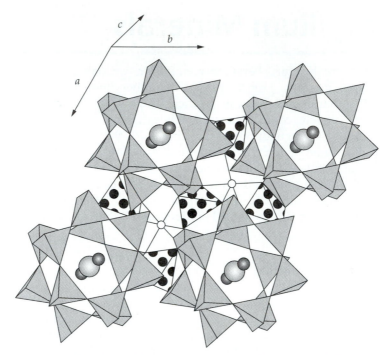

Figure 13   Crystal structure of beryl viewed down the c-axis. Dotted tetrahedra represent BeO$_4$, and shaded are SiO$_4$ in an arrangement of hexagonal rings. AlO$_6$ octahedra are represented by open circles, after Artioli et al. (1993).

2. Bertrandite

Bertrandite has a chemical composition of Be$_4$Si$_2$O$_7$(OH)$_2$ with BeO content in the range from 39.6 to 42.6%, SiO$_2$ between 49.3 and 51.8%, and H$_2$O between 6.9 and 8.9%. The crystal structure of bertrandite was determined by Solovéva and Belov (1965), and refined by Guiseppetti et al. (1992). It consists of a three-dimensional framework of BeO$_3$(OH) and SiO$_4$ tetrahedra in a 2:1 ratio. In the 100 corrugated layers, all tetrahedra share each of the three vertices with two tetrahedra, whereas the fourth vertices of oxygen and hydroxyl groups are shared by tetrahedra from two equivalent layers. Neutron diffraction analysis reveals a zigzag chain of weak H bonds within the structure (Downs and Ross 1987). Bertrandite is orthorhombic with a = 8.716Å, b = 15.255Å, and c = 4.565Å.

Bertrandite has a hardness of 6 and a specific gravity of 2.60. It is optically negative with refractive indices of α = 1.584, β = 1.603, and γ = 1.611. Bertrandite is colorless, white, or yellow, and the crystals usually have tabular habit (Beus 1966).

GEOLOGICAL OCCURRENCE AND DISTRIBUTION OF DEPOSITS

The geological occurrence of beryllium minerals is a direct consequence of the crystal chemistry of beryllium, which, as a small, divalent cation, finds among the common rock-forming cations no suitable ones for it to substitute. Therefore it becomes concentrated in the residual magmatic fluids and crystallizes at late stages.

The most common occurrences of both beryl and bertrandite are in pegmatite. Beryllium-bearing pegmatites have been found in all geological ages, but more than 50% of the known deposits are Precambrian and 37% are Paleozoic (Beus 1966). Beryl occurs in unzoned pegmatites and in all types of units of zoned pegmatites, but only in a few zoned pegmatites are beryls commonly rich enough to be of industrial importance (Griffitts and Norton 1955). The most common sites of beryl concentrations are; (1) deposits that border quartz cores; (2) intermediate zones and cores that contain potash feldspar, and lithium minerals; and (3) wall and intermediate zones that contain mica. Beryl also occurs in nepheline-syenites and in mica schists and marbles. Major deposits in production are in Kazakhstan, the Kola Peninsula, the Urals, Altay, Transbaykal, and the western Ukraine of the former USSR, and Alagoas, Ceará, Paraiba, Bahia, and Minas Gerais of Brazil (Zabdotnaya 1977; Soja and Sabin 1986).

Except for pegmatite veins, bertrandite also occurs in greisens as a late and, in some places, secondary mineral. Wood (1992) proposed fluoride complexes as Be-carrier in the formation of bertrandite. In rocks of Triberg granitic complex in the Schwarzwald, Germany, late magmatic and hydrothermal fluids interacted with beryl-bearing pegmatites and produced bertrandite with or without phenakite in the temperature range of 220° to 230°C (Markl and Schumacher 1997).

The Topaz-Spor Mountain bertrandite deposit, located in Juab County, 150 km south of Salt Lake City, Utah, is the only bertrandite mine in production. The bertrandite occurs in a microcrystalline form intimately associated with fluorite within Tertiary rhyolitic tuffs (Shawe 1968; Schiller 1985). The Thor Lake bertrandite deposit, located 100 km southeast of Yellowknife, Northwest Territory, Canada, is in development and has economic potential for beryllium production (Trueman 1986).

INDUSTRIAL PROCESSES AND USES

1. Milling and Beneficiation

The mining of beryl takes place, in the main, alongside the recovery of other pegmatitic minerals (Griffitts and Norton 1955). Initially, the rock is blasted and then hand- cobbed for any valuable minerals, or crushed in a coarse manner, and then passed along conveyor belts for sorting. The specific gravity of beryl is slightly higher than that of feldspar or quartz, but the difference is not enough to favor a gravity separation.

Flotation has proved to be the most applicable method for the milling of beryllium ores. In the flotation plant, ore is first ground in a ball mill, the fines feeding to a conditioner where flotation reagents are added. After floating off the muscovite, the pulp is conditioned at pH 5 with a tallow amine acetate and its beryl and feldspar are bulk-floated using an alcohol frother. These minerals are separated by conditioning with calcium hypochloride followed by flotation of the beryl with a petroleum sulfonate (Pryor 1965). For successful flotation, it is necessary to recognize the flotation characteristics of associated minerals and to selectively remove the heavy minerals ahead of beryl flotation.

2. Processing of Beryl and Bertrandite

Beryl is treated mainly by sulfate and fluoride processes (Hausner 1965). The chloride process is a direct process for recovering beryllium chloride for electrolysis.

In the sulfate process, beryl is heated at 1,625°C and quenched in cold water to produce a glass. The glass is dried and ground into a −200 mesh powder, which is then leached with sulfuric acid so that the beryllium and aluminum components become sulfates. The sulfates are then extracted with the aid of water. Ammonia is used to separate the aluminum sulfate. Further treatment with sodium hydroxide produces beryllium hydroxide, which is then calcined to BeO.

In the fluoride process, beryl is sintered at 750°C with sodium hexafluosilicate. Only the beryllium oxide is converted to a water-soluble salt; aluminum oxide, silicon dioxide, and other impurities are essentially not affected. The reaction product must be leached at room temperature immediately with water to avoid the precipitation of beryllium salts that are sparingly soluble, and to avoid reaction between SiO_2 and fluoride solutions which will occur at elevated temperatures. To obtain beryllium hydroxide, the solution is made alkaline with NaOH. Beryllium hydroxide precipitates when the solution is diluted with water and boiled. BeO is obtained by calcination of beryllium hydroxide.

In the treatment of bertrandite, the ore is wet ground and screened to form a slurry of −20 mesh particles. It is then leached with a 10% sulfuric acid solution at about 100°C. Because of the different mineralogy between the two beryllium ores, there is no need to melt or sinter the bertrandite in order to access the mineral to acid attack. The sulfate solution is extracted with 0.3m solution of ammonium di-2-ethylhexyl phosphate/di-2-ethylhexyl phosphoric acid/kerosine in a five-stage mixer at 50°C. Beryllium, aluminum, and iron enter the organic phase, while other impurities remain in the sulfate solution. Following the separation of the two liquid phases, the metal ions are removed from the organic compounds by ammonium carbonate solution. The result is the formation of a basic beryllium carbonate solution containing beryllium hydroxide. Pure beryllium hydroxide is obtained from the basic solution by filtration and decomposition at 140° to 160°C (Foos 1968).

A flow diagram of Brush Wellman's process to treat both beryl and bertrandite is illustrated in Fig. 14 (Schiller 1985). The beryl ores are preheated and put into the bertrandite ore metallurgical flow cycle to yield a finished beryllium carbonate and beryllium hydroxide.

3. Industrial Uses

A wide range of beryllium products, metal, copper and other alloys, and oxide, is manufactured and consumed in various industries. Beryllium alloys accounted for 75% of U.S. total production, whereas metal and oxide made up 10% and 15%, respectively.

Industrial uses of beryllia may be summarized in three major areas: (1) electrical/electronic, (2) refractory, and (3) structural (Griffiths 1985; Schiller 1985). The most outstanding characteristic of BeO is its extremely high thermal conductivity. This is most unusual for oxides and makes BeO an especially interesting refractory material (Beaver 1955; Walsh 1979). It is utilized as conduction cooling components in electronic assemblies, and as a heat sink in transistors and semiconductors. Because it has high dielectric strength with low dielectric constant, beryllia also operates well as an electrical

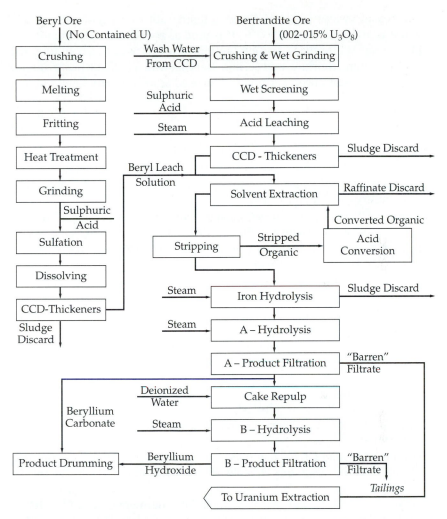

Figure 14 Flow diagram of Brush Wellman's process plant for beryllium products at Delta, Utah. Reproduced with permission of Mining Journal Ltd., London.

insulator. Beryllia, fabricated by hot pressing or firing, results in a wide range of densities from 2.0 to 2.9 gm/cm³. At higher densities, thermal and electrical conductivities as well as mechanical properties such as tensile strength, compressive strength, modulus of rupture, and elastic modulus are all higher.

The combination of high thermal conductivity and relatively high strength gives thermal fracture resistance to beryllia ceramics that is superior to that of most other refractory oxides. The room temperature mechanical properties of beryllia are not affected until a temperature of 1,000°C is reached, whereas the tensile and compressive strengths remain constant to about 700°C. Beryllia ceramics are widely used in applications where high temperature resistance is important (Stonehouse 1986).

Beryllia has an extremely high melting point at 2,550°C, 400°C higher than alumina. It resists corrosion by practically all materials except hydrofluoric acid, fused alkalies and water vapor at elevated temperatures. Reaction between BeO and water vapor is appreciable at 1,300°C. It is used as rocket nozzles, crucibles, and thermocouple tubing (White 1955). In metal refining, BeO is used as a container for molten nickel, iron, platinum, thorium, and cerium. Beryllia with a ballistic performance that compares well with that of boron carbide can be used as armor for vehicles and personnel.

REFERENCES

Artioli, G., R. Rinaldi, K. Stahl, and P. F. Zanazzi. 1993. Structure refinements of beryl by single-crystal neutron and x-ray diffraction. *Amer. Mineral.* 78:762–768.

Aurisicciho, C., G. Fioravanti, O. Grubessi, and P. F. Zanazzi. 1988. Reappraisal of the crystal chemistry of beryl. *Amer. Mineral.* 73:826–837.

Beaver, W. W. 1955. "Refractory Compounds an Cermets of Beryllium." In *The Metal Beryllium,* Eds. D. W. White, Jr. and J. E. Burke, 570–598. Cleveland, Ohio: Amer. Soc. Metals.

Beus, A. A. 1966. *Geochemistry of Bryllium an Genetic Types of Beryllium Deposits.* San Francisco: W. H. Freeman & Co., 461p.

Cerny, P., and F. C. Hawthorne. 1976. Refractive indices versus alkali contents in beryl. General limitations and applications to some pegmatitic types. *Can. Mineral.* 14:491–497.

Downs, J. W., and F. K. Ross. 1987. Neutron-diffraction study of bertrandite. *Amer. Mineral.* 72:979–983.

Duchi, G., M. Franzini, M. Giamello, P. Orlandi, and F. Riccobono. 1993. The iron-rich beryls from Alpi Apune. Mineralogy, chemistry and fluid inclusion. *Neues Jahr. Min., Monat.* 199–207.

Foos, R. A. 1968. Applications of solvent extraction to beryllium oxide manufacturing. *AIME Annual Meeting* New York: abstract.

Griffiths, J. 1985. Beryllium minerals: Demand strong for miniaturisation. *Ind. Minerals* (June):41–52.

Griffitts, W. R., and J. J. Norton. 1955. "Occurrence of Beryllium Ores and Their Treatment." In *The Metal Beryllium,* Eds. D. W. White, Jr. and J. E. Burke, 42–48. Cleveland, Ohio: Amer. Soc. Metals.

Guiseppetti, G., C. Tadini, and V. Mattioli. 1992. Bertranite, $Be_4Si_2O_7(OH)_2$, from Val Vigezzo (NO), Italy: The x-ray structural refinement. *Neues Jahr. Min. Monat.* 13–19.

Harben, P. W., and M. Kuzvart. 1996. Industrial minerals, a global geology. *Industrial Minerals Information, Ltd., Metal Bulletin* (London) 462p.

Hausner, H. H., ed. 1965. *Beryllium: Its Metallurgy and Properties.* Berkeley: Univ. California Press, 322p.

Markl, G., and J. C. Schumacher. 1997. Beryl stability in local hydrothermal and chemical environments in a mineralized granite. *Amer. Mineral.* 82:194–202.

Pryor, E. J. 1965. *Mineral Processing.* Amsterdam: Elsevier, 844p.

Schiller, E. A. 1985. Beryllium—Geology, production and uses. *Mining Magazine* (April): 317–322.

Shawe, D. R. 1968. "Geology of the Spor Mountain Beryllium District, Utah." (Graton-Sales Volume), In *Ore Deposits of the United States, 1933-67,* Ed. J. D. Ridge, 1148–1161. New York: AIME.

Soja, A. A., and A. E. Sabin. 1986. Beryllium availability—Market economy countries—A mineral availability appraisal. *U.S. Bur. Mines Inform. Circ.* 9100:19p.

Solov'eva, L. P., and N. V. Belov. 1965. Precise determination of the crystal structure of bertrandite, $Be_4Si_2O_7 (OH)_2$. *Sov. Phys. Cryst.* 9:458–460.

Stonehouse, A. J. 1986. Physics and chemistry of beryllium. *Jour. Vac. Sci. Technol.* A4:1163–1170.

Trueman, D. L. 1986. The Thor Lake rare-metal deposits, Northwest Territories, Canada. Vol. 1.
Proc. 7th Ind. Minerals Intern. Congr., Ed. G. M. Clarke and J. B. Griffiths, 127–131. Monte Carlo.

Walsh, K. A. 1979. "Extraction." Vol. 2, In *Beryllium Science and Technology,* Eds. D. R. Floyd and
J. N. Lowe, 1–11. New York: Plenum Press.

White, J. F. 1955. "The Refractory Properties of Beryllium Oxide." In The Metal Beryllium, Eds.
D. W. White, Jr. and J. E. Burke, 599–619. Cleveland, Ohio: *Amer. Soc. Metals.*

Wood, S. A. 1992. Theoretical prediction of speciation and solubility of beryllium in hydrothermal solution to 300°C at saturated vapor pressures: Application to bertrandite/phenakite deposits. *Ore Geol. Rev.* 7:249–278.

Zabdotnaya, N. P. 1977. "Deposits of Beryllium." Vol. 3, In *Ore Deposits of the USSR,* Ed. V. I.
Smirnov, 320–371. New York: Pitman Publ.

Borax and Borates

Boron has a great affinity for oxygen, and in nature always occurs in the oxygenated state as borates. There are more than 150 known borate minerals, but only less than a half dozen of them are important in the borax industry. Major uses of borate minerals are in the manufacture of glass, vitreous enamels, and glazes. Others are in fertilizers, detergents, insecticides, and metal welding.

The United States, Turkey, the former USSR, and China are the world's major producers of borax and borates.

MINERALOGICAL PROPERTIES

Among some 150 known borate minerals, sixteen common ones are listed in Table 5. Among them, borax, kernite, ulexite, colemanite, and szaibelyite are important industrial minerals. Sassolite is a natural boric acid found in small quantities in the Lardarello region, Italy and has historic interest only. Probertite has been found only in deposits that were probably subjected to elevated temperatures. Priceite is abundant only in Turkey. All others are secondary or accessory borate minerals.

Borax occurs as short, white prisms, as compact glossy masses, in solution in saline lakes, and as a glistening white efflorescence of alkali soils. It is monoclinic with a=11.858Å, b=10.674Å, c=12.179Å, and β=106°41′ (Morimoto 1956). The hardness, specific gravity, and refractive indices have ranges from 2 to 2.5, from 1.69 to 1.72, and from 1.447 to 1.472, respectively. Borax crushes freely with conchoidal fracture and dissolves readily in water. On heating, borax loses 5 moles of water from 50° to 100°C and loses an additional 3 moles up to 169°C. The remaining 2 moles of water are lost at about

Table 5 Common borate minerals

Mineral	Chemical Composition	B_2O_3 wt%
Borax	$Na_2B_4O_7 \cdot 10H_2O$	36.5
Tincalconite	$Na_2B_4O_7 \cdot 5H_2O$	47.8
Kernite	$Na_2B_4O_7 \cdot 4H_2O$	51.0
Ulexite	$NaCaB_5O_9 \cdot 8H_2O$	43.0
Proberitite	$NaCaB_5O_9 \cdot 5H_2O$	49.6
Inyoite	$Ca_2B_6O_{11} \cdot 13H_2O$	37.6
Priceite	$Ca_4B_{10}O_{19} \cdot 7H_2O$	49.8
Meyerhofferite	$Ca_2B_6O_{11} \cdot 7H_2O$	46.7
Colemanite	$Ca_2B_6O_{11} \cdot 5H_2O$	50.8
Sassolite	$B(OH)_3$	56.4
Szaibelyite	$Mg_2B_2O_5 \cdot H_2O$	41.4
Boracite	$Mg_3B_7O_{13}Cl$	62.2
Suanite	$Mg_2B_2O_5$	46.3
Kurnakovite	$Mg_2B_6O_{11} \cdot 15H_2O$	37.3
Ludwigite	$(Fe,Mg)_4Fe_2B_2O_7$	17.8
Datolite	$Ca_2B_2Si_2O_9 \cdot H_2O$	21.8

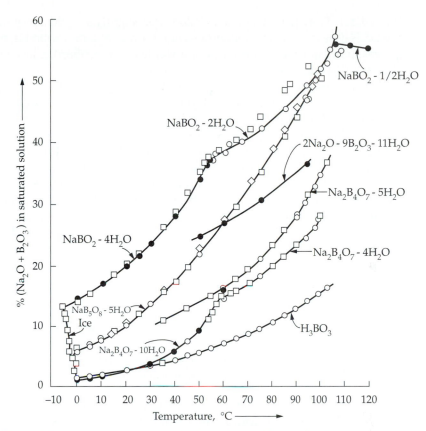

Figure 15   Solubility versus temperature curves for borate hydrates. Reprinted with permission from Nies and Hulbert, *J. Chem. Eng. Data* 12. 1967, American Chemical Society.

400°C (Menzel et al. 1935). This is a direct consequence of its crystal structure, in which two of the ten moles of water exist as hydroxyl. The formula is best represented by $Na_2[B_4O_5(OH)_4]\cdot 8H_2O$ (Morimoto 1956). Rapid heating of borax causes puffing. Puffed borax consists of small glassy hollow spheres with a bulk density of about 0.07 gm/cm³. The solubility of borax and other sodium borates is shown in Fig. 15 (Nies and Hulbert 1967). Borax crystallizes from aqueous solution below 60.8°C, the decahydrate-pentahydrate transition temperature. The pH value of a borax solution increases slightly with increasing concentration, and decreases slightly with increasing temperature.

Kernite, another sodium borate hydrate, occurs in transparent, colorless monoclinic crystals, or in large cleavable masses. Compared with borax, kernite has a higher hardness (H=2.5), a higher specific gravity (1.95), and similar index of refraction. It is slowly soluble in cold water, but readily soluble in hot water. Kernite often occurs in association with borax, but forms at relatively higher temperatures than borax (Christ and Garrels 1959).

Ulexite is a sodium-calcium borate hydrate. It occurs in white, rounded masses, composed of loosely compacted, fine silky fibers that are easily pulverized between fingers. It has a hardness of 1 and a specific gravity of 1.96. Index of refraction ranges from

1.491 to 1.520. Ulexite is insoluble in cold water, slightly soluble in hot water, and soluble in acids. It is the primary sodium-calcium borate mineral formed in borate marshes or playas under normal conditions. The lower hydrate, probertite, has been found only in deposits that were subjected to elevated temperatures.

Colemanite occurs as fine brilliant crystals in vugs at most deposits, where the bulk of the mineral generally is in massive layers and granular aggregates. Colemanite is monoclinic with a=8.743Å, b=11.264Å, c=6.102Å, and β=100°7′ (Christ et al. 1958). Typical colors are white, gray, and yellowish gray. The hardness is 4.0 to 4.5 and specific gravity 2.26 to 2.52. The index of refraction ranges from 1.586 to 1.614. Like borax, there are hydroxyls present in the structure, and the formula is best written as $Ca_2B_6O_8(OH)_6 \cdot 2H_2O$ (Christ et al. 1958). Colemanite dissolves in water very slowly, but is soluble in acid. Solubility in water at 25°C is 0.18 wt% as B_2O_3 and 0.38 wt% at 100°C. For the calcium borate minerals (Table 5), the lowest hydrate is the common mineral and the highest hydrate has been found only in a few small deposits (Aristarain and Erd 1971).

Szaibelyite is the major Russian borate and is also abundant in China. It is a magnesium borate. From structural analysis (Peng et al. 1963), the formula is best represented as $Mg_2(OH)[B_2O_4(OH)]$. Like colemanite, it is very slowly soluble in water and decomposes to boric acid with sulfuric acid at 95°C.

The crystal structure of borate minerals is complex and is the most important factor in the determination of borate's use as a glass-former. It is well known that the borate content and variation in proportion of BO_4 tetrahedra and BO_3 triangular groups have distinct effects upon the physical properties of glass (Hansson 1961). Christ and Clark (1977) proposed six rules that govern the formation of polyanions in hydrated borate minerals:

1. boron links either three oxygens to form a triangle, or four oxygens to form a tetrahedron;
2. polynuclear anions are formed by corner sharing of boron-oxygen triangles and tetrahedra in such a manner that a compact insular group results;
3. in the hydrated borates protonatable oxygen atoms will be protonated in the following sequence: available protons are first assigned to free O^{2-} ions to convert these to free OH ions, and additional protons are assigned to tetrahedral oxygens in the borate ion, next, protons are assigned to triangular oxygens in the borate ion, and finally, any remaining protons are assigned to free OH ions to form H_2O molecules;
4. the hydrated insular groups may polymerize in various ways by splitting out water; this process may be accompanied by the breaking of boron-oxygen bonds within the polyanion framework;
5. complex borate polyanions may be modified by attachment of an individual side group, such as an extra borate tetrahedron, an extra borate triangle, two linked triangles, and so on; and
6. isolated $B(OH)_3$ groups, or polymers of these, may exist in the presence of other anions. Among the five borates considered to be industrial minerals, each has a distinct structural arrangement. Both borax and kernite have tetraborate units of two triangles and two tetrahedra. In borax the tetraborate units are isolated, whereas in kernite they form chains. Ulexite contains isolated pentaborate units of two triangles and three tetrahedra. Both colemanite and szaibelyite have triborate units. One triangle and two tetrahedra join corners to form chains in colemanite, and two triangles link together to form an isolated arrangement in szaibelyite.

Based upon Christ and Clark's scheme, Strunz (1997) proposed another classification for borates. The number of borate atoms in the borate anions is used to define primary divisions such as monoborates, diborates, and tetraborates. Each division is further divided into subdivisions related to the degree of polymerization of the individual borate units, such as, for example, neo-tetraborates, ino-tetraborates, phyllotetraborates, and tecto-tetraborates.

GEOLOGICAL OCCURRENCE AND DISTRIBUTION OF DEPOSITS

Boron is a widespread trace element with an average concentration in the earth's crust of about 0.0003% (Walker 1975). The ocean is a major reservoir. Both marine and non-marine clastic sediments often have a high boron content from detrital tourmaline and illite (Shaw and Bugry 1966). When boron becomes concentrated in geological environments, borate deposits form. There are five major processes for the formation of borate minerals (Barker and Lefond 1985): (1) by precipitation from brine in shallow or deep lakes, (2) as crusts or crystals in mud of playas, (3) by direct precipitation near springs or fumaroles, (4) by direct evaporation of marine water, and (5) by crystallization at or near granitic contacts or in veins.

Major borate deposits of the world are distributed in a well-defined pattern that follows the Cenozoic tectonic-volcanic belts and passes through regions of arid climate (Tageeva 1943), including the southwestern United States, the Andes of South America, Tibet-China, and Turkey. Borate deposits are also found in Permian evaporites in Germany and the former USSR.

United States: In the United States, borate deposits are concentrated in the deserts of California (Harben and Kuzvart 1996). Sodium borate is produced at Boron, sodium-calcium and calcium borate in Death Valley, and several borates from brine at Searles Lake. The Boron deposit is located in the northwestern Mojave Desert about 145 km northeast of Los Angeles. Geologically, the deposit lies near the edge of a large Tertiary basin. The borates are underlain by Tertiary basalt and overlain by a series of continental sands, and are a part of the shale member of the Kramer beds, which is early Middle Miocene in age. The deposit is a buried mass of crystalline borax, kernite, and ulexite (Barnard and Kistler 1966). Sodium borates and claystone form a central core that is enveloped by calcium borates and claystone (Bowser and Dickson 1966). The borates were formed by precipitation from brine in a lake contained in an east-west trending basin, formed by subsidence on the north side of a normal fault. Both Barnard and Kistler (1966), and Bowser and Dickson (1966) conclude that borax crystallized on or slightly beneath the floor of a perennial lake that received the discharge from thermal springs whose waters contained large amounts of dissolved borate, and that ulexite grew as pods beneath the surface of the mud when borate concentrations fell below borax saturation levels. Diagenesis by Mg- and Ca-bearing groundwaters is responsible for the peripheral colemanite, ulexite, and most of the other Ca- and Mg-borate minerals in those zones. Burial temperatures above 60°C are responsible for the conversion of primary borax to kernite. A stratigraphic profile of Boron deposits is shown in Fig. 16a (Kistler and Smith 1983).

Death Valley is a long north-trending valley near the Nevada border of southern California. It was produced by normal faulting that shifted the area near Bad Water to

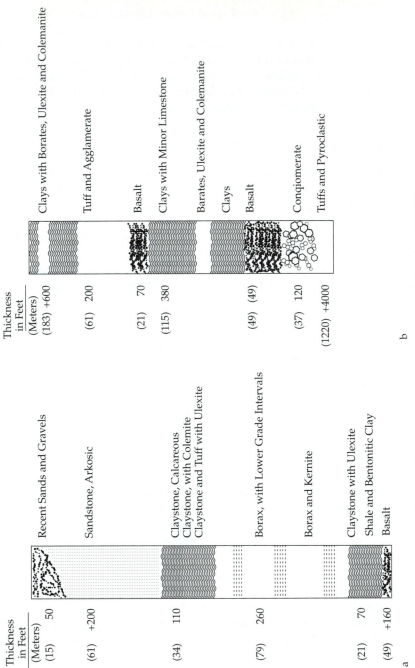

Figure 16 Stratigraphic sections at (a) Boron, and (b) Death Valley, California, after Kistler and Smith (1983).

(a)

Thickness in Feet (Meters)	Lithology
(15) 50	Recent Sands and Gravels
(61) +200	Sandstone, Arkosic
(34) 110	Claystone, Calcareous / Claystone, with Colemite / Claystone and Tuff with Ulexite
(79) 260	Borax, with Lower Grade Intervals
(21) 70	Borax and Kermite
(49) +160	Claystone with Ulexite / Shale and Bentonitic Clay / Basalt

(b)

Thickness in Feet (Meters)	Lithology
(183) +600	Clays with Borates, Ulexite and Colemanite
(61) 200	Tuff and Agglamerate
(21) 70	Basalt
(115) 380	Clays with Minor Limestone
	Barates, Ulexite and Colemanite
(49) (49)	Clays / Basalt
(37) 120	Conqiomerate
(1220) +4000	Tuffs and Pyroclastic

86 m below sea level and the Panamint Range to 3,370 m above. The ore is contained in the lower part of the thick Pliocene Furnace Creek Formation that consists mainly of clastic fluvial and lake sediments (Wilson and Emmons 1977). The three minable borate minerals are distributed in three zones in the following order of abundance: (1) probertite in the core or deeper zone, (2) ulexite next, and (3) colemanite in the surrounding outermost zone. Probertite is a significantly large part of only one of the deposits, whereas ulexite is a large part of many, and colemanite is predominant in most (McAllister 1970). In Fig. 16b, a stratigraphic section is shown.

Various borate minerals are produced from the brine extracted from Searles Lake, including boric acid, borax pentahydrate, and anhydrous borax (Smith 1979). Searles Lake occupies the central portion of a desert valley midway between Death Valley and Boron, California. It is a large evaporite basin formed during the most recent two Sierra Nevada ice stages. The Lake consists of a central salt flat overlying a mass of admixed salts, borax, clays, and interstitial brine, approximately 100 km^2 in extent. There are two producing horizons in the lake, the upper salt which ranges from 9 to 27 m thick, averaging 15 m, with a brine composition averaging 1% B_2O_3, and the lower salt which is 8 to 14 m thick, averaging 12 m with a brine composition averaging 1.2% B_2O_3. These two salt beds are separated by a relatively impervious clay layer with an average thickness of 4 m. Playa-type borax occurs on the shores of Searles Lake.

Christ et al. (1967) studied the borate mineral assemblages of Boron and Death Valley deposits in the system Na_2O-CaO-MgO-B_2O_3-H_2O. They proposed several activity-activity relationships (Fig. 17) which may be used to trace out the paragenetic sequences as a function of changing cation and H_2O activities. Fig. 17b, for example, represents two situations corresponding to the conditions for the stability fields of borax and ulexite, taking into account that the primary hydrated borates were formed originally as chemical precipitates in saline lakes and always as higher hydrates. Borax is formed in lakes high in Na-content, low in Ca-content, and having high pH, whereas ulexite forms in lakes of relatively higher Ca-content. Dehydration or transformation from a primary hydrate to a lower hydrate is illustrated by passing along a horizontal coordinate from right to left in the activity-activity diagram. Upon burial in a lake, the mineral assemblages will change to pass along some of the horizontal coordinates in the diagram, if relative composition of the solution remains the same. If, during burial, the composition of the solution varies, the direction describing the change of the mineral assemblages (from right to left across the activity-activity diagram) will have both horizontal and vertical components.

Turkey: Turkey possesses roughly 70% of the world reserves of boron minerals and is the second largest producer of boron after the United States. Borate deposits are located in five major areas in western Turkey, at Kirka, Emet, Bigadic, Sultancayiri, and Kestelek (Albayrak and Protopapas 1985; Helvaci 1978, 1983). Geologically, the borate deposits of Turkey, like those in the Tertiary rocks of California (Kistler and Smith 1983), consist of lenticular beds in deformed sedimentary rock that were deposited in the continental basins during a period of tectonic-volcanic activity in the late Tertiary. In the Kirka deposit (Inan et al. 1973), Tertiary marl, clay, and tuff underlie and overlie borate ore that is nearly monomineralic borax thinly interlayered and intergrown with dolomitic clay and, in places, ulexite. The thickness of the orebody

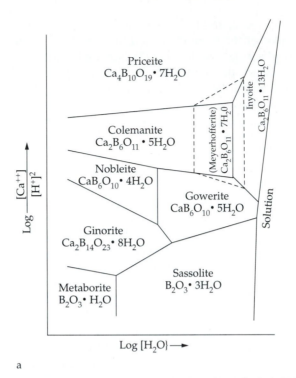

a

Figure 17   Schematic plots of phase relations in (a) the system $4CaO \cdot 5B_2O_3 \cdot 7H_2O - R_2O_3 \cdot H_2O - H_2O$, (b) the system $Na_2O \cdot 2B_2O_3 \cdot 4H_2O - 2CaO \cdot 3B_2O_3 \cdot 5H_2O - H_2O$, and (c) the system $2CaO \cdot 3B_2O_3 \cdot 5H_2O - 2MgO \cdot 3B_2O_3 \cdot 7 - H_2O$. Reprinted from Christ et al. *Borate Mineral Assemblages in the System Na²O-CaO-MgO-B₂O₃-H₂O.* Copyright 1967. 313–32, with permission from Elsevier Science.

reaches 70 m. The Emet deposit (Helvaci 1986) consists of colemanite closely packed in shale and interstratified with late Tertiary marls, clays, limestones, and tuffs. The upper part of the shale averages 14 m thick and is about 75% colemanite, whereas the lower part is 7 m thick and approximately 25% colemanite. The chemical composition of colemanite is 50.9% B_2O_3, 27.2% CaO, and 21.9% H_2O (Albayrak and Protopapas 1985). The deposit at Bigadic (Helvaci 1995; Helvaci and Alaca 1991) consists of priceite and colemanite-ulexite interbedded with late Tertiary marl, clay, and gypsum. The B_2O_3 contents in colemanite and ulexite are, respectively, 40 to 42% and 37.52%. The Kestelek deposit consists of nodular colemanite boulders up to 1 m in diameter, which occur in a 5 m thick horizon. The upper 1.5 m horizon contains very pure, coarsely crystalline colemanite whereas the lower horizon contains many clay partings (Helvaci 1990). In the Sultancayiri borate deposit, the stratigraphic sequence consists of basement conglomerate, tuff, limestone, borates, gypsum, and limestone. Priceite is abundant, and other borates include colemanite and howlite (Helvaci 1995). Borate deposits of colemanite, ulexite, and borax were also found in nearby Samos Island, Greece. Tuffaceous rocks interbedded with borates are rich in authigenic silicate minerals such as zeolites, K-feldspar, and opal-CT (Helvaci et al. 1993).

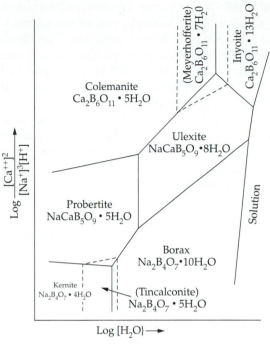

b

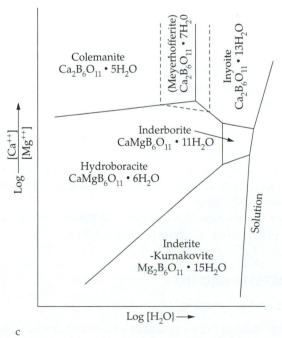

c

Figure 17 (Continued)

The Former USSR: Major borate deposits are concentrated in the Inder district, north of the Caspian Sea in the southern part of the country. Szaibelyite is the dominant borate in the deposits and is associated with hydroboracite rather than the normal sodium or calcium borates. The ores occur along fractures on the north and south margins of a salt dome of about 250 km^2 as a replacement for salt, clay, and gypsum (Kistler and Smith 1983). Borax is also recovered from the brine of Lake Inder, which has an average B$_2$O$_3$ content of 0.23%.

The Primorsky Industrial Amalgamation mines near Dalnegorsk in the district of Dalnegoskoye, some 550 km northeast of Vladivostok, Siberia, is another major producer. The deposits are borosilicate skarns consisting of datolite, a silicate borate mineral, wollastonite, calcite, hematite, garnet, danburite, hedenbergite, and axinite (Harben and Kuzvart, 1996).

South America: Some 40 borate deposits occur in an 885 km long stretch of the Andes along the common borders of Argentina, Bolivia, Chile, and Peru. These are either small aprons or cones containing several thousand tons of ulexite and borax, or beds containing several million tons of ulexite, borax, and inyoite (Norman and Santini 1985; Kistler and Helvaci 1994). Northern Argentina is the only significant producer, mainly from the Tincalayu mine in the north-central region of Salar del Hombre Muerto in the province of Salta. The borax is intercalated with sandstone, claystone, tuff, evaporite, limestone, and conglomerate. In addition, near the upper contact between borax and sediment are horizons of claystone with overlying nodules of ulexite, inyoite, and other borate minerals.

China: China's borate reserves are large, containing perhaps 10% of the world total (Lyday 1991) with major deposits in Liaoning, Qinghai, and Tibet. Liaoning, a northeast province, has the largest borate mines. The ores are szaibelyite, ludwigite, and suanite and occur as veins in early Proterozoic magnesian marbles. The average B$_2$O$_3$ content ranges from 5 to 18%. Associated minerals are magnesite, magnetite, dolomite, and talc (Liu 1988). From geology and field relationships, Peng and Palmer (1995) concluded that the Liaoning deposits are metamorphosed evaporites in which the boron was originally present in primary evaporite borate minerals.

The Chaidam basin, covering an area of 200,000 km^2 in the northwest portion of Qinghai Province, contains twenty large salt lakes. The largest one, Chaerhan Lake (5,000 km^2), contains sylvite, mirabilite, and borax (borax and magnesium-borax), along with bromine and iodine.

Large borate reserves exist on the Tibet Plateau, where fifty-seven lakes have been identified to contain borates (Lock 1991).

INDUSTRIAL PROCESSES AND USES

1. Milling and Processing

The milling of borate ores generally consists of crushing, as the preliminary step, calcination to remove water and screening, and air separation to remove clays before further refining processes.

At Boron, the ore is crushed, screened, and dissolved in weak and hot borax liquor. Large insoluble particles are separated when the dissolved ore solution is passed over vibrating screens. The liquor and the fines are pumped into a series of large countercurrent thickeners. From the thickeners, the clarified liquor is filtered and pumped to vacuum crystallizers. Crystals of borax from the crystallizers are separated from the spent liquor in automatic centrifuges and then dried as refined borax decahydrate or borax pentahydrate. Anhydrous boric acid and anhydrous borax are produced by heat treatment in gas-fired furnaces (Kistler and Smith 1983; Kistler and Helvaci 1994).

In Death Valley, the colemanite ore is crushed, screened, and fed to a washing plant to remove the clay. After being treated in a ball mill, the product is introduced to a bank of six float cells. Float concentrates are thickened, filtered, and dried, then run through cyclones to remove the 325 mesh material. After calcination, the product is again run through a series of cyclones, cooled, and classified for final product (Smith and Walters 1980).

Borax can be floated from halite by oleic acid or with an aromatic amine (Taggart 1945). A froth flotation process was proposed by Yarar (1985) to separate colemanite from a colemanite-calcite mixture. Sodium dodecyl benzene sulfonate is a selective collector for colemanite, whereas sodium oleate is not selective in a flotation pulp containing both calcite and colemanite.

Sodium borates from the brine of Searles Lake are treated by one of two processes: evaporation or carbonation (Garrett 1960). In the evaporation process, the brine is treated by rapid, controlled cooling which selectively removes the crystallized NaCl, Na_2CO_3, and KCl. The remaining borax-containing liquor is fed into large crystallizers, where it is mixed with a thick bed of borax seed crystals. The resulting borax crystals are separated from the liquor in cyclones, filtered, washed, redissolved, and recrystallized in vacuum crystallizers. Products from these crystallizers are again separated in cyclones and then dried to produce the final products. In the carbonation process, carbon dioxide from lime kiln gases or boiler flue gases is bubbled through the brine to convert Na_2CO_3 to $NaHCO_3$, which, being only slightly soluble in the brine, crystallizes and is filtered out. The filtrate is cooled in vacuum crystallizers in the presence of borax seed. The crystallized borax is dewatered in centrifuges and dried to produce borax decahydrate and borax pentahydrate.

It is generally recognized that a technical grade Borax should have greater than 99.5% $Na_2B_4O_7 \cdot 10H_2O$, and a special quality grade greater than 99.9%.

2. Industrial Uses

The largest users of borates are the glass and ceramic industries whose consumption accounts for 60% of the total production in the United States. Borates have very easy fusibility and high fluxing ability, and are major components in borosilicate glasses, insulation fiberglass, textile fiberglass, porcelain enamels, and ceramic glazes. Other uses, as illustrated in Fig. 18 (Kistler and Smith 1983), are many.

Glasses: Glass is a material which can accommodate a large range of chemical compositions resulting in a large range of physical properties for many fields of application. The standard procedure for the manufacture of glass consists of melting to produce a homogeneous molten batch, refining to remove the bubbles and cords,

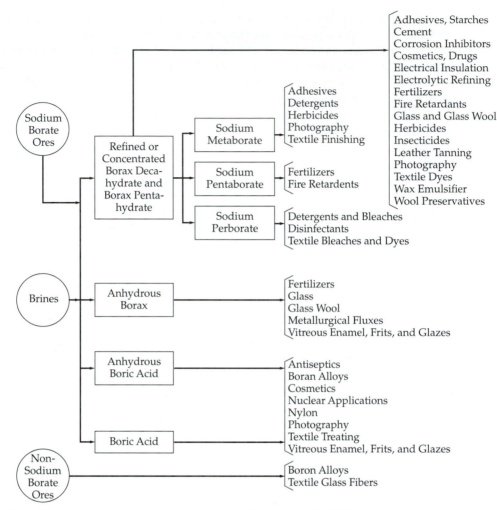

Figure 18   A summary of major uses of borates, after Kistler and Smith (1983).

cooling to solidify the molten batch, and annealing to release stresses and to stabilize glass properties. Almost every chemical element can be used for the production of glass, but about 90% of the worldwide production contains three major compositions, (1) silica, (2) lime, and (3) sodium oxide. Other compositions are added to the glass batch to modify and improve glass properties.

B_2O_3 is added to a lime-soda glass batch as a flux to modify the optical properties of the glass. It also improves chemical resistivity and resistance to thermal shock by lowering the coefficient of expansion of the glass (Baumgart 1984). In borosilicate heat-resistance glass, such as Pyrex glass, B_2O_3 forms an essential part of the glass-forming batch. These glasses possess a low coefficient of expansion and are highly resistant to thermal and mechanical shock and to chemical action. Pyrex glass 774 has the lowest liquidus temperature (774°C) of any known mixture that contains such a high silica content (80% SiO_2). This explains the exceptional ability of B_2O_3 to withstand devitrification (Baumgart 1984). Borosilicate glasses are widely used in the manufacture of laboratory ware, containers for pharmaceutical and medical products, pipelines and equipment for chemical and food processing plants, transparent cooking and oven ware, windows for high temperature environments, and tubing, bulbs, and lenses for illuminators (Deeg 1978). Typical compositions of borosilicate glasses are listed below:

SiO_2	81	76	62	76	74	68	67
B_2O_3	12	11	5	8	10	8	2
Al_2O_3	2	3	17	4	5	9	3
MgO	—	—	7	—	—	—	—
CaO	—	1	8	1	—	6	7
BaO	—	4	—	4	4	—	—
ZnO	—	—	—	—	—	—	7
Na_2O	3	4	1	6	6	9	14
K_2O	1	1	—	—	—	—	—

Borosilicate optical glasses are used mainly for lenses in high quality instruments, for light filters with specific spectra transmission, and for ophthalmic photochromic lenses (Deeg 1978). The prime reason for the introduction of B_2O_3 in optical glasses is its effect on dispersion in the short wavelength range of the visible spectrum. An analysis of approximately 250 optical glasses shows that 55% of them contain more than 5 wt% B_2O_3. Only 27%, most of them in the flint glass group, are boron-free. Not all boron containing optical glasses are borosilicates. Some of them can be classified as rare earth borates, lead borates, and barium borates. For filter glasses, an analysis of approximately 110 glasses shows that 75% contain B_2O_3. Among them, 39% have B_2O_3 of more than 5 wt%. The highest B_2O_3 concentrations (approximately 20 wt%) are found in neutral gray and in gold ruby glasses. Compositions of selected optical borosilicate glasses are as shown below:

SiO_2	55	68	36	46	44	2	5
ZrO_2	—	—	—	—	—	4	—
B_2O_3	7	13	13	14	19	40	53
Al_2O_3	—	—	5	3	1	—	10
Sb_2O_3	—	—	—	21	—	—	—
La_2O_3	—	—	—	—	—	41	—
CaO	—	—	—	—	—	6	—
BaO	—	10	46	—	29	—	—
ZnO	20	—	—	4	—	6	17
PbO	—	—	—	—	—	—	10
Na_2O	1	9	—	—	3	—	—
K_2O	16	—	—	12	4	—	5

Dielectric and seal glasses are lead borosilicate glasses with B_2O_3 content in the range of 14 to 21.5% (Tummala 1978). Those with very steep viscosity as a function of temperature were made in the system $PbO-B_2O_3$ and SiO_2 with additions of CaO and MgO. Five compositions are listed here:

	1	2	3	4	5
PbO	73.5	69.6	63.4	56.0	56.0
B_2O_3	12.7	13.6	17.0	21.5	21.5
SiO_2	13.6	13.6	12.0	12.5	12.5
Al_2O_3	0.2	0.2	0.2	1.0	1.0
CaO	—	—	4.4	5.0	5.0
MgO	—	3.0	3.0	2.0	2.0
Na_2O	—	—	—	2.0	2.0

Glass 1 is a typical lead glass and has a low dielectric deformation temperature (430°C). The deformation temperature increases as MgO is added to the batch composition. However, even adding MgO content to its compositional limit of 3.0% before the glass starts to crystallize is not enough to bring the temperature to the required level (485°C) without the addition of CaO. The dielectric deformation temperatures of glasses 2, 3, 4, and 5, are 463°, 490°, 490°, and 500°C, respectively. Glasses with compositions in the system $PbO-ZnO-B_2O_3$ as modified by the addition of CuO, Al_2O_3, Bi_2O_3, and SiO_2 have resulted in vitreous seal glass with required low-melting properties. A typical composition may be represented by 66% PbO, 14% B_2O_3, 2% SiO_2, 3.5% Al_2O_3, 2.0% CuO, 2.0% Bi_2O_3, and 10.5% ZnO.

Glass Fibers: Glass fiber manifests itself in three major forms: (1) insulating glass fibers, (2) textile glass fibers, and (3) optical glass fibers. Insulating glass fiber is made from a borosilicate glass melt by pouring molten glass into a rapidly spinning dish whose circumference is punctured by hundreds of small holes. The fibers that are forced out of the holes by centrifugal force are sprayed with a binder and blown into random distribution to form a matted wool (Russell 1991). The purity requirements for raw materials are not particularly high. The suitability of a raw composition for fiber formation is determined by the ratio of the viscosity-increasing SiO_2 and Al_2O_3 to the viscosity-decreasing ones such as Na_2O and CaO. Insulating glass fiber is used in acoustic, thermal, and electrical applications.

The manufacture of textile glass fiber is more sensitive to chemical and physical requirements. High purity of chemical components is required and viscosity must be uniform. The raw materials must be well-mixed and then melted in an elongated tank furnace for a prolonged period of time to produce a clarified glass melt. The conversion of melt to fiber is made by the direct melting process that is currently adopted by most producers. The glass melt is fed into spinning jets (bushings) at about 1,250°C. The individual bushings normally have 400 to 2,000 orifices with diameters of 1 to 2 mm. Upon emerging from the jet, the glass melt forms filaments which combine into a spinning strand below the bushings. Different constituted glass fibers are utilized in different application fields. Among borosilicate glass fibers, the E-glass (54.5% SiO_2, 14.5% Al_2O_3, 0.5% Fe_2O_3, 7.5% B_2O_3, 17.5% CaO, 4.5% MgO, and 0.8% Na_2O) is by far the most important textile glass fiber and is utilized for reinforcing plastics, rubber, and cement.

Optical glass fiber is made of two glasses, a core of Ge-doped fused silica glass with high refractive index, and a cladding of borosilicate glass with low refractive index. Light directed into the glass fibers travels through the length of the fibers without loosing much of its energy because of an almost total reflection along the borders of differently reflecting glass layers. These particles of light can travel much faster through fiber than electrons through copper wire, revolutionizing information transmission (Baumgart 1984).

Glazes and Enamels: B_2O_3 is an oxide of strong fluxing power, comparable to that of PbO or Na_2O. It can be used as either a primary or an auxiliary flux in a glaze and covers a large range of temperatures. Because it is a network-forming oxide, B_2O_3 generally contributes to a lowering of the expansion of the glaze. Borosilicate glass is used in both ground-coat porcelain enamels and cover-coat enamels. In the former, small amounts of adherence-promoting oxides, such as cobalt oxide, nickel oxide, and copper oxide are used in the formulation, whereas various amounts of titanium dioxide and color-producing oxides are required in the latter application (Taylor and Bull 1986).

Cleaning Agents: Borax can react with a strong base or a strong acid to form a buffer solution. This characteristic ability is used in the manufacture of soaps and detergents (Lyday 1985). Sodium perborate, marketed as a tetrahydrate or monohydrate, has both a detergent and a mild bleaching action. The tetrahydrate is manufactured by reacting NaOH and borax with H_2O_2, whereas the monohydrate is made by drying the tetrahydrate. Detergent formulations are the largest single use for borate in Europe and Japan (Kistler and Smith 1983).

Fertilizers and Insecticides: Borax is used as an essential gradient of fertilizers on boron-deficient soils to eliminate many forms of plane diseases caused by such a deficiency. It is also used in insecticides and as a weed killer. Small amounts of boric acid applied to wood provide protection against the common house bore (Lyday 1985).

Metal Welding: Boric acid and borax are exceptional fluxes for metal welding. Molten borates have the ability to dissolve metal oxides formed at the surface of the metal under welding, thus providing a clean surface for good metal-to-metal contact. Potassium perborate is used in metallurgy for welding and brazing stainless steel (Lyday 1985).

REFERENCES

Albayrak, F. A., and T. E. Protopapas. 1985. "Borate Deposits of Turkey." In *Borates: Economic Geology and Production,* Eds. J. M. Barker and S. J. Lefond, 71–85. New York: Sym. Proc. Soc. Mining Eng., AIME.

Aristarain, L., and R. C. Erd. 1971. Inyoite, $2CaO \cdot 3B_2O_3 \cdot 13H_2O$ de la Puna Argentina. *Anales de la Sociedad Cientifica Argentina* 191:181–211.

Barker, J. H., and S.J. Lefond. 1985. "Boron and Borates: Introduction and Exploration Techniques." In *Borates: Economic Geology and Production,* Eds. J. M. Barker and S. J. Lefond, 14–34. New York: Sym. Proc. Soc. Mining Eng., AIME.

Barnard, R. M., and R. B. Kistler. 1966. Stratigraphic and structural evolution of the Kramer sodium borate ore body, Boron, California, Second Symposium on Salts. Vol. 1. *The Northern Ohio Geol. Society,* Ed. J. L. Rau, 133–156. (Cleveland, Ohio).

Baumgart, W. 1984. "Glass." In *Process Mineralogy of Ceramic Materials,* Eds. W. Baumgart, A.C. Dunham, and G.C. Amstutz, 28–50. New York: Elsevier Sci. Publ.

Bowser, C. J., and R. W. Dickson. 1966. Chemical zonation of the borates of Kramer, California, Second Symposium on Salt. Vol. 1. *The Northern Ohio Geol. Society,* Ed. J. L. Rau, 122–132. (Cleveland, Ohio).

Christ, A., H. Truesdell, and R. C. Erd. 1967. Borate mineral assemblages in the system Na^2O-CaO-MgO-B_2O_3-H_2O. *Geochem. Cosmochim. Acta* 31:313–337.

Christ, C. L., J. R. Clark, and H. T. Evans. 1958. Borate Minerals III. The crystal structure of colemanite. *Acta Cryst.* 11:761–770.

Christ, C. L., and J. R. Clark. 1977. A crystal-chemical classification of borate structures with emphasis on hydrated borates. *Phys. Chem. Minerals* 2:59–87.

Christ, C. L., and R. M. Garrels. 1959. Relations among sodium borate hydrates at the Kramer deposit, Boron, California. *Amer. J. Sci.* 257:516–528.

Deeg, E. W. 1978. "Industrial Borate Glasses." Vol. 12, In *Borate Glasses,* Eds. L. D. Pye, V. D. Fréchette, and N. J. Kreidl, 587–596. New York: Plenum Press.

Garrett, D. E. 1960. "Borax Processing at Searles Lake." 3rd ed. In *Industrial Minerals and Rocks,* Ed. J.L. Gillson, 119–127. New York: AIME.

Hansson, A. 1961. On the crystal structure of hydrated sodium peroxoborate. *Acta Chem. Scand.* 15:934–935.

Harben, P. W., and M. Kuzvart. 1996. Industrial minerals: A global geology. *Industrial Minerals Information, Ltd., Metal Bulletin* (London) 462p.

Helvaci, C. 1978. A review of the mineralogy of the Turkish borate deposits. *Mercian Geol.* 6:257–270.

Helvaci, C. 1983. Mineralogy of Turkish borate deposits. *Jelogi Muhendisligi* 17:37–54.

Helvaci, C. 1986. Stratigraphic and structural evolution of the Emet borate deposits, Western Anatolia, Turkey. Res. Paper MM/JEO–86, AR 008, *Dokuz Eylul Univ.* (Izmir, Turkey) 28p.

Helvaci, C. 1990. Mineral assemblages and formation of the Kestelek and Sultancayiri borate deposits. *Proc., Intern. Earth Sci. Congr on Aegean Regions* (Izmir, Turkey) 11–12.

Helvaci, C. 1995. Stratigraphy, mineralogy and genesis of the Bigadic borate deposits, western Turkey. *Econ. Geol.* 90:1237–1260.

Helvaci, C., M. G. Stamatakis, C. Zagouroglou, and J. Kanaris. 1993. Borate minerals and related authigenic silicates in northeastern Mediterranean Lat Miocene continental basins. *Exploration & Mining Geol.* 2:171–178.

Helvaci, C., and O. Alaca. 1991. Geology and mineralogy of the Bigadic borate deposits and vicinity. (Foreign ed.) *Bull. Min. Res. & Exploration* 113:31–63.

Inan, K., A. C. Dunham, and J. Esson. 1973. Mineralogy, chemistry and origin of Kirka borate deposit, Eskisehir Province, Turkey. *Trans., Inst. Mining and Metal.,* 82B:114–123.

Kistler, R. B., and C. Helvaci. 1994 "Boron and Borates." 6th ed. In *Industrial Minerals and Rocks,* Ed. D.D. Carr, 171–186. Littleton, Colorado: Soc. Mining, Metal., and Exploration.

Kistler, R. B., and W. C. Smith. 1983. "Boron and Borates." 5th ed. In *Industrial Minerals and Rocks,* Ed. S.J. Lefond, 533–559. New York: Soc. Mining Engineers, AIME.

Liu, D. M. 1988. Boron mineral resources in the northeastern region of China—A technical and economic evaluation. *Ministry of Geol. & Mineral Res.* (Beijing) 5p.

Lock, D. 1991. Notes on the borate resources of the People's Republic of China. M.R.I. Rept., *Australian Nat. Univ.* 53p.

Lyday, P. A. 1985. "End Uses of Boron Other than Glass." Vol. 12, In *Borate Glasses: Structure, Properties, Applications,* Eds. L. D. Pye, V. D. Fréchette, and N. J. Kreidl, 255–267. New York: Plenum Press.

Lyday, P. A. 1991. Boron—1990. Mineral Commodity Summaries, *U.S. Bureau Mines* 133–142.

McAllister, J. F. 1970. Geology of the Furnace Creek Borate Area, Death Valley, Inyo County, California. Map Sheet 14, *California Div. Mines and Geology* 1–9.

Menzel, H., H. Schulz, L. Sieg, and M. Volgt. 1935. The system sodium tetraborate-water. *Zeit. Anorg. Chem.* 224:1–22.

Morimoto, N. 1956. The crystal structure of borax. *Mineral. J.* (Sapporo) 2:1–18.

Nies, N. P., and R. W. Hulbert. 1967. Solubility isotherms in the system sodium oxide-boric oxide-water. *J. Chem. Eng. Data* 12:303–313.

Norman, J. C., and K. N. Saritini. 1985. "An Overview of Occurrence and Origin of South American Borate Deposits with a Description of the Deposit at Laguna Salinas, Peru." In *Borates: Economic Geology and Production,* Eds. J. M. Barker and S. J. Lefond, 53–69. New York: Sym. Proc. Soc. Mining Engineers, AIME.

Peng, C., C. Wu, and P. Chang. 1963. Crystal structure of ascharite. *Sci. Sinica* 12:1761–1764.

Peng, Q. M., and M. R. Palmer. 1995. The paleoproterozoic boron deposits in eastern Liaoning, China: A metamorphosed evaporite. *Precambrian Res.* 72:185–197.

Russell, A. 1991. Minerals in fibre glass. *Ind. Minerals* (November):27–41.

Shaw, D. M., and R. Bugry. 1966. A review of boron sedimentary geochemistry in relation to new analyses of some North American shales. *Can. J. Earth Sci.* 3:49–63.

Smith, G. I. 1979. Surface stratigraphy and geochemistry of late Quaternary evaporites, Searles Lake, California. *U.S.G.S. Bull.* 1181:1–58.

Smith, P. R., and R. A. Walters. 1980. Production of colemanite at American Borate Company's plant near Lathrop Wells, Nevada. *Mining Eng.* (Feb): 199–204.

Strunz, H. 1997. Classification of borate minerals. *European Jour. Mineral.* 9:225–232.

Tagéeva, N. V. 1943. Geochemistry of boron and flurine. *Priroda* 6:25–35.

Taggart, A. F. 1945. *Handbook of Mineral Dressing.* New York: John Wiley & Sons, 12–125.

Taylor, J. R., and A.C. Bull. 1986. *Ceramic glaze technology.* Oxford: Pergamon, 263p.

Tummala, P. R. 1978. "Application of Borate Glasses in Electronics." Vol. 12, In *Borate Glasses: Structure, Properties, Applications,* Eds. L. D. Pye, V. D. Fréchette, and N. J. Kreidl, 617–622. New York: Plenum Press.

Walker, C. T. 1975. *Geochemistry of Boron, Dowden, Hutchinson and Ross.* (Stroudsburgh) 414p.

Wilson, J. L., and D. L. Emmons. 1977. Origin and configuration of the oxidized zone in Tertiary formations, Death Valley region, California. *Geology* 5:696–698.

Yarar, H. 1985. "The Surface Chemical Mechanism of Colemanite-Calcite Separation by Flotation." In *Borate: Economic Geology and Production,* Eds. J. M. Barker and S. J. Lefond, 221–233. New York: Sym. Proc., Soc. Mining Engineers, AIME.

Celestite

Celestite ($SrSO_4$) is the sole strontium mineral of industrial importance, and its principal use is in the ceramic and glass industries. Strontium carbonate occurs in nature as strontianite, but its distribution is restricted and economically workable deposits are scarce.

The strontium industry was dominated by the United Kingdom's production of celestite for more than several decades, but now the major producers are Mexico, Turkey, Spain, and China.

MINERALOGICAL PROPERTIES

Celestite has a barite-type structure ($BaSO_4$) with a=8.377Å, b=5.350Å, and c=6.873Å (Garske and Peacor 1965; Chang et al. 1996). Like barium in barite, strontium occupies a large irregular twelve-coordinated site in the structure, which limits cation substitution. There is a complete series of solid solution in the system $SrSO_4$-$BaSO_4$ (Brower 1973), but $(Ba,Sr)SO_4$ is an inert phase for Ba-Sr exchange in nature, and only in some rare cases does an intermediate member of the series occur in nature (Hanor 1968). A Sr-rich member, $(Sr_{0.87}Ba_{0.13})SO_4$, was found in volcanic clastic rocks in Vicenza, Italy, and has a=8.408Å, b=5.372Å, and c=6.897Å (Brigatti et al. 1997). Therefore, most celestites have compositions close to the ideal $SrSO_4$. The Ba-content in celestite can be estimated by measuring $d_{(121)}$ of the barite-type structure (Goldish 1989).

Celestite is normally white, but in some cases it has a bluish color that was determined by Przibram(1956) to be a radiation color. On heating, the color disappears and weak thermoluminescence is exhibited; on exposure to irradiation the blue color is restored. Celestite usually occurs in tabular or prismatic crystals or as fibrous cleavable masses. It has a hardness of 3.0 to 3.5, and a specific gravity of 3.8. Index of refraction ranges from 1.619 to 1.632. Both basal and prismatic cleavages are perfect with angles of about 104° and 76°.

The solubility of celestite in water as a function of temperature is shown in Fig. 19(a) illustrating a maximum at 20°C (Reardon and Armstrong 1987). Fig. 19(b) shows the solubility of celestite in NaCl solution of various concentrations at 25°C.

GEOLOGICAL OCCURRENCE AND DISTRIBUTION OF DEPOSITS

Celestite occurs mainly in sedimentary rocks, particularly dolomite, limestone, and marl, either as a primary precipitate from aqueous solution or, more usually, by the interaction of gypsum or anhydrite with Sr-rich waters. It may be a primary mineral in hydrothermal veins and is also found filling veins and cavities in basic eruptive rocks.

The United Kingdom was the leader in the celestite industry for several decades, but now the major producers are Mexico, Turkey, Spain, and China. Others are Canada,

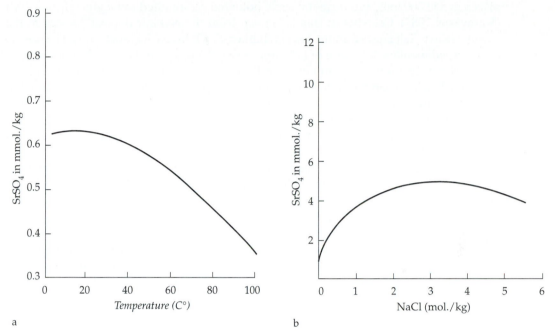

Figure 19   (a) Solubility of celestite as a function of temperature, and (b) solubility of celestite at 25°C in NaCl solutions. Reprinted from Reardon and Armstrong. Copyright 1987, with permission from Elsevier Science.

the United States, and Iran. The annual estimated world production was about 170,000 tons in 1994 and 1995 (Harben and Kuzvart 1996).

In the United Kingdom, celestite deposits are well developed in the Bristol area around Yate (Nickless et al. 1975). Located close to the top of the Triassic Keuper Marl, the celestite beds, in alternation with clay, dolomite, and evaporite, have three types of mineralization. It occurs (1) as stratiform beds of nodular and disseminated celestite, and as thin stringers and veins in Keuper Marl, typically 10 to 15 m below the base of the overlying Tea Green Marl, (2) as nodular and disseminated celestite at the unconformity between the Keuper Marl and the underlying Paleozoic rocks, and (3) as veins, sheets, and infillings of celestite in Paleozoic rocks below the Keuper Marl (Nickless et al. 1975).

In Mexico, the main celestite deposits are distributed in the southern part of the state of Coahuila. The ores are contained in a thick Mesozoic sequence of massive fine-grained limestone (de Brodtkorb 1989; Bearden 1990; Rodriquez Garza, and McAnulty 1996). The celestite stratabound layers are up to 4 m thick and up to 1 km long. Each layer is confined to a single stratigraphic horizon, and some of them may be followed for nearly 30 km (Kesler and Jones 1981). The San Agustin and Bermejillo North and South deposits are nonmineralic, whereas at Matamoros and Escalon, the celestite is in association with calcite.

Turkey has a large celestite deposit near Akkaya, Sivas Province, some 480 km east of Ankara. The celestite occurs within gypsum and clay formations of Neogene age and has a minimum $SrSO_4$ content of 96% (Griffiths 1985a). Identified reserves have been placed at 550,000 tons, and further reserve potential is estimated to be greater than 2 Mt (Karayazici 1987). Celestite in Iran is derived from the Nakhjir deposit, located in the Dasht-e-Kavir Salt Desert, about 200 km southeast of Tehran (Schiebel 1978). The ore occurs in a sedimentary sequence of limestone, marl, shale, salt, and chert of Miocene age.

In Spain, celestite occurs in two parallel east-west belts in the Granada Basin: the short northeastern belt which has the Montevive deposit, and the long southwestern Escuzar belt which has only celestite exposures (Martin et al. 1984; Griffiths 1985b). The Montevive deposit is a sedimentary sequence containing limestone, clay, barite, and gypsum, and celestite occurs as a replacement of the carbonate that is well-stratified in layers 20 to 50 cm thick, and contains horizons of different detrital clays. Two types of celestite occurrences outcrop in the Escuzar belt. One is the product of replacement of stromatolitic carbonates, and the other is karst-filled Miocene breccia with celestite pebbles.

In Canada, celestite deposits were found in Nova Scotia, Newfoundland, New Brunswick, Ontario, and British Columbia (Dawson 1985; Felderhof 1986). The Loch Lomand deposit comprises five distinct horizons interlayered with limestone, shale, siltstone, and gypsum. Celestite is produced in China at Hechuan deposit in Sichuan and Lishui deposit in Jiangshu, and there are celestite deposits in the United States in Texas, Arizona, and California (Fulton 1983).

INDUSTRIAL PROCESSES AND USES

1. Processes and Beneficiation

Because of tonnage requirements and modes of occurrence of celestite, it has been a general practice to treat the crude ore by hand cobbing and washing to reach an approximately 90% $SrSO_4$ (Enck 1960). Although this practice is still in use in some plants, mechanization, including crushing, jigging, and screening, has been installed in most plants, and gravity separation and flotation are widely used.

At Barit Maden in Turkey, the Stripa process is used for the beneficiation of celestite (Svensson 1980; Önal and Dogan 1988). The ore, after being crushed in a jaw crusher, is sent to an attrition tumbler, then to a 15 mm screen, and finally to a stripa heavy medium separation unit as shown in Fig. 20. The stripa preconcentrate is further crushed in an impact crusher in a closed circuit with a vibrating screen. As illustrated, the two size fractions receive different types of treatment, and both jig and table concentrates are combined for drying on a vibrating table. A stripa unit, as illustrated in Fig. 21, includes a cone to serve as a thickener and a surge tank for the medium. There is also a shaking trough in which the separation takes place and a screen on which the medium is washed from the sinks and floats. The higher density celestite sinks down through the bed and settles on the bottom of the trough where the particles are transported at a considerably slower rate than the gangue.

At Loch Lomond, Nova Scotia, low-grade celestite ore is beneficiated by flotation (Crowell 1973). The ore is treated in a series of size reduction processes by using jaw and cone crushers, ball mill, and cyclone before it is fed to the flotation circuit. An anionic flotation method is employed.

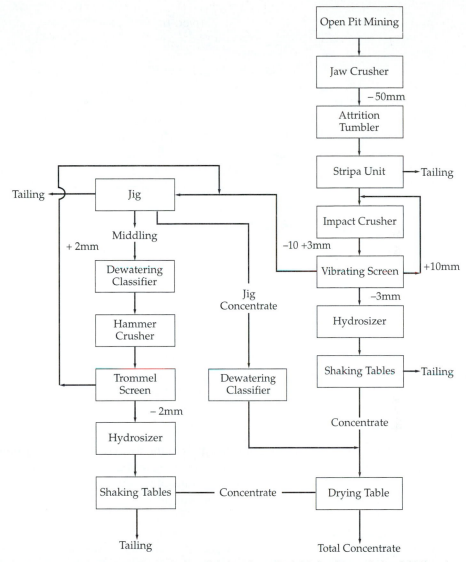

Figure 20   Simplified flowsheet of celestite beneficiating plant at Barit Maden. Source: Industrial Minerals Supplement, March 1998.

2. Specifications

The celestite used in industry is in the form of strontium carbonate and nitrate (Fulton 1983). Typical specifications for the carbonate are 90% $SrSO_4$ (min), 2% $BaSO_4$ (max), and 0.1% F (max), and for the nitrate are 95% $SrSO_4$ (min), 1.5% $CaSO_4$ (max), and 2% $BaSO_4$ (max). Size specifications are six inches (15.24 cm) (max) for the carbonate and minus six inches, plus one-quarter inch (minus 15.24 cm, plus 0.635 cm) for the nitrate. There are two grades in the specifications of manufactured carbonate. The glass

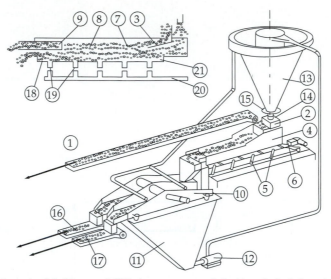

Figure 21   The Stripa unit, after Svensson (1980). 1: ore conveyor, 2: feed box, 3: feed plate, 4: trough, 5: spring rods, 6: tie-rods from eccentric, 7: bed, 8: suspension layer, 9: separating plate, 10: washing screen, 11: pump sump, 12: medium pump, 13: cone thickener, 14: valve, 15: dilution water, 16: conveyor for floats, 17: conveyor for sinks, 18: bottom plates, 19: valves, 20: water pipes, 21: water box.

grade has 96% $SrCO_3$ (min), 3% $BaCO_3$ (max), 0.5% $CaCO_3$ (max), 0.4% Total S (max), 0.01% Fe_2O_3 (max), and 1.0% Na_2CO_3 (max), and the ceramic grade has 96% $SrCO_3$ (min), 1.5% $BaCO_3$ (max), 0.4% Total S (max).

3. Industrial Uses

Unlike barite, the principal use of celestite is in the form of strontium carbonate and nitrate. In the manufacture of strontium carbonate, celestite ore is crushed, ground, and mixed with coke. The mixture is then fed into rotary kilns. At temperatures of 1,100° to 1,200°C, the celestite is reduced to strontium sulfide, known as black ash, according to the reaction (Goodenough and Stenger 1973): $SrSO_4$ (solid) + 2C (solid) → SrS (solid) + $2CO_2$ (gas). The black ash is fine-ground, leached with water, filtered and reacted with carbon dioxide or soda ash to produce the carbonate: SrS (liquid) + CO_2 (gas) + H_2O(liquid) → $SrCO_3$ (solid) + H_2S (liquid) or SrS (liquid) + Na_2CO_3 (solid) → $SrCO_3$ (solid) + Na_2S (liquid). Strontium carbonate produced from this process with or without additional sintering is known as glass grade. Strontium carbonate can also be manufactured by direct conversion. This process involves washing and thickening the celestite, followed by reaction with soda ash and steam for between one and three hours. The strontium carbonate produced is separated from the sodium sulfate liquor by centrifuge. Ceramic grade strontium carbonate is produced by this direct process.

The largest use of strontium carbonate is in the manufacture of glass faceplates for color television tubes (Fulton 1983; Griffiths 1992). It is present in glass at approximately 12 to 14 wt% SrO and functions as an x-ray absorber. Strontium carbonate is an effective x-ray barrier because of its large atomic radius, and its presence is required in the relatively high voltage television sets used in the United States and Japan. For lower voltage television such as that used in Europe, faceplates contain the less expansive barium carbonate as an x-ray absorber. Strontium carbonate, when added to special glasses, glass frits, and ceramic glazes, increases the firing range and lowers acid solubility.

Strontium carbonate is used in the production of high purity, low lead electrolytic zinc. At a use level of 4.4 to 7.7 kg of $SrCO_3$ per ton of metal produced, it removes lead from cathode zinc. Carbonate is also used in the manufacture of strontium ferrite by mixing strontium carbonate with ferric oxide in a ball mill using ethanol liquid as a wetting agent. The resultant slurry is separated by filtration, and strontium ferrite is produced by calcination (Gray and Routil 1973). Strontium ferrites have a higher coercive force than their barium counterparts, which are more easily demagnetized. Consequently, strontium ferrites are used in applications where demagnetization is likely to occur. Strontium titanate, also made from strontium carbonate, is used for electro-technical applications in the form of metal-doped capacitors.

Strontium nitrate is produced by the reaction of milled strontium carbonate with nitric acid. The nitrate slurry is filtered, crystallized, and centrifuged before drying in a rotary dryer for the final product. Strontium nitrate has long been used in pyrotechnics and is a component of signalling devices. It is also used to make red tracer bullets for the military.

REFERENCES

Bearden, S. D. 1990. Celestite resources of Mexico, Proc., 28th Forum on the Geology of Industrial Minerals, *Virginia Div. Mineral. Res. Publ.,* Ed. P. C. Sweet, 119:13–15.

Brigatti, M. F., E. Galli, and L. Medici. 1997. Ba-rich celestite: New data and crystal structure refinement. *Mineral. Mag.* 61:447–451.

Brower, E. 1973. Synthesis of barite, celestite and barium-strontium sulfate solid solution crystals. *Geochim. Cosmochim. Acta* 37:155–158 and 2688–2689.

Chang, L. L. Y., R. A. Howie, and J. Zussman. 1996. Rock-forming minerals. Vol. 5B, *Non-Silicates: Sulphates, Carbonates, Phosphates, Halids,* Essex, England: Longman, 383p.

Crowell, G. D. 1973. "Mineralogy of Strontium." In *Proc. Intern. Conf. on Strontium Containing Compounds,* Ed. T.J. Gray, 23–33. Halifax: Atlantic Industrial Res. Inst.

Dawson, K. R. 1985. Geology of barium, strontium, and fluorite deposits in Canada. *Geol. Surv. of Canada, Econ. Geol. Rept.* 34:136p.

de Brodtkorb, M. K. 1989. "Celestite: Worldwide Classical Ore Field." In *Nonmetalliferous Stratobound Ore Fields,* Ed. M. K. de Brodtkorb, 17–39. New York: Van Nostrand Reinhold.

Enck, E. G. 1960. "Strontium Minerals." 3d ed. In *Industrial Minerals and Rocks,* Ed. J. L. Gillson, 815–818. New York: AIME.

Felderhof, G. W. 1986. Report on celestite deposits of Canada. Canada/Nova Scotia Mineral Development Agreement. *Dept. Suppl. & Serv.* File 6-9031.

Fulton, III R.B. 1983. "Strontium." 5th ed. In *Industrial Minerals and Rocks,* Ed. S. J. Lefond, 1229–1233. New York: Soc. Mining Engineers, AIME.

Garske, D., and D. R. Peacor. 1965. Refinement of the structure of celestite, $SrSO_4$. *Zeit. Krist.* 211:204–210.

Goldish, E. 1989. X-ray diffraction analysis of barium-strontium sulfate (barite-celestite) solid solutions. *Powder Diff.* 14:214–216.

Goodenough, R. D., and V. A. Stenger. 1973. "Magnesium, Calcium, Strontium, Barium and Radium." Vol. 1, In *Comprehensive Inorganic Chemistry,* Ed. J. C. Bailar 591–664.

Gray, T. J., and R. J. Routil. 1973. "Strontium hexaferrites." In *Proc. Intern. Conf. on Strontium Containing Compounds,* Ed. T. J. Gray 11–32. Halifax: Atlantic Industrial Res. Inst.

Griffiths, J. 1985a. Celestite: New production and processing developments. *Ind. Minerals* (November):21–35.

Griffiths, J. 1985b. Spain's industrial minerals. *Ind. Minerals* (July):40.

Griffiths, J. 1992. Celestite and strontianite chemical trade, the Mexican wave. *Ind. Minerals* (October):21–27.

Hanor, J. S. 1968. Frequency distribution of compositions in the barite-celestite series. *Amer. Mineral.* 53:1215–1222.

Harben, P. W., and M. Kuzvart. 1996. Industrial minerals: *A Global Geology. Industrial Minerals Information, Ltd., Metal Bulletin* (London), 462p.

Karayazici, F. 1987. Industry structure, past, present, and future. Proc. Turkey's Ind. Minerals Conf., *Metal Bull.,* PLC (London) 14.

Kesler, S. E., and L. M. Jones. 1981. Sulfur and strontium isotopic geochemistry of celestite, barite and gypsum from the Mesozoic basins of northeastern Mexico. *Chem. Geol.* 31:211–224.

Martin, J. M., M. Ortesa Huerta, and J. Torres Ruiz. 1984. Genesis and evolution of strontium deposits of the Grenada Basin (Southeastern Spain). Evidence of diagenetic replacement of a stromatolitic belt. *Sed. Geol.* 39:281–298.

Nickless, E.F. P., S. J. Booth, and P. N. Mosley. 1975. Celestite deposits of the Bristol area. *Inst. Min. Metal. Trans.* 84B:62–64.

Önal, G., and M. Z. Dogan. 1988. Beneficiation of celestite ore, Stripa application in Turkey. *Ind. Minerals Suppl.* (March):44–46.

Przibram, K. 1956. Irradiation colors and luminescence. *Pergamon* (London), 332p.

Reardon, S. J., and D. K. Armstrong. 1987. Celestite solubilities in water, seawater and NaCl solution. *Geochim. Cosmochim. Acta* 51:63–72.

Rodriques Garza, A., and W. N. McAnulty, Jr. 1996. "Overview of Celestite Deposits in South-Central Coahuila, Mexico—Geology and Exploration." In *Proc., 31st Forum on the Geology of Industrial Minerals,* Eds. G. S. Austin, S. K. Hoffman, J. M. Barker, J. Zidek, and N. Gilson, 154:27–32. New Mexico Mines and Mineral Res. Bull.

Schiebel, W. 1978. New strontium deposits in Iran. *Ind. Minerals* (September):54–57.

Svensson, J. 1980. The Stripa process of dense medium separation. *Mining Mag.* (November):480–487.

Chromite

Chromite is the only Cr-containing mineral of industrial importance and the predominant mineral in chrome ores that are aggregates of chromite, magnesium silicates, and plagioclase. More than 65% of chrome ores are used for the production of ferrochrome, and the remainder is used for refractories and chemicals.

South Africa, the former USSR, Albania, Turkey, India, Zimbabwe, Finland, and the Philippines are the world's major producers of chrome ores, with an average annual production of 10 M tons in the period 1990 to 1995 (Harben and Kuzvart 1996).

MINERALOGICAL PROPERTIES

Spinel minerals with Cr^{3+} as the dominant trivalent cation constitute the chromite series. Divalent cations are largely Fe^{2+} and Mg, and a complete solid solution series exists between $FeCr_2O_4$ and $MgCr_2O_4$. Chromite represents the ferrous members ($Fe^{2+}>Mg$), and magnesian members are magnesiochromite ($Mg>Fe^{2+}$). There is also appreciable replacement of Cr^{3+} by Al and by Fe^{3+} (Deer et al. 1992). A complete representation of the chemical composition of a chromite should be $(Fe,Mg)(Cr,Al)_2O_4$. The Cr_2O_3 content in chrome ores determines their industrial usage. Metallurgical grade has 45 to 56% Cr_2O_3, <10% SiO_2, and Cr/Fe = 2.5 to 4.3, chemical grade Cr_2O_3 > 44%, SiO_2 < 2.5 to 3.5, and Cr/Fe = 1.5, and refractory grade 30 to 40% Cr_2O_3, 25 to 30% Al_2O_3, 6% SiO_2, and Cr/Fe = 2.0 to 2.5. Theoretically, chromite contains 68% Cr_2O_3 and 32% FeO. Analyses of typical chrome ores are shown in Table 6. No. 1 (Transvaal Steelport), which has 47.41% Cr_2O_3 and SiO_2 content below detection, is a chemical grade ore, No. 2 (Philippine Masinloc), which has a low Cr_2O_3 and a high Al_2O_3, is a refractory grade ore, and Nos. 3 to 5 are metallurgical grade ores (Mikami 1983).

Chromite is black to dark gray. In thin section or as fragments it is essentially opaque, but may be yellow-brown or deep brown on thin edges. Its luster ranges from brilliant, shiny metallic to resinous, dull submetallic. Chromite is brittle, lacks cleavage, and fractures unevenly with conchoidal tendencies. $FeCr_2O_4$ has $a_{(cubic)}$ = 8.378Å, D (specific gravity) = 5.09, and n (refractive index) = 2.16. The $a_{(cubic)}$, D, and n of other members of the chromite series $(Fe,Mg)(Cr,Al)_2O_4$ are 8.334Å, 4.43, and 2.00 for $MgCr_2O_4$ (magnesiochromite); 8.396Å, 5.20, and 2.42 for $FeFe_2O_4$ (magnetite); 8.383Å, 4.52, and 2.38 for $MgFe_2O_4$ (magnesioferrite), 8.135Å, 4.40, and 1.835 for $FeAl_2O_4$ (hercynite); and 8.103Å, 3.55, and 1.719 for $MgAl_2O_4$ (spinel) (Deer et al. 1992). The magnetic susceptibility of chromite is highly variable, and the variation is closely related to the percentage of the end-members it contains (Murthy and Gopalakrichna, 1981). Magnetic susceptibilities at 300 K (room temperature) and 77 K, of chromites from Masinloc, Philippines, and Montrose, South Africa, are, respectively, 25.8 and 33.8, and 54.3 and 54.5 ($x10^{-6}$) (Owada and Harada 1985).

Chrome ore generally occurs as massive aggregates of chromite grains of from 0.05 to 30 mm, or more commonly from 1 to 5 mm in size. As gangue increases, chrome ore grades from massive to disseminated ore. In some deposits, gangue can go well beyond 50%, and in others, nodular ore forms that consists of spherical to ovoid aggregates

Table 6 Analysis of chromites and corresponding chrome ores, after Mikami (1983).

	No. 1		No. 2		No. 3		No. 4		No. 5		No. 6	
	C*	O*	C*	O*	C*	O*	C*	O*	C*	O*	C*	O*
Cr_2O_3	47.41	44.52	36.24	32.10	61.44	54.91	57.70	47.00	58.32	48.50	55.90	49.19
Al_2O_3	14.82	15.50	31.86	30.20	11.41	9.92	13.44	12.65	11.06	10.01	11.92	11.37
Fe_2O_3	9.21		2.97		nil		3.42		4.10		2.18	
FeO	16.86		11.32		12.53		11.66		11.10		18.80	
Tot. Fe as FeO†	(25.26)	24.72	(13.99)	12.72	—	12.41	(14.70)	11.93	(14.75)	13.28		18.27
MgO	11.40	10.10	17.10	18.06	13.66	14.92	13.29	15.46	14.23	18.83	10.36	12.35
CaO		0.30		0.44		0.70		1.77		0.40		0.48
SiO_2		2.24		5.00		5.02		5.71		6.94		6.58
TiO_2	0.41	0.43	0.38	0.30	0.47	0.20	0.39	0.32	0.06	0.05	0.76	0.67
MnO	0.09	0.07	0.11	0.08	0.16	0.14	0.24	0.06	0.15	0.12	0.30	0.26
NiO			0.12	0.10					0.20	0.18	0.09	0.08
V_2O_5	0.33	0.30							0.07	0.05	0.18	0.16
L.O.I.				0.35		0.92		3.95		1.20		
Total	100.53	98.17	100.10	99.35	99.37	98.64	100.24	98.85	99.29	99.56	100.49	100.44
Cr:Fe	1.67	1.58	2.30	2.22	4.31	3.90	3.45	3.47	3.45	3.25	2.37	2.37

*C = chromite, O = ore or concentrate.
†Separate FeO and Fe_2O_3 are generally not determined. Generally only Cr_2O_3, Total Fe, Al_2O_3, MgO, SiO_2, and CaO are available in out-turn analyses. These are given on a dried basis. CaO and SiO_2 ranged from nil to 0.17 in separated chromites and were omitted.
No. 1—Trasvaal Steelport. 2—Philippine Masinloc. 3—Kempirsai (USSR). 4—Selukwe, Zimbabwe (talc-dolomite gangue). 5—Turkish. 6—Great Dyke.

mostly from 5 to 40 mm in diameter (Mikami 1983). Depending upon the size and the nature of chrome ores, they are classified as lumpy, friable-lumpy, and run-of-mine fines.

The most common gangue minerals in chrome ore are serpentine, chlorite, talc, and pyroxene. Others commonly found are olivine, amphibole, and calcic plagioclase.

GEOLOGICAL OCCURRENCE AND DISTRIBUTION OF DEPOSITS

In every primary chrome ore district in the world, the chromite occurs in ultramafic rocks or rocks derived from them by alteration. Such rocks are peridotites, pyroxenites, dunites, serpentinites, and talc schists.

Chrome ore deposits are classified by structure of ore bodies and regional pattern of ore distribution into two major types: stratiform and podiform (Stowe 1987a; Harben and Bates 1990). Stratiform deposits occur in parallel layers with great lateral continuity, are remarkably uniform, and are made up of high iron chrome ore. The Bushveld igneous complex (Transvaal, South Africa), the Great Dyke (Zimbabwe), and the Stillwater complex (Montana, United States) are examples. Podiform deposits occur as pods, lens, sackforms, slabs, and other irregular shapes. The ore bodies are not continuous and follow no recognizable systematic distribution within the containing country rock. Examples are deposits in the Ural Mountains, Albania, Zimbabwe, and the Philippines (Wyllie 1967; Thayer 1973; Stowe 1987a).

There are major chrome deposits in South Africa, the former USSR, Albania, Turkey, India, Zimbabwe, Finland, and the Philippines. A brief description for each is given here:

South Africa: The world's largest chrome deposits occur in the Bushveld igneous complex, covering 68,000 km^2 of west-central Transvaal Province, South Africa (Willemse 1969; von Gruenewaldt 1983). The complex is made up of rocks ranging from dunite through peridotite and pyroxenite to norite, gabbro, and anorthosite, and the ore occurs as magmatic segregation seams within the gabbro-norite unit. The mines are located in four main areas on the periphery of the complex, (1) the eastern (Lydenburg) chromite belt, (2) the western (Rustenburg) belt, (3) the Protietersrust district in the northern part of the complex, and (4) the Marico-Zeerrust District, an extension of the Bushveld Complex to the east of the western belt. The ore is generally friable, and the bulk of Bushveld chrome ore is rich in iron (20 to 27% FeO) with a low content of alumina. The Cr_2O_3 content ranges from 42 to 49%, and the chromium to iron ratio averages 1.5:1.

The Former USSR: Chromite deposits occur mainly in the Urals, where twenty-five ore districts are known, including the Donskoy group in the southeastern part of the Kempirsai massif in the southern Urals, and the Saranovskoe deposit on the western slope of the Middle Urals. The massif consists of serpentinized peridotites and dunites, and the mineralization is of late-magmatic origin (Smirnov 1977). The Kempirsai ultrabasic massif extends north-south over a length of 82 km and reportedly contains 160 deposits of various sizes. All minable deposits are situated in the south Kempirsai ore district, where they are aligned in two belts of 20 km length. Large ore bodies have been localized in serpentinized dunites and are accompanied by bands of disseminated ores and fine stringers of massive chromites. Their compositions have ranges of 20 to 63% Cr_2O_3, 7 to 21% Fe_2O_3, 8 to 20% MgO, 8 to 15% Al_2O_3, and 0.2 to 30% SiO_2. In the Saranovskoe deposit, the ore occurs in the gabbro-peridotite massif (Strishkov and Steblez 1985) and aligns in three parallel bodies with the richest deposit in the central body. The ore has a typical composition of 38% Cr_2O_3, 15% Al_2O_3, 18% FeO, and 18% MgO (Smirnov 1977).

Albania: Albania is the world's third largest chromite producer after South Africa and the former USSR. The Upper Jurassic to Lower Cretaceous was an important metallographic epoch in the Dinarides during the course of which ultrabasic massifs were formed. The chromite deposits occur within the Mirdita ophiolite complex and are invariably in association with either olvine- or bronzite-rich rocks (Rabchevsky 1985). The deposits are typically of the podiform type, with much of the ore being hard and lumpy. Four mining areas in operation are (1) the Kukes district of northern Albania, (2) the Bater-Martanesh area of central Albania, (3) the Latai Zone in the east-central region, and (4) the Korce area of southern Albania. Chemical composition of refractory lumpy ore may be represented by 22% Cr_2O_3, 11.8% FeO, 16% SiO_2, 13% Al_2O_3, and 17% MgO.

Turkey: Deposits are associated with ultrabasic rocks and are of a podiform-type occurring in the Alpine-type peridotite-gabbro complex of the Mesozoic Era (Engin 1979; Engin et al. 1987). Three important producing regions are (1) to the south of Bursa, (2) in the Mugla District in the southwest, and (3) in the area around Elazig.

The ores are disseminated, banded, and massive types that display layered, nodular, or net textures. The compositions of chromite vary widely with a Cr_2O_3 content ranging from 38 to 60% and Cr:Fe ratio from 2.6 to 3.7 (Thayer 1964; Engin et al. 1987).

India: The most important deposits occur in the Sukinda ultramafic complex in the Cuttack District (Bannerjee 1971). Chrome ores consisting of talc schist, serpentine, nickeliferous limonite, and chert are concentrated in the lower portion of the complex. Specifications for Indian chromite are 53% Cr_2O_3 (min), 4% SiO_2 (max), 16% Fe_2O_3 (max), and 1% CaO (max) for the refractory grade, and 40% Cr_2O_3 (min), 5% SiO_2 (max), 20% Total iron oxide (max), 14% Al_2O_3 (max), 14% MgO (max), and 3% CaO (max) for the chemical grade (Maihotra 1983).

Zimbabwe: The entire chromite production in Zimbabwe is in the form of metallurgical grades from both the Alpine-type podiform and the Great Dyke layered stratiform deposits (Stowe 1987b). The Cr_2O_3 content of the Great Dyke deposit varies between 36 to 49% with a Cr:Fe ratio of 2.0 to 2.7 of the upper group and between 43 to 54% with a Cr:Fe ratio of 2.7 to 3.9 of the lower group (Prendergast 1987).

Finland: The chromite mineralization is associated with a layered ultrabasic sill-like intrusion between a granite-gneiss basement complex and the schist belt (Isokangas 1978). The deposits are located 7 km north of Kemi on the northern coast of the Gulf of Bothnia and extend northwards for about 10 km. Specifications for chemical concentrate and foundry sand are, respectively, 46% and 46% Cr_2O_3, 2.1 and 1.5% SiO_2, 13.8% and 14.0% Al_2O_3, 10.3% and 10.0% MgO, 25.0% and 25.5% $FeO+Fe_2O_3$, and 0.14% and 0.1% CaO.

Philippines: Although chromite occurs scattered throughout the Philippines, the two main mines, Acoje and Masinloc-Coto, are both located in the Zambales complex in northwestern Luzon (Dela Crus 1962; Leblanc and Violette 1983). The chromite occurs in serpentinized dunite and troctolite in the lower, ultramafic part of the complex. At Acoje mine, the ore bodies are massive layered, disseminated, or podiform. Massive-type ores are high-Cr chromite with an average of 48% Cr_2O_3, and others have 17 to 28% Cr_2O_3 and a Cr:Fe ratio of 2.4. At Masinloc mine, the ore is massive or disseminated with flow banding. It is the high-Al refractory grade ore with an average of 32.3% Cr_2O_3.

INDUSTRIAL PROCESSES AND USES

1. Processes and Beneficiation

Selective mining, hand cobbing and sorting, washing screening, and wet classification have been the routine beneficiation procedures for chrome ores. However, more intensive processes are required in the case of chrome fines. These include heavy medium separation, jibs, spirals, and shaking tables. The introduction of the Bartles Crossbelt concentrator has greatly improved the treatment of chrome fines. Gravity concentration is ideally suited to the upgrading of the lower grade lump ore, as well as the concentration of run of the mine fines. For the latter fines magnetic separation and flotation are both possible alternatives.

There are several chemical methods to enrich the Cr_2O_3-content or Cr:Fe ratio in the fines of chrome ores by chlorination or by reducing Fe^{3+} or iron oxide (Harris 1964).

At Masinloc Mines, the Philippines (Besa and Dela Cruz 1983; Burt 1984), the processing plant consists of three sections: (1) washing and heavy medium separation, (2) crushing, and (3) fine treatment. In the first section, ore coarser than 70 mm is hand picked, whereas the $-70 +8$mm fraction is pre-concentrated in Akins spiral heavy medium separators using ferrosilicon as the medium. The handpicked concentrate is either direct shipped or crushed to 7 mm and sent to fines treatment. This consists of a series of screens and classifiers to size the material into various fractions. The $+1.6$ mm fraction is upgraded in a heavy medium cone separator, the $-1.6 +0.2$ mm fraction in Yuba jigs, the $-0.2 +0.15$ mm fraction by Reichert spirals, and the $-0.15 +0.04$ mm fraction by Humphreys spirals and Wilfley tables.

The Kavak chromite concentrator in Turkey (Ergunalp 1980) has more complex circuitry than is usual in chromite plants. Treatment starts with rod mill grinding, and then the ground product is passed to classification. The complex classification system employed utilizes three sets of sand cones, two hydrosizers, and three stages of cyclones to classify the feed and recycle products into a series of different size fractions. The products are upgraded on banks of spirals and cleaned on banks of sand tables. A grade of 54% Cr_2O_3 is produced from a feed of 28% Cr_2O_3 (Ergunalp 1980).

2. Industrial Uses

The three principal uses of chromite, as stated previously, are (1) metallurgical, (2) refractory, and (3) chemical. In the latter two, chromite is consumed as an industrial mineral. Because of the increasing use of the basic oxygen steelmaking process, the demand for chromite in refractories has declined.

For metallurgical applications, chrome ore of metallurgical grade is reduced in electric carbon arc furnaces to various types of ferrochromium alloys, and pure chromium is prepared by aluminothermic and electrolytic processes from sodium dichromate. Before the introduction of new steelmaking technology only lumpy ores were considered, but today fine material can be employed using various systems of briqueting and pelletization. For the production of stainless steel, only the metallurgical grade with high Cr:Fe ratio could be used in the past. With the introduction of the Argon-Oxygen Decarburization process (Krivsky 1973), it has become possible to use chemical grade chrome ore with high iron content for the production of stainless steel.

For refractory applications, chrome refractories are supplied predominantly as bricks and shapes, mortar, plastic, and gunning (Hayburst 1988; Papp 1994). They are basic refractories and made of chromite, dolomite, magnesite, and various combinations of chromite and magnesite. There are generally four principal types of pressed bricks: (1) chromite and magnesia composite bricks, chrome-magnesia brick with 70% chromite, and magnesia-chrome brick with 30 to 40% chromite; (2) co-sintered magnesia-chrome clinker product for brick making grain; (3) fused magnesia-chrome brick; and (4) magnesia-chrome brick from fused, crushed, and rebonded material (McMichael 1989). In the production of chromite-magnesia composite brick, the required proportions of relatively coarse grains of chrome ore (-10 mesh), and fine grains of magnesite, are mixed with a binder, pelletized, and fired at high temperatures. For the chrome-magnesia mixture, serpentine and other silicates are converted to the refractory mineral forsterite during firing at high temperatures, thus increasing the refractoriness of the brick. For the magnesia-chrome mixture, the chromite grains dissolve and solid solutions are formed in the periclase above 1780°C, the dissolution temperature of chromite. The co-sintered clinker is produced by firing pellets of finely mixed grains of chrome ore and magnesite. Fused magnesia-chrome brick is produced by mixing magnesia clinker with chrome ore in specific proportions and melting the mixture in an electric arc furnace (Power 1985).

The chrome-based refractories are used in the cement and glass industries as well as in the nonferrous processing industries. The principal user of magnesia-chrome refractories in the past, the open hearth furnace for steelmaking, has been totally replaced by the basic oxygen furnace in western industrialized countries that require no chrome-based refractories (Papp 1994).

Chromite is used in the specialized foundry industry and competes well with the established nonsiliceous sands such as zircon and olivine. Its characteristic properties are excellent thermal conductivity, resistance to metal corrosion, reduced wetting action, low rate of thermal expansion, and resistance to thermal shock (Heine 1989).

For chemical applications, the key chemical from which all major industrial chromium compounds are made is sodium dichromate ($Na_2Cr_2O_7$) or sodium dichromate dihydrate ($Na_2Cr_2O_7 \cdot 2H_2O$). The basic procedure in the manufacture of sodium dichromate includes (1) roasting of chrome ore with soda ash or with soda ash and an additive such as lime at 1,100° to 1,200°C, (2) leaching of the calcine to produce a solution of sodium chromate, (3) conversion of the sodium chromate to dichromate with acid: $2Na_2CrO_4 + H_2SO_4 \rightarrow Na_2Cr_2O_7 + Na_2SO_4 + H_2O$, and (4) purification of the sodium dichromate. The operation is shown diagrammatically in Fig. 22 (Büchner et al. 1989).

Chromic acid (CrO_3) is prepared by reacting sodium dichromate with sulfuric acid in aqueous solution as well as in the molten state, in a rotary kiln according to the reaction: $Na_2Cr_2O_7 \cdot 2H_2O + 2H_2SO_4 \rightarrow 2NaHSO_4 + 2CrO_3 + 3H_2O$. Chromium oxide is manufactured by reducing sodium dichromate with charcoal or sulfur in an exothermic reaction in rotary kiln: $2Na_2Cr_2O_7 + 3C \rightarrow 2Cr_2O_3 + 2Na_2CO_3 + CO_2$ or $Na_2Cr_2O_7 + S \rightarrow Cr_2O_3 + Na_2SO_4$. The product is leached with water to remove soluble products, filtered, dried, and ground. Potassium and ammonium dichromates are prepared by metathesis between sodium dichromate and either potassium or ammonium chloride: $2KCl$ or $2(NH_4)Cl + Na_2Cr_2O_7 \rightarrow K_2Cr_2O_7$ or $(NH_4)_2Cr_2O_7 + 2NaCl$.

Pigments are an important single use of chromates. Reds, yellows, and oranges can be produced by the interaction of sodium dichromate with lead salts, blues by oxidizing ferrocyanide with the dichromate, and greens by the combination of chrome yellow pigment with iron-based Prussian blue pigments. A summary of chromate pigments is as follows: (1) medium chrome yellow: an orange yellow, pure $PbCrO_4$; (2) lemon and primrose chrome yellow: $PbCrO_4$ with up to 40% $PbSO_4$; (3) molybdate oranges: solid solutions of $Pb(Cr,S,Mo)O_4$; (4) chrome orange: basic lead chromate, $PbCrO_4 \cdot PbO$; (5) lead silicochromate: essentially medium chrome yellow precipitated on silica; (6) chromium oxide green: Cr_2O_3; and (7) zinc chromate: zinc yellow, a major corrosion inhibiting pigment (Heinrich Heine 1987).

Chromic acid is used for chrome plating of metals for either decorative effect or for reducing wear and corrosion. It is produced by chrome precipitation in an anodiz-

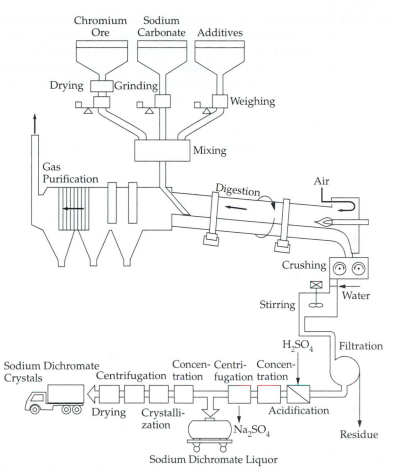

Figure 22 Flowsheet of a rotary kiln plant for the manufacture of sodium dichromate. Source: Industrial Inorganic Chemistry, 1989.

ing bath by means of electrolysis (Hartford 1981; Power 1985). By applying the plating directly on a substrate and with appreciable thickness, hard chromium plating is widely used in cylinder liners and piston rings for internal combustion engines. Chromates are used to inhibit metal corrosion in recirculating water systems. Steel immersed in dilute chromate solutions does not rust, and corrosion of other metals is similarly prevented. Oxide films on aluminum are produced by anodizing in a chromic acid solution. They impart exceptional corrosion resistance and paint adherence.

The chrome tanning of leather is one step in a complicated series of operations leading from the raw hide to the finished product. The tanning formulations, $Cr(OH)_x (SO_4)_{(3-x)/2}$, are produced from sodium dichromate by reacting with a reducing agent such as carbohydrates contained in the raw hides, and the tanning can take place in a matter of hours in the one-bath process (Covington 1985). In the textile industry, sodium chromate is used as an oxidant and source of chromium to dye wool and synthetics.

Wood preservation shares with hard chromium plating the distinction of being the fastest-growing use of chromium chemicals. The preservation is achieved by using chromated copper arsenate made from sodium dichromate. Its functions are (1) to prevent leaching of the water soluble preservatives, (2) to reduce corrosion, and (3) to reinforce the preservative value of other ingredients (Hartford 1973). Wood treated with this compound requires special impregnation under pressure, but after this process the wood may be fabricated and painted as readily as untreated wood. The treated wood becomes less flammable because of the capability of chromium to prevent leaching of the fire retardant from the wood.

Chromium compounds are also used in the formulations of drilling mud, in high-fidelity magnetic tapes, in fused-salt batteries, and in photography and photoengraving.

REFERENCES

Bannerjee, P. K. 1971. The Sukinda chromite field, Cuttack District, Orissa. *India Geol. Sur. Rec.* 96:140–170.

Besa, R. A., and S. D. Dela Cruz. 1983. "Masinloc Chromate Operation: Today and the Future." In *Minerals in the Refractory Industry—Assessing the Decade Ahead,* Eds. E. M. Dickson and P.W. Harben, 41–45. Ind. Minerals Suppl. (March).

Büchner, W., R. Schliebs, G. Winter, and K. H. Büchel. 1989. *Industrial Inorganic Chemistry.* VCH Verlagsgesellscha mbH, New York, 614p.

Burt, R. O. 1984. *Gravity Concentration Technology.* Amsterdam: Elsevier, 605p.

Covington, A .D. 1985. Chromium in the leather industry: Chromium review. No. 8. *Intern. Chromium Development Assoc.* (Paris, France), 2–9.

Deer, W. A., R. A. Howie, and J. Zussman. 1992. *An Introduction to Rock-Forming Minerals.* Essex, England: Longman, 695p.

Dela Cruz, S. "Masinloc Chromite Operation of Benquet Corporation. Republic of Philippines." In *Proc., The Raw Materials for Refractories Conference,* Tuscaloosa: University of Alabama, 264–285.

Engin, T. 1979. Nature of podiform chromite deposits, exploration problems and mining practices in Turkey. *10th World Mining Congress* (Istanbul, Turkey), II–42, 17p.

Engin, T., O. Özkocak, and U. Artan. 1987. "General Geological Setting and Character of Chromite Deposits in Turkey." In *Evolution of Chromium Ore Fields,* Ed. C.W. Stowe, 194–219. New York: Van Nostrand Reinhold.

Ergunalp, F. 1980. Chromite mining and processing at Kovak mine, Turkey. *Trans. Inst. Min. Metal* (October) 89A:179–184.

Harben, P. W., and M. Kuzvart. 1996. Industrial minerals—A global geology. *Industrial Minerals Information, Ltd., Metal Bulletin Plc.* (London), 462p.

Harben, P. W., and R. L. Bates. 1990. Industrial minerals—Geology and world deposits. *Industrial Mineral Div., Metal Bulletin Plc.* (London), 312p.

Harris, D. L. 1964. Chemical upgrading of Stillwater chromite. *Trans. AIME.* 229:267–281.

Hartford, W. H. 1973. "Chemical and Physical Properties of Wood Preservatives and Wood-Preservative System." Vol. 2., In *Wood Deterioration and Its Prevention by Preservative Treatments,* Ed. J. Nicholas, 1–120. New York: Syracuse University.

Hartford, W .H. 1981. "Trends and Technology in Industrial Chromium Chemicals." No. 40, In *Specialty Inorganic Chemicals,* 311–345. London: Royal Soc. Chemistry Publ.

Hayburst, A. 1988. Chromite in modern refractories: Chromium review. No. 9. *Intern. Chromium Development Assoc.* (Paris, France) 13–19.

Heine, H. G. Heinrich. "Inorganic Pigments." Vol. A20, In *Ullmann's Encyclopedia of Inorganic Chemistry* Weinheim: VCH Publishing Verlagsgesellschaft mbH, 243–361.

Heine, H. J. 1989. Key fundamentals for sand preparation and reclamation. *Foundry Management and Technology* (April):34–41.

Isokangas, P. 1978. "Finland." Vol. 1, In *Mineral Deposits of Europe,* 77–79. In *Northwest Europe,* Eds. S.H.U. Bowie, A. Kralheim, and, H. W. Gaslam, London: IMM/Min. Soc.

Krivsky, W.A. 1973. The Linde argon-oxygen process for stainless steel: A case study of a major innovation in a basic industry. *Trans. TMS/AIME,* 4:1439–1447.

Leblanc, M., and J. F. Violette. 1983. Al and Cr rich chromite pods in peridotites. *Econ. Geol.,* 78:293–301.

Malhotra, V. 1983. Chromites in India—A review. *Ind. Minerals* (September):53–59.

McMichael, B. 1989. Chromite—Ladles refine demand. *Ind. Minerals* (February):25–45.

Mikami, H. M. 1983. "Chromite." 5th ed. In *Industrial Minerals and Rocks,* Ed. S.J. Lefond, 567–584. New York: Soc. Mining Engineers, AIME.

Murthy, I. V. P., and G. Gopalakrishna. 1981. Remanence hysteresis property of chromites. *Proc. Indian Acad. Sci.,* 91:159–165.

Owada, S., and T. Harada. 1985. Grindability and magnetic property of chromites from various locations in relation to their mineralogical properties. *Jour. Min. Metall. Inst.* (Japan), 101:781–787.

Papp, J. J. 1994. "Chromite." 6th ed. In *Industrial Minerals and Rocks,* Ed. D. D. Carr, 209–228. Littleton, Colorado: Soc. Mining, Metal., and Exploration.

Power, T. 1985. Chromite—The non-metallurgical market. *Ind. Minerals* (April):17–51.

Prendergast, M. D. 1987. "The Chromite Ore Field of the Great Dyke, Zimbabwe." In *Evolution of Chromite Ore Fields,* Ed. C. W. Stowe, 89–108. New York: Van Nostrand Reinhold.

Rabchevsky, G.A. 1985. Chromium deposits of Albania. *Chromium Review* (March):14–17.

Smirnov, V.I. 1977. "Deposits of Chromium." Vol. 1, In *Ore Deposits of the USSR,* 179–236. New York: Pitman Publishing.

Stowe, C. W. 1987a. "The Mineral Chromite." In *Evolution of Chromium Ore Fields,* Ed. C. W. Stowe, 1–22. New York: Van Nostrand and Reinhold.

Stowe, C. W. 1987b. "Chromite Deposits of the Shurugwi Greenstone Belt, Zimbabwe." In *Evolution of Chromium Ore Fields,* Ed. C. W. Stowe, 297–320. New York: Van Nostrand Reinhold.

Strishkov, V. V., and W.G. Steblez. 1985. The chromium industry of the USSR. *U.S.B.M. Mineral Issues Series,* 33p.

Thayer, T. P. 1964. Principal features and origin of podiform chromite deposits and some observations on the Guteman-Sordaz District, Turkey. *Econ. Geol.,* 59:1497–1524.

Thayer, T. P. 1973. "Chromium." In *United States Mineral Resources,* Eds. D. A. Brobst and W.P. Pratt, 820:247–259. U.S.G.S. Prof. Paper.

Thayer, T. P. 1977. The mineral resources of the Bushveld Complex. *Min. Sci. Engin.* 9:93–95.

Von Gruenewaldt, G. 1983. Chromium deposits of the Bushveld Complex. *Chromium Review* (April):8–11.

Willemse, J. 1969. "The Geology of the Bushveld Igneous Complex, the Largest Repository of Magmatic Ore Deposits in the World. In *Magmatic Ore Deposits—A Symposium,* Ed. H. D. B. Wilson, Monograph, 4:1–22.

Wyllie, P. J. 1967. *Ultramafic and Related Rocks.* New York: John Wiley, 464p.

Diatomite

Diatomite is a siliceous, sedimentary rock consisting mainly of the fossilized remains of the diatom, a microscopic one-cell aquatic plant. It is also known as diatomaceous earth or kieselguhr. Because of the highly porous structure of diatom, the major industrial uses of diatomite are as a filter aid in clarifying liquids and as a functional filler in paint, paper, rubber, and plastics.

The world's annual production for each of the last twenty years varied within a range of from 1.2 to 1.9 M tons. The United States leads the world in diatomite production (Lemons 1996).

MINERALOGICAL PROPERTIES

Diatomite consists primarily of accumulated frutules of diatoms, each in the shape of the original diatom plant. Their designs are as intricate and varied as snow flakes. Over 10,000 varieties have been classified (Calvert 1930). Various diatom shapes were well illustrated by Schrader (1974).

Chemically, diatomite consists primarily of silica with combined water. It is essentially inert and reacts only with strong alkalies and hydrofluoric acid. The silica is a variety of opal ($SiO_2 \cdot nH_2O$) that appears to be amorphous under the light microscope. Cristabolite has also been identified by x-ray diffraction to be a crystalline phase in some diatomites. Diatomite from economic deposits usually contains a silica content of over 86%, an alumina content of from 2 to 10%, and variable but small amounts of Fe_2O_3, CaO, MgO, Na_2O, and K_2O (Cressman 1962). The chemical composition of diatomite from Lompoc, California, has the following values: 89.70% SiO_2, 3.72% Al_2O_3, 1.09% Fe_2O_3, 0.10% TiO_2, 0.10% P_2O_5, 0.30% CaO, 0.55% MgO, 0.31% Na_2O, and 0.41% K_2O with 3.70% of ignition loss (Kadey 1983). Diatomite deposits commonly contain the contaminants clay, volcanic ash, silt, carbonate, and organic matter. The types and amounts of contaminants are highly variable and depend upon the conditions of sedimentation at the time of diatom deposition.

The color of pure diatomite is usually white, or nearly white, but commonly varies from light tan or grey to brown, greenish, or nearly black. The hardness of diatom skeleton is between 4.5 and 5, but the hardness of massive diatomaceous earth is about 1.5 because of the comparative friability of the porous mass. The refractive index of diatomite is comparable to that of opaline silica and ranges from 1.41 to 1.48 (Kadey 1983).

Low density and high porosity are imparted to diatomite by the diatom frustules, which are typically perforated by numerous holes, the combined area of which may constitute 10 to 30% of the frustule (Lewin and Guiland 1963; Kamatani 1971). Dry unconsolidated diatomite has a bulk density between 0.12 gm/cm^3 and 0.25 gm/cm^3 or less than half that of water, whereas the density of the opaline silica of diatom frustules is about twice that of water (Clark 1978). It has been estimated that a cubic inch of diatomite from the Monterey Formation of California may contain as many as 21 million diatom frustules (Bramlette 1946).

The thermal conductivity of diatomite is low and increases with the increase in percentage of contaminants and the weight per unit volume. The fusion point depends on purity, averaging about 1,590°C for pure material, slightly less than that of pure silica. Diatomite has weak adsorption powers, but shows excellent absorption of acids, liquid fertilizers, alcohol, water, and other fluids (Calvert 1930; Kamatani 1971).

High temperatures and pressures will damage the delicate diatom frustule and convert the opaline silica (opal-A) to opal containing cristobalite and tridymite (opal-CT). They also reduce the surface area and specific gravity of diatomite. Results from calcination of diatomite show that surface area decreases from 10 to 30 m^2/gram to 0.5 to 5 m^2/gram, hardness increases from 4.5 to 6, specific gravity increases from 2.0 to 2.3, and refractive indices increase from 1.40 to 1.49 (Calacal and Whittemore 1987).

GEOLOGICAL OCCURRENCE AND DISTRIBUTION OF DEPOSITS

Diatomites are formed in environments that favor the growth of diatoms, under depositional conditions that have restricted input of terrigenous, volcanic, and biogenic carbonate materials, and from a geologic history that has allowed little diagenesis. Diagenetic processes cause the diatomite to lose its characteristic low density and high porosity. Diatoms flourish in marine environments having upwellings that bring nutrients to the surface where they can be utilized by diatoms (Lisitzin 1972; Heath 1974; Barron 1987). The association of marine diatoms with regions of upwelling has existed since the Cretaceous, when diatoms first became abundant. In the Pacific, diatomaceous ooze is found in regions of biologic productivity north of 40°N, between 45° and 65°S, and along the equator at water depths below the calcium carbonate compensation depth. Nonmarine diatomaceous sediment is typically lacustrine in origin. As in the marine setting, lacustrine diatoms require an environment that (1) is rich in nutrients that promote diatom growth, (2) is not excessively alkaline in order to preserve the diatomite that has accumulated, and (3) does not have a steady influx of terrigenous material. Nonmarine diatomites can be as old as late Eocene and are commonly associated with volcanic rocks.

Deposits of diatomite are known to exist on every continent. Over half of the states in the United States reportedly contain diatomaceous deposits (Cummins 1960; Kadey 1983). California has the largest formation of diatomite, the Monterey formation that extends from Point Arena in Mendocino County in the north to San Onofre in the South. The most extensive deposit in the area is near Lompoc, Santa Barbara County, and is of marine origin (Bramlette 1946; Dibblee 1950, 1982). In that area, the high quality diatomite is characterized by rhythmic bedding in thin laminae. The ore is coherent but tends to separate along bedding planes. It is variably light in color, highly porous, and contains up to 60% moisture in the deposit. Impure coarsely bedded and impure massive diatomites also occur in Lompoc deposits, but they have limited commercial value. Diatomites of freshwater origin and of Miocene to Pliocene age occur in western Nevada, southern Oregon, and central Washington. In Canada, large freshwater diatomite deposits exist in British Columbia, and small deposits are distributed in Nova Scotia, New Brunswick, Quebec, and Ontario (Hora 1981).

The French deposits are mainly of freshwater origin and widely distributed in Tertiary and Quaternary lake beds in the district of the Massif Central in the Departments of Cantal, Puy-de-Dôme, and Ardèche (Cummins 1960; Clarke 1980; Loughbrough 1994). The Lüneburger Heide deposits in Germany, formed in the last interglacial period, lie in water-soaked beds covered with sand and soil. In Spain, productive diatomite deposits are distributed in the Albacete Province. The high-grade diatomite is interbedded in a thick series of lacustrine rocks (Requeiro et al. 1993). There are several occurrences of diatomaceous rocks in both lake and marine sediments in Greece. The saline deposits of the Mytilinii basin, Samos Island, and the marine deposits of the islands of Crete and Zakyathes are of economic interest (Stamatakis and Tsipoura-Vlachou 1990). Other important deposits are reported to exist in Algeria, Kenya, South Africa, Denmark, Mexico, and the former USSR.

INDUSTRIAL PROCESSES AND USES

1. Milling and Beneficiation

The particulate shape and structure of diatom skeleton is the physical property that distinguishes diatomite from other forms of silica and characterizes diatomite in its industrial applications. In milling processes, therefore, every possible care must be taken to ensure the preservation of this property. Fig. 23 is a flow sheet showing the processing of diatomite at a plant at Lompoc, California (Neu and Alciatore 1987).

The crude diatomite is first crushed in hammer mills, followed by simultaneous milling and drying at relatively low temperatures. Because of the high moisture content of the ore, all processes are done while the suspended particles of diatomite are carried in a stream of hot air. Coarse and gritty nondiatomaceous earth is removed in separators, and preliminary size separation is made in cyclones. Products requiring no further treatment are known as natural milled diatomite. The coarse fractions produced are generally used for filter aids, and the finer fractions are used for filler (Breese 1994).

When adjustment of particle size distribution is required, such as for filter aids, diatomite powder is calcined in rotary kilns at about 980°C, followed by further milling and classifying to remove coarse agglomerates as well as extreme fines. Products resulting from this process are known as the calcined grades. Further adjustment of particle size is made by the addition of a flux, usually soda ash, before the calcination. The use of sodium chloride as a flux is avoided because of its corrosive action. Such processing reduces the surface area of the particles, changes their color from buff to white, and renders various impurities. Products resulting from this process are known as the flux-calcined grades (Pettifer 1982; Breese 1994).

2. Industrial Uses

As a result of its unusual particulate structure and chemical inertness, processed diatomite finds its way into a wide range of industrial applications. Of these, three major uses are (1) filter aid that is used as an expandable processing aid, (2) filler that is used as a component of the manufactured product, and (3) insulation that is used as a structural part of the heated equipment.

Filtration: By far, the largest use of processed diatomite is as a filter aid for the separation of suspended solids from fluids. Some common applications are the

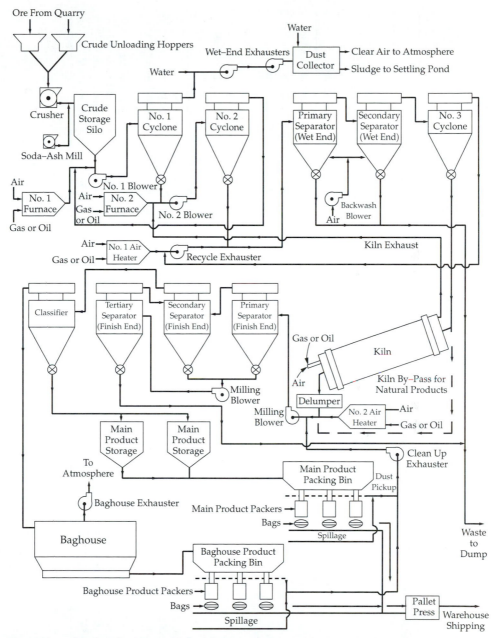

Figure 23 A flowsheet for the processing of diatomite at a Lompac plant. Source: Ullman's Encyclopedia of Industrial Chemistry, 1987.

Table 7 Physical properties of diatomite used as commercial filter aids after Kadey (1983).

	Grade*	Color	Density, Lb per Cu Ft		Screen Analysis, % Retained, 150 Mesh	pH	Specific Gravity	Water Absorption	Relative Flow Rate	Typical Applications, Filtration
			Dry	Wet						
Natural	Filter-Cel	Gray	7.0	15.9	1.0	7.0	2.10	235	100	Vegetable oil catalyst
Calcined	Celite 505	Pink	8.0	21.0	—	7.0	2.15	170	135	Apple juice
	Standard Super-Cel	Pink	8.0	17.2	3.0	7.0	2.15	255	200	Beer and wine
	Celite 512	Pink	8.0	17.9	4.0	7.0	2.15	250	300	Sugar
Flux Calcined	Hyflo Super-Cel	White	9.0	17.2	5.0	10.0	2.30	245	500	Dry cleaning solvents, chemicals, etc. (most widely used filter aid)
	Celite 501	White	9.5	16.9	8.0	10.0	2.30	250	750	Grape juice
	Celite 503	White	9.5	17.2	9.0	10.0	2.30	240	900	Industrial and potable water
	Celite 535	White	12.0	17.6	9.0	10.0	2.30	245	1350	Industrial wastes
	Celite 545	White	12.0	18.0	12.0	10.0	2.30	240	2160	Swimming pools
	Celite 550	White	18.1	21.0	20.0	8.0	2.30	220	2380	Pharmaceuticals
	Celite 560	White	19.5	20.0	50.0	10.0	2.30	220	7500	Phosphoric acid

*Grades listed are Manville registered trademarks.

filtration of pharmaceuticals; organic and inorganic chemicals; beer, whiskey, and wine; raw sugar liquors; antibiotics; industrial and municipal waters; fruit and vegetable juices; lube, rolling mill, and cutting oils; jet fuels; dry cleaning solvents; and varnishes and lacquers (Halvorgen 1952; Jackson 1961; Baumann 1965; Cummins 1973; Kadey 1983). Table 7 lists the physical properties of diatomites used as commercial filter aids.

The principal requirements for filter aid suitability are skeleton structure, porosity, density, and soluble impurities of diatom. The interstices and chambers within the structure, and the relationship and interplay between variously shaped particles in multi-species assemblages, characterize the effectiveness of diatomite as a filter aid (Kiadey 1983).

Fillers: Diatomite is used as an important mineral filler, extender, and diluent for many fabricated materials. A generally recognized order of importance is in protective coatings (paints), paper, insecticides, plastics, asphaltic compositions, and fertilizers (Frissell 1956; Kranich and Zilli 1966; Coombs 1970; Kranich 1973). In the protective coating of paint, diatomite is used as a function filler to control gloss and sheen. Rigorous requirements regarding brightness, pH, absorption properties, refractive indices, and chemical stability must be met (Breese 1994).

Insulation: Diatomite makes an efficient thermal insulator because of its high resistance to heat and its high porosity. Materials in the form of powders, aggregates, and bricks are most commonly used in heating equipment, board and pipe covering, and loose-fill.

Others: Other industrial uses of diatomite include (1) a pozzolanic admixture in concrete mixes, (2) a carrier for catalysts in petroleum refining, (3) a natural insecticide to protect seeds and grain, (4) an anticaking agent in fertilizers, (5) a mild abrasive for finish polishing, and (6) a fluffing agent for heavier dusts.

In addition, diatomite is a reactive form of silica that can be used in the manufacture of a variety of silicates that have wide industrial applications.

REFERENCES

Barron, J. A. 1987. "Diatomite: Environmental and Geological Factors Affecting Its Distribution." In *Siliceous Sedimentary Rock-Hosted Ores and Petroleum,* 164–178. New York: Van Nostrand Reinhold.

Baumann, E. R. 1965. Diatomite filters for municipal use. *Jour. Amer. Water Works Assoc.* 57:157–180.

Bramlette, M. N. 1946. The Monterey Formation of California and the Origin of its siliceous rocks. *U.S.G.S. Prof. Paper* 212:57p.

Breese, R. C. Y. 1994. "Diatomite." 6th ed. In *Industrial Minerals and Rocks,* Ed. D.D. Carr, 397–412. Littleton, Colorado: Soc. Mining, Metal., and Exploration.

Calacal, E. L., and O. J. Whittemore. 1987. The sintering of diatomite. *Bull. Amer. Ceramic Soc.* 66:790–795.

Calvert, R. 1930. *Diatomaceous Earth.* New York: Amer. Chem. Soc. Monograph Series, the Chemical Catalog Co., 251p.

Clark, W. B. 1978. Diatomite industry in California. *California Geol.,* 31:3–9.

Clarke, G. 1980. Industrial minerals of France. *Ind. Minerals* (December):31–33.

Coombs, G. 1970. Pesticide formulation as seen by a producer of carriers. *Agricultural Chemicals* 15:47–105.

Cressman, E. R. 1962. "Nondetrital Siliceous Sediments." 6th ed. In *Data of Geochemistry,* 11–23. U.S.G.S. Prof. Paper 440-T.

Cummins, A. B. 1960. "Diatomite." 3d ed. In *Industrial Minerals and Rocks,* Ed. J. L. Gillson, 303–319. New York: Soc. Mining Engineers, AIME.

Cummins, A. B. 1973. Development of diatomite filter aid filtration. *Filtration and Separation* (London) 10:53–60, 215–219, 317–320.

Dibblee, Jr., T. W. 1950. Geology of southwestern Santa Barbara County, California. *California Div. Mines Bull.,* 150:95p.

Dibblee, Jr., T. W. 1982. "Geology of the Santa Ynez-Topatopa Mountains, Southern California." In *Geology and Mineral Wealth of the California Transverse Range,* Eds. D.L. Fyfe and J.A. Minch, 45–56. Santa Ana, California: South Coast Geol. Soc.

Frissell, W. J. 1956. Effect of fillers in polyethylene plastics. *Plastic Tech.* 2:723–729.

Halvorgen, G. G. 1952. Diatomaceous filter aids and their applications in the brewing industry. *Brewers Jour.* 107:70–74.

Heath, G. R. 1974. "Dissolved Silica and Deep Sea Sediments." In *Studies in Paleo-Oceanography,* Ed. W.W. Hay, 20:77–93. S.E.P.M. Special Paper.

Hora, I. D. 1981. "British Columbia." *In Industrial Minerals of Canada,* Eds. P.W. Harben and E.M. Dickson, (August):35.

Jackson, T. M. 1961. Filter aids speed up difficult filtration. *Chem. Eng.,* 68:141–146.

Kadey, F. L. 1983. "Diatomite." 5th ed. In *Industrial Minerals and Rocks,* Ed. S.J. Lefond, 677–707. New York: Soc. Mining Engineers, AIME.

Kamatani, A. 1971. Physical and chemical characteristics of biogeneous silica. *Marine Biology,* 8:89–95.

Kranich, H. 1973. "Diatomaceous Silica." Vol. 1, In *Pigment Handbook,* Ed. T.C. Patton, 141–155. New York: John Wiley.

Kranich, H., and R. J. Zilli. 1966. Diatomite in chemical specialties. *Soap and Chem. Specialties,* 42:73–74.

Lemons, J. F. 1996. Diatomite. *U.S.G.S. Annual Review* (July):4p.

Lewin, J. C., and R. R. L. Guiland. 1963. Diatoms: Annual review. *Microbiology,* 17:373–429.

Lisitzin, A. 1972. Sedimentation in the World Ocean. *S.E.P.M. Special Publication,* 17:218p.

Loughbrough, R. 1994. The industrial minerals of France—Enroute to recovery. *Ind. Minerals* (July):26–27.

Neu, E. L., and A. F. Alciatore. 1987. "Diatomite." Vol. 7, In *Ullmann's Encyclopedia of Industrial Chemistry,* 603–614. Weinheim: Verlag Chemie.

Pettifer, L. 1982. Diatomite—Growth in the face of adversity. *Ind. Minerals* (April):48–58.

Requeiro, M., J. P. Calvo, E. Elizagn, and V. Calderon. 1993. Spanish diatomite—Geology and economics. *Ind. Minerals* (March):57–67.

Schrader, H. J. 1974. Revised diatom stratigraphy of the experimental Mohole drilling, Guadalupe site. *Proc. California Acad. Sci.* 4th series, XXXIX, 517–562.

Stamatakis, M. S., and M. Tsipoura-Vlachou. 1990. "Diatomaceous Rocks in Greece." In *Minerals, Materials and Industry,* Ed. M. J. Jones, 185–192. London: Inst. Mining & Metal.

Dolomite

The term "dolomite" is used for both the mineral, $CaMg(CO_3)_2$, and the rock composed predominantly of the mineral. Although the term "dolostone" has been proposed for the rock in order to avoid confusion with the mineral, it is not widely used. Dolomite rocks are common sedimentary rocks, and most of them are mono-mineralic consisting more than 90% of the mineral dolomite.

End uses for dolomite are many, but they can essentially be broken down into the following four major categories: (1) constructional, (2) agricultural, (3) chemical, and (4) metallurgical.

MINERALOGICAL PROPERTIES

Normally, the composition of dolomite is fairly close to its formula $CaMg(CO_3)_2$, but many dolomites contain some Fe^{2+} in place of Mg. The Fe^{2+} substitution produces a complete series to ankerite, $Ca(Mg,Fe)(CO_3)_2$. Mn and Zn may also replace Mg in dolomite. This is a direct consequence of the structural relationship between calcite- and dolomite-type carbonates (Chang et al. 1996). In general, the crystal structure of dolomite may be described as retaining the calcite structure, but substituting every other calcium layer with a magnesium layer, producing an ordering arrangement. A random distribution of divalent cations in both layers will produce a disordered dolomite. Dolomite with calcium in excess of its formula is known. Selected chemical compositions of dolomites from a list of chemical analyses compiled by Chang et al. (1996) are shown below to illustrate the limits of Fe, Mn, and Zn in dolomite:

CaO	33.83	32.69	28.81	33.15	22.17
MgO	18.95	19.67	13.40	12.39	2.16
MnO	—	—	6.37	9.08	0.07
ZnO	—	—	—	—	25.15
FeO	—	0.15	5.98	6.07	0.57
CO$_2$	47.23	47.21	44.80	39.31	50.12

The thermal decomposition of dolomite takes place in two steps: $CaMg(CO_3)_2 \rightarrow CaCO_3 + MgO + CO_2$ at 775°C and 300 mmHg, and $CaCO_3 \rightarrow CaO + CO_2$ at 870°C and 300 mmHg. With decrease in CO_2 pressure, the temperature of the first reaction is shifted to a higher temperature, and the second one moves in the opposite direction, until, at a pressure of <260 mmHg, the decomposition of dolomite proceeds in one step (Chang et al. 1996).

Dolomite when pure is white or colorless, but it often has a yellow or brown color due to the presence of Fe^{2+} or a rose-pink color from the Mn content. Refractive indices of dolomite are $1,500(\varepsilon)$ and $1.679(\omega)$, and increase with the substitution of Fe^{2+} for Mg. Dolomite has a hardness of 3.5, slightly higher than that of calcite. Its specific gravity is 2.86, and it is increased by the substitution of Fe^{2+} and/or Mn^{2+} for Mg. Dolomite is one of the strong luminescing carbonate minerals (Gies 1975).

To distinguish between dolomite and calcite in carbonate rocks, differential staining methods are commonly used and found to be very effective. The method is based upon the fact that organic dye stains calcite in acid solution and dolomite in alkaline solution (Adams et al. 1984). Inorganic staining solutions have also been widely used. In the Lemberg method, the carbonate sample is immersed in a 2.5% ferric chloride solution for a few seconds, thoroughly washed, and then immersed in an ammonium sulfide solution for a few seconds. Calcite will stain black, whereas dolomite remains unaffected (Chang et al. 1996). Dolomite and calcite can be distinguished, and their ratio when present together can be determined, by x-ray powder diffraction using the principal reflections for calcite at 3.03Å and for dolomite at 2.88Å (Royse et al. 1971). Calibration curves given by Tennant and Berger (1957), Weber and Smith (1961), and Royse et al. (1971) agree well with each other. However, systematic error in the method can be caused by the variation in crystalline size, lattice deformation, and interference by quartz. Because all three effects may occur together, correction can become complex (Chang et al. 1996).

The solubility of $MgCO_3$ in calcite is on the average about 15% in recent sediments, and no dolomite is formed; in older rocks, dolomite may occur when the $MgCO_3$ is only 2 to 3%. After a prolonged period of burial or at equilibrium, the high-Mg calcites lose their magnesium to produce dolomite (Chang et al. 1996). Dolomite is also formed from an assemblage of carbonate minerals of aragonite and high-magnesian calcite by burial reaction (Hatch and Rastall 1965).

In carbonate rocks, the mineral dolomite has a very strong tendency to form euhedral crystals, whereas the occurrence of calcite euhedral habit is rare. Dolomite crystals, particularly those of sand size or larger, commonly contain concentric, alternating zones of iron-rich (red) and iron-poor (clear) dolomite that marks the stages of growth. Zonation is also present in dolomite without the presence of iron minerals (Blatt 1982).

Dolomite rocks are carbonate rocks predominantly composed of the mineral dolomite. Rocks intermediate in composition between pure dolomite and limestone may be classified conveniently as following (Pettijohn 1957):

Dolomite: containing <10% calcite, >90% dolomite
Calcitic dolomite: 10 to 50% calcite, 50 to 90% dolomite
Dolomitic limestone: 50 to 90% calcite, 10 to 50% dolomite
Magnesian limestone: 90 to 95% calcite, 5 to 10% dolomite
Limestone: containing >95% calcite, <5% dolomite

Impurities in dolomite rock vary considerably in type and amount. Two important features that will determine the usefulness of the rock are how much impurity is present and how evenly an impurity is distributed in the rock. The most common impurity in dolomite rocks is clay minerals, including mainly kaolinite, illite, smectite, and chlorite. They are either disseminated throughout the rock or concentrated in laminae. Chert is another common impurity in dolomite rock and may be disseminated as grains throughout the rock or concentrated in nodules, lenses, or beds. Compact cherts have a Mohs hardness of 7 and high impact toughness, which make them particularly abrasive to processing facilities (Carr and Rooney 1983).

GEOLOGICAL OCCURRENCE AND DISTRIBUTION OF DEPOSITS

Dolomite typically occurs as the major constituent of extensive bedded sedimentary formations in association with calcite. The variety of dolomite types has been illustrated by numerous dolomite formations ranging from Paleozoic to Recent in age. The types are distinguished by field relations, crystal form and chemistry, and geologic age. The variety of types itself demands that no single set of conditions can be called upon to explain the genesis of dolomite. Most dolomite is formed from calcite and aragonite by the action of magnesium-rich solutions from marine or hydrothermal sources. A few sedimentary rocks are known to contain dolomite precipitated directly from solution, such dolomite being termed primary. Highly saline, isolated marine waters or saline lakes are the normal environments for primary dolomite precipitation (Chang et al. 1996).

Occurrences of dolomite rock are extensive and widespread throughout the world. In the United States, all fifty states except Delaware, Louisiana, New Hampshire, and North Dakota, all of the Canadian Provinces, and twenty-one Mexican states have dolomite resources (Carr and Rooney 1983). Spain is a country rich in carbonate rocks and is host to some of the purest dolomite consumed by industry. The main producing areas are Cantabria-Asturia and the provinces of Malaga and Granada in Andalucia. Norway has extensive dolomite resources near Bodo in northern Norway, where dolomite of 99.5% purity is produced. Belgium is one of the largest producers of dolomite in Europe with a major deposit at Neau. The United Kingdom, Germany, France, Brazil, and China also have extensive dolomite resources (O'Driscoll 1988).

INDUSTRIAL PROCESSES AND USES

Because of its extensive distribution and economic value, industrial processes to beneficiate dolomite rock are at a minimum. Generally, a primary crusher is all that is needed, and in many uses, it has to go through a secondary crusher and has to be sized. Grinding and heat treatment are sometimes used to prepare dolomite for special uses. The purity of dolomite determines its use as a constructional material, in agriculture, and for the chemical and metallurgical industries. Chemical and metallurgical dolomites require high purity, whereas constructional and agricultural uses do not. The size of carbonate rocks is also a major factor in determining their uses in various industries. A broad base classification made by Scott and Dunham (1984), and modified by Harben and Bates (1990), is illustrated below:

>1 m	Cut and polished stone
>30 cm	Building stone, riprap, and armor stone
1 to 20 cm	Aggregate for concrete, roadstone, railway ballast, roofing granules
0.2 to 5 cm	In chemicals and glass
3–8 cm	Filter-bed stone, poultry grit
<4 cm	Agriculture stone
<3 cm	Foundry and fluxstone
<0.2 cm	Filler in plastics, paint, paper, rubber, putty, asphalt; mild abrasive; in glazes and enamels; carrier for fungicide and insecticide

<0.1 cm	In flue gas desulfurization
Various	Bulk fill

Constructional Uses: The crushed stone and aggregate industries are the single largest consumers of dolomite rock (O'Driscoll 1988). The physical properties that determine the quality of dolomite to be used for these purposes are high strength, good hardness and toughness, low porosity and permeability, and low dust and fines. The chemical composition is immaterial. Various uses of dolomite aggregates depend upon size, and the U.S. Bureau of Mines has established a code to define the range of sizes for aggregates as (1) large coarse aggregate, (2) graded coarse aggregate, (3) fine aggregate, and (4) combined coarse and fine aggregate. They are used, respectively, for (1) macadam, riprap, jetty stone, and filter stone; (2) concrete aggregate, bituminous aggregate, bituminous surface treatment aggregate, and railroad ballast; (3) stone sand; and (4) graded roadbase and unpaved road surfacing. Each category is subject to specifications drawn up by ASTM and some state agencies.

   In common cement applications, use of dolomite is rather limited. Adding up to 40% dolomite by weight to cement accelerates the rate of cement hydration (Soroka and Setter 1977). The dolomite filler should have a surface area ranging between 1.150 and 10.300 cm^2.gm, and is added in the cement/sand ratio of 1:2.75 to mortars made from portland cement and siliceous sand. A dolomite clinker, prepared from dolomite (32.51% CaO and 20.59% MgO) and calcium fluoride additive at 1500°C, has been shown to be effective in the retardation of hydration and in the increased rate of crystallization of MgO (Anani 1984).

   Semicalcined dolomite is used to manufacture magnesium oxychloride cement and magnesium oxysulphate cement (Taylor 1964). In the preparation of magnesium oxychloride cement, a mixture of $MgO + CaCO_3$ is obtained by selective calcination of dolomite. The equivalent of 100 parts MgO is added to 100 parts of magnesium chloride dissolved in 30 ml of water in the presence of 1 part sodium hexametaphosphate. This reaction produces magnesium oxychloride cement of the composition $5MgO \cdot MgCl_2 \cdot 9H_2O$. Desirable properties of magnesium oxychloride cement include its resistance to solvents, good compressive and flexible strength, fire-proof, and low cost. It is used as a thickener in the manufacture of polyester sheet moulding material that is widely used in the manufacture of automobile body components (Anani 1984). Magnesium oxysulphate cement is prepared by adding sodium hexametaphosphate and MgO from semicalcined dolomite to a 50% concentrated solution of magnesium sulphate heptahydrate. This reaction produces magnesium oxysulphate of the composition $5MgSO_4 \cdot MgSO_4 \cdot 8H_2O$.

Agricultural Uses: Dolomite is used as a neutralizer for soil acidity by a base exchange reaction in which Ca^{2+} and Mg^{2+} replace H^{1+} in the soil (Kamprath and Foy 1971). This usage represents dolomite's second largest consumption. Dolomite products used for this purpose are under the inclusive term "aglime." They are produced by crushing and grinding dolomite rock and limestone to achieve a free flowing powder, usually able to pass through sieves in the 8 to 100-mesh range or finer. The purity of aglime is measured by a calcium carbonate equivalent (CCE) and is determined by the amount of pure calcium and magnesium carbonate present in the carbonate rock (O'Driscoll 1988).

Chemical Uses: Magnesium oxide, magnesium hydroxide, and magnesium carbonate can all be produced from dolomite. By treating dolomite at high temperatures, dolime or dolomitic lime is produced that is mainly composed of either CaO or MgO depending upon the original carbonate contents of the raw materials. By reacting with magnesium chloride from seawater, magnesium hydroxide is precipitated. Magnesium oxide is produced by dehydration of $Mg(OH)_2$ at high temperatures. Viswanathan et al. (1979) established a procedure for the separation of magnesium carbonate and calcium carbonate from dolomite. The process consists of treating crushed dolomite under a set of temperatures and carbon dioxide pressures, during which only magnesium carbonate will decompose to produce semicalcined dolomite. Through wet grinding, magnesium hydroxide is produced from the semicalcined dolomite. Recarbonation regenerates magnesium carbonate, which is separated from the remaining calcium slurry.

Dolomite is added to glass batches as a flux to reduce the tendency for devitrification, inhibit chemical attack by atmospheric moisture, increase the fracture resistance of the glass caused by thermal shock, and improve workability by lowering the setting rate.

Metallurgical Uses: Dead-burned dolomite is used in brick and shape manufacture, for monolithic refractories, and as a fluxing agent in steelmaking. Dead-burned dolomite or "doloma" is produced by the following reactions: $MgCa(CO_3)_2 \rightarrow CaCO_3 + MgO + CO_2$ and $CaCO_3 + MgO + CO_2 \rightarrow CaO + MgO + CO_2$. In the temperature range of 600° to 900°C, the dissociation of dolomite results in the formation of $CaCO_3$ and MgO. Calcination above 900° to 1,000°C drives off the CO_2 completely from the solid assemblage and produces MgO+CaO oxides. This product requires further heat treatment to reduce its reactivity and its porosity. Further calcination at about 1,700°C is necessary to produce dense grained, dead-burned dolomite (Clancy and Benson 1983). Dead-burned dolomite refractories are mainly used in basic oxygen furnaces, ladle linings, argon oxygen decarburization vessels, cement kilns, and electric arc furnaces. Grades of dead-burned dolomite generally require high purity with dolomite composition close to its formula.

Dolomite is one of the primary source materials for the manufacture of magnesium by electrolytic or pyrometallurgical processes. In the electrolytic process (Gilchrist 1989), $Mg(OH)_2$, produced by reacting calcined dolomite with seawater, is converted to $MgCl_2$ with hydrochloric acid, followed by a chloridizing roast under reducing conditions. The $MgCl_2$ solution in a mixture of $CaCl_2$ and NaCl is electrolyzed at high temperatures in a special cell. The Mg floats on the electrolyte and the anodic Cl_2 is recycled. The pyrometallurgical process uses dolomite and ferrosilicon as raw materials (Hughes et al. 1968). The calcined dolomite is briquetted with powdered ferrosilicon and the reduction carried out under vacuum in small retorts heated externally at 1,100° to 1,200°C. The reaction takes place in two stages: calcined dolomite or CaO and ferrosilicon first react to produce a liquid Ca-Si-Fe alloy that then reduces MgO with evolution of magnesium that is strongly endothermic. Both cathodes obtained from the electrolytic process, or condensates from the pyrometallurgical process, must be remelted. Iron may be reduced in the melt by adding small amounts of Mn and Zr that would form insoluble phases that sink to the bottom of the melt.

REFERENCES

Adams, A. F., W. S. Mackenzie, and C. Gulford. 1984. *Atlas of Sedimentary Rocks under the Microscope.* London: Longman, 104p.

Anani, A. 1984. Applications of dolomite. *Ind. Minerals* (October):45–55.

Blatt, H. 1982. *Sedimentary Petrology.* San Francisco, W.H. Freeman & Co., 564p.

Carr, D. D., and L. F. Rooney. 1983. "Limestone and Dolomite." 5th ed. In *Industrial Minerals and Rocks,* 833–868 New York: Soc. Mining Engineers, AIME.

Chang, L. L. Y., R. A. Howie, and J. Zussman. 1996. Rock-forming minerals. Vol. 5B—*Nonsilicates: Sulphates, Carbonates, Phosphates, Halides.* Essex, England: Longman Group Ltd., 383p.

Clancy, T. A., and D. J. Benson. 1983. Refractory dolomite raw materials. *Proc. Raw Materials for Refractories Conference,* 119–139. Tuscaloosa, Alabama: Amer. Ceramic Soc., Columbus, Ohio.

Gies, H. 1975. Activation possibilities and geochemical correlations of photoluminescing carbonates, particularly calcite. *Mineral Deposita* 10:216–227.

Gilchrist, J. D. 1989. *Extractive Metallurgy.* 3rd ed. Oxford: Pergamon Press, 431p.

Harben, P. W., and R. L. Bates. 1990. Industrial minerals: Geology and world deposits. *Industrial Minerals Division, Metal Bulletin Plc.* (London) 312p.

Hatch, F. H., and R. H. Rastall. 1965. *Petrology of the Sedimentary Rocks.* 4th ed. New York: Hafner Publ. Co., 408p.

Hughes, W. T., C. E. Ransley, and E. F. Emley.1968. "Reaction Kinetics in the Production of Magnesium by the Dolomite-Ferrosilicon Process." In *Advanced in Extractive Metallurgy,* 429–454. London: Symposium Proceedings, Institute of Mining and Metallurgy.

Kamprath, E., and C. Foy. 1971. Lime fertilizer—Plant interactions in acid soils. 2nd ed. *Fertilizer Technology and Use,* 105–151.

O'Driscoll, M. 1988. Dolomite: More than crushed stone. *Ind. Minerals* (September):37–63.

Pettijohn, J. 1957. *Sedimentary Rocks.* New York: Harper, 718p.

Royse, C. E., J. S. Wadell, and L. E. Petersen. 1971. X-ray determination of calcite-dolomite: An evaluation. *J. Sed. Petr.* 41:483–484.

Scott, P. W., and A. C. Dunham. 1984. Problems in the evaluation of limestone for diverse markets. *Proc. 6th Industrial Minerals Inter. Congress* (Toronto) 1–21.

Soroka, I., and N. Setter. 1977. The effect of fillers on strength of cement mortars. *Cement and Concrete Research* 6:73–83.

Taylor, H. F. W. 1964. *The Chemistry of Cements.* New York: Academic Press, 460p.

Tennant, C. B., and R. W. Berger. 1957. X-ray determination of dolomite-calcite ratio of carbonate rock. *Amer. Mineral* 42:23–29.

Viswanathan, V. N., D. V. Ramana Rao, K. Kumar, and S. J. Raina. 1979. Simultaneous production of magnesium carbonate and calcium carbonate from dolomites and dolomitic limestones. *J. Miner. Oricess* 6:73–83.

Weber, J. N., and F. G. Smith. 1961. Rapid determination of calcite-dolomite ratios in sedimentary rocks. *J. Sed. Petr.* 31:130–131.

Feldspar

Feldspars are a group of aluminosilicate minerals. As a group, they are the most abundant minerals in the earth's crust. The composition of common feldspars is defined by two series: the alkali feldspar series $KAlSi_3O_8$ - $NaAlSi_3O_8$ and the plagioclase series $NaAlSi_3O_8$ - $CaAl_2Si_2O_8$. Industrial applications rest upon these compositions in the manufacture of glass and ceramics.

Nepheline syenite and phonolite are feldspathoidal felsic rocks consisting mainly of feldspathoids and alkali feldspars, and have industrial uses identical to those of feldspar.

MINERALOGICAL PROPERTIES

Chemically, feldspars are aluminosilicates containing sodium, potassium, and calcium that may be represented by two series: the alkali series, $KAlSi_3O_8$ - $NaAlSi_3O_8$, and the plagioclase series, $NaAlSi_3O_8$ - $CaAl_2Si_2O_8$. Generally, each of these two series can take less than 5 to 10% of the other into its solid solution. The sodium member that is common to both series may be richer in K- or Ca-content than the alkali series and plagioclase, respectively (Ribbe 1983). Structurally, feldspars are framework-type silicates with both Al^{3+} and Si^{4+} in the tetrahedral coordination. Therefore, to characterize a feldspar, not only the chemical composition but also the structural state must be defined. The latter depends upon the temperature of crystallization and the subsequent thermal history. Each of the end members as well as the members of the solid solutions all have polymorphic forms. Sanidine, anorthoclase, and plagioclase are mineral names representing disordered monoclinic potassium-rich, triclinic sodium-rich, and triclinic sodium-calcium solid solutions, respectively. A purely chemical definition of plagioclase is given with special names for each of the six compositional ranges into which the plagioclase series is divided. Thus albite, oligoclase, andesine, labradorite, bytownite, and anorthite refer to An percentages (An: $CaAl_2Si_2O_8$) 0 to 10, 10 to 30, 30 to 50, 50 to 70, 70 to 90, and 90 to 100, respectively. For ordered feldspars in the compositional range from 85 to 100% Or (Or: $KAlSi_3O_8$), the names orthoclase and microcline are used. Microcline is triclinic when the Al^{3+} and Si^{4+} in it are fully ordered, but with the gradual disordering of Al/Si, it may attain a monoclinic symmetry. Orthoclase represents potassium-rich feldspars of partial ordering. The transition from orthoclase to microcline involves the development of distinct domain microstructures. Ordering in feldspars causes exsolution in the disordered solid solutions of sanidine, anorthoclase, and plagioclase. Three textural terms are commonly used in the description of feldspars: (1) perthites are intergrowths of sodium-rich feldspar in a potassium-rich feldspar, (2) an intergrowth of potassium-rich feldspar in a plagioclase is named antiperthitic, and (3) one with intermediate proportions is referred to as mesoperthite (Deer et al. 1992).

Feldspars are white, yellow, or salmon in color, with vitreous to pearly luster. Hardness and specific gravity of feldspars range from 6 to 6.5, and from 2.56 to 2.76, respectively. Refractive indices in the alkali series increase from a range of 1.518 to 1.522 for microcline, or 1.522 to 1.530 for sanidine, to a range of 1.532 to 1.541 for albite. In plagioclase, the Ca-rich end member (anorthite) has a higher range, 1.580 to 1.590 (Philips and Griffen 1981). Microperthite may produce iridescent luster in orthoclase (moonstone), and green microcline is known as amazonite. Both are valued as gemstones.

At one atmosphere, $KAlSi_3O_8$ melts incongruently at about 1,150°C, to $KAlSi_2O_6$ (leucite) and liquid that has a very high viscosity at 1,400°C (Schairer and Bowen 1955). This results in a slow fusion of feldspar and still slower formation of leucite. The liquidus temperature is 1,530°C. Because of its high viscosity, the melt is readily undercooled to produce glass, with softening temperatures at about 950°C. The ability of the melt to crystallize at temperatures in the feldspar stability range is so poor that a complete recrystallization is hardly achieved even after a prolonged period of treatment. With the presence of silica, the melting temperature of potassium feldspar is lowered slightly to about 1,000°C, the eutectic temperature between $KAlSi_3O_8$ and SiO_2. The liquidus, leucite + liquid → liquid, decreases sharply with sodium feldspar content to the lowest temperature of about 1,100°C and a composition of 35 mol% $KAlSi_3O_8$ (Bowen and Tuttle 1950). Sodium feldspar, $NaAlSi_3O_8$, has a congruent melting point at 1,118°C. The melt shows a lower viscosity compared to that of potassium feldspar, but the fusion process is also slow. Increasing pressure has a marked effect on lowering the melting point in the series $KAlSi_3O_8$-$NaAlSi_3O_8$. Melting in the plagioclase series is represented by a liquidus without maximum or minimum from $CaAl_2Si_2O_8$ at 1,550°C to $NaAlSi_3O_8$ at 1,110°C (Smith 1983; Smith and Brown 1988). The distinction in viscosity between the $KAlSi_3O_8$ and $NaAlSi_3O_8$ melts segregates these two feldspars in their industrial applications. The former is preferred in ceramics, whereas the glass industry requires the latter.

X-ray diffraction is a standard analytical procedure for the identification of feldspars and for the measurement of cell dimensions. The a-dimension is not sensitive to the (Al,Si) order, whereas the b- and c-dimensions depend on both composition and (Al,Si) order. The β-angle is not sensitive to the (Al,Si) order; the γ-angle varies little with composition, but is uniformly sensitive to the (Al,Si) order for the entire alkali series; and the α-angle has a lesser and more variable sensitivity to the (Al,Si) order than the γ-angle (Kroll and Ribbe 1983). Unlike the alkali feldspars, there is no procedure that can be used to obtain both composition and (Al,Si) order of a plagioclase. This is because composition and ordering are structurally interrelated, and sodium and calcium are similar in size (Kroll 1983).

There are a number of procedures for staining potassium feldspar in thin sections so that it may easily be distinguished from plagioclase and quartz (Williams et al. 1982). In comparison of physical properties, alkali feldspars have lower specific gravity and lower refractive indices than plagioclase feldspars. The presence of lamellar twinning in plagioclase feldspars and the presence of perthitic texture in alkali feldspars may also be used to distinguish members of these two series.

GEOLOGICAL OCCURRENCE AND DISTRIBUTION OF DEPOSITS

Feldspar of industrial value is generally found in pegmatites, in association with quartz and micas. Minor quantities of tourmaline, beryl, garnet, spodumene, pyrite, and magnetite are also present. Although pegmatites are widely distributed in the world, feldspars occurring in large minable bodies and sufficiently free from impurities are not so common. Besides pegmatite, there is a series of granitic rocks mined for feldspars including (1) graphic granite: a granite with intergrowth between a single K-feldspar crystal and enclosed small, angular skeletal quartz; (2) alaskite: a coarse-grained granitic rock consisting mainly of orthoclase, microcline, and quartz; and (3) aplite: a fine-grained igneous rock with the same mineralogical composition as granite. Beach sand and alluvial deposits are additional sources of feldspar.

Nepheline syenite (plutonic) and phonolite (volcanic) are feldspathoidal felsic rocks, with feldspathoids and alkali feldspars as their principal constituents. Their SiO_2 contents are usually between 52 and 60% (Robbins 1986a).

United States: North Carolina is by far the largest feldspar-producing state with 70% of U.S. production (Harben and Bates 1990). Most of this comes from the Spruce Pine district in the mountainous Blue Ridge region of the state. The regional geological setting consists of mica and amphibole gneiss and schist, which is cut by small bodies of dunite, pegmatite, and alaskite, all of Paleozoic age, and basalt and diabase dikes of Triassic age (Babst 1962). The Spruce Pine alaskite is characterized by exceptional mineral purity, uniformity, coarse-grained texture, and large size of deposit. The alaskite is composed approximately of 45% plagioclase, 25% quartz, 20% microcline, and 10% muscovite (Parker 1952).

The feldspathic aplite of Virginia occurs in a belt-shaped deposit in northeastern Amherst and southwestern Nelson counties. Aplite is essentially an intrusive mass with variable texture, composed mainly of plagioclase of andesine composition with quartz, biotite, and some heavy minerals (Brown 1962). Feldspathic sand is beneficiated as a by-product of a granite quarry near Pacolet, Spartanburg, South Carolina (Rogers and Neal 1983).

Representative analyses of feldspars from various sources, including Canadian nepheline syenite, are given in Table 8 (Rogers and Neal 1983).

Canada: The Blue Mountain nepheline syenite deposit in Methuen, Ontario, is the only commercial deposit of its kind in North America (Harben and Bates 1990). The orebody intrudes and partially replaces the Blue Mountain metasedimentary rocks of the Grenville Series and is a uniform foliated fine- to medium-grained rock, consisting of 20 to 25% nepheline, 48 to 54% albite, and 18 to 23% microcline (Hewitt 1961; Harben and Bates 1990).

Russia: Nepheline syenite is produced in the Khibiny Mountain region in the Kola Peninsula, and also in central Siberia, southeastern Ukraine, and the Urals (Allen and Charsley 1968). At the Kola Peninsula, the orebody is of a lenticular form within the Khibiny pluton and is divided into two zones. The upper zone contains 65% apatite, 30% nepheline, and 10% amphibole; and the lower zone has corresponding figures of 45%, 30%, and 18%.

Europe: In Germany, feldspars are produced from feldspar-quartz pegmatite in the Hagendorf area, near Munich, from weathered kaolinite deposits in the

Table 8 Chemical analyses of feldspar and feldspathic rocks modified after Rogers and Neal (1983).

Soda Flotation Feldspar, Spruce Pine, NC		Potash Flotation Feldspar, Kings Mountain, NC		Dry Ground Feldspar, Custer, SD	
SiO_2	67.54%	SiO_2	67.04%	SiO_2	71.84%
Al_2O_2	19.25	Al_2O_2	18.02	Al_2O_2	16.06
Fe_2O_2	0.06	Fe_2O_2	0.04	Fe_2O_2	0.09
CaO	1.94	CaO	0.38	CaO	0.48
MgO	Trace	MgO	Trace	MgO	Trace
K_2O	4.05	K_2O	12.10	K_2O	7.60
Na_2O	6.96	Na_2O	2.12	Na_2O	3.72
Loss	0.13	Loss	0.30	Loss	0.20

Feldspathic Sand, Bessemer City, NC		Low Iron Aplite, Montpelier, VA		Canadian Nepheline Syenite Nephton, Ont. Canada	
SiO_2	79.20%	SiO_2	63.71%	SiO_2	61.40%
Al_2O_2	12.10	Al_2O_2	21.89	Al_2O_2	22.74
Fe_2O_2	0.06	Fe_2O_2	0.09	Fe_2O_2	0.06
CaO	0.52	CaO	5.70	CaO	0.70
MgO	Trace	MgO	Trace	MgO	Trace
K_2O	2.62	TiO_2	0.43	K_2O	4.95
Na_2O	4.80	K_2O	2.37	Na_2O	9.54
Li_2O	0.35	Na_2O	5.60	Loss	0.60
Loss	0.35	Loss	0.21		

Hirschau-Schnaittenbach area of Bavaria, and from feldspathoid-rich phonolite in the Eifel district (Lüttig 1980). In Italy, the northern Alpine arc in the provinces of Vercelli, Novara, Brescia, Como, and Trento has the main deposit. The ore consists of aplite layers between layers of gneiss and mica schist, and albite is the major feldspar mineral in the ore body (Robbins 1986b).

In southern Norway, rich pegmatite dikes and bodies occur within gneiss and amphibolite in the Fenno-Scandinavian Shield. The pegmatite has a typical composition of 29% K-feldspar (85% orthoclase and 15% albite), 40% Na-feldspar (87% albite and 13% anorthite), 28% quartz, and from 2 to 4% other minerals (Harben and Bates 1990). Nepheline syenite is mined on the southern shore of Stjernoy Island, Norway (Geis 1979). Within the ore body, a hornblende-pyroxene and a biotite facies were identified. Both contain 56% perthite feldspar, 34% nepheline, and 10% accessory minerals, including hornblende and biotite. The feldspar produced in Sweden is from a graphic granitic body that is found in two varieties. The white granite contains 24.1% quartz and 75.1% feldspar, of which 72.8% is K-feldspar and 27.2% Na-feldspar. The pink granite has 27.5% quartz and 71.4% feldspar, and lower K-feldspar content than the white body. Feldspar deposits in pegmatites are also very extensive in Finland (Watson 1982).

Potash feldspar from pegmatite is mined in the Porto and Braga areas of Portugal, and Na-feldspar from feldspathic sand deposits around Segovia in Spain. South Africa's feldspar comes from pegmatite in the eastern and northern Transvaal.

Cornish stone, or Cornwall stone, is a type of decomposed granite mined in England (Ladoo and Myers 1951). Four varieties are produced, based on varying degree of alteration: (1) hard purple, (2) mild purple, (3) dry white, and (4) buff stone. The feldspar, quartz, and kaolin contents in the hard purple and buff stone are 77.2% and 55.5%, 16.1% and 30.9%, and 6.7% and 14.6%, respectively.

INDUSTRIAL PROCESSES AND USES

1. Beneficiation and Milling

When the rock consists of large crystals of potassium feldspar, and quartz, and not much else, it is hand-cobbed for the feldspar. In the presence of other minerals and alteration products, the rock requires beneficiation by grinding, separation, and flotation. In grinding, when the crude feldspars are of different types, but each is fairly pure and uniform, storing and grinding are done separately. Subsequently, blending is done for different grades. A generalized procedure for milling and beneficiation is described below (Taggart 1945; Ladoo and Meyers 1951; Rogers and Neal 1983). Crude feldspar is separated into oversize and fines by Grizzley or bar screens. Both sizes are introduced to the jaw crusher after the fines are dried in a rotary dryer and further treated in a magnetic separator to remove particles of iron minerals. Grinding is done in Hardinge pebble mills in a closed circuit, with screens for the coarse and air separators for the fine products. The Virginia aplite is beneficiated by a wet method in which a series of treatments is made in hammer mill, rod mill, and screen, scrubber, spiral separator, rotary dryer, and magnetic separator.

Currently, most crude feldspars are refined by froth flotation and high intensity magnetic separation (Taggart 1945; Rogers and Neal 1983). In flotation, with cationic collectors quartz tends to float selectively in slightly alkaline pulps, whereas in acidic pulps, feldspar floats selectively, especially in those acidified with hydrofluoric acid. A routine procedure is to remove micaceous minerals with amine as collector in an acid environment, assisted by pine oil and fuel oil, and to remove iron-bearing minerals with petroleum sulfonate as a collector. The feldspar is floated away from quartz in hydrofluoric acid environments using amine as a collector once again. The final steps consist of dewatering the feldspar concentrate in drain bins or over a vacuum filter or centrifuge, and drying it in rotary dryers.

A high intensity magnetic separator may be used either alone to remove iron-bearing minerals or to do supplementary refinement with flotation (Svoboda 1987). There are two magnetic separation procedures. The selection of dry and wet procedures depends upon the size of the mineral particles. If the particles are, for example, 75 μm, both separation procedures may be used, but if they are <75 μm, the wet procedure is more appropriate, although this threshold particle size can be smaller or greater than 75 μm, depending on the amount of fines present and on the physical properties of the feed. A further choice of magnetic separation procedures depends on the magnetic properties of the mineral to be treated. For strongly magnetic minerals, such as the iron-bearing minerals in feldspathic rocks, low-intensity drum magnetic separators can be used in either dry or wet modes of operation.

2. Industrial Uses

Feldspar and feldspathic rocks, including aplite and nepheline syenite, have two major industrial uses. More than 60% of feldspar produced is used in glass and 35% in glaze and enamel manufacture (Robbins 1986a).

Alumina for Glass: The principal ingredients of glass are silica, soda ash, and lime, but other oxides are mixed with the principal ingredients to modify the properties. Feldspar is added to the glass compositions mainly for its alumina content, as well as for the alkalis it contains. There are several important functions for the added alumina. It increases the resistance of glass to breakage from impact and bending, the resistance to scratching, the tensile strength, and the chemical durability. In the meantime the added Al_2O_3 content decreases the tendency of the glass toward devitrification. The alkali content of feldspar added to the glass compositions aids in fluxing and serves to replace a portion of the other forms of alkali needed. Sodium feldspar, with at least 7% Na_2O and 19% Al_2O_3, is generally preferred (Harben 1991).

The grain size distibution of feldspar used in glassmaking is very critical. A similar sizing improves blending of the materials, and the elimination of coarser fractions aids melting. Typical chemical and physical working specifications on feldspar, nepheline syenite, and aplite are listed below (the specification on grain size distribution is based on % retained on U.S. Series screens) (Mills 1983):

	Chemical Specifications	*Physical Specifications*
Feldspar	Al_2O_3-over 19%	+16 mesh-0
	Alkali-over 11%	+20 mesh-1% max.
	Fe_2O_3-0.10% max.	−100 mesh-25% max.
Nepheline	Al_2O_3-over 22%	+30 mesh-0
syenite	Alkali-over 13%	+40 mesh-3.5% max.
	SiO_2-62% max.	−100 mesh-25%
Aplite	Al_2O_3-over 22%	+16 mesh-0
	Fe_2O_3-0.10% max.	+20 mesh-2% max.
		+30 mesh-20% max.
		−100 mesh-10% max.

The amount of alumina is generally of the order of 1.5 to 2.0% Al_2O_3 in container glass, around 2.0% in flat glass, and of variable amount in glass fibers, depending on the intended use. In some fibers, the Al_2O_3 may reach 15%.

Fluxes for Ceramics: Feldspar that contains alkali is added to ceramic mixes as a flux. It melts before quartz and clay, thus producing a liquid phase that wets the surface of solid particles of other constituents and promotes bonding between them upon solidification. Since different ceramic bodies require different degrees of vitrification, the amount and type of flux needed for particular ceramics varies accordingly (Watson 1981). Among a number of factors that can affect the fluxing properties of a mineral, the most important factors are the total alkali content and ratio of the alkali oxides Na_2O and K_2O. The greater the alkali content, the better the fluxing action, and the lower the melting point. Therefore, only feldspars of the alkali series are used for ceramics. The feldspar flux content in porcelain is 25 to 40%, 18 to 35% in tableware, 30 to 36% in sanitaryware, and 15 to 35% in whiteware, pottery, and tile (Harben and Bates 1990).

Feldspar is widely used in enamel and glaze formulations (Maskall and White 1986; Taylor and Bull 1986). A typical batch composition of ground-coat enamel for sheet metal is 30 to 35% silica sand, 15 to 25% feldspar, 5 to 10% fluorspar, 25 to 30% borax, and 10 to 15% soda ash. Cobalt oxide (0.3 to 0.5%) and nickel oxide (0.5 to 1.0%) are added to promote adhesion. Raw and fritted glazes are based on mineral fluxes, such as feldspar, with the addition of clays, quartz, calcium carbonate, dolomite, zinc oxide, and zirconium silicate as common components. Compositions for a yellow sanitaryware glaze and a white opaque glaze for vitreous china are as follows:

Yellow raw glaze		*White opaque glaze*	
Feldspar	29.4	Potash feldspar	26.3
Ground quartz	19.4	Silica	28.4
Wollastonite	24.4	Calcium Carbonate	3.2
Kaolin	11.3	Barium Carbonate	5.9
Zinc Oxide	2.4	Dolomite	11.4
Zirconium silicate	12.3	Zinc oxide	4.7
Zircon praseodymium	3.0	Kaolin	10.3
		Zirconium silicate	9.8

Filler for Paints: Feldspar is used as a replacement for calcium carbonate in paints subjected to acid environments. The use of feldspar filler in plastics has great potential (Robbins 1986a).

REFERENCES

Allen, J. B., and T. J. Charsley. 1968. *Nepheline syenite and phonolite.* London: Her Majesty's Stationary Office, 169p.

Babst, D. A. 1962. Geology of the Spruce Pine district, Avery, Mitchell and Yancey Counties, North Carolina. U.S.G.S. Bull., 1122A, 26p.

Bowen, N. L., and O. F. Tuttle. 1950. The system $NaAlSi_3O_8$-$KAlSi_3O_8$-H2O. *J. Geol.* 58:489–511.

Brown, W. R. 1962. *Mica and feldspar deposits of Virginia.* Virginia Div. Mineral Res., Mineral Res. Rept., 3, 195p.

Deer, W. A., R. A. Howie, and J. Zussman. 1992. *An Introduction to rock-forming minerals.* Essex, England: Longman, 696p.

Geis, H. P. 1979. Nepheline syenite on Sterney, northern Norway. *Econ. Geol.* 74:1268–1295.

Harben, P. 1991. Glass raw materials: Aspects of quality, quantity, and prices. *Ind. Minerals* (July): 31–43.

Harben, P. W., and R. L. Bates. 1990. *Industrial minerals: Geology and world deposits.* London: Industrial Minerals Division, Metal Bulletin plc, 212p.

Hewitt, D. F. 1961. Nepheline syenite deposits of southern Ontario. Vol. 69, pt. 8. Ontario Dept. Mines.

Kroll, H. 1983. Lattice parameters and determination methods for plagioclase and ternary feldspars. Vol. 2 *Reviews of Minerology.* 2nd ed. Ed. P. H. Ribbe, 101–119. Washington, D.C.: Mineral Soc. Amer.

Kroll, H., and P. H. Ribbe. 1983. Lattice parameters, composition and (Al,Si) order in alkali feldspars. Vol. 2 *Reviews of mineralogy.* 2nd ed. Ed. P. H. Ribbe, 57–99. Washington, D.C.: Mineral. Soc. Amer.

Ladoo, R. B., and W. M. Meyers. 1951. *Nonmetallic minerals,* 2nd. ed. New York: McGraw-Hill Corp. 601p.

Lüttig, G. W. 1980. *General geology of the Federal Republic of Germany: Proc. 26th Intern. Geol. Congress, Paris,* 39–40. Stuttgart: E. Schweiz. Verlags-buchhandlung.

Maskall, K. A., and D. White. 1986. *Vitreous enamelling.* Oxford: Pergamon Press, 112p.

Mills, H. N. 1983. Glass raw materials. In *Industrial minerals and rocks.* 5th ed. Ed. S. J. Lefond, 339–347. New York: Soc. Mining Engineers, AIME.

Parker, III, J. M. 1952. *Geology and structure of part of the Spruce Pine district, North Carolina.* North Carolina Div. Mineral Res. Bull. 65, 26p.

Philips, W. R., and D. T. Griffen. 1981. Optical mineralogy. San Francisco: W. H. Freeman & Sons, 677p.

Ribbe, P. H. 1983. The chemistry, structure and nomenclature of feldspar. Vol. 2 *Reviews of mineralogy.* 2nd ed. Ed. P. H. Ribbe, 1–20. Washington, D.C.: Mineral. Soc. Amer.

Robbins, J. 1986a. Feldspar and nepheline syenite. *Ind. Minerals* (September): 69–101.

———— 1986b. Italy's industrial minerals. *Ind. Minerals* (December): 33–34.

Rogers, C. P., and J. P. Neal. 1983. Feldspar. In *Industrial minerals and rocks.* 5th ed. Ed. S. J. Lefond, 709–722. New York: Soc. Mining Engineers, AIME.

Schairer, J. F., and N. L. Bowen. 1955. The system $K_2O\text{-}Al_2O_3\text{-}SiO_2$. *Amer. J. Sci.* 253:681–746.

Smith, J. V. 1983. Phase equilibria of plagioclase. Vol. 2 *Reviews of mineralogy.* 2nd ed. Ed. P. H. Ribbe, 223–239. Washington, D.C.: Mineral. Soc. Amer.

Smith, J. V., and W. L. Brown. 1988. *Feldspar minerals l. crystal structures, physical, chemical and microstructural properties.* Berlin: Springer-Verlag, 828p.

Svoboda, J. 1987. *Magnetic methods for the treatment of minerals.* Amsterdam: Elsevier, 692p.

Taggart, A. F. 1945. *Handbook of mineral dressing.* 2nd. ed. 3-26–3-30, New York: John Wiley & Sons.

Taylor, J. R., and A. C. Bull. 1986. *Ceramics glaze technology.* Oxford: Pergamon Press, 263p.

Watson, I. 1981. Feldspathic fluxes—the rivalry reviewed. *Ind. Minerals* (April): 21–42.

———— 1982. The industrial minerals of Finland. *Ind. Minerals.* (January): 23–33.

Williams, H., F. J. Turner, and C. M. Gilbert. 1982. *Petrography: An introduction to the study of rocks in thin sections.* 2nd ed. San Francisco: W. H. Freeman, 626p.

Fluorspar

Fluorine is a relatively abundant element in nature with concentrations up to 2,000 ppm in some alkali rocks. It is also the most electronegative element and has a strong affinity toward many metals, but most of fluorine is fixed in nature by combining with calcium to form fluorite, CaF_2, commercially known as fluorspar. As an industrial mineral, fluorspar is used widely in the chemical, metallurgical, and ceramic industries.

Other fluorides include cryolite (Na_3AlF_6), which is a rare mineral with economic deposits found only in Greenland and used in the manufacture of aluminum, and fluorapatite ($Ca_5(PO_4)_3F$), a calcium phosphate with fluorine content between 3 and 4%.

China, Mexico, and Mongolia are the world's major fluorite producers. Average annual world production was estimated to be 4 M tons in the period from 1990 to 1995 (Miller 1996).

MINERALOGICAL PROPERTIES

The crystal structure of fluorite (CaF_2) is based upon two penetrating face-centered cubic lattices of fluorine that provide eight-coordinated positions per unit. Calcium occupies four of these positions, with a 50% occupancy. This arrangement can accommodate appreciable amounts of rare earths such as Y and Ce for Ca, extending to the fluorite-type YF_3 and CeF_3. However, most fluorite is at least 99% CaF_2 (Chang et al. 1996).

Fluorite as pure CaF_2 is colorless; however, it is usually colored, most commonly yellow, green, blue, and purple. The causes suggested for the coloration include (1) defect structures, (2) radioactive inclusions or emanations, (3) presence of rare earths or MnO_2, and (4) carbonaceous inclusions. Colorless fluorite changes to purple after irradiation, but resumes its original color on heating to 200°C (Chang et al. 1996).

Fluorite has a hardness of 4 and a specific gravity of 3.18. Its refractive index is between 1.433 and 1.435, but increases slightly with substitution of Ca by Y. Purple fluorite is associated with an increase in refractive index and a decrease in specific gravity. Fluorite usually occurs in nature as euhedral crystals, and it is sometimes coarse to fine granular. Octahedral {111} cleavages are distinctive and commonly produce fragments such as trigonal pyramids or triangles.

The solubility of fluorite is 1.7 mg/100 g H_2O at 25°C. In NaCl solution the solubility of fluorite shows an increase of up to 100°C and starts to decrease above it (Richardson and Holland 1979). In the system CaF_2-$MgCl_2$-H_2O at 150° to 300°C and 4.5 to 85 bar, Schaefer and Strubel (1979) found a maximum solubility of 5.16×10^{-1} $gCaF_2/kgH_2O$ at 150°C, 4.5 bar, and 8.6 $gCaF_2/kgH_2O$ at 300°C, 85 bar. CaF_2 melts at 1,414°C and forms eutectics with most metallic oxides, making it a useful flux. CaF_2 is only slightly soluble in dilute, cold acids, but is attacked by hot sulfuric acid.

GEOLOGICAL OCCURRENCE AND DISTRIBUTION OF DEPOSITS

Fluorite commonly occurs in carbonate rocks as stratiform replacement deposits, and as replacements along contacts with acid igneous intrusive. It is also found in fissure veins and stockworks in shattered zones in granitic rocks, greisens, and pegmatites. Deposits associated with carbonatites and alkaline igneous rocks are becoming of increasing economic importance. Residual concentrations resulting from the weathering of primary deposits, and the occurrence as gangue in base metal deposits, are also found (Grogan et al. 1974; Hodge 1986; Fulton and Montgomery 1994).

Stratiform replacements or bedded fluorite deposits occur in carbonate rocks. The replacements are generally along or adjacent to joints or faults, but in some deposits no such relationship exists. The Illinois-Kentucky ore district is the main fluorite producing region in the United States (Baxter et al. 1963; Baxter and Desborough 1965; Baxter et al. 1967; Baxter and Bradbury 1989). In the Lower Carboniferous limestones of the Cave-in-Rock district of southern Illinois, mineralization concentrates are found in metasomatic stratiform ore bodies, collapse breccias, and diatremes in association with an uplift structure and no igneous activity. The associated minerals are sphalerite, galena, and calcite, with accessory barite, strontianite, and quartz. The formation temperature is about 150°C. Some of the fluorite ore has an 85% CaF_2 concentration that can be used directly for metallurgical purposes. In the Encantada district in northern Coahuila, Mexico, the presence of rhyolite plugs in the general vicinity of the fluorite deposits, and the association of fluorite with rhyolite injections along bedding planes, suggests a connection between the formation of the deposit and igneous activity. Fluorite from Coahuila has a 60 to 85% CaF_2 content (Kesler 1977; Kendall 1995). At the Ottoshoop district in South Africa, fluorite replaces dolomite underlying impervious cherts, shales, and quartzites. The nearby Oogvan-Malmanie deposit has the shape of a mushroom, the foot of which is filled with dolomite fragments that are cemented with fluorite, whereas the cap consists of pure fluorite (Vaneček 1994).

Replacement deposits in carbonate rocks along contact with intrusive rhyolite are well developed in the Rio Verde fluorite district of Central Mexico (Ruiz et al. 1980). The Las Cuevas deposit, which is one of the largest known high-grade fluorite deposits, is at the contact between Lower Cretaceous limestone and Tertiary rhyolite breccia, and was formed by replacement of the limestone followed by open-space filling. It consists almost entirely of fluorite with small amounts of silica and calcite. There is a distinct compositional zonation within the deposit, with silica increasing toward the rhyolite breccia, and calcite increasing toward the limestone (Ruiz et al. 1980).

Fissure veins along faults or shear zones are a common occurrence in fluorite deposits. Silica, calcite, barite, and iron-lead-zinc sulfides are the associated minerals. In Spain, fluorite occurs in veins in granite with xenolites of argillites in Osor in Catalania, and in veins in Carboniferous limestones on the contact with Triassic schist near Collada in Asturia (Wicke and Guimon 1983). The major fluorite deposits in England are in the fissure veins in Durham, Northern Pennine Orefield, and in Derbyshire, Southern Pennine Orefield (Mason 1973; Greenwood 1977; Harben and Kuzvart 1996). In the Northern Orefield, fluorite deposits in narrow, near vertical fissure veins are confined to rocks consisting of thin alternating limestone, sandstone, and shale, the intruded quartz-dolerite sill, and the Coal Measure. The fluorite has a 70 to 80% CaF_2 content. In the Southern Orefield, mineralization occurs in the upper part of the Carboniferous limestone and was controlled by fracture patterns. The ore has a lower CaF_2 content (50 to 70%). The fluorite deposits of the St. Lawrence District, Newfoundland, Canada, form veins in Lower Carboniferous granite and are associated with quartz, calcite, barite, and sulfides (Collins and Strong 1985). Some vein deposits of fluorite also occur in the Rosiclare district of southern Illinois. Veins may be up to several kilometers in length, and ore may be present to a depth of 300 meters or more below the surface. There are widespread fluorite deposits in China concentrated in the provinces of Zhejiang, Jiangxi, Shangdong, and Hunan (Harben and Kuzvart 1996). The leading producing deposit at Wu-Yi, in Zhejiang province, south of Shanghai, occurs in a series of east-west striking veins in a porphritic acidic tuff host rock. Fluorite vein deposits also exist in other provinces, and limestone replacement deposits are found in the province of Jianxi. In Inner Mongolia, the Da Gai Tang fluorite, a north-south vein deposit, and the Hei Shao Ton fluorite, an east-west vein deposit, are the principal producing mines (Kilgore et al. 1985).

Fluorite often occurs as stockworks and fillings in shear and breccia zones. The deposits are usually wide, but have low CaF_2 content. The fluorite deposits in the Zuni Mountains of New Mexico, and near Jamestown, Colorado, in the United States, and the Zwartkloof and Buffalo deposits in the Transvaal Province of South Africa are examples (Fulton and Montgomery 1994). Most of the fluorite deposits in Thailand occur as fillings in shear and breccia zones at the contact of granite with limestone, sandstone, and shale (Rachdawong 1982).

INDUSTRIAL PROCESSES AND USES

1. Milling and Beneficiation

There are three principal industrial grades of fluorite: (1) metallurgical, (2) ceramic, and (3) acid. The specifications for each grade are fairly well established, and beneficiation is generally done accordingly (Grogan 1960; Fulton and Montgomery 1983, 1994). Acid grade fluorite contains not less than 97% CaF_2, and limits for silica, calcium carbonate, sulfur, alumina, and ferric oxide are specified in each particular use with a maximum value of 1%, 1%, 0.05% (as sulfides), 1.5%, and 1.5%, respectively. Moisture content in dried concentrates is usually set at 0.10% or less. For some uses, a specific size is required. Ceramic grade fluorite generally contains 93 to 94% CaF_2, but it may require a concentration as high as 96% CaF_2 and as low as 85% CaF_2 for some specific uses. The limits are specified for silica (<2.5 to 3.0%), calcite, iron oxides, and sulfides. User specifications are the common practice. Metallurgical grade fluorite contains a CaF_2 content in the range of 75 to 85% and less than 0.5% sulfur or sulfide. An "effective percentage" of CaF_2 has been used to specify the CaF_2 in relation to the SiO_2 content. The effective percentage is calculated by multiplying the silica percentage in the analysis by a factor of 2.5 and subtracting this number from the calcium fluoride percentage in the analysis.

Most fluorite must be beneficiated for industrial use. For high-grade lump or gravel ore, hand-sorting is commonly used, followed by crushing and screening to remove the finer fraction. For ores of lower grade, and with relatively coarse interlocking mineral grains, gravity concentration processes are used for producing metallurgical grade fluorite, based upon the specific gravity of 3.2 for fluorite and less than 2.8 for most gangue minerals. Ore is crushed, screened, and sized. Fine fractions are sent to flotation, while the coarse particles go to a heavy media separation. At Buffalo fluorite deposits, South Africa, crushing is done in three stages, and sizing is made by a preparation screen at 1.9 mm. Oversize is subjected to heavy medium (ferrosilicon) separation using Dyna Whirlpool separators (Ryan 1982). A carousel high-gradient separator is also used at Buffalo fluorite to reduce the phosphate content in the final concentrate (Svoboda 1987). Magnetic separation is effective to remove iron oxide and arsenic contaminants from fluorite concentrates.

Ceramic and acid grades of fluorite are produced by flotation (Fulton and Montgomery 1983, 1994). The conventional reagent plan controls pH by the use of soda ash or sodium hydroxide, and the collector agent is oleic acid or one of its modifications. Flotation temperatures are kept above 25°C and in the range of 80°C to avoid freezing the oleic acid. Fatty acids are used as collectors for the fluorite. Depressants used are (1) Quebracho or Tannin for calcite and dolomite, (2) sodium silicate for iron oxides and silica, (3) starch and dextrin for barite, and (4) cyanide for sulfides. In the procedure, the ore is crushed and ground to proper size. Fluorite is removed in a quick pass through a rougher flotation circuit and sent on to the cleaner circuit; the rougher tailing is then discarded. The middling product is reground to further separate the finely interlocked grains of fluorite and gangue, and these are then separated in one or more cleaner circuits. The final products consist of an acid grade concentrate and one or more concentrates of lower grade that are used as ceramic grade, or pelletized and used as metallurgical grade.

2. Industrial Uses

The chemical and steel industries are the major users of fluorite, consuming an average of 98% of the total production in the last twenty-five years.

Fluorite is used to make hydrofluoric acid by reacting with sulfuric acid in rotary kilns. The acid grade fluorite used should have a minimum of 97% CaF_2 with specified limits for silica, calcium carbonate, sulfides, alumina, and ferric oxide contaminates and a grain size maintained below 150 μm. The reaction, which takes place in a rotary furnace at 300°–350°C, is (Klott and Short 1990):

$$CaF_2 + H_2SO_4 \rightarrow CaSO_4 + 2HF$$

and produces (1) a gas phase containing HF with impurities such as SiF_4 from silica reacting with HF, and CO_2, SO_2, and H_2S from $CaCO_3$, aluminum oxide and iron oxides reacting with H_2SO_4, and H_2O, and (2) a solid phase assemblage containing calcium sulfate with impurities of iron and aluminum sulfates and some reacted raw materials. The raw HF is scrubbed with sulfuric acid. After multistage condensation and refrigeration at 120°C, pure HF (>98%) is produced from the gaseous HF. Further purification can be made by distillation. Aqueous solutions of HF of various concentrations up to 60 wt% are also produced. Anhydrous HF acid is kept in steel containers because iron is a metal that, by becoming passivated, offers good resistance towards the acid. Calcium sulfate is a by-product of this process. The other by-product, silicon tetrafluoride, is reacted with water to produce hexafluorosilicic acid, either in the absence of hydrofluoric acid with silicon dioxide precipitation: $3SiF_4 + 2H_2O \rightarrow 2H_2SiF_6 + SiO_2$, or with added hydrofluoric acid, in the presence of water without silicon dioxide precipitation: $SiF_4 + 2HF \rightarrow H_2SiF_6$. The reactors used in various plants for the production of HF can be grouped into five main types: (1) simple rotary furnace, (2) rotary furnace with pre-mixer, (3) rotary furnace with prereactor, (4) rotary furnace with recycle, and (5) rotary furnace with recycle and prereactor. Hydrogen fluoride is used (1) in the manufacture of inorganic fluorides such as aluminum fluoride, boron fluoride, uranium tetrafluoride, and ammonium hydrogen fluoride; (2) in the manufacture of organic fluoro-compounds; (3) for etching and polishing in the glass industry; (4) as a catalyst in alkylation reactions; (5) for pickling stainless steel; and (6) in the manufacture of semiconductors (Klott and Short 1990).

Organic fluorides are volume leaders in the fluorine chemical industry. Fluorinated chlorocarbons and fluorocarbons are prepared by the interaction of anhydrous HF with chloroform, perchlorethyeelene, and carbon tetrachloride. They are used as refrigerants, aerosol propellants, solvents, and cleaning agents, and as intermediates for polymers such as fluorocarbon resins and elastomers (Fulton and Montgomery 1983). However, the usage of CFC has been declining at accelerated rate because of environmental restrictions (Griffiths 1990).

HF is used in the manufacture of cryolite (Na_3AlF_6), whose natural occurrence is limited to one deposit (Ivigtut, Greenland). The main hydrofluoric acid process may be summarized by the equation:

$$12HF + Al_2O_3 \cdot 3H_2O \rightarrow 2AlF_3 \cdot 3HF + 6H_2O$$

followed by one of the three equations:

$$2AlF_3 \cdot 3HF + 6NaOH \rightarrow 2AlF_3 \cdot 3NaF + 6H_2O$$
$$2AlF_3 \cdot 3HF + 3Na_2CO_3 \rightarrow 2AlF_3 \cdot 3NaF + 3H_2O + 3CO_2$$
$$2AlF_3 \cdot 3HF + 6NaCl \rightarrow 2AlF_3 \cdot 3NaF + 6HCl$$

These operations are generally carried out by a batch process. After the alumina is dissolved in 40 to 60% HF solution at 70°C, cryolite is precipitated by the addition of NaOH, Na_2CO_3, or NaCl. After the slurry is filtered, the solids are calcined at 600°C. These processes have been placed on an industrial scale by Kuhlmann (Patent FR 1519691 1967; Patent FR 2076577 1970). Cryolite is used as an electrolyte in the manufacture of aluminum from its oxide.

Metallurgical grade fluorite is used in the steel industry as a flux in basic open hearth and electric furnaces, where it is added to the heats in amounts ranging from 5 to 20 pounds per ton of steel produced. It promotes fluidity of the slag and thus facilitates the removal of sulfur and phosphorus from the steel into the slag. Fluorite serves the same function in iron foundries, where it is added to the cupola charge in a proportion of around 15 to 20 pounds per ton of metal melted (Grogan 1960).

Fluorite of ceramic grade is mainly used to produce white or colored opal glasses and flint glass mixtures which commonly contain 3% fluorspar. It is the most common fluoride material used in enamel or glaze, with its addition being of particular value in low-temperature glazes (Taylor and Bull 1986). Opal glasses are used in containers and ornamental glassware, and opaque enamels are used to cover steel fixtures (Fulton and Montgomery 1983).

In the self-shielding flux-cored process, the fluxes must provide the shielding gas. Four types of self-shielding flux-cored wires are in use. They are (1) fluorspar-alumina, (2) fluorspar-titania, (3) fluorspar-lime-titania, and (4) fluorspar-lime (Lawrence 1983). Among them, fluorspar-alumina and fluorspar-titania are most commonly used.

Minor uses of fluorspar include the manufacture of magnesium, calcium, and zinc metals, portland cement, calcium carbide, and others.

REFERENCES

Baxter, J. W., P. E. Potter, and F. L. Doyle. 1963. *Areal geology of the Illinois fluorspar district. Pt. 1—Saline mines, Cave-in-Rock, Dekoven and Repton quadrangles.* Illinois State Geol. Surv.: Circular 342, 43p.

Baxter, J. W., and G. H. Desborough. 1965. *Areal geology of the Illinois fluorspar district. Pt. 2— Karberg Ridge and Rosiclare quadrangles.* Illinois State Geol. Surv.: Circular 385, 40p.

Baxter, J. W., J. C. Bradbury, and C. W. Shaw. 1967. *Areal geology of the Illinois fluorspar district. Pt. 3—Herod and Shetlerville quadrangles.* Illinois State Geol. Surv.: Circular 413, 41p.

Baxter, J. W., and J. C. Bradbury. 1989. Illinois-Kentucky fluorspar mining district. In *Precambrian and Paleozoic geology and ore deposits in the midcontinent region.* 28th Intern. Geol. Congr., Field Trip T147. Ed. J. W. Baxter et al., 3–36. Washington, D.C.: Amer. Geophys. Union.

Chang, L. L. Y., R. A. Howie, and J. Zussman. 1996. Vol. 5B of *Rock-forming minerals.* 2nd ed. Essex, England: Longman, 383p.

Collins, C. J., and D. F. Strong. 1985. The behavior of fluorite in magmatic system: A case study of Canada's largest fluorspar deposit. In *Granite-related mineral deposits,* Eds. R. P. Taylor and D. F. Strong. 183–197. Halifax: CIM Geol. Divi.

Fulton, R. B., and G. Montgomery. 1983. Fluorspar and cryolite. In *Industrial minerals and rocks.* 5th ed. Ed. S. J. Lefond, 723–744. New York: Soc. Mining Engineers, AIME.

———— 1994. Fluorspar. In *Industrial minerals and rocks.* 6th ed. Ed. D. D. Carr, 509–522. Littleton, Colorado: Soc. Mining , Metal. and Exploration.

Greenwood, D. A. 1977. Fluorspar mining in the northern Pennines. *Trans. Inst. Mining and Metal.* 86:B181–B190.

Griffiths, J. 1990. Fluorspar markets in transition. *Ind. Minerals* (July): 21–31.

Grogan, R. M. 1960. Fluorspar and cryolite. In *Industrial minerals and rocks.* 3rd ed. Ed. J. L. Gillson, 363–382, New York: AIME.

Grogan, R. M., P. K. Cunningham-Dunlap, H. F. Bartlett, and I. J. Czel. 1974. The environments of deposition of fluorspar. Spe. Publ., Kentucky Geol. Surv. x, 22, 4–9p.

Harben, P. W., and M. Kuzvart. 1996. Industrial minerals—A global geology. London: Industrial Minerals Information, Ltd., Metal Bulletin Plc, 462p.

Hodge, B. L. 1986. Occurrence and exploitation of fluorite. In *Geology in the real world—The Kingsley Dunham volume.* Ed. R. W. Nesbitt and I. Nichol. Inst. Mining Metall.

Kendall, T. 1995. Mexico: promises of prosperity. *Ind. Minerals* (September): 27–55, 165p.

Kesler, S. E. 1977. Geochemistry of Monto fluorite deposits, northern Conhuila, Mexico. *Econ. Geol.* 72:204–218.

Kilgore, C. C., S. R. Kramer, and J. A. Bekkala. 1985. Fluorspar availability—Market economy countries and China. U.S.B.M. Circular 9060, 57p.

Klott, K. A., and E. L. Short. 1990. Vol. 1 of *Industrial Chemistry,* 312–347. New York: Ellis Horwood.

Lawrence, F. V. Fluxes. 1983. In *Industrial minerals and rocks.* 5th ed. Ed. S. J. Lefond, 259–270. New York: Soc. Mining Engineers, AIME.

Mason, J. E. 1973. Geology of the Derbyshire fluorspar deposits, United Kingdom. *Proc. 9th Forum on the Geology of Industrial Minerals,* Ed. D. W. Hutchson, 10–22. Kentucky Geol. Surv. Spe. Publ., 22.

Miller, M. M. 1996. Fluorspar. *U.S.G.S. Annual Review* (August 1995), 11p.

Rachdawong, S. 1982. The fluorite industry of Thailand, 1976–1981. *Ind. Minerals* (August): 53–55.

Richardson, C. K., and H. D. Holland. 1979. The solubility of fluorite in hydrothermal solutions, an experimental study. *Geochim. Cosmochim. Acta* 43:1313–1325.

Ruiz, J., S. E. Kesler, L. M. Jones, and J. F. Sutter. 1980. Geology and geochemistry of the Las Cuevas fluorite deposit, San Luis Potosi, Mexico. *Econ. Geol.* 75:1200–1209.

Ryan, P. J. 1982. A review of the fluorspar-mining industry in South Africa. *Proc. 12th CMMI Congress 1,* 228–247. Johanesburg: South Africa Inst. Min. Metall.

Schaefer, H., and G. Strubel. 1979. Hydrothermal solubilities of fluorite in the system CaF_2-$MgCl_2$-H_2O: *Neues Jahrb. Min., Montat.,* 233–240.

Svoboda, J. 1987. *Magnetic methods for the treatment of minerals.* Amsterdam: Elsevier, 692p.

Taylor, J. R., and A. C. Bull. 1986. *Ceramic glaze technology.* Oxford: Pergamon Press, 263p.

Vaneček, M. 1994. *Mineral deposits of the world, developments in economic geology 28.* Amsterdam: Elsevier, 519p.

Wicke, F. L., and A. B. Guimon. 1983. A review of the Spanish fluorite industry and its role in world markets. *Proc. 5th Ind. Minerals Intern. Congress.* Eds. B. M. Coope and G. M. Clarke, 51–59.

Garnet

Garnet represents an isomorphous group of orthosilicates with similar physical properties and varying chemical compositions. As an industrial mineral, garnet is mainly used as an abrasive, and among the members of the group, almandine, an iron-aluminum garnet, is used most often.

The United States has long been the dominant leader in world garnet production.

MINERALOGICAL PROPERTIES

Garnet is cubic with eight $X^{2+}_3Y^{3+}_2(SiO_4)_3$ formula units per unit cell. The crystal structure consists of alternating (SiO_4) tetrahedra and (YO_6) octahedra that link together by sharing corners to form a three-dimensional network. The interstices are occupied by X-polyhedra in the form of a distorted cube (Novak and Gibbs 1971). The arrangement of various polyhedra is nearly ideal, and the cubic edge may be expressed by an addition function of ionic radius of the cations in the structure: $a(Å) = 9.04 + 1.61(r)_X + 1.89(r)_Y$.

The garnet structure is formed by numerous chemical compounds, some of which are magnetic. The rare-earth yttrium iron garnet $Y_3Fe_5O_{12}$ is prototypical of the rare-earth ferromagnetic insulators. The compound is cubic and isostructural with garnet. A wide range of cation substitution by transition metal is possible. Consequently, the saturation magnetization, magnetocrystalline anisotropy, and other magnetic properties may be modified over a wide range (Reynolds and Buchanan 1991).

The six major garnet minerals are generally divided into two groups: Pyralspite garnets and Ugrandite garnets. They are listed below with their hardness (H), specific gravity (G), refractive index (n), and color (Deer et al. 1992):

Pyralspite Garnet:

Pyrope	$Mg_3Al_2(SiO_4)3$	H=7.5	G=3.582	n=1.714
	Deep red, pink, purple.			
Almandine	$Fe_3Al_2(SiO_4)_3$	H=7–7.5	G=4.318	n=1.830
	Dark red, brown, black.			
Spessartine	$Mn_3Al_2(SiO_4)_3$	H=7–7.5	G=4.190	n=1.800
	Dark red, brown, black.			

Ugrandite Garnet:

Andradite	$Ca_3Fe_2(SiO_4)_3$	H=6.5–7	G=3.859	n=1.887
	Brown, yellow, green, black.			
Grossular	$Ca_3Al_2(SiO_4)_3$	H=6.5–7	G=3.594	n=1.734
	Yellow, brown, green, colorless.			
Uvarovite	$Ca_3Cr_2(SiO_4)_3$	H=7.5	G=3.83	n=1.865
	Dark green, emerald green.			

Within the pyralspite group, complete solid solution among the three end members exists, and in the ugrandite group, there are extensive solid solutions among the end members. However, ranges of solid solutions between pyralspite garnets and ugrandite gar-

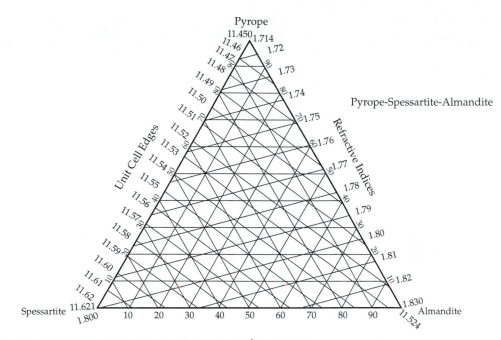

Figure 24   Refractive index (n) and cell dimension (a, Å) relationships among garnet.

nets are limited. Garnets of pure end member composition rarely occur in nature. Among all garnet minerals, almandine is the most abundant.

The chemical composition of garnet, consisting of no more than three end members, can be determined by measuring the unit cell dimension "a" using x-ray diffraction, and the index of refraction "n" using optical microscopy. An example of the linear relationship between "a" and "n" as established by Sriramadas (1957) is shown in Fig. 24.

The garnet appears most commonly as euhedral or nearly euhedral crystals of dodecahedral {110} or trapezohedral {112} form, or as some combination of the two forms as shown in Fig. 25. The dodecahedron exhibits twelve faces with four sides, whereas the trapezohedron consists of twenty-four faces with eight sides. Garnets have no distinct cleavage, but a {110} parting is frequently present. The crystal shape appears to be blocky and nearly equidimensional. The crystal form, and the parting, as well as the crystal shape, are major factors in determining the potential industrial uses of garnet.

GEOLOGICAL OCCURRENCE AND DISTRIBUTION OF DEPOSITS

Garnet commonly occurs as an accessory mineral in many metamorphic rocks, including gneiss, schist, and recrystallized limestone, and is found as well in some granites and pegmatites. Because garnet is fairly resistance to weathering, it is often deposited in detrital sediments. Almandine is the garnet in schists and gneisses of regional metamorphism. It forms in the amphibolite through granulite facies, and intermediate almandine-pyrope garnets form in the eclogites. Almandine also appears in hornfels in contact metamorphic

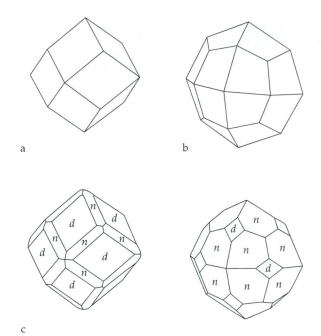

Figure 25 Common crystal forms of garnet (a) The dodecahedron, {110} (b) The trapezohedron, {112} and (c) Combined forms of dodecahedron, d, and trapezohedron, n.

rocks, and as a late-stage mineral in granitic aplites and pegmatites. Pyrope occurs in ultrabasic rocks such as peridotites and kimberlites and in associated serpentinites. Spessartine is found in skarn deposits and in pegmatites. The grossular and andradite characteristically occur in contact metamorphic rocks and are also formed from impure limestones by regional metamorphism. Uvarovite occurs in peridotite or serpentinite in association with chromite (Deer et al. 1992).

Among all known garnet deposits, the largest and most productive one is in Gore Mountain, near North Creek, in the Adirondack Mountains, New York (Fig. 26), and is the source of 95% of the world's production of technical garnet.

The Gore Mountain deposit is a lenticular body and lies along the contact between gabbro on the north and syenite on the south, with anorthosite on the east and west as shown in Fig. 27 (Levin 1950). The garnet ores can be classified into four types:

1 The dark ore: This garnet is associated with olivine gabbro and is common in the Gore Mountain deposit. It is spherical in shape and large in size (usually 5 to 10 cm in diameter). The garnet grains are usually surrounded by a thick rim of hornblende, and the matrix, made of plagioclase and hornblende, is homogeneous and massive.
2 The porphyroblastic gabbroic anothosite gneiss: This type of garnet ore, which is the most abundant, is strikingly different from the dark ore. The garnets are small (5 cm across) and embedded in a gneissic matrix of gabbroic anorthosite composition. In the feldspar-rich variety, the garnets are euhedral and do not have any rims, whereas in the more mafic varieties, they are surrounded by dactylic intergrowth of hypersthene and/or hornblende, with plagioclase.

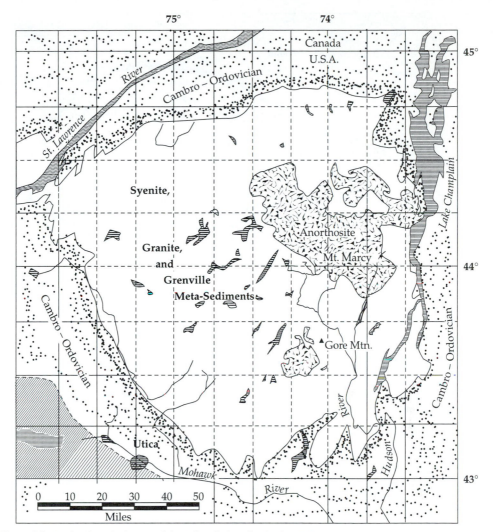

Figure 26   Regional geologic map of the Adirondacks, after Levin (1950).

3 Porphyroblastic garnetiferous amphibolites: The metasedimentary belt north and north-east of Gore Mountain contains lenses and layers of garnetiferous amphibolites. The garnets are irregular in shape.

4 The light ore: This ore is characterized by its nongneissic matrix, and by the rims of pure plagioclase that surround the garnets (Bartholomé 1960).

   The Gore Mountain garnets are a combination of almandine and pyrope, and possess many desirable properties for abrasive usage. They exhibit lamellar parting planes and have a hardness close to the higher end of garnet's limit of 8. When crushed to a fine size, the garnet grain retains its natural morphological shape and produces a fresh, sharp set of edges (Hight 1983).

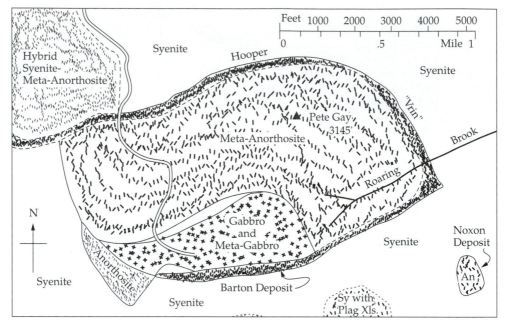

Figure 27   Areal geologic map, Gore Mountain District, after Levin (1950).

Garnet deposits are also found in the United States, in Maine, Idaho, North Carolina, and Montana. The Wing Hill deposit in western Maine is one of the largest and produces the highest grade garnet in the world (O'Connor 1985). Garnet is mined from both crystalline rock and alluvial deposits in Australia, Canada, China, and Norway (McMichael 1990).

INDUSTRIAL PROCESSES AND USES

1. Milling and Beneficiation

Industrial garnet is a sized and graded product. Sizings generally conform to standards established for quality criteria and are based upon standard methods of testing, in cooperation with the American National Standards Institute (ANSI). Grading standards are documented under the titles "Grading of Abrasive Grain on Coated Abrasive Products (B74.18–1977)," and "Grading of Abrasive Grains for Grinding Wheels (B74.12–1977" (Austin 1994).

The treatment of garnet normally consists of (1) primary crushing and screening, (2) a sink-float process with ferrosilicon as the medium, and (3) secondary screening with screens divided longitudinally to serve both float and sink. The product is further treated in rolls and classified, with middling being returned to the crusher. A concentrate of 98% garnet-content (minimum) is achieved for the various grade sizes (Hight 1983; Burt 1984). In addition, an induced magnetic roll separator is also used in the beneficiation of garnet. The milled garnet is fed to the top of the roll by a vibratory feeder. As the roll revolves, the garnet passes through a narrow gap between the pole of the magnet and the roll. The nonmagnetic particles are discharged from the roll, while the garnet particles are attracted to the roll (Taggart 1945; Svoboda 1987).

2. Industrial Uses

Garnet is mainly used as an abrasive grain in the manufacture of coated abrasive products, and as an airblasting medium. In addition, garnet is also used as bedding material for water filtration. Some varieties of garnet are considered to be semiprecious stones.

Abrasive Grains: Garnet grains are used for sandblasting large structures and buildings. They are a low-cost, silica-free abrasive that avoid the silicosis problem associated with silica sand blasting. They are also heavier than silica sand, and, thus, grains of equal size deliver harder blow. In comparison with copper slag for sandblasting, garnet has faster cleaning and lower consumption rates, and causes fewer environmental problems. Garnet grains are more expensive than copper slag (McMichael 1990).

Coated Abrasives: Garnet-coated papers and cloths are used primarily for wood sanding, but also for finishing leather, rubber, plastics, glass, and softer metals. Paper and cloth are generally used for the backings, although vulcanized fiber can also be used. Three types of adhesives are used in the manufacture of coated abrasives: (1) glue, (2) resin, and (3) varnish. The glue is generally animal hide glue. The resins and varnishes are basically phenols or ureas, and depending on their use, they are modified in various ways to give shorter or longer drying times, greater strength, or more flexibility. There are two types of abrasive grain coatings made in coated abrasives: closed coat and open coat. A closed coat is one in which the abrasive grains completely cover the coatside surface of the backing. An open coat is one in which the individual grains are spaced at predetermined distances from one another, resulting in 50 to 70% coverage of the surface of the backing. The closed coat is designed for severe service and is used for most applications, whereas the open coat has greater flexibility and resistance to filling (Anon. 1958). Electrostatic coating is now a commonly used manufacturing process. The applied electrostatic charge causes the individual grains to imbed upright in the wet bond on the backing and to be spaced more evenly due to grain repulsion (Hight 1983).

Water Filtration: Garnet is used as a bedding material for water filtration, and its clarifying power is comparable to sand as a deep bed filter (Tien 1989). Because of its hardness and toughness, garnet more effectively resists physical and chemical breakdown during filtering and backwashing (Dickson 1982).

Lapping and Grinding: Garnet is used for lapping and grinding of glass, ceramics, rubber, plastics, and metals.

REFERENCES

Anon. 1958. *Coated abrasives—modern tool of industry.* New York: McGraw-Hill, 426p.

Austin, G. T. 1994. Garnet. In *Industrial minerals and rocks,* 6th ed., Ed. D. D. Carr, 523–533. Littleton, Colorado: Soc. Mining, Metal., and Exploration.

Bartholomé, P. 1960. Genesis of the Gore Mountain garnet deposit, New York. *Econ. Geol.* 55:255–277.

Burt, R. O. 1984. *Gravity concentration technology.* Amsterdam: Elsevier, 605p.

Deer, W. A., R. A. Howie, and J. Zussman. 1992. *An introduction to the rock-forming minerals.* Essex, England: Longman, 696p.

Dickson, E. M. 1982. Garnet—cutting into filtration. *Ind. Minerals* (February):35–39.

Hight, R. P. 1983. Abrasives. In *Industrial minerals and rocks.* 5th ed. Ed. S. J. Lefond, 11–32. New York: Soc. Mining Engineers, AIME.

Levin, S. B. 1950. Genesis of some Adirondack garnet deposits. *Bull. Geol. Soc. Amer.* 61:519–565.

McMichael, B. 1990. Abrasive minerals. *Ind. Minerals* (February):19–33.

Novak, G. A., and G. V. Gibbs. 1971. The crystal chemistry of silicate garnets. *Amer. Mineral.* 56:791–825.

O'Connor, M. P. 1985. The Wing Hill garnet deposit of Rangeley, Maine and its relationship to world market: Preprint 85–29, 5p, *Soc. Mining Engineers, AIME Annual Meeting, February, 1985,* New York.

Reynolds, T. G., and R. C. Buchanan. 1991. Ferrite (magnetic) ceramics. In *Ceramic materials for electronics.* 2nd ed. Ed. R. C. Buchanan, 207–248. New York: Marcel Dekker, Inc.

Sriramadas, A. 1957. Diagrams for the correlation of unit cell edges and refractive indices with the chemical composition of garnets. *Amer. Mineral.* 42:294–298.

Svoboda, J. 1987. *Magnetic methods for the treatment of minerals.* Amsterdam: Elsevier, 692p.

Taggart, A. F. 1945. *Handbook of mineral dressing,* 13–19 and 25. New York: John Wiley & Sons.

Tien, C. 1989. *Granular filtration of aerosols and hydrosols.* Boston: Butterworths, 365p.

Graphite

Graphite is one of the two crystalline forms of carbon occurring in nature. The other one is diamond. Graphite is a good conductor of electricity and is a superior refractory. It has high thermal conductivity, high thermal shock resistance, excellent high temperature strength, excellent machinability, high sublimation temperature, a low modulus of elasticity, and a hardness of 1 on the Moh's scale. The combination of these properties has led to its diversified uses in the metallurgical, chemical, electrical, and nuclear industries.

Two types of graphite, natural and synthetic, are used. China, the former USSR, South and North Korea, Austria, Sri Lanka, Mexico, and Madagascar are the leading producers of natural graphite, whereas synthetic graphite is mainly produced by the United States, Japan, Germany, France, Canada, and Italy. Annual production of natural graphite worldwide was estimated to be 0.7 to 1.0 M tons between 1990 and 1995 (Harben and Kuzvart 1996).

MINERALOGICAL PROPERTIES

The mineralogical properties of graphite are distinguished by its crystal structure (Fig. 28), which is made up of parallel sheets of carbon atoms, each sheet being a network of hexagonal, planar C_6 rings (Taylor 1992). Each carbon atom is joined to three neighboring carbon atoms, arranged at 120° angles in the plane of the sheet. The carbon-carbon distance in the rings is 1.415Å, and the width of each C_6 ring is 2.456Å. The sheets of graphite are arranged parallel to one another with an intersheet distance of 3.354Å. Weak van der Waals forces join the carbons in adjoining sheets. The rings of a given sheet do not lie directly above those of the next sheet below, but rather are displaced laterally by a distance of 1.415Å. Half of the carbon atoms of a given sheet lie above the atoms of the sheet below, and half lie above the centers of the hexagonal rings. Repetition of this lateral shift causes the third sheet of rings to lie directly over the first. This arrangement produces a hexagonal form of graphite containing the most common stacking order of ABABABAB .... The distance between the sheets is large and the bonding force is weak. Therefore, cleavage readily takes place between the sheets (Mantell 1968; Wege 1984).

1. Natural Graphite

Natural graphite is a very soft, black, greasy feeling mineral occurring in disseminated flakes or in scaly, granular, compact, or earthy masses. It is easily sectile and flexible, but not elastic. Its hardness is 1 on the Moh's scale, and its specific gravity ranges between 1.9 and 2.3 depending upon the perfection of the crystal structure. Optically, graphite is strongly pleochroic and uniaxial, and the sign of birefringence is negative. Thin flakes are normally opaque to light, but in extremely thin sections they transmit a greenish to blue light (Cameron 1960; Graffin 1983). Natural graphite has a low absorption coefficient for X-rays and electrons because of its low atomic number (Mantell 1968).

Graphite is a good conductor of heat and electricity (Reynolds 1968). The thermal conductivity shows a maximum near room temperature with κ (thermal conductivity)

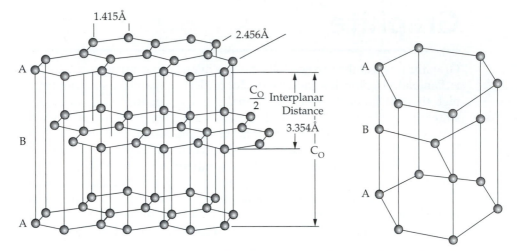

Figure 28   Crystal structure of graphite. *Kirk-Othmer/Encyclopedia of Chemical Technology,* Taylor. Copyright 1992. Reprinted with permission of John Wiley & Sons, Inc.

= 10 to 15 W/cm·K in the direction of a-axis, and κ = 0.02 to 0.04 in the c-axis direction, and a/c > 300. The specific heat increases markedly with temperature. There is a sharp rise in C_p above 3,225°C that is probably caused by the formation of defects. The coefficient of linear expansion in the direction of the a-axis changes from slightly negative below 383°C to slightly positive above this temperature; its average in the direction of the c-axis is 238×10^{-7} between 15°C and 800°C. The specific electric resistivity of natural graphite is 10^{-4} Ω · cm at room temperature in the direction of a-axis parallel to the sheet and is 1 Ω · cm in the direction of c-axis. Therefore the c/a axis anisotropy ratio is 10^4. Any increase in sheet separation may cause higher anisotropic ratio. Graphite is strongly diamagnetic because of its abundance of π electrons. The value of the specific magnetic susceptibility for natural graphite from Sri Lanka at 20°C is -6.5×10^{-6} in the direction of c-axis and -22×10^{-6} in the direction of the a-axis. Plastic deformation occurs in graphite above 2,500°C (Taylor 1992).

The graphite is resistant to attack by most chemical reagents, and is inert to hydrochloric and hydrofluoric acids. Upon heating, graphite decomposes slowly in air at temperatures above 450°C, and the rate increases with increasing temperature and surface area. Graphite reacts with oxygen to form CO and CO_2, with metals to form carbides, with oxides to form metals and CO, and with many other substances to form laminar compounds (Roberts et al. 1973). Phase relations in the system of carbon show that a triple point, graphite+diamond+liquid, exists at 4,000°C and 12 GPa. Graphite sublimes at 3,730°C.

2. Synthetic Graphite

Synthetic graphite, sometimes called manufactured graphite, artificial graphite, electrographite, or graphitized carbon, refers to a bonded granular carbon body whose matrix has been subjected to a thermal process known as graphitization (Marsh 1989). Two additional types of synthetic graphite are manufactured for industrial uses. They are pyrolytic graphite and graphite fibers. The former is produced by chemical vapor deposi-

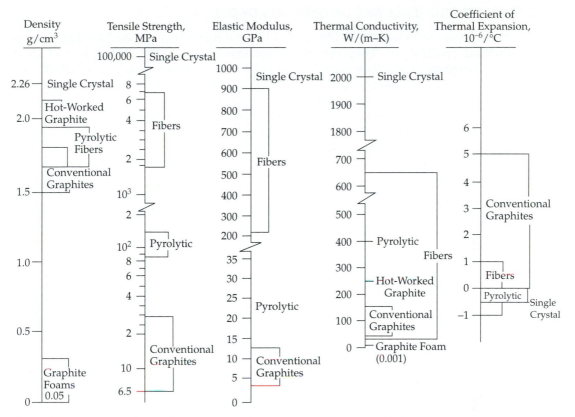

Figure 29   Mechanical and thermal properties of synthetic graphite (with-grain). *Kirk-Othmer/Encyclopedia of Chemical Technology,* Hupp. Copyright 1992. Reprinted with permission of John Wiley & Sons, Inc.

tion at 800° to 2,300°C (Moore 1973), whereas the latter refers to a variety of filamentary products prepared from polyacrylonitrile (PAN) or pitch, by pyrolysis (Bennett and Johnson 1979; Turk 1992).

Synthetic graphite is strongly anisotropic as the direct result of its crystal structure. In the formation of synthetic graphites, the long axis of the carbon particles tends to align perpendicularly to the molding force in molded graphites, and parallel to the extrusion force in extruded graphites. By the selection of raw materials and processing conditions, graphites can be manufactured with a wide range of properties of degree of anisotropy (Hutcheon 1970; Hupp 1992). There are two distinct cuts of manufactured graphite, based upon their relation to the direction of the forming force. A cross-grain is designated as samples cut parallel to the molding force for molded graphites, or perpendicular to the extrusion force for extruded graphites. A with-grain is designated as samples cut parallel to the molding plane of molded graphites, or parallel to the extrusion axis for extruded graphites. ASTM has a set of test procedures for the determination of properties of graphite (Annual Book of ASTM Standards, v.15 - 01 1989). A series of properties is shown in Fig. 29. In general, synthetic graphites have lower density, higher electrical resistance, and higher porosity than their natural counterparts (Pettifer 1980a).

GEOLOGICAL OCCURRENCE AND DISTRIBUTION OF DEPOSITS

Natural graphites have three physically different forms that are used for different industrial applications. They are (1) disseminated flake, (2) crystalline vein, and (3) amorphous. Each form is further subdivided into grades based upon carbon content, particle size, and types of impurities (Harben and Bates 1990). Flake graphite consists of flat, plate-like particles that occur in a disseminated form through layers of metamorphosed carbonaceous sediments such as quartz-mica schists, micaceous or feldspathic quartzites, gneisses, and marbles. Common associated minerals are mica, quartz, calcite, feldspar, amphibole, and garnet. The size of the flakes can vary greatly, from less than 1 mm in diameter to as large as 5 cm with an average size of 0.5 cm. Flake graphite is further subdivided into "coarse flake" (-20 to $+100$ mesh), and "fine flake" (-100 to $+325$ mesh). There are specifications for special grades, such as Madagasy crucible flake graphite, which must have a minimum of 85% carbon and be essentially all -8 to $+60$ mesh, and North Korean Grade 1 flake graphite, which must have a minimum of 90.0% carbon and a granulation of $+48$ mesh (Pettifer 1980b).

Vein graphite is found in well-defined veins or as pockety accumulations along fissures, fractures, or cavities traversing igneous and metamorphic rocks, commonly Precambrian. Vein graphite is typically massive with particle size ranging from very fine, known as "amorphous," or "amorphous lump," to coarse and flaky, usually called "crystalline lump." Vein graphite exhibits a wide variation in width from a few millimeters to over 2 meters and is subdivided into two types: foliated or bladed, and columnar or fibrous. Minerals that are commonly present in vein graphite include feldspar, quartz, mica, pyroxene, zircon, rutile, apatite, pyrite, and pyrrhotite. The amounts of impurities in vein graphite are comparatively lower than in other types of graphite (Pettifer 1980b).

Amorphous graphite is not amorphous. It occurs as minute microcrystalline particles more or less uniformly distributed in feebly metamorphic rocks such as slates, or in beds consisting almost entirely of graphite. The former are usually altered carbonaceous sediments with carbon content ranging from 25 to 60%, whereas the latter are metamorphosed coal seams and carry as much as 80 to 85% carbon. In common with vein and flake, amorphous graphites vary considerably in size and purity.

Deposits of Flake Graphite: Deposits of flake graphite are widely distributed in China, Russia, and India. Other countries reported to have economic deposits of flake graphite are the United States, Norway, Czechoslovakia, Germany, Canada, and Madagascar. In China, graphite deposits are distributed in the Tarim belt of the North China platform in eastern Shandong, the southern part of Inner Mongolia, in Heilongjian along the Sino-Russia border, and in northern Shanxi. In the Nan-Shu and Bai-Shu districts, Shandong, the host rocks are gneiss and schist, with an average 6% of flake graphite, and at the Liu-Mao and Lin mines, Heilonjiang, graphite occurs in metamorphosed coal seams intercalated with metasedimentary rocks. The average content of flake graphite is 20% (Xu and Cheng 1988; Su 1992).

Graphite deposits are concentrated around Kuldzhuktau in Uzbekistan and at Zavalye mine in the Ukraine. In the former, flake graphite occurs in limestone alternated by a gabbroic intrusive body, whereas in the latter, disseminated flake graphite is contained in schist and limestone with 18 to 20%C in four separate seams (Harben and Bates 1990). The Botogol deposit in the eastern Sayan Mountain is considered by Kuzvart (1984) to be of an early magmatic origin. Flake graphite occurs in schist, gneiss, quartzite, and marble intruded by nepheline syenite and microcline granite. In India, the major producing mines are located in the Titlagarh graphite belt, in the Balangir and Kalahandi districts of Oriss. Flake graphite occurs in garnetiferous schist, gneiss, and quartzite, and in intimate association with pegmatite and quartz veins (Kanungo 1976; Russell 1991).

The flake graphite deposits of the highest grade in the world are on the island of Madagascar. The deposits occur in an extended belt of mica gneiss and mica schist from the port town of Tamatave to the southern end. Manampotsy, Ambatolampy, and Ampanihy are the principal production districts. The graphite content has reported ranges up to 60%, and the flake is characterized by its uniform thickness, toughness, and cleanness (Murdoch 1967; Fogg and Boyle 1987).

Deposits of Vein Graphite: Sri Lanka has the most extensive vein graphite deposits (Erdosh 1970; Graffin 1983; Dissanayake 1994). Graphite veins and stockworks, pockets, lenses, and cavity fillings are distributed in Precambrian gneiss, granulite, quartzite, and dolomite, which are intruded by quartz veins, pegmatite, and granite. Important deposits are found in the Kahatagaha-Kolengaka and Bogata districts (Dissanayake 1981). Graphite is coarse-grained and crystalline, with the best grade containing 97%C. Vein graphite deposits are also found in the United States, Brazil, India, China, Korea, and Japan.

Deposits of Amorphous Graphite: Amorphous graphite deposits are known to exist in Mexico, Korea, Austria, and other countries (Graffin 1983; Kendall 1995). The most important deposits of amorphous graphite in Mexico occur in the State of Sonora. The graphite beds are in Triassic sandstone that has been folded and intruded by granite. Six beds of graphite have been reported, with the thickest bed averaging approximately 3 meters. The graphite ore is soft and massive, and is considered to have formed through the metamorphism of coal beds (Kuzvart 1984). In Austria, large deposits of amorphous graphite are present in a 50 km long belt of folded metasediments extending from Leoben to Rottenmann in the Styrian Alps (Dickson 1981). The metasediments are mainly limestone and slate, and the contained graphite, with an average of 50 to 60%C, resulted from the metamorphism of coal seams. The graphite deposits at Niederösterreich in Lower Austria occur in graphite-bearing slate sandwiched between gneiss and mica schist. The graphite is generally harder and more crystalline than that of Styria. In Korea, graphite deposits are confined to the Gyeonggi Massif (Fogg and Boyle 1987). Large amorphous graphite deposits were formed by contact metamorphism of coal beds and occur as irregular lenses parallel to the structure of enclosing schists and phyllites.

The origin of graphite has been a subject of discussion for many years. It is generally recognized that organic matter in sediments is the source, and the conversion by metamorphism is the process, for the formation of syngenetic graphite. As metamorphism advances from zeolite to amphibolite facies, crystallinity of graphite progresses from amorphous through intermediate stages, to fully ordered crystalline graphite (Erdosh 1970; Landis 1971; Hollister 1980; Mancuso and Seavoy 1981). Controversy exists over the origin of the epigenetic graphite commonly found in Precambrian regionally metamorphosed schist and gneiss. A stable carbon isotope studied by Weis et al. (1981) suggests that epigenetic graphite is produced from syngenetic graphite or carbonaceous detritus, by means of a water-gas reaction that converts carbon to CO and precipitates it by the Boudouard reaction. Possible causes influencing high crystallinity of epigenetic graphite from C-O-H fluids suggested by Luque and Rodas (1999) include (1) large graphite occurrences that are generated from large volumes of fluids that maintain their temperature for long periods of time, (2) high temperature conditions that are required for a fluid to precipitate a major portion of its dissolved carbon during a small temperature decrease, (3) the incorporation of carbon into C-O-H fluids mainly through devolatilization reactions, and (4) the generation of highly crystalline graphite at high temperature and high pressure conditions that is less susceptible to resorption as pressure decreases or by subsequent fluid flow.

INDUSTRIAL PROCESSES AND USES

1. Milling and Beneficiation of Natural Graphite

The milling of graphite varies from plant to plant. In general, it involves a series of crushing and classification procedures. Flotation is the common milling procedure used to beneficiate graphite, and hand-sorting is performed in some mines (Cameron 1960; Graffin 1983). In Sri Lanka, the ore is roughly picked over at the mine. This reduces the gangue impurities to about 5 to 10%. At the processing plant, hatchets are used to break up larger samples, and sizing is made by screens. The finished product is graded as large lump, ordinary lump, chip, and chippy dust. In Madagascar, the crude ore is crushed by a primary crusher and then conveyed to a series of roll crushers and classifiers to remove the oversizes and gangue minerals. Flotation is used to take off the graphite, and classification of the concentrations into various flake sizes is made by screening.

In the Kaiserburg mines at Styria, Austria, hand sorting and air classification are used, whereas in Korea, sorting, crushing, and washing of ore by hand is the general practice. In Mexico, gangue minerals are removed partly by hand-sorting and partly by screening. Further milling by roller mills and air classifiers is also used.

Flotation is usually used for final beneficiation of graphite at most processing plants (Crozier 1992). The usual treatment starts with a pulp of about pH 8 that is made alkaline with soda ash. Despite the natural floatability of graphite, the separation of feldspar, quartz, mica, and marble is normally improved by the addition of a small amount of kerosene and floated with pine oil as the frother. If the product is insufficiently clean, it can be upgraded by a treatment such as is shown in Fig. 30. Cleaning with regrinding usually is not practical because graphite tends to smear gangue minerals and makes them floatable. When interleaving of mica is present, it is not possible to produce graphite of over 95% by flotation.

Many commercial grades of graphite are marketed, and specifications for graphite products are varied and often complex. Although both the ASTM and the U.S. Gov-

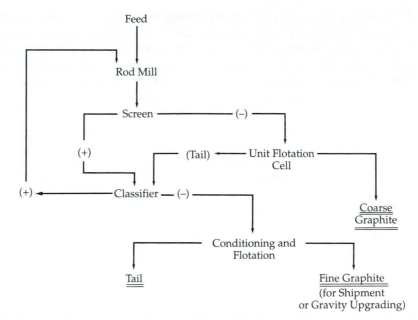

Figure 30   Flow sheet for the recovery of coarse and fine graphite. Reprinted from Pryor, International Journal of Processing. Copyright 1965, p. 764, with permission from Elsevier Science.

ernment have published specifications and recommended practices, there are no standard specifications. In general, the common factors used for grading graphite are (1) carbon content, (2) flake or particle sizes, and (3) kinds and amounts of impurities. Flake graphite of high grade contains 95 to 96% carbon, and low grade 90 to 94% carbon. High purity crystalline flake contains 99% carbon. Amorphous graphite is classified according to locality and carbon content.

2. Production of Synthetic Graphite

Raw Materials and Binders: The raw materials used for the production of synthetic graphite are mostly derived from petroleum and coal. Petroleum coke, including both needle or premium coke and regular coke, is the most commonly used raw material. Premium coke is anisotopic and has a very low coefficient of thermal expansion, whereas regular coke is isotropic and has a high coefficient of thermal expansion. Both have low sulfur content, volatile matter, and ash. The raw materials used in its production have an important bearing on the properties and applications of synthetic graphite. For the two types of coke used, the needle or premium coke is mainly used for the production of graphite electrodes, whereas regular coke is used for other graphite products (Mantell 1968; Pettifer 1980a; Lewis 1992).

A binder is required in the production of synthetic graphite to fluidize the coke particles, enabling them to flow into an ordered alignment during the forming process, and to preserve the shape of the green carbon. The binder should (1) have high carbon yield, (2) have good wetting adhesion properties to bind the coke particles together,

(3) be reasonably priced and readily obtainable, (4) possess a low level of impurities and ash, and (5) produce binder coke that can be graphitized to improve electrical and thermal properties (Lewis 1992). Small quantities of additives may also be added to the filler and binder mix for the purpose of improving extrusion rates, as well as the structure of extruded products. Light extrusion oils and lubricants, including petroleum oils, waxes, fatty acids, and esters, are often used (Lewis 1992).

Production Processes: The petroleum coke is calcined in rotary kilns in order to compact the particle and to remove hydrocarbons and other volatiles. The run-of-kiln coke is broken down by crushers and mills, and sized into a series of fractions through screens. For metallurgical and nuclear applications, particles as small as micrometers in diameter are required. If the product is to be used for the production of electrodes, a high yield of particles of up to 25 mm in diameter is necessary.

After the coke is crushed and sized, it is mixed with the binder and additives to produce a uniform distribution of binder over the surface of the filler grains. A dense mass is formed either by molding, or more usually by extrusion of the mixture. At this stage in the process, it is desirable to produce graphite of maximum density by minimizing void volume in the formed product. It is also the purpose of this forming step to produce a shape and size as near to that of the finished product as possible. The product produced at this stage is known as green bar or green stock (Turk 1992).

In forming, the anisotropic coke particles tend to be orientated with their long axes parallel to the direction of extrusion, or perpendicular to the direction of molding, and this gives rise to a difference in the with-grain and cross-grain properties (Pettifer 1980a).

The next processing step is to coke the pitch binder in the green bar to develop an infusable carbon bond at temperatures between 800° and 1,000°C. Approximately one-third of the pitch binder is driven off through destructive distillation. There is a loss in apparent density, and a development of porosity and permeability to liquids and gases. These characteristics can be improved by impregnation with coal tar or petroleum pitch. By means of this process, a hard and corrosion resistant carbonaceous body, known as baked coke or baked carbon, is produced.

The conversion of the carbon to graphite is performed at temperatures between 2,600° and 3,000°C. Graphitization requires a period of two weeks.

Pyrolytic Graphite: Pyrolytic graphite is produced from the gas phase by the pyrolysis of hydrocarbons. It is known as the chemical vapor deposition (CVD) process. Deposition occurs on some suitable substrate, usually graphite, that is heated at temperatures in excess of 1,500°C, in the presence of a hydrocarbon such as methane, propane, acetylene, or benzene (Mantell 1968). The density of pyrolytic graphite is dependent only on the temperature of deposition. Minimum density is 1.14 g/cm^3 for deposition at 1,700°C, and a maximum density of 2.22 g/cm^3 is obtained at 2,100°C.

The properties of pyrolytic graphite exhibit a great degree of anisotropy. The tensile strength in the ab-direction is five to ten times greater than that of graphite prepared by conventional processes, and the strength in the c-direction is proportionately lower. Similarly, the thermal conductivity of pyrolytic graphite in the ab-direction ranks among the highest of the elementary materials, whereas in the c-direction it is quite low.

Graphite Fibers: Graphite fibers are produced from polyacrylonitrile (PAN) or pitch. The processes consist of (1) fiber formation, (2) spinning, (3) stabilization to thermoset the fiber, (4) carbonization, (5) graphitization, (6) surface treatment, and (7) sizing. Thermal conversion of PAN starts at temperatures of between 200° and 300°C in air. Graphitization occurs in inert gas at temperatures between 1,000° and 3,000° (Wege 1984).

3. Industrial Uses of Natural Graphite

Because of their distinct properties, natural graphite and synthetic graphite serve different industries and overlap only in a limited number of applications. Natural graphite is mainly used in refractories, foundries, batteries, lubricants, brushes, pencils, brake linings, and paint (Mantell 1968; Pettifer 1980b; Taylor 1992).

Refractories and Foundries: Natural graphite refractories are either formed ware such as crucibles, shapes such as carbon-magnesia brick and continuous casting ware, or ramming mixes (Pettifer 1980b). The carbon content should be greater than 85%, and particle size large enough to allow good bonding with other ingredients. The refractory ware consists chiefly of crucibles for melting metals and alloys. The bonding materials used in the manufacture of crucibles are refractory clays, mixtures of tars and pitches, and silicon carbide. Clay-graphite crucibles require 40 to 50% graphite, as compared with 18 to 25% graphite in silicon carbide-graphite crucibles, but the clay-graphite crucibles have a greater resistance to oxidation. The flake graphite is preferred because it has a flexible and interlocking texture that withstands greater thermal stress and has longer use life than crucibles made of other varieties of graphite (Mantell 1968; Taylor 1992).

Carbon-magnesite bricks, which are made from fused magnesia, flake graphite, and a binder, are mainly used to line basic oxygen converters, in the slag-lines, water-cooled side walls of electric arc furnaces, and steel ladles. The graphite content of carbon-magnesite brick ranges from 8 to 30%. Alumina-graphite continuous casting ware has excellent characteristics in thermal shock resistance and corrosion resistance. It is used for holding and moving hot metals, and for stoppers, nozzles, slabs, rods, and other uses. Flake graphite is also required in the manufacture of continuous casting ware.

The largest single use of natural graphite, primarily amorphous graphite, is foundry facings that provide molds and cores with a smooth surface and prevent molten metal from penetrating into or reacting with the foundry sand.

Lubricants: Graphite is widely used as a lubricant in various industries because of its softness, low friction, inertness, and heat resistance. Graphite lubricants include (1) loose powder, (2) powder mixed with liquid lubricants or greases, (3) bonded dry films produced with film-forming volatile liquids, (4) composites formed with synthetic resins and powder metal for bearings, and (5) colloidal graphite (Taylor 1992). The adhesives used for the bonded graphite films are (1) air-drying thermoplastic resins,

(2) thermosetting resins, (3) inorganic and ceramics, and (4) metal matrix bonds. In the preparation of metal graphite composites, molten metal is impregnated into the pores of graphite or carbon-graphite in solid molded form. The usual metals are babbitt, copper, bronze, cadmium, and silver. The metal content of these composites generally varies from 35 to 70%. For metal-rich composites, fabrication is normally done by powder metallurgy techniques (Clauss 1972). Colloidal graphite is dispersed in water, oil, and various solvents of fine graphite particles, in micron to submicron size (Taylor 1992). The colloidal particles have a electronegative charge that assists in stabilizing the particles in suspension and in attaching the particles to metal surfaces of opposite polarity (Clauss 1972).

Batteries: In dry cells, graphite is used to render pyrolusite conductive through intimate admixture, and it is also used in alkaline batteries. The degree of graphitization is a factor, in that graphites with the same carbon content and from the same locality give different results.

Brushes: Natural graphite is used in the manufacture of some electrical brushes. Different end products require different types of graphite, so no single type is appropriate. Natural graphite is particularly useful in the manufacture of brushes for d-c equipment because of its high conductivity, high contact drop, and anisotropy (Pettifer 1980b).

Brake Linings: Crystalline flake, lump, and amorphous graphite are used in brake and clutch linings, mostly for heavy-duty vehicles as opposed to passenger automobiles.

Pencils: The so-called lead pencil is actually a baked ceramic rod of clay-bonded graphite encased in wood. The quality of the pencil is determined by the type of graphite used. The better ones are made from a mixture of crystalline and amorphous graphite. The hardness of the pencil depends upon the clay to graphite ratio.

Paint: Graphite is used as a reinforcing pigment in paint to provide tough, durable, and flexible protective coatings for structural steel and metal surfaces that are exposed to rigorous conditions.

4. Industrial Uses of Synthetic Graphite

Synthetic graphite is mainly used in electrode, metallurgical, aerospace, nuclear, chemical, and mechanical applications (Mantell 1968; Pittifer 1980a).

Electrode Applications: The largest use of synthetic graphite is as electrodes in electric-arc furnaces (Taylor 1985). For iron and steel production, the average consumption is 2 to 5 kg/ton. Typical properties of synthetic graphite used for electrodes are as follows:

	Regular Grade	Premium Grade
Bulk density, g/cm^3	1.60	1.70
Resistivity, $\mu\Omega\cdot$m	7.3	5.5
Flexible strength, kPa		
with-grain	6,900	9,100
cross-grain	5,800	7,000
Elastic modulus, GPa		
with-grain	5.3	7.6
cross-grain	3.5	5.0
Coefficient of thermal expansion, $10^{-6}/°C$		
with-grain	0.60	0.40
cross-grain	1.40	1.10
Thermal conductivity, W/(m·K)		
with-grain	134	168
cross-grain	67	101
Thermal shock parameter, 10^3		
with-grain	290	490
cross-grain	80	101

The thermal shock parameter is a measurement of the combined characteristics of (thermal conductivity × strength) divided by (coefficient of thermal expansion × elastic modulus). The ability of graphite electrodes to withstand thermal shock has been improved significantly in the past decade as a consequence of enhancement in the elastic modulus and coefficient of thermal expansion, by both refined raw materials and advanced manufacturing technology. The current-carrying capacity of electrode columns was greater than 60,000 A in 1990, for the 510 mm diameter electrode that was designed and introduced in 1938 with a capacity of 26,000 A. The premium grade electrode is used where very high performance is required, such as in ultrahigh powered arc furnaces. The approximately 30% lower electrode resistivity is essential for the successful operation (Nafzier and Tress 1976; Taylor 1985).

Metallurgical Applications: Structural graphite shapes are used in a great number of metallurgical applications at high temperatures, because graphite (1) does not melt, (2) does not fuse with common metals and ceramics, (3) increases its strength with temperature, and (4) has a high thermal conductivity. These shapes include various forms of dies, molds, rods, plates, crucibles, trays, boats, and many others for metallurgical processes (Mantell 1968; Criscione et al. 1992). For frequent exposure to high temperatures, graphite must be protected from oxidation. During prolonged periods of contact with metals at high temperatures, graphite may react to form carbides.

Machined graphite shapes are widely used as electric heating elements that may produce temperatures up to approximately 3,000°C in protective atmospheres.

Aerospace Applications: The excellent properties of graphite make it possible to sustain an aerospace environment that is characterized by high operating temperatures and rapid temperature rise. The entrance cap, throat, and exit cone sections in rocket nozzles are usually made or lined with graphite. Nose cones and leading edge components fabricated of graphite are used on both ballistic and glider-type reentry vehicles (Criscione et al. 1992).

Nuclear Applications: The use of graphite in nuclear technology is based upon its moderator and reflector qualities, combined with its structural strength and thermal stability (Mantell 1968).

A higher level of purity of graphite is often required for nuclear applications. Halogen purification is used to remove stable carbides, especially those of boron, that melt at 2,350°C, but do not boil until 3,500°C.

Chemical Applications: Graphite is used in process equipment where corrosion is a problem because of its excellent resistance to attack by acids, alkalies, and many organic and inorganic compounds (Ford and Greenhalgn 1970). For example, self-supported graphite vessels are used for the direct chlorination of metal and alkaline-earth oxides, and water-cooled graphite towers serve as reaction chambers for the burning of phosphorus in air (Criscione et al. 1992).

Since graphite products are manufactured by consolidating particles with a hydrocarbon bond, it is obvious that the final products will be porous. For applications where fluids under pressure must be retained, impervious graphite is made by blocking the pores of the graphite with thermosetting resins. Many types of impervious graphite shell and tube, cascade, and immersion exchangers are in service (Mantell 1968; Criscione et al. 1992). Impervious graphite heat exchangers are used for (1) evaporation of phosphoric acid, (2) cooling electrolytic copper cell liquor, (3) heating pickle liquor for descaling sheet steel, (4) boiling, heating, cooling, and absorbing hydrochloric acid and hydrogen chloride, and (5) heating and cooling exchange of chlorinated hydrocarbons and sulfuric acid.

Mechanical Applications: Graphite possesses lubricity, strength, dimensional stability, thermal stability, and ease of machining, a combination of properties that has led to its use in a wide variety of mechanical applications for supporting, rotating, or sliding loads in contact. The principal applications are in bearings, seals, and vanes.

REFERENCES

Bennett S. C., and D. J. Johnson. 1979. Electron microscope studies of structural heterogeneity in PAN-based carbon fibers. *Carbon* 17: 25–39.

Cameron, E. N. 1960. Graphite. In *Industrial minerals and rocks*. 3rd ed. Ed. by J. L. Gillson, 455–469. New York: Soc. Mining Engineers, AIME.

Cheng, A. and H. Xu. 1998. Some characteristics of the Precambrian mineralization in China. Krystalinikum 19: 21–29.

Clauss, F. J. 1972. *Solid lubricants and self-lubricating solids*. New York: Academic Press., 260p.

Criscione, J. M., and others. 1992. Applications of baked and graphitized carbon. In Vol. 4 of *Kirk-Othmer encyclopedia of chemical technology.* 982–1011. 4th ed. New York.

Crozier, R. D. 1992. *Flotation—theory, reagents and ore testing.* Oxford: Pergamon Press, 343p.

Dickson, E. M. 1981. The industrial minerals of Austria. *Ind. Minerals* (February): 27–29.

Dissanayake, C. B. 1981. Origin of vein graphite in Sri Lanka. *Organic Geochem.* 3:1–7.

———— 1994. Origin of vein graphite in high-grade metamorphic terrains: Role of organic matter and sediments subduction. *Mineral. Deposita* 29:57–67.

Erdosh, G. 1970. Geology of the Bogala mine, Ceylon, and the origin of vein-type graphite. *Mineral. Deposita* 5:375–382.

Fogg, C. T., and E. H. Boyle, Jr. 1987. Flake and high-crystalline graphite availability—market economy countries: A mineral availability appraisal. U.S.B.M., Inform. Circ., 9122, 40p.

Ford, A. R., and E. Greenhalgn. 1970. Industrial applications of carbon and graphite. In *Modern aspect of graphite technology,* Ed. L. C. F. Blackman, 258–292. London: Academic Press.

Graffin, G. D. 1983. Graphite. In *Industrial minerals and rocks.* 5th ed. Ed. S. J. Lefond, 757–773. New York: Soc. Mining Engineers, AIME.

Harben, P. W., and R. L. Bates. 1990. Industrial minerals—geology and world deposits. London: Industrial Minerals Division, Metal Bulletin Plc, 312p.

Harben, P. W., and M. Kuzvart. 1996. Industrial minerals: A global geology. London: Industrial Mineral Inform., Ltd., Metal Bulletin Plc, 462p.

Hollister, V. F. 1980. Origin of graphite in the Duluth Complex. *Econ. Geol.* 75:764–766.

Hupp, T. R. 1992. Properties of manufactured graphite. In Vol. 4 of *Kirk-Othmer encyclopedia of chemical technology.* 974–981. 4th ed. New York: Interscience Publ.

Hutcheon, J. M. 1970. Polycrystalline carbon and graphite. In *Modern aspects of graphite technology,* Ed. L. C. F. Blackman, 1–48. London: Academic Press.

Kanungo, S. C. 1976. Graphite mineralisation in Titlagarh graphite belt, Bolangir and Kalahandi districts, Orissa. *Indian Minerals* 30:1–5.

Kendall, T. 1995. Mexico: Promises of prosperity. *Ind. Minerals* (September):27–55.

Kuzvart, M. 1984. *Industrial minerals and rocks.* Amsterdam: Elsevier, 454p.

Landis, C. A. 1971. Graphitization of dispersed carbonaceous material in metamorphic rocks. *Contr. Mineral. Petrl.* 30:34–45.

Lewis, I. C. 1992. Baked and graphitized carbon. In Vol. 4 of *Kirk-Othmer encyclopedia of chemical technology,* 953–960. 4th ed. New York: Interscience Publ.

Luque, F. J., and M. Rodas. 1999. Constrains on graphite crystallinity in some Spanish fluid-deposited occurrences from different geologic settings. *Mineral. Deposita* 34:215–219.

Mancuso, J. J., and R. E. Seavoy. 1981. Precambrian coal or anthraxolite: A source for graphite in high-grade schists and gneisses. *Econ. Geol.* 76:951–954.

Mantell, C. I. 1968. *Carbon and graphite handbook.* New York: Interscience Publ, 538p.

Marsh, H. 1989. *Introduction to carbon science.* Boston: Butterworths, 256p.

Moore, A. W. 1973. Highly oriented pyrolytic graphite. *Chem. Phys. Carbon* 11:69–187.

Murdoch, T. M. 1967. Mineral resources of the Madagasy Republic. U.S.B.M. Information Circular 8196, 112–121.

Nafzier, R. H., and J. E. Tress. 1976. Electric furnaces melting of by-product metallurgical slags. *Can. Inst. Metall. Bull.* 69:73–78.

Pettifer, L. 1980a. Synthetic graphite—electrodes continue to lead. *Ind. Minerals* (November):19–30.

———— 1980b. Natural graphite—the dawn of tight market. *Ind. Minerals* (September):19–32.

Pryor, E. J. 1965. *Mineral processing.* 3rd ed. Amsterdam: Elsevier, 844p.

Reynolds, W. N. 1968. *Physical properties of graphite.* New York: Elsevier Publ, 197p.

Roberts, M. C., M. Oberlin, and J. Mering. 1973. Lamellar reactions in graphitizable carbon. *Chem. Phys. Carbon* 10:141–211.

Russell, A. 1991. India's industrial minerals: Completing the picture. *Ind. Minerals* (September):45–55.

Su, F. 1992. Marketing and production of Chinese natural flake graphite, Proc. 10th Ind. Minerals Intern. Congr., Ed. G. B. Griffiths, 63–64. San Francisco: Metal Bull.

Taylor, C. R. 1985. *Electric furnace steelmaking.* Warrendale, Pennsylvania: The Iron and Steel Soc., AIME.

Taylor, H. A. 1992. Natural graphite. In Vol. 4 of *Kirk-Othmer encyclopedia of chemical technology*, 4th ed., 1097–1117. New York: Interscience Publ.

Turk, D. L. 1992. Processing of baked and graphitized carbon. In Vol. 4 of *Kirk-Othmer encyclopedia of chemical technology*, 4th ed., 960–974. New York: Interscience Publ.

Wege, E. 1984. Graphite and paracrystalline carbon. In *Process mineralogy of ceramic materials*, Ed. W. Baumgart, A. C. Dunham, and G. C. Amstutz, 125–149. New York: Elsevier.

Weis, P. L., I. Friedman, and J. P. Gleason. 1981. The origin of epigenetic graphite: Evidence from isotopes. *Geochim. Cosmochim. Acta* 45:2325–2332.

Gypsum

Gypsum, $CaSO_4 \cdot 2H_2O$, is an important industrial mineral with an abundant occurrence in the earth's crust. It is produced by more than seventy countries in the world. With an annual production on the average of 100 M tons from 1990 to 1995, the United States, China, Canada, Thailand, and Mexico occupy the top positions.

Based upon its unique property of readily giving, or taking on, water of crystallization, gypsum's largest consumption is in the manufacture of gypsum board for the construction industry. It is also used as a retarder for Portland cement, as a fertilizer in agriculture, and as a filler material.

Gypsum is a by-product of several industrial processes. Three important ones are (1) desulphogypsum, (2) phosphogypsum, and (3) fluorogypsum. Japan is the world's leader in the utilization of desulphogypsum.

MINERALOGICAL PROPERTIES

Gypsum is the dihydrate form of calcium sulfate, $CaSO_4 \cdot 2H_2O$, and in its natural occurrence most gypsum shows very little variation in chemical composition. In its structure, two sheets of SO_4 ions parallel to (010) are bonded together by Ca ions. These double sheets are separated by layers of H_2O molecules. Each calcium is coordinated by eight oxygens, six from SO_4 tetrahedra, and two from H_2O molecules (Cole and Lancucki 1974). The relatively weak hydrogen bonding between the successive pairs of sheets explains the perfect (010) cleavage of gypsum. Ca-O bonds have the shortest distance along the direction [001]. This corresponds approximately with the direction of fibrous cleavage, small expansion coefficient, and high refractive index. Gypsum is monoclinic with a=5.68Å, b=15.18Å, c=6.29Å, and β=113°50′ (Chang et al. 1996).

Gypsum is usually colorless and transparent, or white, but is sometimes gray, yellow, brown, and blue. It has a hardness of 2, specific gravity of 2.31, and a vitreous to pearly luster. Refractive indices are α = 1.519 to 1.521, β = 1.522 to 1.1526, and γ = 1.529 to 1.931, and birefringence is weak.

Most gypsum occurs as massive rock gypsum with associated impurities of clays, iron oxides, and other minerals. When gypsum is fine grained and white, it is used as an ornamental stone known as "alabaster." Granular or compact masses of gypsum are sometimes called "sand" or "seed" gypsum; "kopi" is an earthy variety. Lenticular crystal forms are commonly intergrown as rosette-like aggregates. Gypsum often occurs as euhedral transparent crystals or broad transparent plates, and as fibrous forms elongated along the c-axis. The former is known as "selenite," whereas the latter is called "Satin-spar" (Chang et al. 1996).

The dehydration and hydration of gypsum are important in the mineralogy of gypsum. Four solid phases exist in the system $CaSO_4$–H_2O at room temperature. They are (1) $CaSO_4 \cdot 2H_2O$ (gypsum), (2) $CaSO_4 \cdot \frac{1}{2}H_2O$ (hemihydrate), (3) γ - $CaSO_4$ (soluble anhydrite), and (4) β-$CaSO_4$ (anhydrite). A fifth phase, α-anhydrite, is present in the system only at temperatures above 1,180°C. Of the four phases present at room temperature, hemihydrate and γ -CaSO$_4$ are metastable. Thus, under equilibrium conditions, the reaction, gypsum = anhydrite + water, occurs without intermediate phases. The temperature of this reaction in H$_2$O is in the range of 40° to 42°C, but gypsum may persist metastably above this temperature. The transition temperature is lowered considerably in the presence of NaCl (Chang et al. 1996).

A combined thermogravimetric analysis (TGA), differential thermal analysis (DTA), and x-ray diffraction analysis by Khalil and Gad (1972) shows that gypsum is converted to hemihydrate between 100°C and 200°C when the heating rate is 10°C/min. Complete conversion below 100°C occurs when the heating rate is slower. Heating above 250°C for a few minutes produces γ -CaSO$_4$; β-CaSO$_4$ forms around 360°C; and α-CaSO$_4$ forms at 1,239°C. The decomposition of CaSO$_4$ occurs at 1,300° to 1,350°C.

The hemihydrate is reported to have two morphological forms (Morris 1963; Fowler et al. 1969; Miyazaki and Takagi 1970). The α-form has well-shaped crystals, whereas crystals of the β-form are needle-like. The α-form is less reactive and has a slower rate of strength development than the β-form, but the rehydrated α-form produces a denser, stronger plaster. The conditions of synthesis of α- and β-form occur, respectively, at >45°C from aqueous solutions, and between 45° and 200°C in dry air.

The dehydration of gypsum (synthetic) at temperatures above 110°C and pressures at P$_{(water)}$ = 10^{-5} mm and P$_{(water)}$ = 45 mm is controlled by diffusion (Ball and Norwood 1969). At lower temperatures, nucleation, formation of boundaries, and migration of boundaries are important. The activation energy of the diffusion-controlled mechanism at P$_{(water)}$= 17 mm is 34.6 kcal/mole. The dehydration of hemihydrate (β-form) also has a diffusion-controlled mechanism with an activation energy of 6.4 kcal/mole at P$_{(water)}$ = 10^{-5} mm, and 49 kcal/mole at P$_{(water)}$ = 42 mm (Ball and Norwood 1969, 1970). Somewhat different results (15.6 and 35 kcal/mole) were obtained for the α-form (Ball and Urie 1970).

Although hemihydrite is unstable at room temperature, it nevertheless occurs naturally as bassanite (β-hemihydrate). Yamamoto and Kennedy (1969) examined the effect of pressure upon the reactions:

$$CaSO_4 \cdot 2H_2O = CaSO_4 \cdot 1/2H_2O + 1\ 1/2H_2O$$

and

$$CaSO_4 \cdot 1/2H_2O = CaSO_4 + 1/2H_2O$$

Results illustrate that the two curves representing the above reactions intersect at 85°C and 2 kbar, indicating that below this pressure hemihydrate does not exist as a stable phase, and suggesting that natural bassanite is metastable under the conditions reported for its occurrence.

The hydration of hemihydrate under both isothermal and adiabatic conditions was found to be greatly self-accelerated and to follow similar courses, either when accelerated by sodium chloride (1,000 mg), or inhibited by borax (0.4000 m), or unmodified (Ridge and Surkevicius 1962).

Heats of hydration of hemihydrate and anhydrite, and dehydration of gypsum have been determined to be 25°C (Kelley et al. 1941) and are listed below:

Hydration of hemihydrate and anhydrite:

$$\beta\text{-}CaSO_4 \cdot 1/2H_2O + 3/2H_2O \rightarrow CaSO_4 \cdot 2H_2O \qquad 19{,}300 \text{ J/mol}$$

$$\alpha\text{-CaSO}_4 \cdot 1/2H_2O + 3/2H_2O \rightarrow CaSO_4 \cdot 2H_2O \qquad\qquad 17200 \text{ J/mol}$$
$$\text{CaSo}_4 \text{ (anhydrite)} + 2H_2O \rightarrow CaSo_4 \cdot 2H_2O \qquad\qquad 16{,}900 \text{ J/mol}$$

Dehydration of gypsum:

$$CaSO_4 \cdot 2H_2O \rightarrow \beta\text{-CaSO}_4 \cdot 1/2H_2O + 2/3H_2O \qquad 86{,}700 \text{ J/mol}$$
$$CaSO_4 \cdot 2H_2O \rightarrow \alpha\text{-CaSO}_4 \cdot 1/2H_2O + 2/3H_2O \qquad 84{,}600 \text{ J/mol}$$
$$CaSO_4 \cdot 2H_2O \rightarrow CaSO_4 \text{ (anhydrite)} \qquad\qquad 108{,}600 \text{ J/mol}$$

The solubility of gypsum in H_2O has been a subject of many studies (Bock 1961; Marshall and Slusher 1966; Christoffersen and Christoffersen 1976), and the results obtained range from 1.47 to $1.52 \times 10^{-2}M$. The solubilities of gypsum and anhydrite, as a function of temperature, were determined by Innorta et al. (1980). The results can be illustrated by the equations below:

$$S \text{ (wt\% } CaSO_4) = 8.69 \times 10^{-6}T^2 - 1.73 \times 10^{-3}T + 0.2624$$

for gypsum, and

$$S \text{ (wt\% } CaSO_4) = 1.14 \times 10^{-5}T^2 - 3.667 \times 10^{-3}T + 0.3517$$

for anhydrite.

The dehydration of gypsum, $CaSO_4 \cdot 2H_2O \rightarrow CaSO_4 \cdot 1/2H_2O + 2/3H_2O$, produces hemihydrate, the major component of Plaster of Paris, and the hydration of hemihydrate, $CaSO_4 \cdot 1/2H_2O + 2/3H_2O \rightarrow CaSO_4 \cdot 2H_2O$, reforms the dihydrate that is a solid mass, set hard through the formation of interlocking crystals of gypsum. These processes form the basis of gypsum technology.

GEOLOGICAL OCCURRENCE AND DISTRIBUTION OF DEPOSITS

Gypsum mainly occurs as sedimentary deposits associated with limestones, dolomites, shales, marls, or clays, and as evaporite deposits, from the early Paleozoic to Recent eras (Sonnenfeld 1984; Warren 1989). The principal modes of formation of gypsum are direct deposition by the evaporation of brine, and by hydration of anhydrite. Others are the oxidation of sulphides, and sulphatization of calcium-bearing rocks.

As the solubility data discussed in the previous section show, gypsum should be precipitated first from saturated solution in H_2O when evaporation takes place below 42°C; above this temperature anhydrite is the stable form. The presence of sodium chloride should lower the transition temperature, so that in concentrated NaCl solution anhydrite can form at lower temperatures. However, in modern evaporite deposition, as well as in ancient evaporites, gypsum is always the primary mineral, and anhydrite is a secondary mineral produced by the dehydration of gypsum (Bundy 1956; Morris and Dickey 1957; Warren 1989). From their study on the mechanism of gypsification, Conley and Bundy (1958) illustrate the gypsum-anhydrite relations by the following characteristics:

1. In dilute solution, the hydration of anhydrite takes place through the medium of transient surface complexes. Concentrated solutions may precipitate double salts.
2. Double salts and/or gypsum are stable phases below the gypsum-anhydrite transition temperature in the dehydration by concentrated salt solutions; above the transition temperature double salts may replace anhydrite as the stable phase.

3. Gypsum may remain a metastable phase indefinitely in its saturated solution below the hemihydrate transition temperature (98°C).

Conley and Bundy (1958) concluded that the precipitation of anhydrite from sea water is unlikely.

Gypsum deposits are widely distributed in North America. Principal producing districts are shown in Fig. 31. The deposits, in order of economic importance, are (1) Mississippian gypsum deposits in southwestern Canada, Michigan, Indiana, and Virginia; (2) Permian deposits in Texas, Oklahoma, Kansas, Colorado, Wyoming, Nevada, Arizona, and Mexico; (3) Tertiary deposits in California, Jamaica, Arizona, and Mexico; and (4) Silurian deposits in New York, Ontario, and Ohio (Appleyard 1983a). Most gypsum and anhydrite deposits contain clastic sediments, commonly clay minerals and fine sand, as well as chemical sediments such as limestones and dolomites, as illustrated in the evaporites in Texas, Oklahoma, and New Mexico. Soluble evaporite minerals are also frequently found in gypsum deposits (Braitsch 1971). Individual gypsum deposits show great variation in the quantity and type of associated soluble chloride and sulfate salts. In general, chloride salts are predominant in gypsum deposits of eastern North America, whereas sulfate salts exceed chloride in western North American deposits (Appleyard 1983a).

The Permian salts and related strata in the Texas Panhandle and western Oklahoma, and the upper Permian Ochoa series of the Delaware Basin, west Texas, and southeastern New Mexico, are good examples to illustrate an evaporation sequence. As shown in Fig. 32 (Harben and Bates 1990), the Permian salts consist of four major evaporite units: (1) Wellington, (2) Cimarron, (3) Beckham, and (4) Cloud Chief in ascending order, which make up an 8,000 km sequence of gypsum, anhydrite, and rock salt, interbedded with shale, silt, sandstone, and dolomite. Gypsum of Beckham and Cloud Chief evaporites are of commercial importance with 95 to 99% and 90 to 98% gypsum, respectively.

In the Ochoa series, the Castle, the basal formation of the series, is composed of anhydrite, calcite, salt (NaCl), limestone, minor amounts of other evaporites, and minute quantities of fine clastics. Above the Castle, and with an uncomformity with the Castle, is another thick section of evaporites of the Salado formation. In the Salado evaporite, salt is more abundant than anhydrite, and both NaCl and KCl are present. Deposition of the Rustler, the third formation of the Ochoa series, was preceded by a period of uplift and erosion in the area. The Rustler formation consists of gypsum interbedded with redbeds, gray sandstone, and gypsiferous dolomite. A conformable contact exists between the Rustler and the overlying Dewey Lake formation, that is a redbed, in sharp contrast to the evaporites of the lower Ochoa. The gypsum that is present with the red sand and shale occurs as cement, secondary crystals, and veins (Adams 1944).

Outside of North America, Tertiary deposits occur in Australia, France, and Italy, Permian and Jurassic deposits in England, and Triassic deposits in France, Germany, and England. Many gypsum deposits are also Quaternary in age and were formed by the direct precipitation and conversion of anhydrite (Appleyard 1983a).

INDUSTRIAL PROCESSES AND USES

1. Ore Processing and Beneficiation

For many years, the main preparatory process for gypsum ore was the primary crushing done at most quarries before transportation to the plant, followed by little or no

Figure 31   Locations of gypsum-producing districts in North America, after Appleyard (1983a).

beneficiation (Harvard 1960). Crude ore, consisting of any appreciable amount of impurities that would have to be separated, is not used. The impurities that are commonly found in gypsum and ores may be classified as three types, based on their effect on the manufacturing process.

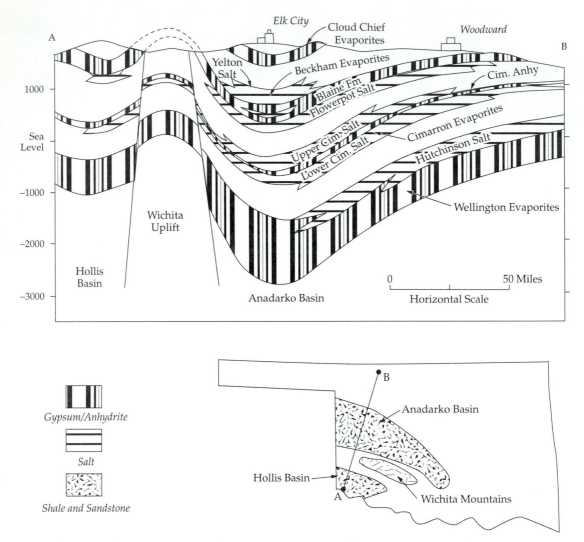

Figure 32 Cross section of Permian salts and associated strata in the Texas Panhandle and western Oklahoma.
© P.W. Harben & R. L. Bates.

1. Insoluble or relatively insoluble minerals such as dolomite, limestone, anhydrite, and silica may constitute up to 10 to 15%. Their main effect is upon the strength of rehydrated stucco and the weight of the finished products.
2. Soluble salts including halite, sylvite, and others, whose main effect is on the calcining temperature. Their concentration is generally limited to 0.02%.
3. Hydrous minerals consisting of sulfate salts and clays, whose main effects are on the absorption of moisture of the finished products and the bonding characteristics of the gypsum stucco core of wall-board to its paper covering. The limit of sulfate salts is 0.03% and of clays to 2.0% (Appleyard 1983a).

In recent years, the demand for higher quality gypsum has led to ore processing and beneficiation of gypsum as a common practice. This process usually consists of four steps.

1. Primary crushing is generally made by gyratory, jaw, or impact crusher. The size of the mine-run ore dictates the type of crusher.
2. Secondary crushing is done in either hammer mills or cone-type crusher. It is usually conducted with vibrating screens in the circuit to reduce the fines.
3. Drying is necessary to add to the circuit if the moisture of the mine run ore is high. This is generally done in a rotary dryer, and drying temperature must not be allowed to exceed the temperature of dissociation of gypsum.
4. Fine grinding is achieved by air-swept roller mills with integral air separator or high-energy impact mills with air classifier.

Gypsum to be used for Portland cement is recovered at the end of secondary crushing, whereas that for fillers and fertilizers is produced at the end of fine grinding (Appleyard 1983b).

Upgrading gypsum ore by further beneficiation is now done, but only in a limited number of processing plants. This process includes (1) dry screening and air classification, (2) washing and wet screening, and (3) heavy-media separation and froth flotation.

2. By-Product Gypsum

By-product gypsum or chemical gypsum, or both, synthetic gypsum, phosphogypsum, desulfogypsum, fluorogypsum, titanogypsum, and other names used are synonymous. It has been suggested that the definition of by-product is the chemical end-product of industrial processes, consisting primarily of calcium sulphate dihydrate. Many of those processes do not produce a commercially useful by-product because of contaminants, particle size, or volume produced (Petersen et al. 1992). There are, however, two chemical processes that do produce sufficient volume to have potential industrial value.

Desulphogypsum (flue-gas gypsum, FDG-gypsum) is produced by scrubbing sulfur dioxide out of flue gases of fossil fuels such as anthracite, bituminous coal, lignite, and oil, in large combustion plants, especially power stations. Among numerous processes that have been developed, the scrubbing process with limestone consists of three steps.

1. The formation of $CaSO_3 \cdot 1/2H_2O$ at pH 7 to 8 by the reaction of $SO_2 + CaCO_3 + 1/2H_2O \rightarrow CO_2 + CaSO_3 \cdot 1/2H_2O$, which is insoluble.
2. The conversion of calcium sulfite to calcium bisulfite at pH 5 by the reaction of $2CaSO_3 \cdot 1/2H_2O + 2SO_2 + H_2O \rightarrow 2Ca(HSO_3)_2$, which is soluble.
3. The formation of desulphogypsum at pH 5 by the reaction of $Ca(HSO_3)_2 + O_2 + 2H_2O \rightarrow CaSO_4 \cdot 2H_2O + H_2SO_4$.

The sulfuric acid produced reacts with limestone to produce additional gypsum. The gypsum produced is separated from the aqueous suspension in hydrocyclones and vacuum drum filters. The product is a moist, fine, fairly pure powder with a free water content of less than 10%. About 5.4 tons of gypsum are produced per ton of sulfur in the fuel (Wirsching 1985).

Phosphogypsum is produced when phosphate ore is acidulated to extract phosphoric acid according to a reaction that may be illustrated in the general form:

$$Ca_3(PO_4)_3F + 5H_2SO_4 + 10H_2O \rightarrow 5(CaSO_4 \cdot 2H_2O) + 3H_3PO_4 + HF$$

(Wirsching 1985). By far the largest quantity of by-product gypsum is produced by this process. For every ton of phosphorus pentoxide manufactured, about 5 tons of gypsum are produced. However, because of a high percentage of impurities and water content, the cost for beneficiation and purification makes phosphogypsum uneconomical.

By-product gypsum is also produced from fluorspar acidulation and is used to (1) produce hydrofluoric acid, (2) treat sulfate containing waste waters, (3) manufacture TiO_2 pigment, and (4) process zinc ores (Petersen et al. 1992; Pressler 1984).

3. Calcination

Calcination is a process used to produce hemihydrate and anhydrite from gypsum by dehydration (Appleyard 1983a). The products are known as (1) α-plaster or molding plaster of α-hemihydrate, (2) β-plaster, stucco, or plaster of Paris of β-hemihydrate, and (3) overburnt plaster or deadburned gypsum of anhydrite. Gypsum begins to lose its water of crystallization when it reaches a temperature of 42° to 49°C. When the temperature reaches 116° to 120°C, the water released provides a vapor pressure that equals the atmospheric pressure. Hemihydrate is produced after the boiling ceases, and drying at 149° to 166°C produces what is known as "first settle" stucco containing a water content of 5 to 6% as against 20.9% in the original gypsum. The "second settle" stucco is made by further heating the first settle stucco at 177°C to boil off the remaining water, and the product produced is in the form of a soluble anhydrite.

β-plasters are produced by dry calcination in a temperature range between 120° and 180°C from gypsum in either directly fired rotary kilns or indirectly heated kettles. In the rotary process (Fig. 33a) (Wirsching 1985), granular gypsum ore is fed continuously from the silo by a weigh-belt feeder to the kiln where hot gases are produced in a brick-lined combustion chamber. The heat is transferred directly from the hot gases to gypsum to produce a complete calcination. The start-up and shutdown bin collects the gypsum that is not dehydrated. Fine grinding and air classifying are used to achieve a product of uniform quality. The continuous kettle process is illustrated in Fig. 33b (Wirsching 1985). Predried and finely ground gypsum is fed continuously from the kettle-fed bin to a kettle that is externally heated. As the process occurs, the water vapor set free keeps the gypsum bed fluidized. The hemihydrate produced settles at the bottom of the kettle and is continuously discharged through a pipe connected to the side of the kettle.

α-plasters are produced by a wet calcining process at 120° to 135°C, in a sealed retort under a steam pressure of about 117 kPa for 5 to 7 hours. The hot moist product is quickly dried at 105°C and pulverized. α-plasters are also produced at atmospheric pressure in acids or aqueous salt solutions between 80° and 150°C.

Overburnt plasters are produced in a dry calcining process in beehive kilns, rotary calciners, or flash calciners, at temperatures up to 500°C.

4. Industrial Uses

The industrial uses of gypsum are generally divided into four categories: (1) construction (70%), (2) industrial (23%), (3) agricultural (6%), and (4) miscellaneous (1%) (Appleyard 1983b; Wirsching 1985).

Figure 33 Production of ß-hemihydrate plaster by (a) the rotary kiln process and (b) the continuous kettle process, after Wirsching (1985). In (a): (A) silo for gypsum, (B) weight-belt feeder, (C) rotary kiln with combustion chamber, (D) primary mill, (E) start-up and shut-down bin, (F) fine mill, (G) air classifier, (H) electrostatic precipitator, and (I) ß-hemihydrate plaster. In (b): (A) silo for gypsum rock, (B) drying-grinding unit, (C) cyclone, (D) kettle feed bin, (E) continuous kettle, (F) cooling bin, and (H) ß-hemihydrate plaster.

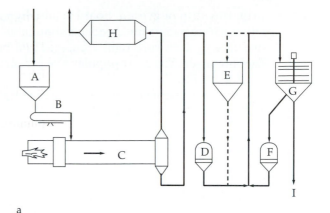

a

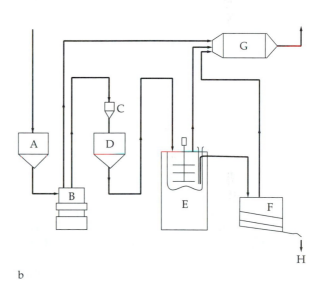

b

Construction Uses: Gypsum products are used in the construction industry as covering and finishing materials. Their distinct features are (1) they are noncombustible and among the most effective fireproofing materials known; (2) they can be used in various forms such as board, tile, block, and others, and their properties may be varied by using different calcining techniques in the production of the plaster; and (3) they are economic and compete favorably in their applications with other covering and finishing materials.

The largest end use for gypsum in the United States is in the manufacture of wallboard or gypsum board. These boards are light, porous, dry, and nonbrittle, with β-hemihydrate as their main component. Gypsum boards are manufactured by feeding β-hemihydrate plaster into a continuous mixer from controlled feeding devices, and mixing it continuously with water and adhesives, in order to form a homogeneous and rapidly setting slurry. The slurry is dropped onto a moving strip of cardboard, covered with a second strip of cardboard, and passed over a molding platform to be shaped into a completely encased strip. This strip of gypsum board is initially soft, but hardens within minutes. It is cut into separate panels and dried in a continuous tunnel dryer. The finished product consists of a gypsum core tightly encased and bonded to cardboard. A wide variety of sizes are made. The most popular size gypsum board is 1.2 m × 2.4 m × 12.7 mm (4ft × 8ft × 1/2 in).

Other construction materials made of gypsum are partition panels, ceiling tiles, and fiber boards (Wirsching 1985).

Agricultural Uses: Gypsum or anhydrite is used in agriculture mainly as a soil conditioner for the following purposes: (1) to improve soil physical conditions by breaking up compacted clays, and by increasing porosity for better drainage; (2) to supply neutral, soluble calcium, and sulfate sulfur; (3) to neutralize alkali soil and correct high sodium irrigation water; (4) to improve utilization of nitrogen; and (5) to stimulate soil micro-organisms (Appleyard 1983b).

Other Industrial Uses: For other uses, gypsum can be divided into three usage categories: (1) calcined, (2) anhydrous, and (3) uncalcined, each of which utilizes distinct properties of the mineral (Appleyard 1983a).

Calcined gypsum is used extensively by industry for molds used in the manufacture of sanitary ware, pottery, and metal casting. Special uses include art plasters, orthopedic plasters, and dental plasters. Molding plaster is made from high purity gypsum, at least 95%, and is characterized by its water content, color, strength, setting time, and expansion qualities. It may be made from either β- or α-hemihydrate, or a combination of the two.

Anhydrous gypsum, or second settle stucco, is calcium sulphate without water of crystallization, and is often referred to as "soluble anhydrite." Its outstanding property is its extreme affinity for any moisture, which makes it a very efficient drying agent. In ambient moisture-laden air, it readily hydrates to hemihydrate.

Uncalcined gypsum, sometimes referred to as "raw gypsum," is added to portland cement clinker to stop the rapid reaction of calcium aluminate. Because of its ability to accelerate strength development, uncalcined gypsum is properly termed a set regulator, rather than a retarder. Terra alba is finely ground gypsum of high purity and very white color. It is manufactured with the end use in mind as either an inert filler or a diluent. When used in food products or in pharmaceuticals, terra alba must meet Food Chemicals specifications.

REFERENCES

Adams, J. E. 1944. Upper Permian Ochoa Series of Delaware Basin, west Texas and southeastern New Mexico. *Amer. Assoc. Petroleum Geol.* 28:1596–1625.

Appleyard, F. C. 1983a. Gypsum and anhydrite. In *Industrial minerals and rocks.* 5th ed. Ed. S. J. Lefond, 775–792. New York: Soc. Mining Engineers, AIME.

—— 1983b. Construction Materials. In *Industrial minerals and rocks.* 5th ed. Ed. S. J. Lefond, 183–198. New York: Soc. Mining Engineers, AIME.

Ball, M. C., and I. S. Norwood. 1969. Studies in the system calcium sulphate-water, Part I. Kinetics of dehydration of calcium sulphate dihydrate. *J. Chem. Soc.* A:1633–1637.

—— 1970. Studies in the system calcium sulphate-water, Part III. Kinetics of dehydration of α-calcium sulphate hemihydrate. *J. Chem. Soc.* A:1476–1479.

Ball, M. C., and R. G. Urie. 1970. Studies in the system calcium sulphate-water, Part II. The kinetics of dehydration of β-CaSO$_4$ · 1/2H$_2$O. *J. Chem. Soc.* A:528–530.

Bock, E. 1961. On the solubility of anhydrous calcium sulphate and of gypsum in concentrated solutions of sodium chloride at 25°C, 30°C, 40°C and 50°C. *Can. J. Chem.* 39:1746–1751.

Braitsch, O. 1971. *Salt Deposits, Their Origin and Composition.* New York: Springer-Verlag, 338p.

Bundy, W. M. 1956. Petrology of gypsum-anhydrite deposits in southwestern Indiana. *J. Sedi. Petr.* 26:240–252.

Chang, L. L. Y., R. A. Howie, and J. Zussman. 1996. *Rock-forming minerals.* Vol. 5B, *Non-silicates: Sulphates, carbonates, phosphates, halides.* 2nd ed. Essex, England: Longman, 383p.

Christoffersen, J., and R. Christoffersen. 1976. The kinetics of dissolution of calcium sulphate dihydrate in water. *J. Cryst. Growth* 35:79–88.

Cole, W. F., and C. J. Lancucki. 1974. Refinement of the crystal structure of gypsum CaSO$_4$ · 2H$_2$O. *Acta Cryst.* B30:921–929.

Conley, R. F., and W. M. Bundy. 1958. Mechanism of gypsification. *Geochim. Cosmochim. Acta* 15:57–72.

Fowler, A., H. G. Howell, and K. K. Schiller. 1969. The dihydrate-hemihydrate transformation in gypsum. *J. Appl. Chem.* 18:366–372.

Harben, P. W., and R. L. 1990. Bates. *Industrial minerals, geology and world deposits.* London: Industrial Minerals Division, Metal Bulletin Plc, 312p.

Harvard, J. F. 1960. Gypsum. In *Industrial minerals and rocks.* 3rd ed. J. L. Gillson, Ed., 471–486. New York: AIME.

Innorta, G., E. Rabbi, and L. Tomadin. 1980. The gypsum-anhydrite equilibrium by solubility measurements. *Geochim. Cosmochim. Acta* 44:1931–1936.

Kelley, K. K., J. C. Southard, and C. T. Anderson. 1943. *Thermodynamic properties of gypsum and its dehydration products.* U.S. Bureau Mines, Tech. Rept., 625, 73p.

Khalil, A. A. A., and G. M. Gad. 1972. Thermochemical behavior of gypsum. *Build. Intern.* 5:145.

Marshall, W. L., and R. Slusher. 1966. Thermodynamics of calcium sulfate dihydrate in aqueous sodium chloride solutions, 0°-110°C. *J. Phys. Chem.* 170:4015–4027.

Miyazaki, H., and N. Takagi. 1970. Difference between α and β forms of calcium sulphate hemihydrate. *J. Chem. Soc. Japan* 73:1766–1769.

Morris, R. C., and P. A. Dickey. 1957. Modern evaporite deposition in Peru. *Bull. Amer. Assoc. Petroleum Geol.* 41:2467–2474.

Morris, R. J. 1963. X-ray diffraction identification of the alpha- and beta-forms of calcium sulphate hemihydrate. *Nature* 198:1298–1299.

Petersen, D. J., N. W. Kaleta, and L. W. Kingston. 1992. Calcium Sulfate. In Vol. 4 of *Kirk-Othmer Encyclopedia of Chemical Technology,* 4th ed., 812–826. New York: Interscience Publ.

Pressler, J. W. 1984. Byproduct gypsum. In *The Chemistry and Technology of Gypsum,* Ed. R. A. Kuntze, 105–115. ASTM Special Technical Publication 861.

Ridge, M. J., and H. Surkevicius. 1962. Hydration of calcium sulphate hemihydrate, I. Kinetics of the reaction. *J. Appl. Chem.* 12:246–252.

Sonnenfeld, P. 1984. *Brine and Evaporites.* New York: Academic Press, 630p.

Warren, J. K. 1989. *Evaporite Sedimentology.* Englewood, N.J.: Prentice Hall, 285p.

Wirsching, F. 1985. Calcium sulfate. Vol. A4 In *Ullmann's Encyclopedia of Industrial Chemistry,* 555–585. Weinheim: VCH Publ.

Yamamoto, H., and G. C. Kennedy. 1969. Stability relations in the system CaSO$_4$−H$_2$O at high temperatures and pressures. *Amer. J. Sci.* 267A:550–557.

Industrial Diamond

Industrial diamonds are diamonds that do not have gem quality. They are used in industries as an abrasive because of their distinct property of hardness. Among the world's major producers, Australia, Zaire, the former USSR, Botswana, and South Africa are ranked as the top five producers, accounting for 95% of the world's volume (58 M carats in 1995).

Industrial diamonds are also produced synthetically. It is estimated that they serve 80% of the market, amounting to more than 250 M carat/year (Harben and Nötstaller 1991). The United States, South Africa, and Japan are leaders in the production of synthetic diamonds.

MINERALOGICAL PROPERTIES

Diamond is one of the two naturally occurring forms of carbon. The other one is graphite. Crystallographically, it is cubic, with a cell dimension ranging from 3.56683Å to 3.56725Å at 25°C (Skinner 1957; Kaiser and Bond 1959). The crystal structure of diamond is a simple one. Each of the carbon atoms forms four equivalent covalent bonds, the angles between them being 109°28′. There are eight carbon atoms in the unit cell, and the calculated density is 3.51525 g/cm^3, which correlates well with measured density (3.51524 g/cm^3).

Diamond, although it is commonly pure, contains other elements in addition to carbon. Of the various elemental impurities detected in diamond, some twenty-five are present in concentrations exceeding 10^{-4} atom %. They are H, B, N, O, Na, Mg, Al, P, Ca, Sc, Ti, Cr, Mn, Fe, Co, Cu, Sr, Ba, Zr, La, Lu, Pt, Au, Ag, and Pb, of which only nitrogen and boron are known with certainty to be incorporated in the diamond structure. Their presence may exert a considerable effect on the physical properties and structure of the crystals (Orlov 1973). Harris and Gurney (1979) have identified twenty-five mineral inclusions in diamond. The relatively abundant ones are biotite, pyroxenes, olivine, rutile, orange and purple garnets, spinels, and pyrrhotite.

Diamond does not react with common acids, but oxidizes readily at high temperatures in air and in an oxygen-containing atmosphere. Strong oxidizers, such as chlorates, perchlorates, hydroxids, and nitrates, cause characteristic etched figures on the surface of diamond at temperatures as low as 375°C. The rate of diamond oxidation by liquid salts containing potassium is twice as high as that by sodium-containing salts. Diamond reacts with metals to form either carbides (W, Ti, Ta, Zr) or solid solutions (Fe, Co, Ni, Mn, and Cr) at moderate to high temperatures. The fact that diamond dissolves in iron and iron alloys above 550°C restricts the use of diamond tools for most machining operations on ferrous metals (Bakon and Szymanski 1993).

Fig. 34 is a phase diagram of carbon showing the temperature-pressure conditions of diamond stability. Under normal conditions diamond is not a stable form of carbon, and transformation to graphite should occur in terms of thermodynamics (Bundy 1980). In the course of diamond-graphite transformation, the carbon atoms are displaced and the interatomic bonds change their character and orientation. Pure natural diamond and

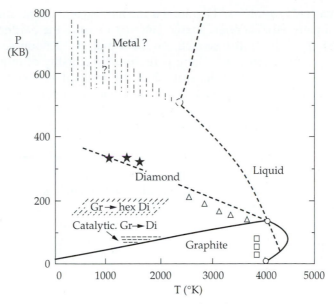

Figure 34   Pressure-temperature phase diagram of carbon modified, after Gardinier (1988). In the diagram, "Gr" and "Di" refer to gragraphite and diamond, respectively, and "hex" to hexagonal. Solid stars, open triangles, and open squares represent experimental data points of various investigators.

synthetic diamond of good quality begin to graphitize at about 1,500°C. The presence of aggregates of impurities and large inclusions adversely affects the stability of diamond. The temperature of diamond-graphite transformation decreases in proportion to the number of inclusions. Crystals containing inclusions of metallic elements are especially susceptible to graphitization. Elements which react with carbon or those which readily dissolve in it have distinct catalytic effects upon the transformation (Evans 1979).

Diamond crystals are classified into four types based principally on optical properties, but also on nitrogen content (Collins 1997):

1. Type Ia diamonds are characterized by the presence of appreciable amounts of nitrogen of the order of 0.1%, which appears to be distributed into small aggregates, the concentration of which can be determined from the intensity of the infrared absorption at 7.8 μm (Kaiser and Bond 1959). Most natural diamonds are of this type.
2. Type Ib diamonds contain nitrogen on isolated substitutional lattice sites at concentrations of up to 500 ppm. Because nitrogen produces paramagnetic resonance, its concentration can be determined by electron spin resonance (Berman et al. 1975). Almost all synthetic diamonds are of this type, but this is very rare in natural diamonds.
3. Type IIa diamonds contain the least amount of nitrogen (10 to 100 ppm) and have the highest conductivities (Slack 1973). This type is very rare in natural diamonds.
4. Type IIb diamonds have such a low concentration of nitrogen that some of the boron acceptors are not compensated and the crystal is a p-type semiconductor. This type is extremely rare in natural diamonds.

Synthetic semiconducting diamond can be manufactured by incorporating boron and excluding nitrogen. Some diamonds have been shown to consist of more than one type.

Mechanical Properties: Diamond is the hardest and the most abrasion resistant of all known materials. There are many scales to measure the hardness of crystalline solids. The Mohs scale or the scratch hardness and the Knoop scale or the indentation hardness are mostly used. There is reasonable equality of intervals between the first nine integers on the Mohs scale, but the interval between 9 (corundum) and 10 (diamond) represents a much larger difference in indentation hardness than a single unit on the Mohs scale would suggest (Brookes 1979). The ten hardness units on the Mohs scale are 1 (talc), 2 (gypsum), 3 (calcite), 4 (fluorite), 5 (apatite), 6 (orthoclase), 7 (quartz), 8 (topaz), 9 (corundum), and 10 (diamond). For indentation hardness measured on (111) surface, in the [110] direction, and with a 500 gm load, the values are (in kg/mm^2) 9,000, 4,500, 2,600, 2,250, 2,190, 2,190, and 2,000 for diamond, cubic boron nitride (CBN), silicon carbide, boron carbide, tungsten carbide, titanium carbide, and aluminum oxide, respectively.

The carbon atoms in the diamond structure show different arrangements in different planes, and these are responsible for the anisotropy in mechanical properties of diamond. The highest density of atoms is in the octahedral plane {111}, followed by the {100} plane. The varying diamond cleavage behavior in different directions is caused by the directional differentiation of atomic density. Diamond cleaves preferentially in the {111} plane. This fact must be allowed for in the manufacture and use of many types of diamond tools (Field 1979; Gielisse 1997).

It is generally recognized that diamond is anisotropic in its wear resistance and shows so-called soft and hard directions (Bakon and Szymanski 1993). In the case of {100} face, the "soft" and "hard" directions are bivectorial. Of the {110} face, the optimal "soft" cutting direction is parallel to the crystallographic axes, and the direction perpendicular to it is "hard." Of the {111} face, the "soft" direction coincides with the direction of the height of the equilateral triangle formed by the octahedron wall, is vectorial, and runs towards the octahedron vertex. Wilks and Wilks (1972) illustrated the dependence of hardness of diamond crystal on the crystallographic directions in a series of directions. With increasing hardness, the directions are (1) dodecahedron parallel to crystal axis, (2) cube parallel to crystal axis, (3) octahedron towards dodecahedron, (4) octahedron towards cube, (5) dodecahedron 90° to axis, and (6) cube 45° to axis.

Elastic moduli of diamond have been measured in many studies, and results from Prince and Wooster (1953), McSkimin and Bond (1957), and Grimsditch and Ramdas (1975) are C_{11}=11.0, 10.76, and 10.764, C_{12}=3.3, 1.25, and 1.252, and C_{44}=4.4, 5.76, and 5.774 (all $\times 10^{12}$ dynes/cm^2), respectively. Bulk modulus calculated from data of Prince and Wooster (1953) and of McSkimin and Bond (1957) are, respectively, 5.9×10^{12} and 4.42×10^{12} dynes/cm^2.

Both tensile strength and shear strength of diamond are difficult to measure because of the nature of the crystal, which consists of defects, inclusions, impurities, and other features. Therefore, results show a wide range of variations (Field 1979; Gielisse 1997).

Thermal Properties: Diamond is the best heat conductor of any known material, at normal temperatures. At 20°C (room temperature), the thermal conductivity of type IIa diamond is 25 watts/cm · K, more than six times the value of copper (4 watts/ cm · K). It increases with decrease in temperature and reaches a maximum of 125 watts/cm · K at 80 to 100 K. The thermal conductivity of type I diamond is 9 watts/cm · K at room temperature (Klages 1997).

Linear thermal expansion of diamond at 20°C is $(0.8 \text{ to } 1.2) \times 10^{-6} \text{ K}^{-1}$, and specific heat is 516 to 550 J/Kg·K (Skinner 1957; Berman 1979). The thermal expansion coefficient calculated from lattice measurements by x-ray diffraction is from $1.3 \times 10^{-6} \text{K}^{-1}$ at 293 K to $7 \times 10^{-6} \text{K}^{-1}$ at 1673 K. The thermal expansion at 293 K is $(1.0 \text{ to } 1.2) 10^{-6} \text{K}^{-1}$ for both synthetic and natural diamonds (Bakon and Szymanski 1993).

Electric Properties: At normal temperatures, diamond is essentially dielectric, although it may also be considered as a semiconductor, with a very wide forbidden gap of $\Delta E = 5.7$ eV. The calculated resistivity for a perfect diamond is $10^{70} \, \Omega \cdot$ cm (Champion 1963; Collins 1997). Impurities considerably reduce this figure. In most natural diamond crystals, which are classed as type I, the resistivity attains values ranging from 10^{14} to $10^{16} \, \Omega \cdot$ cm, while type II diamonds have resistivities ranging from 1 to $10^8 \, \Omega \cdot$ cm. Type IIb diamonds have been shown by Custers (1955), and Bakon and Szymanski (1993), to be p-type semiconductors, and a characteristic of natural diamonds with semiconducting properties is their blue color. The conductivity of synthetic diamonds shows anisotropy (Collins and Lightowlers 1979). Boron-doped diamonds have conductivity 10^6 to 10^8 times higher in the {111} direction than in the {100} direction. This is attributed to the different spatial distribution of boron atoms in the structure.

The dielectric constant (ϵ) of diamond is 5.7 ± 0.05 at 300 K, and its temperature dependence can be described by the equation: $\epsilon = 5.70111 - 5.35167 \times 10^{-5}T + 1.6603 \times 10^7 T^2$ (Bakon and Szymanski 1993). Irradiated diamonds exhibit photoconductivity (Fontanella et al. 1977).

Optical Properties: Pure diamond is colorless. The presence of impurities and crystal defects affects absorption in the visible region and produces various shades of color. Diamond crystals with yellow, blue, and green colors are relatively common.

The refractive indices of diamond at 5,461Å (green) and 6,563Å (red) are 2.4237 and 2.4099, respectively. They decrease when the crystal is subjected to pressure and increase when it is heated (Orlov 1973).

Pure diamond crystals are transparent in the ultraviolet up to 2,200 to 2,350Å. Impurities strongly affect their absorption in the ultraviolet. Diamond crystals with a relatively high nitrogen content (type I) are opaque to ultraviolet light at wavelengths below 3,200 to 3,000Å. The infrared absorption spectrum of a pure diamond shows absorption only in the 3 to 6μ region (2,030, 2,450, and 3,200 cm^{-1}) due to thermal vibration of the carbon atoms in the structure. The absorption spectrum is affected by the presence of nitrogen, and by the nature of its occurrence in the diamond structure. The absorption coefficient at 1,280 cm^{-1} (7.8μ) is directly proportional to the nitrogen content (Orlov 1973).

Many diamond crystals become luminescent when exposed to ultraviolet light, cathode rays, x-rays, or gamma rays, or when heated or placed in an electric field (Collins 1997).

GEOLOGICAL OCCURRENCE AND DISTRIBUTION OF DEPOSITS

The primary source of diamonds, kimberlite, is characterized by its confinement to stable continental cratons, notably in eastern and southern Africa, the Siberian platform, the Grenville province of Quebec, the Scandinavian Shield, Western Australia, and an inlier in Arkansas. Kimberlites occur as pipes, dikes, veins, and sills. They constitute an ultrabasic rock with an alkali trend, possessing porphyritic texture. Studies of kimberlites have suggested that they were primarily associated with alkali-ultrabasic magma originating in the upper mantle and welling up through deep faults to the upper zones of the Earth's crust. The distribution of kimberlite pipes on platforms shows that they are usually located near the marginal region of bending of the Earth's crust, where conditions are most favorable for the formation of deep faults, and for the upwelling of magma from the upper mantle (Orlov 1973; Harben and Bates 1990; Meyer and Seal 1997).

In the diamond-producing region in South Africa, the filling of kimberlite pipes is weathered to oxidized "yellow ground," passing downward into "blue ground." At greater depths, the unweathered kimberlite is a holocrystalline rock consisting of magnesian olivine, phlogopite, chromian diopside, enstatite, pink magnesian garnet, and ilmenite, all set in a matrix of serpentinized olivine and calcite. The distribution of kimberlites in South Africa is illustrated by Harben and Bates (1990) as shown in Fig. 35. Only a few kimberlites are diamond-bearing, and still fewer may be considered of commercial significance.

The most widespread sources of diamonds are placer deposits. Their source is diamond-bearing kimberlite found in volcanic pipes that have been attacked by weathering processes. Diamonds and other weather-resistant minerals, released from the disintegrated kimberlite, were carried away by river systems and deposited to form alluvial, fluviatile, and beach deposits.

There are three major types of natural industrial diamonds: (1) Bort includes off-color, flawed, or broken fragments of diamonds unsuitable for gems, (2) Carbonado, or black diamond, is a very hard and extremely tough aggregate of small diamond crystals, and (3) Ballas is a very hard, tough, globular mass of diamond crystals radiating from a common center. Carbonadoes come only from Bahia, Brazil; ballas mainly from Brazil and South Africa; and Bort comes from all diamond-producing districts (Hight 1983).

INDUSTRIAL PROCESSES AND USES

1. Processing Techniques

The diamonds present in the yellow ground at the upper, oxidized zone of the pipe are recovered easily by inspection, washing, screening, and hand-sorting. In many mines, the diamond-bearing rocks from dumps and heaps are reworked several times to ensure a complete recovery. The blue ground, mined at greater depths, is a hard and unweathered rock. For years the general practice was to expose the ore on the ground at the surface and allow it to weather slowly with frequent stirring. After months of exposure the

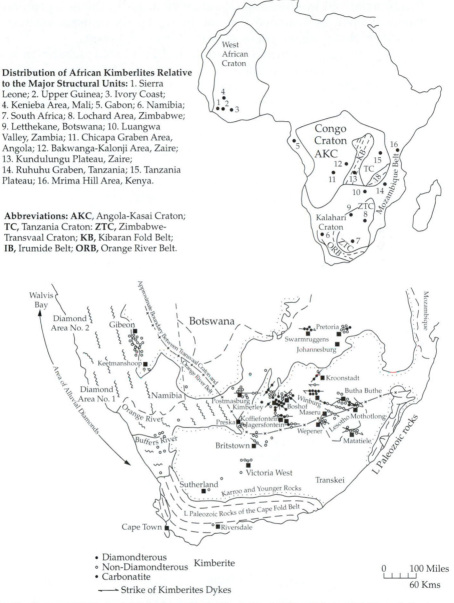

Distribution of African Kimberlites Relative to the Major Structural Units: 1. Sierra Leone; 2. Upper Guinea; 3. Ivory Coast; 4. Kenieba Area, Mali; 5. Gabon; 6. Namibia; 7. South Africa; 8. Lochard Area, Zimbabwe; 9. Letthekane, Botswana; 10. Luangwa Valley, Zambia; 11. Chicapa Graben Area, Angola; 12. Bakwanga-Kalonji Area, Zaire; 13. Kundulungu Plateau, Zaire; 14. Ruhuhu Graben, Tanzania; 15. Tanzania Plateau; 16. Mrima Hill Area, Kenya.

Abbreviations: AKC, Angola-Kasai Craton; **TC**, Tanzania Craton: **ZTC**, Zimbabwe-Transvaal Craton; **KB**, Kibaran Fold Belt; **IB**, Irumide Belt; **ORB**, Orange River Belt.

Figure 35 Cratons of Africa and distribution of kimberlites in South Africa. © *Industrial Minerals Geology and World Deposits,* P. W. Harben & R. L. Bates, 1990, published by Industrial Minerals.

ore would disintegrate down to the extent that the diamonds could be liberated without much effort (Reckling et al. 1983; Meyer and Seal 1997).

As mining has progressed deeper, and operations have grown bigger, mechanical crushing and concentration have been adopted by most operations. The ore is crushed in either jaw or gyrator crushers or by corrugated spring rolls, generally in two or more stages, separated by screening (Reckling et al. 1983). Primary concentration of diamonds is done by Diamond Pan and by jig. Heavy medium separation and heavy medium cyclones are used in most modern treatment plants.

As with all heavy medium separators, effective feed preparation is a prerequisite for efficient separation (Chaston and Napier-Munn 1974). Excessive fines must be removed, otherwise they will accumulate and cause difficulties with the density of the medium and will increase the viscosity of the pulp. This can result in lower efficiency of separation in the cyclone. Feed and medium are mixed in the constant head box and then gravitate to the cyclone. Sinks and float screens are used for the concentrates and rejects, respectively. Ground ferrosilicon is the preferred medium in all diamond plants. Heavy medium separation is capable of reducing the bulk of the ore to 0.1% of the feed, with up to 98% recovery of diamonds, giving a concentrate in the range of 100 to 1,000 carats per tonne (Chaston and Napier-Munn, 1974).

Secondary concentration is generally done by the process of grease tabling. The rough concentrate from the heavy medium separation is passed over included vibrating tables, which are covered in a thick layer of petroleum grease. The diamonds are attracted to and become embedded in the grease due to their hydrophobicity, whereas the gangue particles are washed off the table. The diamonds are collected within the first inches of the table (Wills 1981; Meyer and Seal 1997).

Diamonds are recovered from placer deposits by washing, dry winnowing, screening, panning, jigging, tabling, and handpicking, generally in some combination. For consolidated gravel, some crushing is required. Further treatments include the grease belt process, ball-milling, electrostatic separation, and x-ray luminescence separation.

2. Synthesis and Production of Industrial Diamond

In the conventional diamond synthesis, a base metal and graphite are mixed in a cell and subjected to high pressures and temperatures within the stability field of diamond, as shown in Fig. 34 (Gardinier 1988). The five-stage process from carbon to diamond abrasive is

1. graphitization, in which carbon material changes to graphite,
2. metal melting and graphite dissolution, in which graphite changes to a metal-carbon solution,
3. nucleation and crystal growth, in which diamond precipitates from the solution and forms a eutectic mixture with the metal,
4. dissolution, in which graphite and metals are dissolved in acids, and
5. grading and sorting, in which diamond abrasive grains are manufactured.

In the metal-graphite systems, the temperatures and pressures typically employed for diamond production are $1,400°$ to $2,100°C$ and 4.5 to 7.0 GPa (Wedlake 1979). For Ni+graphite, Co+graphite, and Fe+graphite, they are $1,460°C$ and 5.5 GPa, $1,450°C$ and 5.0 GPa, and $1,475°C$ and 5.7 GPa, respectively. Among high-pressure equipment (Johnson et al. 1979), a belt apparatus and multianvil press are commonly used for the synthesis and production of diamond.

A number of nonconventional techniques are available for diamond synthesis under metastable conditions, the most commonly used being chemical vapor deposition (CVD) (Wedlake 1979; Bakon and Szymanski 1993). The process involved may proceed in accordance with the reaction: $(a+b+c)C_W \rightarrow aC_D + bC_G + cC_X$, where C_W is carbon from a gas phase; C_D, C_G, and C_X are carbon in the form of diamond, graphite, or some other allotrope; and a, b, c are proportionality coefficients. The conditions under which the process is conducted should be such as to terminate crystallization at the point when the maximum quantity of diamond is formed. When methane is used, the reaction is as follows: $CH_4 \rightarrow C_D$ or $C_G + 2H_2$. The substrate for diamond growth is generally fine diamond powder, and substrate temperatures in the range of 1,200 to 1,400 K are preferred. Growth rates are slow, and after a period of growth graphite may nucleate.

Polycrystalline diamond (PCD) may be produced by sintering fine diamond powders to form a polycrystalline mass. An important characteristic of synthetic polycrystalline diamond, which differentiates it from natural and other synthetic diamond crystals, is the isotropy of its physical and chemical properties.

3. Synthesis of Cubic Boron Nitride (CBN)

Diamond and CBN are the only two abrasives considered as superabrasives, and their applications are widely overlapping. CBN is isostructural with diamond and is synthesized by the same techniques. It has excellent thermal stability and can be heated in air up to 1,400°C before it changes to the hexagonal boron nitride, a graphite analog. However, CBN reacts with water at temperatures in excess of 800°C (DeVries 1972).

4. Industrial Use

The diamond is by far the most important industrial abrasive (Reckling et al. 1983) and, along with cubic boron nitride, is considered a superabrasive (Metzger 1986; Subramanian and Shanbhag 1997). Its industrial application falls into six distinct areas: (1) truing, (2) drilling, (3) dressing, (4) sawing, (5) grinding, and (6) lapping and polishing. Each application requires a certain particle size. The largest range of sizes is used in truing (typically 0.5 carat whole stone), followed, as the size scale decreases, by drilling, sawing, grinding, and finally lapping and polishing applications. In most of these applications the diamond grains are bonded into a solid matrix, except for lapping and polishing, where the grains are suspended in a supporting fluid such as oil (Subramanian and Shanbhag 1997).

The larger diamonds of higher quality are used as dressers to true and dress abrasive tools. To obtain maximum efficiency from a tool diamond, it is important to orient it so that the hard direction bears the workload.

In the drilling application that accounts for 10% of all industrial diamond used, blast-hole drill bits are usually made with small crystals of regular shape. Mineral exploration bits require diamonds ranging in size from 15 to 100 stones per carat, masonry and concrete bits make use of smaller diamonds ranging from 50 to 200 stones per carat, and oil well bits require larger sizes, usually 12 stones per carat. The relationship between the size of the stones and the weight in carats of the diamonds in a drill bit is given in the following examples (Meinert 1964):

Diamonds/Carat	Stones/Bit	Carats/Bit
10	100	10.00
18	144	8.00
27	162	6.00
35	190	5.50
45	225	5.00

Diamond drill can be subdivided into two types (1) surface set bits, where each diamond is set and positioned by hand, and used mainly for oil and exploration drilling; and (2) impregnated bits, where the diamond grits are mixed with a metal powder and then sintered, and are used mainly in civil engineering (Busch 1979).

Approximately 75% of all industrial diamonds are consumed in the form of grit and powder. Diamond grit is used primarily in saw blades and grinding wheels. The grit ranges in size from very coarse (8/10, +8 −10) to fine (325/400, −325 +400), and the grains are sized by sieving on woven wire mesh or electroformed screens. Screens are designated according to the number of meshes per inch, and screened grits are defined in terms of the screens between which they fall. A −325 +400 grit is one that has passed through a 325-mesh screen and will not pass through a 400-mesh screen (Ruckling et al. 1983). Diamond saw tools are generally subdivided into two categories, namely (1) segmented saw blades, where the diamond impregnated segments are brazed onto either the periphery of a steel blank or a straight steel blank, and which are used mainly in the cutting of concretes, stones, and refractories; and (2) continuous rim blades, which are used mainly in the cutting of ceramics, glass, and gemstones (Bakon and Szymanski 1993).

Diamond grinding wheels are used for hard, brittle materials, such as tungsten carbides, glasses, and ceramics, that are difficult to machine with conventional abrasives such as aluminum oxide or silicon carbide (Metzger 1986). The grinding wheel consists of a thin diamond impregnated rim or surface, with grits of selected size and type dispersed in a bonding matrix. There are four bond types: (1) resin, (2) metal, (3) vitrified, and (4) electroplated. The first three derive their names from the binder material. Common to all three is the fact that, during the manufacturing process, the bonding agent becomes fluid in sintering and wets to some degree all the other components present in the wheel rim or surface. The bonding agent is some form of synthetic polymer or resin in the case of resin bonds, a metal or an alloy in the case of metal bonds, and a glass or a ceramic in the case of vitrified bonds (Bakon and Szymanski 1993). The fourth bonding agent, in the case of electroplated wheel, is made by covering the blank in nickel electrolyte with a thin layer of diamond grit (Daniel 1967). The primary categories of diamond wheels and grinding modes are (1) peripheral grinding with peripheral wheels, and (2) face grinding with cup wheels. Both can be used for surface grinding and cylindrical grinding. The former is specifically for internal grinding, profile grinding, slot grinding, and cut-off grinding, whereas the latter is for tool and cutter grinding, and free-hand grinding (Bakon and Szymanski 1993; Hay and Galimberti 1997).

Very fine diamond powder, known as graded micron powder, is used mainly in the manufacture of lapping and polishing compounds, and as fine finishing tools. Fine diamond powders are graded by a combination of sedimentation and centrifuging. Table 9 lists the standard grade number for these powders.

Diamond has several nonabrasive industrial uses (Caveney 1979; Bakon and Szymanski 1993) that can be described by five categories:

1. Diamond transmits light from the absorption edge at 0.22 μm in the ultraviolet to about 3.7 μm in the infrared, and then for wavelengths greater than 6 μm. These exceptional optical transmission properties make diamond a good window material for infrared experiments.
2. Diamond is a high strength material with a low scattering factor for X-ray; hence a diamond cell is used for high pressure x-ray studies.

Table 9   Standard grade numbers for diamond powder, after Ruckling et al. (1983).

Grade Nos.	μ Range	Approximate Mesh Equivalent
1/2	0-1	50,000
1	0-2	14,000
3	2-4	8,000
6	4-8	3,000
9	8-12	1,800
15	12-22	1,200
30	22-36	800
45	36-54	500/600
60	54-80	400/500

3. Diamond has a low coefficient of friction with biological tissue. A knife made of an oriented diamond can cut tissue specimens into very thin sections without producing knife tracks and other artifacts.
4. Diamonds fluoresce under the action of ionizing irradiation and, thus, can be used as a radiation detector.
5. Diamond is used to make wire-drawing dies, especially for fine wire of diameters below 1 and 2 mm. A tolerance close to 1 to 2μm can be achieved.

REFERENCES

Bakon, A., and A. Szymanski. 1993. *Practical Uses of Diamond.* New York: Ellis Horwood, 237p.

Berman, R. 1979. Thermal properties. In *The properties of diamond,* Ed. J. E. Field, 3–22. London: Academic Press.

Berman, R., P. R. W. Hudson, and M. Martinez. 1975. Nitrogen in diamond, evidence from thermal conductivity. *Solid State Phys.* 8:L430–434.

Brookes, A. 1979. Indentation hardness. In *The properties of diamond,* Ed. J. E. Field, 383–402. London: Academic Press.

Bundy, F. P. 1980. The P-T phase reaction diagram for elemental carbon. *J. Geophys. Res.* 85B:6930–6936.

Busch, D. M. 1979. Industrial uses of diamond. In *The properties of diamond,* Ed. J. E. Field, 595–618. London: Academic Press.

Caveney, R. J. 1979. Non-abrasive industrial uses of diamond. In *The properties of diamond,* Ed. J. E. Field, 619–639. London: Academic Press.

Champion, F. C. 1963. *Electronic Properties of Diamond.* London: Butterworths, 132p.

Chaston, T. R. M., and T. J. Napier-Munn. 1974. Design and operation of dense-medium cyclone plants for the recovery of diamond. *Joou. South Afr. Inst. Min. Metall.* 74:120–133.

Collins, A. T. 1997. The electronic and optical properties of diamond. In *The physics of diamond, Proc. Intern. School of Physics, Course CXXXV, 288–354,* Ed. A. Paoletti and A. Tucciarone, Italian Physical Soc. Amsterdam: IOS Press.

Collins, A. T., and E. C. Lightowlers. 1979. Electrical properties. In *The properties of diamond,* Ed. J. E. Field, 79–106. London: Academic Press.

Custers, J. F. H. 1955. Semiconductivity of a type IIb diamond. *Nature* 176:173–174.

Daniel, P. 1967. Making diamond tools—The ingredients for a successful formula. *Industrial Diamond Review* 324:466–470.

DeVries, R. C. 1972. Cubic boron nitride. *Handbook of properties,* Rept. 72CRD178, Technical Information Series, General Electric Research & Development 17p.

Evans, T. 1979. Changes produced by high temperature treatment of diamond. In *The properties of diamond,* Ed. J. E. Field, 403–424. London: Academic Press.

Field, J. E. 1979. Strength and fracture properties of diamond. In *The properties of diamond,* Ed. J. E. Field, 281–324. London: Academic Press.

Fontanella, R., L. Johnston, J. H. Colwell, and C. Andeen. 1977. Temperature and pressure variation of the refractive index of diamond. *Appl. Optics* 16:2949–2951.

Gardinier, C. F. 1988. Physical properties of superabrasives. *Ceramic Bull.* 67:1006–1009.

Gielisse, P. J. 1997. Mechanical properties of diamond, diamond films, diamond-like carbon, and like diamond materials. In *Handbook of industrial diamonds and diamond films,* Ed. M. A. Prelas, G. Popvici, and L. K. Bigelow, 49–88. New York: Marcel Dekker Inc.

Grimsditch, M. H., and A. K. Ramdas. 1975. Brillouin scattering in diamond. *Phys. Rev.* B11:3139–3148.

Harben, P. W., and R. L. Bates. 1990. *Industrial Minerals, Geology and World Deposits.* London: Industrial Minerals Division, Metal Bulletin Plc 312p.

Harben, P. W., and R. Nötstaller. 1991. Diamonds—scintillating performance in growth and prices. *Industrial Minerals* (March):35–47.

Harris, J. W., and J. J. Gurney. 1979. Inclusions in diamond. In *The properties of diamond,* Ed. J. E. Field, 556–591. London: Academic Press.

Hay, R. A., and J. M. Galimberti. 1997. Cutting and wear applications. In *Handbook of industrial of diamonds and diamond films,* Ed. M. A. Prelas, G. Popvici, and L. K. Bigelow, 1135–1147. New York: Marcel Dekker.

Hight, R. P. 1983. Abrasives. In *Industrial Minerals and Rocks.* 5th ed. Edited by S. J. Lefond, 11–39. New York: Soc. Mining Engineers, AIME.

Johnson, W., A. G. Mamalis, M. C. de Malherbe, and H. K. Tönshoff. 1979. Ultra-high pressure equipment and techniques mainly for synthesizing diamond and cubic boron nitride. *Fortschritt-Berichte, Zeit.* 2:1–74.

Kaiser, W., and W. L. Bond. 1959. Nitrogen, a major impurity in common type I diamond. *Phys. Rev.* 115:857–863.

Klages, C. P. 1997. Thermal conductivity of diamond and diamond thin films. In *The physics of diamond, Proc. Intern. School of Physics, Course CXXXV,* 355–372. Ed. A. Paolette and A. Tucciarone, Italian Physical Soc. Amsterdam: IOS Press.

McSkimin, H. J., and W. L. Bond. 1957. Elastic moduli of diamond. *Phys. Rev.* 105:116–121.

Meinert, H. J. 1964. Diamond bits. In *The industrial diamond, a symposium,* 63–64. New York: Industrial Diamond Association of America.

Metzger, J. L. 1986. *Superabrasive grinding.* London: Butterworths 226p.

Meyer, H. O. A., and M. Seal. 1997. Natural diamond. In *Handbook of industrial diamonds and diamond films,* Ed. M. A. Prelas, G. Popvici, and L. K. Bigelow, 481–526. New York: Marcel Dekker.

Orlov, Yu. L. 1973. *The mineralogy of the diamond.* New York: Wiley-Interscience Publ 235p.

Prince, E., and W. A. Wooster. 1953. Determination of elastic constants of crystals from diffuse relexion of x-ray. III. Diamond. *Acta Cryst.* 6:450–454.

Reckling, K., R. B. Hoy, and S. J. Lefond. 1983. Diamonds. In *Industrial minerals and rocks.* 5th ed. Ed. S. J. Lefond, 653–676. New York: Soc. Mining Engineers, AIME.

Skinner, B. J. 1957. The thermal expansions of thoria, periclase and diamond. *Amer. Mineral.* 42:39–55.

Slack, G. A. 1973. Nonmetallic crystals with high thermal conductivity. *J. Phys. Chem. Solids* 34:321–335.

Subramanian, K., and V. R. Shanbhag. 1997. Abrasive applications of diamond. In *Handbook of industrial diamonds and diamond films,* Eds. M. A. Prelas, G. Popvici, and L. K.Bigelow, 1023–1134. New York: Marcel Dekker.

Wedlake, R. J. 1979. Technology of diamond growth. In *The properties of diamonds,* Ed. J. E. Field, 501–535. London: Academic Press.

Wilks, E. M., and J. Wilks. 1972. Variation in diamond hardness dependent on its crystallographic direction. *Sin. Almazy* 3:52–53 (Chem. Abs., 76:146021).

Wills, B. A. 1981. *Mineral process technology.* Oxford: Pergamon Press, 629p.

Iron Oxide Pigments

Iron oxide pigments, both natural and synthetic, represent the most important group of mineral pigments. Natural iron oxides can be classified into four categories and are produced in a relatively small number of countries. They are (1) ochres that are yellow with some red variants (France, Spain, South Africa, and the United States), (2) umbers that are generally brown (Cyprus), (3) siennas that are yellow with an accentuated orange hue (Cyprus and Italy), and (4) red iron oxides (Spain and India). Synthetic iron oxide pigments can be red, yellow, orange, or black. Major producers are Germany, the United States, Japan, Mexico, and Brazil (Toon 1986).

Micaceous iron oxide pigment consists of a flake-shaped iron oxide and has a metallic gray color. It is mainly mined in Austria and is used as a corrosion protection coating.

MINERALOGICAL PROPERTIES

The iron oxide minerals which form the major constituents of pigments are hematite, magnetite, geothite, maghemite, and lepidocrocite. Limonite, often used in the description of pigments, is a hydrated iron oxide of poor crystallinity and consists mainly of goethite or lepidocrocite. The general physical properties of iron oxide and hydroxide minerals are listed in Table 10.

Hematite, α-Fe_2O_3, and magnetite, Fe_3O_4 or $FeO \cdot Fe_2O_3$, are two stable iron oxides. Both structures consist of layers of oxygen ions, but the layer stacking sequence in Fe_2O_3 is a hexagonal close packing arrangement, whereas that in Fe_3O_4 is a cubic close packing arrangement. Consequently, hematite is hexagonal, with a = 5.03Å and c = 13.772Å, and magnetite is cubic, with a = 8.396Å. Magnetite can take an excess amount of Fe_2O_3 into its structure and grades toward the end member maghemite, a cubic (γ) form of Fe_2O_3, forming a complete series of solid solution of the spinel-type structure. Maghemite is metastable and inverts to hematite on heating in a temperature range from 200° to 700°C. Magnetite is stable in air above 1,388°C, below which hematite is the stable iron oxide (Deer et al. 1992).

Goethite (α-FeOOH) and lepidocrocite (γ-FeOOH) are two stable iron hydroxides. Goethite is isostructural with diaspore (α-AlOOH, Fig. 6, p. 28) with a hexagonal close packing arrangement of oxygens. Ferric ions occupy octahedrally coordinated positions between layers in such a way as to form strips of octahedra, the direction of which defines the c-axis. The strips have the width of two octahedra and produce an orthorhombic unit cell with a = 4.59Å, b = 9.94Å, and c= 3.02Å. The structure of lepidococite is similar to that of boehmite (Fig. 6, p. 28). It contains double sheets of octahedra of FeO_6 linked together in chains parallel to the a-axis. The orthorhombic unit cell has a = 3.87Å, b = 12.53Å, and c = 3.06Å. On dehydration, goethite gives hematite, whereas lepidocrocite gives maghemite.

In natural occurrences, Mn is a common substitute for Fe in both oxides and hydroxides, and has an effect upon the pigmentary character of the mineral. At temperatures above 1,000°C, synthetic Fe_2O_3 (hematite) can take approximately 10 wt% Al_2O_3

Table 10   Physical properties of iron oxide and hydroxide minerals.

Mineral	H	Sp. Gr.	R.I.	Color
Hematite	5–6	5.25	2.87–3.22	Light red to dark violet
Goethite	5–5 1/2	4.3	2.26–2.51	Green yellow to brown yellow
Lepidocrocite	5	4.09	1.94–2.51	Yellow to orange
Magnetite	7 1/2–8	5.20	2.42	Black
Maghemite	7 1/2–8	4.88	2.52–2.74	Brown to orange brown

into its solid solution and forms a complete series with $FeTiO_3$ (ilmenite). The spinel-type iron oxides, magnetite and maghemite, have extensive substitutions with Mg, Cr^{3+}, Mn^{2+}, Zn, and Al. All iron oxides and hydroxides are insoluble in water and common organic solvents, but HCl is able to dissolve them. Magnetite and maghemite are ferro-magnetic (Deer et al. 1992).

Among the four major iron oxide pigments, ochres contain limonite predominantly and are characterized by a yellow color. Dehydration converts limonite to hematite and produces a color change in the product from yellow to red. Ochres have a relatively low Fe_2O_3 content, typically in the 20% (French ochre) to 50% (African ochre) range. Yellow ochre produced in the United States and red ochre produced in the United Kingdom have Fe_2O_3 levels in the 55 to 60% and 40 to 50% range, respectively (Benbow 1989). Umbers have a yellow and yellow-brown color due to the admixture of yellow iron oxide hydrate and black manganese oxides, generally in the range of 10 to 15%. The Fe_2O_3 content in umbers varies from deposit to deposit, but typically is in the 40 to 65% range (Cyprus umber). Calcined at temperatures of 250° to 425°C, the umber changes its color from yellow to dark brown as a combined effect of iron and manganese oxides, and the product is known as burnt umber.

Siennas consist mainly of goethite with a Fe_2O_3 content at about 60% and a small amount (1 to 2%) of manganese oxide. Calcination changes the color of sienna from yellow and yellow-orange to bright red-brown. Natural reds such as Persian red and Spanish red are characterized by their high hematite content, of the order of 90% Fe_2O_3, and fine particle size, with 99.5% passing a 325 mesh sieve. The Spanish reds have a brown undertone, whereas the Persian reds are distinguished by a pure hue (Buxbaum 1993). The micaceous iron oxide pigment consists mainly of hematite, and a typical composition has 85 to 90% Fe_2O_3, 4 to 6% SiO_2, 2 to 3% Al_2O_3, 1 to 2% MgO, and 0.5 to 1.0% CaO. The individual hematite flakes vary in size in the 10 to 100 μm range with a thickness of only 5 μm (Benbow 1989; Buxbaum 1993).

GEOLOGICAL OCCURRENCE AND DISTRIBUTION OF DEPOSITS

Hematite occurs in sediments as weathering products and diagenetic alteration of iron-bearing minerals, and is also produced by direct precipitation in the colloidal form. Magnetite occurs typically as an accessory mineral in many igneous and metamorphic rocks. Maghemite is a rare mineral that usually results from the supergene alteration of magnetite. Some of the major banded iron formations contain both hematite and magnetite in association with quartz, carbonates, and iron-bearing silicates. Both goethite and lepidocrocite are weathering products, typically under oxidizing conditions, of iron-bearing minerals such as siderite, magnetite, and pyrite.

The island of Cyprus in the Mediterranean Sea is the world's major producer of the iron oxide pigments umber and ochre (Griffiths 1984). As described by Robertson (1975), the umbers are pale, chestnut colored or almost black, fine grained mudstones with no calcium carbonate. Goethite and poorly crystalline manganese oxides dominate the mineralogy of the umbers. They exist in association with the upper Pillow Lavas, the uppermost igneous unit of the Troodos Massif (Fig. 36). Umbers occur in hollows, underlain by thin lava breccias and by deeper zones of intensely veined and fragmented pillow lavas. Occasional thicker umbers are located in elongated fault-controlled depressions, within and above thick lava breccias, that are restricted to the south margin of the Troodos Massif. Good examples of umber hollows occur at Kinousa, Margi, Pyrga, and Zaxharia (Fig. 36). Interlava umbers occur near Kambia, Kataliondas, and Zaxharia. Lymbia and Troulli pits represent the occurrence of fault-controlled umbers. On the basis of field evidence, umbers are interpreted as rapid chemical precipitates in seawater from submarine thermal springs active during the latest stages of volcanism on the Tethyan ocean ridge. Upwards, the umbers and the Pillow Lavas are succeeded by chert, radiolarian mudstone, chalks, and reef limestone. Downwards, there is the sulfide ore zone of mainly pyrite and ochres, followed by premineralization lava. The two lava flows, the sulfides, and the ochre form the Troodos volcanic series of Cyprus. At Mathiati, brown-yellow ochres occur in direct contact with the sulfide ores and contain blocks of porous pyrite. A bed of ochre up to 3.5 m thick lies between sulfide ore below, and unmineralized lavas above, at Mousoulos. The massive ochre deposits include brown and orange-yellow bands alternating with clayey tuffs and massive sulfides, alternating bands of red hematite ochre, sulfide-rich ochre, sulfide-poor ochre, and yellow ochre of goethite with massive sulfide (Lilljequist 1969; Robertson 1975).

The Catresville mining district, Georgia (the United States), contains residual deposits of barite, manganese oxides, brown iron, ochre, umber, and primary bedded deposits of micaceous iron oxide (Kesler 1950). The district is underlain by a series of metasedimentary rocks divided, in ascending order, into the Weisner, Shady, and Rome formations of Lower Cambrian age, and the Conasauga formation of the Middle and Upper Cambrian age. The Weisner formation consists principally of finely micaceous metashale, whereas the Shady formation is made of variably siliceous specular hematite interbedded with dolomite. The lithology of the Rome formation is not uniform, varying from crystalline dolomite and limestone in the western portion to metashale in the southeastern portion. The Conasauga formation consists mainly of metashale. The deposits of ochre and umber occur in the weathered, surfacial portion of the Shady formation and are thin-bedded with a conformable relation to the underlying Weisner rocks. They consist of finely intermixed limonite and clay, with smaller and variable proportions of fine-grained quartz, and moscovite. Finely disseminated manganese oxide is present in variable quantity, and its proportion determines whether the iron oxide is classed as ochre or as umber. Ochre has less than 2% MnO_2 showing a bright orange-yellow color, and umber has a MnO_2 content, as much as 5%, with a chocolate-brownish color.

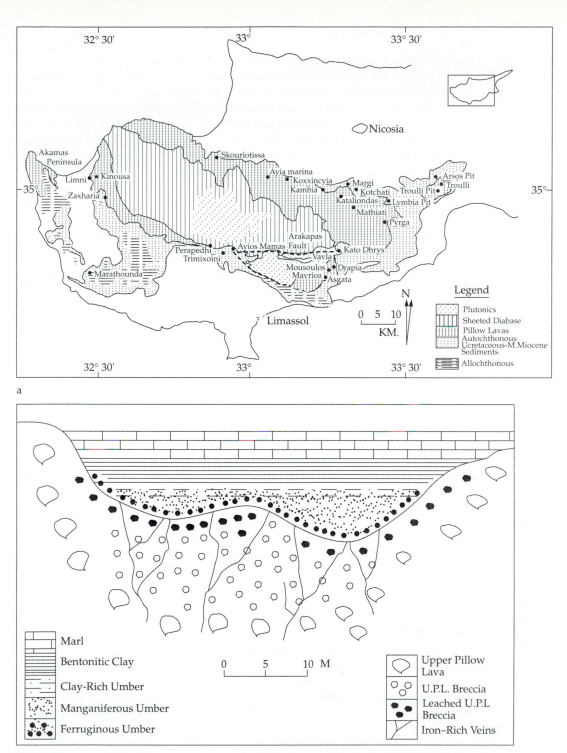

Figure 36 (a) Geological map of Troodos Massif, Cyprus, and (b) Cross section of a typical deposit of umber in a small hollow, after Robertson (1975).

The well-known Spanish red ochre occurs around Málaga and Jaén in the southwest part of Spain. Both red and yellow ochres are mined at Tierga in northeastern Spain. In addition, micaceous iron oxide occurs at La Aparecida in Sierra Nevada with 89 to 96% Fe_2O_3, and low sulfur content. Red ochre also occurs at Bellary, Karnataka, India, but its quality is inferior to the Spanish red ochre. In South Africa, yellow and red ochres are produced in the Riversdale district, Cape Province, and red ochre from the Dundee, Newcastle, and Utrecht districts in Natal. Italy is the leading producer of sienna, a product that takes its name from the town of Siena in Tuscany. However, siennas appear to have suffered more than other natural iron oxide pigments from the impact of synthetics, and production at Siena has fallen accordingly (Griffiths 1985; Toon 1986).

Austria's micaceous iron oxide is mainly mined in the Waldenstein region, a part of the mountain chain known as the Koralpe, which is underlain by high-graded metamorphic rocks. The micaceous iron oxide occurs at the contact between volcanics and the crystalline limestone (Harben and Bates 1990).

INDUSTRIAL PROCESSES AND USES

1. Beneficiation and Milling

The processing of natural iron oxide pigments depends on their composition. They are either washed, slurried, dried, ground, or dried immediately, and then ground in ball mills, or more often in disintegrators or impact mills. Siennas and umbers are calcined, and the hue of the products is determined by the calcination period, temperature, and raw composition. Magnetic beneficiation is used in some cases to remove undesirable siliceous minerals, which can produce a concentrate with purity in excess of 99.5% Fe_2O_3 or Fe_3O_4 from lower grade concentrates. The use of fluid energy mills, micronizers, and others has improved the quality of natural iron oxide pigments and made them competitive with synthetic pigments in many applications.

Summarized in Fig. 37 (Hancock 1983) are alternate processing procedures for natural iron oxide pigment ores that require grinding with and without calcination.

2. Synthetic Iron Oxide Pigments

Synthetic iron oxide pigments have become increasingly important because of their pure hue, consistent properties, and tinting strength. They are produced with red, yellow, orange, and black colors corresponding to that of hematite, goethite, lepidocrocite, and magnetite. Brown pigments are usually made from red, yellow, or black. Three major methods for the manufacture of synthetic iron oxide pigments are (1) solid state reaction of iron compounds, for red, black, and brown pigments; (2) precipitation and hydrolysis of solutions of iron salts for yellow, red, orange, and black pigments; and (3) reduction of organic compounds for black, yellow, and red pigments (Büchner et al. 1989; Buxbaum 1993).

Solid State Reactions: Red iron oxide pigments are obtained by calcination of all decomposable iron compounds in an oxidizing atmosphere to produce a wide range of different red colors, depending upon the starting material. The "Venetian reds" are

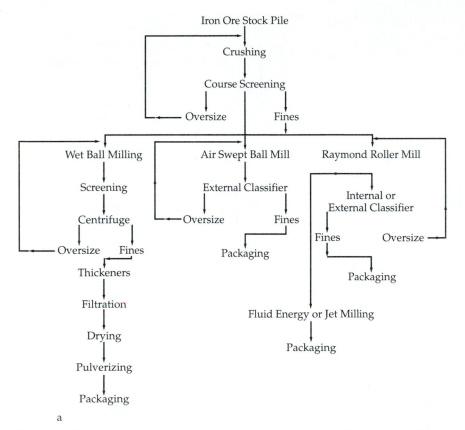

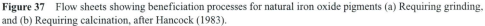

Figure 37   Flow sheets showing beneficiation processes for natural iron oxide pigments (a) Requiring grinding, and (b) Requiring calcination, after Hancock (1983).

produced by calcining an intimate mixture of ferrous sulfate and lime or $Ca(OH)_2$. The reactions may be expressed as follows (Love and Ayers 1942; Buxbaum 1993):

$$4[FeSO_4 \cdot 7H_2O] + 4CaO + O_2 \rightarrow 2Fe_2O_3 + 4CaSO_4 + 28H_2O$$

or

$$2[FeSO_4 \cdot 7H_2O] + CaO \rightarrow Fe_2O_3 + CaSO_4 + SO_2 + 14H_2O$$

The difference between these two processes is the amount of oxygen present in the calcining reaction. High-quality "Copperas reds" are obtained by dehydration of $FeSO_4 \cdot 7H_2O$:

$$FeSO_4 \cdot 7H_2O \rightarrow FeSO_4 \cdot H_2O + 6H_2O$$

and thermal decomposition above 650°C:

$$6[FeSO_4 \cdot H_2O] + 1\frac{1}{2}O_2 \rightarrow Fe_2O_3 + 2Fe_2(SO_4)_3 + 6H_2O$$

and

$$2Fe_2(SO_4)_3 \rightarrow 2Fe_2O_3 + 6SO_3$$

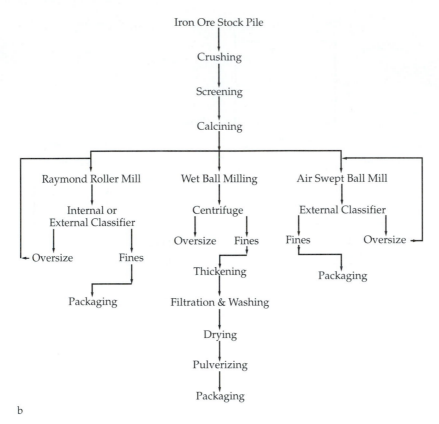

b

Figure 37 *(Continued)* Flow sheets showing beneficiation processes for natural iron oxide pigments (a) Requiring grinding, and (b) Requiring calcination, after Hancock (1983).

The SO_3 formed can be processed to produce sulfuric acid.

Black Fe_3O_4 pigments with a high tinting strength can be prepared by calcining iron compounds under reducing conditions. Brown pigments with compositions of $(Fe,Mn)_2O_3$ are made by the calcination of α-FeOOH with small quantities of manganese compounds. Micaceous iron oxide is obtained in high yield by reacting iron chloride ($FeCl_3$) and iron at 500° to 1,000°C in an oxidizing atmosphere in a tubular reactor (Buxbaum 1993).

Precipitation Processes: In principle, α-FeOOH (yellow), γ-FeOOH (orange), Fe_2O_3 (red), and Fe_3O_4 (black) can all be prepared from aqueous solution of iron salts with the correct choice of reaction conditions and the use of certain seed crystals. However, precipitation with alkali produces neutral salts such as NaCl or Na_2SO_4 as by-products that enter the wastewater as generally used (Büchner et al. 1989).

In the manufacture of yellow pigment, the raw materials are $FeSO_4 \cdot 7H_2O$ and alkali [NaOH, $Ca(OH)_2$, ammonia, or magnesite]. The NaOH solution is added, for example, to a $FeSO_4$ solution to precipitate basic ferrous sulfate, which is converted to α-FeOOH by oxidation producing a seed suspension according to the following reaction:

$$2FeSO_4 + 4NaOH + \tfrac{1}{2}O_2 \rightarrow 2FeOOH \ (\alpha) + 2Na_2SO_4 + H_2O$$

The α-FeOOH seed crystals produced are very small in size, and upon which pigments with hiding properties are obtained by crystal growth. The well-known Penniman-Zoph Process consists of growing α-FeOOH by adding scrap iron and oxidizing it with air without the consumption of NaOH, $Ca(OH)_2$, or $MgCO_3$:

$$2FeSO_4 + 3H_2O + \tfrac{1}{2}O_2 \rightarrow 2FeOOH + 2H_2SO_4$$

and

$$Fe + H_2SO_4 \rightarrow FeSO_4 + H_2$$

In the manufacture of black pigments, ferrous salt solutions are neutralized and oxidized with NaOH at 90°C to a Fe^{2+}/Fe^{3+} ratio of 0.5 for the formation of Fe_3O_4. Red iron oxide pigments can be obtained by a similar procedure to produce yellow pigment on the production of basic ferrous sulfate. The basic ferrous sulfate is converted to α-Fe_2O_3 seed suspension instead of α-FeOOH seed suspension by a complete oxidation, followed by crystal growth of α-Fe_2O_3. Orange iron oxide pigment of lepidocrocite is obtained if dilute solution of the iron salt is precipitated with NaOH solution or other alkalis until neutral. The suspension is then heated for a short period, rapidly cooled, and oxidized.

Reduction of Organic Compounds: By modifying the industrial process for the manufacture of aniline, a number of iron oxide pigments are obtained. This is known as the Laux Process. The raw materials are ground and sieved into largely degreased cast iron or wrought iron chips. By reacting with an aromatic nitro compound such as nitrobenzene in the presence of various chemicals, the iron is oxidized and various iron oxides are produced as expressed by the following reactions:

$$C_6H_5NO_2 + 2Fe + 2H_2O \ (+AlCl_3) \rightarrow C_6H_5NH_2 + 2FeOOH$$

or

$$4C_6H_5NO_2 + 9Fe + 4H_2O \ (+FeCl_3) \rightarrow 4C_6H_5NH_2 + 3Fe_3O_4$$

In the presence of $AlCl_3$, a high-quality yellow pigment is obtained, whereas in the presence of $FeCl_3$, a black pigment with very high tinting strength is produced. Addition of phosphoric acid leads to the formation of brown pigments, and calcination of these products gives a light red pigment of Fe_2O_3.

A systematic tabulation of synthetic iron oxide pigments, prepared by Hancock (1983), is shown in Table 11. It illustrates relationships among chemical composition, mineralogical composition, particle size, and color.

3. Industrial Uses

Iron oxide pigments are used to provide body color for paint, plastics, ceramics, cements, bricks, wood stains, and paper (Benbow 1989). They are suitable in all respects for use in coloring concrete, mortar, shingles, bricks, and similar building materials (Love 1968). The construction industries represent the largest consumer of iron oxide pigments. In paint and coating applications, iron oxide pigments provide excellent properties for many types

Table 11 Physical and chemical properties of synthetic iron oxide pigments, after Hancock (1983).

Chemical Formula:	Fe_2O_3	Fe_2O_3	Fe_2O_3	Fe_2O_3	Fe_2O_3	Fe_2O_3	Fe_2O_3	Fe_2O_3	$CaSO_4$	Fe_2O_3
Mineral Class:	Hematite	Hematite	Hematite	Hematite	Hematite	Hematite	Hematite	Hematite	Mixed	Hematite
Method of Mfg.:	Decomp. $FeSO_4$	Decomp. $FeSO_4$	Ppt.	Ppt.	Calc. of $Fe_2O_3 \cdot H_2O$	Calc. of Fe_3O_4	Pickle Liq. Decomp. (HCl)	B.O.F. Flue Dust	Decomp. $FeSO_4$ & $Ca(OH)_2$	Organic Reduction (Calcined)
Chemical Properties										
% Fe_2O_3	99.5	99.5	97.0	98.0	97.5	99.3	98.4	82.2	40.0	96.5
% FeO	—	—	—	—	—	—	—	—	—	[2.5]
% SiO_2	0.03	0.03	0.05	0.05	0.10	0.01	0.07	9.10	—	—
% Al_2O_3	0.02	0.03	0.01	0.01	0.01	0.05	0.0005	3.10	—	—
% MgO	0.01	0.005	0.01	0.01	0.01	0.015	0.03	1.00	—	—
% CaO	0.02	0.03	0.10	0.10	0.05	0.006	0.05	0.40	25.0	0.10
% Mn	0.02	0.25	0.02	0.02	0.02	0.10	0.30	0.36	—	—
% SO_3	0.05	0.02	0.80	0.80	0.60	0.10	—	—	15.0	—
% Cl_2	—	—	—	—	—	—	—	—	—	—
% LOI @ 1000° C	0.20	0.05	2.50	1.50	0.70	0.60	—	2.50	—	0.50
% Soluble salts	0.10	0.10	0.10	0.10	1.00	0.12	0.10	0.22	—	0.30
Other Properties										
Color	Light red	Dark red	Light red	Dark red	Medium yellow red	Red	Dark red (poor color)	No value	Light red	Light red
Oil absorption*	24	13	24	22	48	—	24	—	20	26
Particle size† Range (10–90%)	0.1–1	2–6	0.1–6	0.5–1.5	0.5–1	0.3–1.2	0.3–7	Varies widely 0.3–5	0.3–5	—
median, μ	0.25	3.0	0.35	0.90	0.75	0.50	1.5	Varies	1.2	0.3‡
Particle shape	Nodular	Nodular	Spheroidal	Spheroidal	Acicular	Cubical	Spheroidal system	Spheroidal	Mixed	Spheroidal
% Retention +325	<0.10	<0.10	<0.05	<0.10	<0.10	—	<0.10	20.0	<0.50	0.05
Surface area¶	10.0	2.2	9.4	3.7	9.1	5.4-	6.3	—	8.8	—
Specific gravity	5.15	5.15	4.90	4.9	4.9	5.15	5.18	—	3.55	5.2

$Fe_2O_3 \cdot xH_2O$	$Fe_2O_3 \cdot H_2O$	$Fe_2O_3 \cdot H_2O$	$Fe_2O_3 \cdot H_2O$	$Fe_2O_3 \cdot xH_2O$	$ZnFe_2O_4$	$xFeO \cdot \gamma Fe_2O_3$		γFe_2O_3	Fe_3O_4
—	Goethite	Goethite	Goethite	—	Zinc Ferrite	Hematite Magnetite	Goethite Hematite Magnetite	Gamma Hematite	Magnetite
Ppt.	Ppt.	Ppt.	Organic Reduction	Ppt.	Calcination	Ppt.	Blend	Reduction & Oxidation	Ppt.
94.0	87.0	87.5	86.5	83.0	(99.6 As) ($ZnO \cdot Fe_2O_3$)	93.3#	94.0	98.5	98.8#
—	—	—	—	—	—	9.0	Balance depends on blended components	—	23.2
0.10	0.05	0.05	[2.0]	—	0.15	0.01		0.06	0.12
—	0.04	0.04	—	—	0.02	0.01		0.01	0.07
—	0.02	0.02	—	—	—	1.12		0.03	0.03
—	—	—	—	—	—	0.04		—	—
0.14	0.05	0.05	—	0.18	0.01	0.19		0.05	0.20
—	1.00	1.00	—	—	0.03	0.20		1.00	0.60
—	—	—	—	—	—	—		1.00	—
5.50	11.00	11.00	12.5	16.2	0.10	—	0.20	0.40	0.15
—	0.15	0.10	8.30	—	0.07	—		—	—
Transparent red	Light yellow	Dark yellow	Yellow	Transparent yellow	Tan	Brown	Brown	Brown	Black
84	55	52	42	55	34	28	26	60	27
—	0.1–0.4	1.2–5	—	—	0.4–1.2	0.2–0.8	—	—	0.2–0.6
<0.10	0.20	1.5	0.2 × .7‡	<0.10	—	0.4	0.5§	0.2§	0.4§
Acicular	Acicular	Acicular	Acicular	Acicular	Acicular	Cubical	Mixed systems	Acicular	Cubical
0.20	<0.10	<0.10	0.10	0.16	<0.10	0.20	<0.20	<1	<0.10
—	15.2	7.4	—	—	—	11.8	—	15.4	7.2
4.56	4.03	4.03	3.97	3.88	5.24	4.77	4.70	4.60	4.96

* Oil absorption—lbs of oil per 100 lbs of pigment.

† As determined by Andressen pipette or Micromeritics sedigraph, except as noted. Range is the weight percentage between 10% and 90%. 50% weight point is median size.

‡ By electron microscopy, acicular, width × length.

¶ Square meters per gram—by B.E.T.

§ Magnetic, by electron microscopy.

Total iron expressed as Fe_2O_3.

of paints and coatings, and can be incorporated into many types of binders (Hancock 1983). Hematite, with varying degrees of pigment loading, is an ingredient in the "primers" that function as bases for various top coats (Podolsky and Keller 1994). Umbers and siennas have characteristic colors and are commonly used in stains. For plastics, low-manganese pigments are used in vinyls, phenolics, polyurethanes, and epoxies (Woernle 1967). The ceramics industries are another important consumer of iron oxide pigments for frits and glazes. The wide applications rely upon their pure hue, good hiding power, good abrasion resistance, and low setting tendency.

Wood stains and exterior metallic coatings are two principal uses for transparent iron oxide pigments, whereas micaceous pigment is used in the protective coating. Magnetic tape recording and magnetic ink oxide represent a growing market for iron oxide (Benbow 1989; Buxbaum 1993).

REFERENCES

Benbow, J. 1989. Iron oxide pigments. *Ind. Minerals* (March): 21–42.

Büchner, W., R. Schliebs, G. Winter, and K. H. Büchel. 1989. *Industrial inorganic chemistry.* Weinheim: VCH Verlagsgesellschaft mbH, 614p.

Buxbaum, G. 1993. ed. *Industrial inorganic pigments.* Weinheim: VCH Verlagsgesellschaft mbH, 281p.

Deer, W. A., R. A. Howie, and J. Zussman. 1992. *An introduction to rock-forming minerals.* Essex, England: Longman, 696p.

Griffiths, J. 1984. The industrial minerals of Cyprus. *Ind. Minerals.* (December): 21–35.

———— 1985. Spain's industrial minerals. *Ind. Minerals.* (October):23–63.

Hancock, K. R. 1983. Mineral pigments. In *Industrial minerals and rocks.* 5th ed. Ed. S. J. Lefond, 349–372. New York: Soc. Mining Engineers, AIME.

Harben, P. W., and R. L. Bates. 1990. *Industrial minerals, geology and world deposits.* London: Industrial Minerals Division, Metal Bulletin Plc, 312.

Kesler, T. L. 1950. Geology and mineral deposits of the Cartersville District, Georgia. *U.S.G.S. Prof. Paper,* 224, 97p.

Lilljequist, R. 1969. The geology and mineralization of the Troulli Inlier. *Geol. Surv. Cyprus Bull.,* n.4, 45–87.

Love, C. H. 1968. Guidelines for coloring concrete products. *Concrete Products* 71:34–37.

Love, C. H., and J. W. Ayers. 1942. The iron oxide pigments. Vol. 2 of *Protective and decorative coatings,* Ed. J. J. Mattiello, 317–318. New York: John Wiley.

Podolsky, G., and D. P. Keller. 1994. Pigments, iron oxide. In *Industrial minerals and rocks.* 6th ed. Ed. D. D. Carr, 765–781. Littleton, Colorado: Soc. Mining, Metal., and Exploration.

Robertson, A. H. F. 1975. Cyprus umbers: basalt-sediment relationships on a Mesozoic ocean ridge. *Jour. Geol. Soc. London* 131:511–531.

Toon, S. 1986. Coloured pigments, news of the hues. *Ind. Minerals* (October):19–39.

Woernle, A. K. 1967. Iron oxide pigments. In *The encyclopedia of basic materials for plastics,* 261–264. New York: Reinhold Publ. Co.

Kaolinite and Kaolinite-Bearing Clays

Kaolinite is a hydrous aluminum silicate mineral with a well-defined chemical composition and crystal structure. It is the essential constituent of kaolin or china clay, ball clay, fire or refractory clay, and certain flint clay. All have major industrial uses, but the consumption of kaolin by papermaking is the largest because of its high brightness, suitable particle size, and low abrasion.

Kaolinite-bearing clay deposits are widely distributed worldwide. The United States, England, the former USSR, India, Czechoslovakia, China, and Brazil are the major producers. As estimated by the U.S. Bureau of Mines, annual world production of kaolin varied between 23 M and 25 M tons in the period from 1989 to 1993.

MINERALOGICAL PROPERTIES

1. Kaolinite

The crystal structure of kaolinite, $Al_4Si_4O_{10}(OH)_8$, consists of a SiO_4 tetrahedral sheet and an AlO_6 octahedral sheet that are combined into a single layer. The atomic distribution and links are illustrated in Fig. 38 (Brindley and Robinson 1947). Only two out of each set of three octahedral sites are occupied by Al ions. Successive layers are superimposed so that, in general, oxygens at the base of one layer are hydrogen-bonded to hydrogen ions at the top of its neighbor, resulting in a triclinic symmetry. Cell dimensions from single-crystal synchotron refinement are a=5.15Å, b=8.93Å, c=7.37Å, α=91.8°, β=104.8°, and

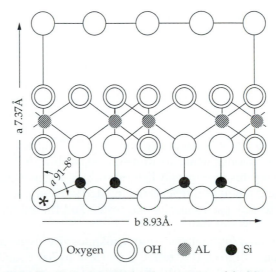

Figure 38 Crystal structure of kaolinite. Published with permission of the Mineralogical Society of Great Britain and Ireland.

$\gamma=90°$ (Neder et al. 1999). Stacking disorders are common, involving both translation of layers by nb/3 and rotations of 120°. The degree of order in the structure can be detected by x-ray diffraction, because it affects the intensities of certain reflections. Various ratios of reflections are often used as a "crystallinity index." Brindley and Robinson (1946) published a schematic chart for distinguishing between well-crystallized kaolinite, common kaolinite, fire-clays, and hydrokaolinite. Murray and Lyons (1956, 1960) proposed a set of thirteen x-ray diffractograms in which a progressive decrease in crystallinity is shown (Fig. 39). Grinding has a direct effect upon the crystallinity of kaolinite. After grinding for one hour in a planetary mill, the crystalline order is destroyed, but amorphization is not complete, even after grinding for a period of 10 hours (Kristóf et al. 1993).

The minerals dickite and nacrite are chemically identical with kaolinite, but have layers stacked in different ordered sequences. Thus dickite has a two-layered and nacrite a six-layered monoclinic unit cell. Structural refinement of dickite and nacrite give cell dimensions of a=5.1474Å, b=8.9386Å, c= 14.390Å, and $\beta=96.483°$ (Bish and Johnston 1993) and a=8.906Å, b=5.146Å, c=15.664Å, and $\beta=115.58°$ (Zhang and Bailey 1994), respectively. Halloysite [$Al_4Si_2O_{10}(OH)_8 \cdot 4H_2O$] has a single layer of water molecules in each interlayer that increases $d_{(001)}$ from the characteristic 7.15Å of kaolinite to about 10Å. The dehydration of halloysite produces a series of intermediate haloysite or hydrohalloysite between two extreme forms, generally referred to as halloysite-7, with c=14.9Å and $\beta=102°$, and halloysite-10, with c=20.7Å and $\beta=99.7°$ (Brindley et al. 1963). The presence of water molecules in the interlayers causes the morphological change of particles from platy to tabular. The structural layers scatter X-rays in a wide and incoherent manner from layer to layer and produce diffused reflections (Brindley and Robinson 1947; Deer et al. 1992).

Little chemical variation of kaolinite from the ideal composition, $Al_4Si_4O_{10}(OH)_8$, is observed, but analytical results from kaolinites that give the exact composition of 46.54% SiO_2, 39.5% Al_2O_3, 13.96% H_2O are seldom obtained. Minor Mg, Fe, or Ti may replace Al in the octahedral sites, and most cations are absorbed by clay layers with broken bonds. Kaolinite is chemically stable over a wide pH range (3 to 9), but some excess SiO_2 or Al_2O_3 contained in kaolinites is readily soluble (Langston and Pask 1969; Prasad et al. 1991). Follett et al. (1965) analyzed kaolinite from Cornwell that showed 3.1 to 4.9% of easily soluble SiO_2 and 1.5 to 5.9% of easily soluble Al_2O_3. Both are presumably present as amorphous cementing materials. Nagelschmidt et al. (1949) demonstrated that the TiO_2 in Georgia kaolinite was present as anatase, and it increased in abundance as the particle size decreased.

Kaolinite has a hardness of from 2 to 2 1/2 and a specific gravity of from 2.61 to 2.68. It is commonly white to iron-stained, with a dull luster and greasy feel. Fine-grained aggregates are strongly light-absorbing and may appear dark in color. Electron microscopic examination reveals the foliated structure of kaolinite, dickite, and nacrite, commonly as tiny pseudohexagonal plates, and the tubular forms of halloysite (Beutelspacher and Marel 1968).

Differential thermal analysis of kaolinites gives two endothermic peaks in the temperature ranges of from 100° to 150°C and 550° to 700°C, and one exothermic peak in the temperature range of from 950° to 1,020°C. The peaks represent the removal of hygroscopic water, the removal of OH links to the octahedral sheet, and the recrystallization of

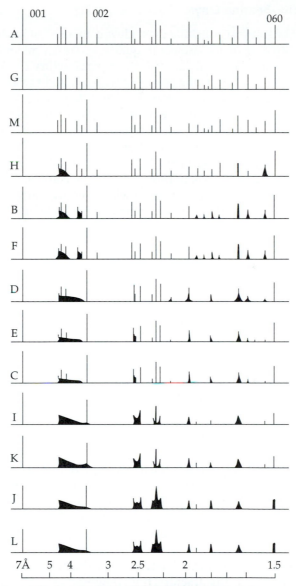

Figure 39   Degrees of crystallinity in kaolinite. © P. W. Harben & R. L. Bates.

α-Al_2O_3 and mullite (Douillet and Nicolas 1969). The infrared spectra of kaolinite, dickite, nacrite, and halloysite are very similar, and the most specific differences appear in the high frequency region of 3,800 to 3,000 cm^{-1} and are related to the O-H stretching vibrations. The resolution of the IR spectra of structural OH increases with decreasing temperature (Farmer and Russell 1964; Prost et al. 1989). In response to HF attack, nacrite is the most stable, dickite the second, and kaolinite the least. This process can be used to enrich both nacrite and dickite in a kaolin clay (Shen 1994).

2. Kaolinite-Bearing Clays

Kaolinite-bearing clays include those clays composed of kaolinite as their essential constituent, with illite or mica, smectite, quartz, and other minerals. The important ones are kaolin (or china clay), ball clay, fire clay (or refractory clay), and flint clay. Kaolin is a white and soft clay, consisting chiefly of well-ordered kaolinite with a low iron content. It is an aggregate of kaolinite flakes in the forms of stacks, packets, and sheaves with various degrees of plasticity. Kaolin may form either as a residual or primary deposit, or as a sedimentary or transported deposit. The terms "china clay" and "kaolin" are often used synonymously (Highley 1984).

Ball clay is a fine-grained, mostly disordered, mixture of kaolinite, illite, mica, quartz, smectite, and carbonaceous matter. Its color varies with the amount of carbonaceous matter and changes from white to gray, blue, brown, or black. Ball clay is highly plastic (Highley 1975). In a comparison with kaolin, both chemical and physical differences were noted as follows:

1. Ball clays have a higher silica-to-alumina ratio than is found in most kaolin, 53.50 to 51.88% SiO_2 and 29.84 to 32.42% Al_2O_3 of ball clays from Tennessee and Kentucky, versus 46.86 to 45.17% SiO_2 and 37.00 to 38.25% Al_2O_3 of kaolins from North Carolina and Georgia.
2. Ball clays are generally finer than kaolins, containing higher percentages of particles of 0.25 micron or smaller in diameter. This promotes dense body structure and develops higher strength in the green ware.
3. Ball clays are considerably more viscous than kaolins, thus they require higher dosages of dispersing agents to attain maximum fluidity. Suspensions containing ball clays generally exhibit thixotropic tendencies.
4. All major deposits of ball clay are sedimentary.

Fire clay or refractory clay is a sedimentary clay that can stand temperatures above 1,500°C. Patterson and Murray (1983) classified refractory clays on the basis of their thermal service. A low refractory clay has a pyrometric cone equivalent (PCE) of 19 to 26, a moderate refractory clay has a PCE of 26 to 31.5, a high refractory clay has a PCE of 31.5 to 33, and a superduty refractory clay has a PCE of 33 to 34. With an increase in alumina content, decrease in the order of the kaolinite structure, and decrease in illite and smectite content, refractory clays range from plastic kaolin and ball clay through soft flint and hard flint clays, to diaspore- or boehmite-rich clays (Stack and Schnake 1983).

Flint clay is a non-plastic, hard, dense refractory clay, having an appearance much like flint, and a shell-like fracture. It is composed mainly of kaolinite, and the kaolinite grains are tightly interlocked, producing a texture similar to that of monomineralic igneous rocks. Flint clay represents a mineralogical intermediary between kaolin plastic clays and high-alumina clays such as boehmite or diaspore (Keller 1968).

GEOLOGICAL OCCURRENCE AND DISTRIBUTION OF DEPOSITS

Kaolinite occurs principally as primary residual deposits formed by the weathering of feldspars, muscovite, and other Al-rich silicates. The hydrolysis of feldspar can be expressed as:

$$2KAlSi_3O_8 + 3H_2O \rightarrow Al_2Si_2O_5(OH)_4 + 4SiO_2 + 2KOH$$

the silica and KOH being leached away. If potassium is kept in the environment of re-action, then illite instead of kaolinite is likely to form (Deer et al. 1992). Kaolinite may be formed from the hydrolysis of illite if the leaching of KOH takes place at a subsequent stage, as:

$$2KAl_2(AlSi_3)O_{10}(OH)_2 + 5H_2O \rightarrow 3Al_2Si_2O_5(OH)_4 + 2KOH$$

Kaolinite may enter the erosion cycle and be deposited in deltaic, lagoonal, or other nonmarine environment, forming a sedimentary or transported deposit.

Hydrothermal alteration may produce kaolinite. The host rocks must contain high-Al minerals and be permeable to allow the circulation of fluids. Granites are the most common host rocks for hydrothermal alteration.

Kaolin deposits are widely distributed in the world. Selected locations in the United States and England are described below:

Kaolin Belt, the United States: The United States is the main source of kaolin, accounting for 35% of world production. Most of the kaolin is derived from the so-called kaolin belt that extends from South Carolina, across Georgia, and into Alabama, a distance of almost 400 km (Kesler 1963). As shown in Fig. 40, the kaolin deposits occur in the Cretaceous and Lower Tertiary sediments along a Fall line that marks where Coastal Plain sediments meet the crystalline rocks of the Piedmont Plateau. In Georgia, the Cretaceous sediments are composed of horizontally stratified beds of quartz sands, kaolinitic sand, and kaolin; and in South Carolina, the Tertiary sediments contain important deposits of kaolin, along with quartz sands, smectitic clay, and limestone. The kaolin bodies are of a lenticular shape, several meters to 1.5 km long, and several centimeters to 15 meters thick (7 meters on average). Mineralogical components of the deposits are quartz, limonite, anatase, muscovite, biotite, magnetite, tourmaline, zircon, kyanite, monazite, goethite, and graphite. Major deposits are the Aiken-Augusta district, Wrens district, and Macon-Inwinton district. In Andersonville, Georgia, and Eufaula, Alabama, large deposits of refractory clay of Tertiary age occur in association with bauxite. In comparison, the Cretaceous kaolins have a whiter color, higher brightness, better crystal perfection, lower TiO_2 content, and more vermicular crystals than the Tertiary kaolins. Commonly, more than 90% of the Tertiary kaolin particles are less than 2 μm in diameter, whereas the Cretaceous kaolins rarely have more than 65% in that size range (Patterson and Murray 1984; Harben and Bates 1990). This kaolin belt represents a good example of transported or sedimentary deposits (Kesler 1963).

Kentucky-Tennessee, the United States: Major ball clay deposits occur in the Jackson Purchase region, in the western parts of Kentucky and Tennessee (Olive and Finch 1969). Lenses of ball clay occur in the Wilcox, Claiborne, and Jackson Paleocene-Eocene age formations. The ball clays characteristically have a high kaolinite and low montmorillonite content. In the eastern Kentucky coal field, deposits of refractory clays in association with thin coal beds occur in the Olive Hill area (Patterson and Hosterman 1962).

Missouri, the United States: In central Missouri, a 48 km-wide belt stretching for almost 160 km is a classic area for the flint clay facies (Keller 1968, 1981). The deposits

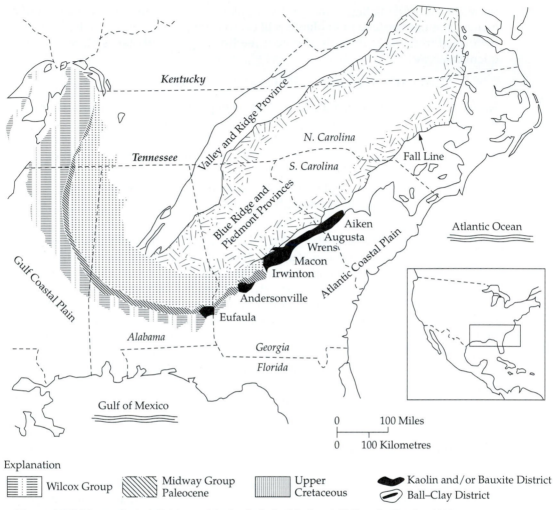

Figure 40   Major geological divisions of the kaolin belt. © Industrial Minerals, October 1990.

occur in the Cheltenham Formation of Pennsylvanian age, and the clays change from plastic and semiplastic in the north, through semiflint and flint clays, to diaspore clay in the south. Along with this change the Al_2O_3 content increases from 30% to almost 80%. This change coincides with the rise of the Cheltenham Formation up the north flank of the Ozark Dome. The increase in elevation in the southern portion of this belt promotes intensive leaching that results in the formation of flint clay mixed with diaspore and boehmite (Keller 1981).

Cornwall and Devon, England: The southwest peninsula of England is underlain by a major granite batholith. The granite exposes found in Dartmoor, Bodmin Moore, St. Austell, Carnmenellis, and Land's End are all altered by greisenization and kaolinization. The most extensive deposit is found at St. Austell. The kaolin deposits in

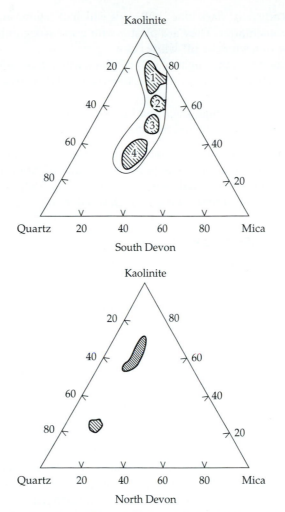

Figure 41 Mineralogical composition of Devon ball clays. © Industrial Minerals, October 1998.

Cornwall are considered to be good examples of the hydrothermal type (Edmonds et al. 1969). A typical kaolin deposit in Cornwall is funnel-shaped, narrowing downwards to a stem as deep as 400 m.

Deposits of ball clay are located in three Tertiary basins in southern England (1) the Bavey Basin in south Devon, (2) the Petrockstow Basin in north Devon, and (3) the Wareham district in southeastern Dorset. The clay deposits, which are of variable thickness, consist of sand, silt, silty clay, ball clay, lignitic clay, and lignite. The mineralogical composition of the Devon ball clays is shown in Fig. 41 (Russell 1988). Of the south Devon clays, four groups are recognized:

1. Clays known as "black" ball clays, which have a dominant kaolinite content with an appreciable amount of carbonaceous matter. They are easily deflocculated, produce fluid stable casting slips, and have a high quality of firing color.

2. Clays known as "dark blue" ball clays, which also have an appreciable amount of carbonaceous matter. They are plastic, with good workability, have high dry strengths, and fire to a white or off-white color.
3. Clays known as the "light blue" ball clays, which have a lower carbonaceous content (less than 0.5%), and thus tend to be extremely thixotropic and to fire to an off-white or ivory color.
4. Clays that have a high silica and low carbonaceous content. They fire to a buff color but have good casting and binding characteristics.

The north Devon ball clays are divided into two types. Type 1, which has a higher kaolinite content, contains an appreciable amount of montmorillonite. This gives the clay higher deflocculant demand values and a relatively high slip viscosity. Type 2, which has a lower kaolinite content, contains much smaller amounts of montmorillonite.

INDUSTRIAL PROCESSES AND USES

1. Industrial Processes

For most fire clays and other low-grade kaolinite-bearing clays, the cost of beneficiation, beyond the limits of selective mining, thorough mixing and blending, if necessary, plus, possibly, crushing or screening out pebbles, is not cost effective.

Preparation of ball clays and certain fire clays consists essentially of shredding, blending, drying, and grinding (Watson 1982). In the shredding process raw clay is fed into a circular, revolving plate with sharp knives attached, which cuts the clay into small pieces. The clay pieces then fall onto a conveyor, which throws them into the designated storage bay. Most of the blending of clays is done at this stage. The usual moisture content of shredded clays is around 15%, but further drying may be needed for some clays. Powdered clays are produced by shredding and milling from 200 to 325 mesh and drying to around 2% moisture. Some ball and plastic clays are calcined to produce chamottes for use in refractories and ceramic bodies. This is usually done in rotary kilns at temperatures around 1,350°C.

The processing of kaolin is accomplished by two basic methods: (1) dry process or air-flotation for the production of rubber and paper filler clays, and (2) wet process or classification in water suspension for the production of filler, coating, and rubber specialty grades (Watson 1984).

In the dry process, the crushed crude kaolin, with about 20 to 25% moisture, is fed in a uniform continuous stream into a rotary dryer from which it emerges with a moisture content of from 1 to 2%. The dried clay is elevated to a storage bin that feeds directly to the Raymond Roller Mill with a whizzer separator. The particles of desirable fineness are lifted from the grinding chambers by air currents, while the coarse particles are rejected from the upward stream by the whizzer separator. Pulverization of the clay is accomplished by means of the centrifugal pressure of the mill roller exerted against a bull ring.

The wet process is accomplished in five stages: (1) blunging, for the initial disintegration of the crude clay lumps in water, (2) fractionation, for the separation of fractions with definite particle size ranges, (3) bleaching to remove discoloration caused by iron compounds, (4) filtration, and (5) drying of the filter cake. Dispersing agents are commonly used in the fractionation stage to promote the formation of a deflocculated suspension, from which coarse clay particles, mica, and quartz are settling. This is carried out in large settling tanks, hydroseparators, or by mechanical apparatus such as the Bird centrifuge. Bleaching is generally done by the use of either zinc or sodium salts of hydrosulfurous acid to reduce the insoluble iron oxides to soluble ferrous compounds. The clay is filtered from its water suspension by a filter press, centrifuge, rotary vacuum filter, or tube filter, and then dried by a band, apron, rotary, drum, spray, or Buell turbo drier.

Further beneficiation is done by (1) flotation to remove fine iron and titanium minerals, (2) delamination to break down the books of kaolin into individual platelets through attrition grinding or extrusion, (3) calcining, and (4) magnetic filtration. Micronization of kaolinite may be done by both ball milling and oscillatory milling. The former is more effective in dry grinding than under wet conditions, and the latter process was found to produce complete destruction of kaolinite structure (amorphization) within one hour (Suraj et al. 1997).

2. Industrial Uses

The papermaking industry is the largest user of kaolin, consuming about 60% of the total domestic production in the United States. The ceramics, rubber, and plastics industries use 10% each, and the remaining 10% is used for various purposes, including the production of cement. Ball and plastic clays are used in earthware, porcelain, tile, and other products, as a binder for china, and to impart plasticity. Fire clays and flint clays of good quality are used mainly for refractories (Murray 1963; Prasad et al. 1991).

Kaolin: Kaolin is used in the paper industry as both a filler and coater. Generally, filler is used in the papermaking industry to

1. improve the surface characteristics of paper,
2. improve whiteness, opacity, and brightness,
3. improve color, and
4. increase dimensional stability.

Some special reasons for coating are

1. to improve the receptivity of the surface to printing,
2. to mask the original surface characteristics of some papers,
3. to upgrade the texture of some papers,
4. to apply a moisture-resistant or moisture-proof surface,
5. to reduce abrasion, and
6. to produce an attractive surface.

Kaolin has excellent physical and chemical properties, including (1) brightness, (2) particle size and distribution of particle shape, (3) viscosity, (4) percentage of grit and abrasion to increase smoothness, (5) gloss, (6) opacity, (7) brightness, and (8) printability (Higham 1983). The 2 micron point is generally used as the commercial particle size control point. Kaolins of the coarse particle size are normally used as filler clay and those of fine particle size are used as coating clay. Usually filler kaolins contain an abundance of stacks, and plates of kaolin are the dominant crystal shape in the coating clay (Stepetys and Cleland 1993).

Of the two basic types of kaolin, air-floated and water-washed, many grades are specified for use as fillers in papermaking. The common filler grades are

1. standard waterwashed filler with 82 to 84 brightness (GE), and 60 to 70% 2μ particle size (nonstokes);
2. premium wasterwashed filler with 84 to 85 brightness (GE), and 60 to 65% 2μ particle size (nonstokes);
3. standard air-floated filler with 76 to 79 brightness (GE), and 50 to 60% 2μ particle size (nonstokes); and
4. premium air-foated filler with 79 to 83 brightness (GE), and 50 to 60% 2μ particle size (nonstokes). All have a viscosity of 400 cpe @ 70% solution (Brookfield viscosity @ 10 RPM, #3 disc)(Pickering and Murray 1994).

Five kaolin coating grades are generally used, and each grade is defined by the same set of parameters as those for fillers. They are

1. No. 1 coating, standard with 87 to 88 brightness, 90 to 94% 2μ and 500 cpe, and premium with 89 to 91 brightness (GE), 90 to 94% 2μ and 500 cpe.
2. No. 2 coating, standard with 86 to 87 brightness (GE), 80 to 84% 2μ and 400 cpe, and premium with 88 to 90 brightness (GE), 80 to 84% 2μ and 400 cpe.
3. High glossing, standard with 86 to 88 brightness (GE), 95 to 98% 2μ and 700 cpe, and premium with 89 to 90 brightness (GE), 98 to 100% 2μ and 700 cpe.
4. Delaminated, standard with 87 to 89 brightness (GE), 78 to 82% 2μ and 300 cpe, and premium with 89 to 90 brightness (GE), 78 to 82% 2μ and 300 cpe.
5. Calcined, high opaque with 80 to 85 brightness (GE), 78 to 80% 2μ and 500 cpe, standard with 90 to 92 brightness (GE), 84 to 86% 2μ and 500 cpe, and premium with 92 to 94 brightness (GE), 90 to 94% 2μ and 500 cpe (Pickering and Murray 1994).

In general, No. 1 grade is for paper of high gloss, brightness, and good smoothness, whereas No. 2 grade is for general publication and magazine paper. Delaminated grades are used for lightweight paper, and high brightness grades are used for premium papers that require high brightness and whiteness, and good smoothness and printability. Calcined grades are for paper that requires matte finish.

Kaolin is one of the minerals used in paint extension and is commonly used as a substitute for the expensive TiO_2. In water-based paint, kaolins are used as a functional extender pigment. Coarse kaolins produce a matte finish to the paint, while fine ones impart a high gloss. Kaolin is used in plastics to smooth the surface finish, reduce cracking and shrinkage during curing, and obscure fiber reinforcement patterns. Loading levels vary from 15 to 60%. As a functional filler, kaolins are used in nylons, polyesters, and other common thermoplastics to improve their mechanical, electrical, and thermal properties. Kaolin filler greatly enhances the insulation properties of PVC coatings on cables. Calcined kaolin improves the infrared absorption characteristics of polyethylene film, when used at a loading level of from 7 to 8% by weight (Pickering and Murray 1994).

Fine-grained kaolins are used in rubber for reinforcing and stiffening, especially in nonblack rubber. In ceramic products, earthware, porcelain, bone china, and vitreous china, each contain, respectively, 25, 60, 25, and 20 to 30% kaolin (Anon. 1955).

Ball and Plastic Clays: The major and also the traditional use of ball clays is in ceramics (Watson 1982). These clays are generally used in varying proportions to give certain properties to particular ceramic bodies during manufacture. These properties include (1) high plasticity to facilitate the shape of the body, (2) high dry-strength to make the handling of the dry or green body of the product easier, and (3) good rheological properties to enable the body to build on the wall of the mould during slip-casting. Ball clays fire to a white, cream, or ivory color, and aid vitrification during firing because of the presence of mica as a flux. Typical compositions of ceramic bodies that contain ball clays, kaolin, flux, quartz, and others, are given below:

Weight% of total solids

	Ball clays	Kaolin	Flux	Quartz	Others
Wall tiles	30	20	—	40	10
Earthwares	25	25	15	35	—
Vitreous china					
Sanitarywares	20–30	20–30	15–25	30–40	0–3
Porcelain	10	60	15	15	—
Insulator					
Porcelain	30	20	25	25	—
Engobe	5–15	30–50	20–35	15–30	—

The production flow of a typical plant for body preparation is shown in Fig. 42 (Russell 1989). Prepared bodies take several forms including plastic bodies, spray dried granulate, and blended slips. Developments in tableware manufacture, as summarized by Russell (1989), are (1) the introduction of dustpressing of flatware, (2) improvement in roller-head processes, (3) rapid firing and fast once-firing, (4) increase in the use of offset printing, and (5) advancement in the use of automation. Developments in tile manufacture are (1) high pressure dry presses, (2) built up granulation, (3) roller dry, and (4) fast once-firing of glazed wall tiles. The most distinct development in sanitary-ware manufacture is pressure casting in place of slip casting. Others are (1) microwave drying, (2) rapid firing, (3) improvements in battery casting, and (4) the use of CAD/CAM techniques in design and manufacture of moulds.

Fire Clays: Fire clays are mainly used for refractories. They are composed essentially of fire clay or calcined fire clay in varying proportions. There are five classes of fire clay bricks. These are designated as (1) superduty, (2) high duty, (3) semisilica, (4) medium duty, and (5) low duty. The semisilica class is specified on its load bearing ability at high temperatures and its silica content, whereas the other four classes use, as the primary criterion for classification, the degree of refractoriness of pyrometric cone equivalent (PCE). The superduty, high duty, medium duty, and low duty fire clay bricks must have a minimum value of 33, 31.5, 26, and 9, respectively. The bulk of fire clay refractories used are in the super and high duty classes (Crookston and Fitzpatrick 1983).

Flint Clays: Like fire clays, flint clays are used mainly in refractories.

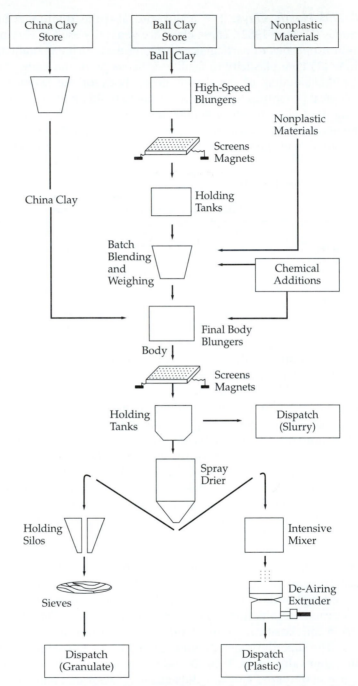

Figure 42   A general scheme to illustrate the production flow of whiteware bodies, after Russell (1989).

REFERENCES

Anon. 1955. *Kaolin clays and their industrial uses.* 2nd ed. New York: J. M. Huber Corp, 214p.

Beutelspacher, H., and J. W. van der Marel. 1968. *Atlas of electronmicroscopy of clay minerals and their admixtures.* Amsterdam: Elsevier 396p.

Bish, D. L., and C. T. Johnston. 1993. Rietveld refinement and Fourier-transform infrared spectroscopic study of the dickite structure at low temperatures. *Clays and Clay Minerals* 41:297–304.

Brindley, G. W., and K. Robinson. 1946. Randomness in the structure of kaolinitic clay minerals. *Trans. Faraday Soc.* 42B:198–205.

———— 1947. The structure of kaolinite. *Mineral. Mag.* 27:242–253.

Brindley, G. W., P. de Souza Santos, and H. de Souza Santos. 1963. Mineralogical studies of kaolinite-halloysite clays. Part I. Identification problems. *Amer. Mineral.* 48:897–910.

Crookston, J. A., and W. D. Fitzpatrick. 1983. Refractories. In *Industrial minerals and rocks.* 5th ed. Ed. by S. J. Lefond, 373–385. New York: Soc. Mining Engineers, AIME.

Deer, W. A., R. A. Howie, and J. Zussman. 1992. *An introduction to the rock-forming minerals.* Essex, England: Longman, 696p.

Douillet, P., and J. Nicolas. 1969. Les minéraux du kaolin-historique—réflexions concernant les diverses classifications et nomenclatures—proposition d'une nomenclature nouvelle. *Bull. Soc. fr. ceram.* 83:87–114.

Edmonds, E. A., M. C. Mckeown, and M. Williams. 1969. *British regional geology: South-west England.* 3rd. ed. London: Her Majesty's Stationary Office, 130p.

Farmer, V. C., and J. D. Russell. 1964. The infrared spectra of layer silicate. *Spectrochim. Acta* 20:1149–1173.

Follett, E. A. C., W. J. McHardy, B. D. Mitchell, and B. F. L. Smith. 1965. Chemical dissolution techniques in the study of soil clays, I.. *Clay Minerals* 6:23–34.

Harben, P. W., and R. L. Bates. 1990. *Industrial minerals, geology and world deposits.* London: Industrial Minerals Division, Metal Bulletin Plc, 312p.

Higham, R. R. A. 1983. *A handbook of papermaking.* London: Oxford University Press, 294p.

Highley, D. E. 1975. Ball clay. *Mineral dossier, no. 11, British geological survey.* London: Her Majesty's Office, 32p.

———— 1984. China clay. *Mineral dossier, no. 26, British geological survey.* London: Her Majesty's Office, 65.

Keller, W. D. 1968. Flint clay and a flint clay facies. *Clays and Clay Minerals* 16:113–128.

———— 1981. The sedimentology of flint clay. *Jour. Sedi. Petr.* 51:233–244.

Kesler, T. L. 1963. Environment and origin of the Cretaceous kaolin deposits of Georgia and South Carolina. *Georgia Mineral Newsletter* 16:2–11.

Kristóf, É., A. Z. Juhász, and I. Vasányi. 1993. The effect of mechanical treatment on the crystal structure and thermal behavior of kaolinite. *Clays and Clay Minerals* 41:608–612.

Langston, R. B., and J. A. Pask. 1969. The nature of anauxite. *Clays and Clay Minerals* 16:425–436.

Murray, H. H. 1963. Industrial applications of kaolin. *Clays and Clay Minerals* 10:291–298.

Murray, H. H., and S. C. Lyons. 1956. Correlation of paper-coating quality with degree of crystalline perfection of kaolinite. *Clays and Clay Minerals* 4:31–40.

———— 1960. Further correlation of kaolinite crystallinity with chemical and physical properties. *Clays and Clay Minerals* 8:11–18.

Nagelschmidt, G., H. F. Donnelly, and A. J. Morcom. 1949. On the occurrence of anatase in sedimentary kaolin. *Mineral. Mag.* 28:492–495.

Neder, R. B., M. Burghammer, T. Grasl, H. Schulz, A. Bram, and G. Fiedler. 1999. Refinement of the kaolinite structure from single-crystal synchrotron data. *Clays and Clay Minerals* 47:487–494.

Olive, W. W., and W. I. Finch. 1969. *Stratigraphic and mineralogic relations and ceramic properties of clay deposits of Eocene age on the Jackson Purchase region, Kentucky, and in adjacent parts of Tennessee.* U.S.G.S. Bull., 1282, 35p.

Patterson, S. H., and J. W. Hosterman. 1962. *Geology and refractory clay deposits of the Haldeman and Wrigley quadrangles, Kentucky.* U.S.G.S. Bull., 1122-F, 113p.

Patterson, S. H., and H. H. Murray. 1983. Clays. In *Industrial minerals and rocks.* 5th ed. Ed. S. J. Lefond, 585–651. New York: Soc. Mining Engineers, AIME.

——— 1984. *Kaolin, refractory clay, ball clay, and halloysite in North America, Hawaii and the Caribbean region.* U.S.G.S. Prof. Paper, 1306, 56p.

Pickering, Jr., S. M., and H. H. Murray. Kaolin. 1994. In *Industrial minerals and rocks.* 6th ed. Ed. D. D. Carr, 255–277. Littleton, Colorado: Soc. Mining, Metal., and Exploration.

Prasad, M. S., K. J. Reid, and H. H. Murray. 1991. Kaolin: processing, properties and applications. *Appl. Clay Sci.* 6:87–119.

Prost, K. R., A. Dameme, E. Huard, J. Driard, and J. P. Leydecker. 1989. Infrared study of structural OH in kaolinite, dickite, nacrite and poorly crystalline kaolinite at 5 to 600 K. *Clays and Clay Minerals* 37:464–468.

Russell, A. 1988. Ball and plastic clays. *Ind. Minerals* (October):27–47.

Russell, A. 1989. Whiteware developments. *Ind. Minerals* (September):29–43.

Shen, Z. 1994. An experimental study of enriching dickite and nacrite from kaolinite. *Jour. Mineral. & Petrol.* 14:18–21 (in Chinese).

Slepetys, R. A., and A. J. Cleland. 1993. Determination of shape of kaolin pigment particles. *Clay Minerals* 28:495–508.

Stack, C. E., and M. A. Schnake. 1983. Refractory clays. In *Minerals in the refractory industry—assessing the decade ahead,* Eds. P. W. Harben and E. M. Dickson, Ind. Minerals, n.187:69–77.

Suraj, G., C. S. P. Iyer, S. Rugmini, and M. Lalithambika. 1997. The effect of micronization on kaolinites and their sorption behavior. *Appl. Clay Sci.* 12:111–130.

Watson, I. 1982. Ball and plastic clays—shaping up to ceramic doldrums. *Ind. Minerals* (August):23–45.

——— 1984. Kaolin—low growth markets put emphasis on quality. In *Raw materials for the pulp and paper industry,* Eds. P. Harben and T. Dickson, 27–39. London: Metal Bulletin, Inc.

Zhang, H., and S. W. Bailey. 1994. Refinement of the nacrite structure. *Clays and Clay Minerals* 42:46–52.

Limestone

Limestone is a common sedimentary rock consisting principally of calcium carbonate in the form of calcite. It occurs widely throughout the world and is an essential raw material for the chemical, metallurgical, construction, and agricultural industries. The most readily available and cost effective alkali, lime or quicklime, is produced by thermal decomposition from limestone.

Several common types of limestone are used. They are (1) chalk, a soft, fine-grained, fossiliferous type of carbonate rock; (2) marl, an impure, soft, earthy carbonate rock; (3) marble, a metamorphic, highly crystalline carbonate rock; (4) travertine, a carbonate rock precipitated from hot springs; and (5) iceland spar, the purest form of limestone, white in color, transparent to light, and always of large crystals.

MINERALOGICAL PROPERTIES OF LIMESTONE

Limestone contains principally calcite, with or without dolomite. Aragonite and magnesian calcite are abundant primary minerals in recent marine carbonate sediments and limestones. Both are metastable, tending to alter to stable calcite or low-magnesian calcite with burial, as well as with time.

1. Calcite

Crystallographically, calcite, $CaCO_3$, is either hexagonal with a=4.989Å and c=17.061Å, or rhombohedral with a=6.375Å and $\alpha=46°05'$. The structure of calcite can be described as a distorted NaCl-type structure with Ca and CO_3 groups in place of Na and Cl atoms, respectively. The face-centered rhombohedral unit cell is produced by compression along one of the three-fold axes of the NaCl unit cell, until the edges that meet it make an angle of $101°55'$ with each other. The shape of the unit cell corresponds well with the cleavage rhombs of calcite (Chang et al. 1996).

Most calcite is relatively pure and fairly close to its formula composition. Common substitutions for calcium in calcite are magnesium, manganese, and ferrous iron. Magnesian calcite, which contains up to 30 mole%$MgCO_3$, is an important constituent of the skeletons of marine organisms and of carbonate cements. Two to three mole% $FeCO_3$ is very common in calcite. Indices of refraction of calcite are $\varepsilon = 1.486$ and $\omega = 1.658$, and vary in an approximately linear manner with $MgCO_3$, $FeCO_3$, and $MnCO_3$ contents. Its maximum birefringence is extreme (Chang et al. 1996).

The solubility of calcite is 0.015 gm/liter at 25°C in distilled water free of oxygen, and increases with temperature to 0.05 gm/liter at 100°C. When the distilled water is in equilibrium with atmospheric CO_2, the solubilities (in gm/liter) are 1.30 at 9°C, 0.943 at 25°C, and 0.765 at 75°C. Calcite shows a decrease in solubility with temperature that overshadows the positive effect of CO_2 on solubility (Boynton 1980). Direct solubility measurements for calcite (Plummer and Busenberg 1982) in CO_2-H_2O solutions between 0° and 90°C give the following values of equilibrium constant for the reaction

$CaCO_{3(s)} = Ca^{2+} CO_3^{2-}$: log K = $-171.9065 - 0.077993T + (2839.319/T) + 715951$ logT, where T is in Kelvins. The rate of dissolution of calcite is proportional to the activity of H^{1+} at low pH and becomes constant in the neutral range. At high pH, the rate of dissolution decreases due to backward precipitation (Chou et al. 1989).

The decomposition temperature of calcite at one atmosphere is 894.4°C. For the reaction $CaCO_3 = CaO + CO_2$ between 7 and 35 bars, the univariant pressure (CO_2)-temperature curves pass through 985° and 1,100°C (Chang et al. 1996).

2. Limestone

Most limestones rarely have only carbonate in their occurrences. Silica, clays, iron oxides, iron carbonates, and iron sulfides are the other common minerals present. The clays include illite, smectite, chlorite, and kaolinite; and the iron minerals are hematite, magnetite, limonite, pyrite, and siderite. The iron carbonate may be incorporated into the calcite as a solid solution. Silica is present in the form of fine-grained chert, fibrous chalcedony, and silt- or sand-sized grains of quartz. These noncarbonate minerals are either disseminated throughout the limestone or concentrated in laminae or thin partings (Pettijohn 1957; Hatch and Rastall 1971). A general classification of sedimentary rocks on the basis of calcium carbonate, sand, mud, and chert defines limestone, sandy limestone, argillaceous limestone, siliceous limestone, and marl (Williams et al. 1982). A mineralogical classification of carbonate rocks in terms of their relative contents of calcite, dolomite, and other noncarbonate minerals is shown in Fig. 43. For practical use, a chemical classification is made on the basis of %$CaCO_3$ and %$MgCO_3$. Ultrahigh and high calcium limestones are defined as having, respectively, more than 97%$CaCO_3$ and 95%$CaCO_3$, whereas a high-purity carbonate rock and a high-magnesian dolomite are defined as having, respectively, more than 95% combined $CaCO_3$ and $MgCO_3$ and 43%$MgCO_3$ (Carr et al. 1994). Trace elements commonly present in limestones are manganese, copper, titanium, fluorine, arsenic, barium, strontium, sulfur, and phosphorus. In addition, finely disseminated organic matter is a common constituent of limestone (Carr et al. 1994).

Pure limestones, including chalk, marble, and travertine, are white, but the presence of impurities can produce a variety of colors and patterns. The gray color, which is caused by carbonaceous matter, often becomes dark gray, and even approaches black. The presence of iron minerals may introduce buff, red, and brown colors. For pure calcite, the density is 2.71 gm/cm^3 at 20°C; and it is 2.93 gm/cm^3 and 3.0 gm/cm^3 for aragonite and dolomite, respectively. The porosity of limestone results in apparent densities of 2.1 to 2.5 gm/cm^3. Pure calcite has a hardness of 3 on the Mohs scale, whereas the hardness of aragonite is 3.5. The value for most limestones varies between 2 and 4. Cherty limestones have a higher hardness because of the silica content, which makes them abrasive to crushers in the beneficiation processes (Chang et al. 1996).

There is wide variance in the strength of limestones. Generally, marble and travertine have the highest values, and chalk and marl have the lowest. Measured compression, shear, and tensile strength range between 8.27 and 195.8 MPa, between 4.14 and 20.27 MPa, and between 2.41 and 6.21 MPa, respectively (Boynton 1980).

The average linear thermal expansion coefficient for limestones is $8 \pm 4 \times 10^{-6}$ and for marble is $7 \pm 2 \times 10^{-6}$. Values of heat capacity in joule/gm are 1.00 at 58°C for lime-

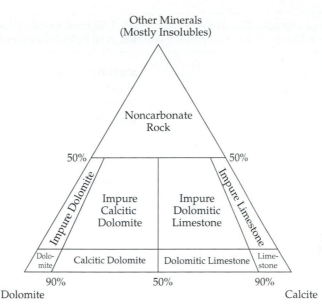

Figure 43   A general mineralogical classification of carbonate rocks, after Carr and Rooney (1983).

stone, and 0.79 at 0°C for marble, increasing to 1.00 at 200°C. Measured specific heats and calculated heat conductivities of high calcium and dolomitic limestones are 817 and 880 joule/kg-K at 20°C, and 0.0039 and 0.0034 cal/cm^3-sec-°C at 125°C, respectively. The heat of formation for the reaction $CaO+CO_2 \rightarrow CaCO_3$, is 48,000 cal/mole at 25°C (Birch et al. 1950).

All limestones are crystalline. The grain size increases with the degree of crystallization that occurred during formation. Shell limestones have a grain size of 1 μm, marls and chalks from 2 to 5 μm, dense limestones from 5 to 25 μm, and marbles and spars from 25 to 1,000 μm. Based upon grains, lime mud, cement, and pores, Leighton and Pendextor (1962) proposed a textural classification for limestone and established a nomenclatural system from the ratio of the relative proportions of grains to micrite (Table 12).

Differential staining is widely used to identify calcite, dolomite, and magnesite in limestone. This method is based on the fact that organic dyes stain calcite in acid solutions, and stain dolomite and magnesite in basic solutions. When both calcite and dolomite are present in a rock, their relative proportions may be determined by x-ray diffraction (Chang et al. 1996).

GEOLOGICAL OCCURRENCE AND DISTRIBUTION OF DEPOSITS

Limestone is widely distributed throughout the world in deposits of varying sizes and degrees of purity. Along with dolomite, limestone comprises about 15% of all sedimentary rocks. Most limestones of industrial importance are partly or wholly formed through organic processes. Marine life extracted the dissolved calcium salts to build shells and skeletons of calcium carbonate. These gradually accumulated on the seabed to produce deposits, many of which were massive (Rooney and Carr 1971). The major environments

Table 12 Textural classification of limestone. M. W. Leighton and C. Pendextor AAPG © 1962, reprinted by permission of the American Association of Petroleum Geologists.

Grain/ Micrite Ratio	% Grains	Grain Type					Organic Framebuilders	No Organic Frame-builders
		Detrital Grains	Skeletal Grains	Pellets	Lumps	Coated Grains		
9:1	~90%	Detrital Ls.	Skeletal Ls.	Pellet Ls.	Lump Ls.	Oolitic Ls. Pisolitic Ls. Algal En-crusted Ls.	Coralline Ls. Algal Ls. Etc.	Caliche Travertine Tufa
1:1	~50%	Detrital-Micritic Ls.	Skeletal-Micritic Ls.	Pellet-Micritic Ls.	Lump-Micritic Ls.	Oolitic-Pisolitic-Etc.) Micritic Ls.	Coralline-Micritic Ls. Algal-Micritic Ls. etc.	
1.9	~10%	Micritic-Detrital Ls.	Micritic-Skeletal Ls.	Micritic-Pellet Ls.	Micritic-Lump Ls.	Micritic-Oolitic (Pisolitic Etc.) Ls.	Micritic-Coralline Ls. Micritic-Algal Ls. Etc.	
		⟶ Micritic Limestone ⟶						

of carbonate formation are epeiric seas, shelf margins, and deep sea basins, probably in that order of abundance in the geologic record. Carbonate sediments in the central parts of the epeiric seas tend to be micrites. Toward the margins of the basin, sand-sized fossils become more abundant, and peloidal muds occur in the more saline, shallower water near the basin edge. Examples of limestones formed in such epeiric seas include the Ordovician and Mississippian rocks of the midcontinental United States. Shelf margins are the most favorable site for reef development because of the supply of nutrients and the agitation of the waters. Modern examples of such sites include (1) the reef tract along the southern tip of Florida, (2) the Great Barrier Reef along the northeastern coast of Australia, and (3) the circular reef growths that partially surround many volcanoes in the Pacific Ocean in the lower latitudes. Carbonate sediments are also found in bathyal and abyssal depths in modern oceanic basins. Limestones formed in such depths have been distinguished from rocks of Mesozoic and Cenozoic age (Blatt 1982). Limestones that form in high-energy zones generally contain few noncarbonates and may be the source of highly pure limestone. Limestones that accumulate in zones of low energy (lagoons or deep water) are more likely to be contaminated with clay and noncarbonates.

The Cretaceous Period is known as the age of chalk. Extensive Cretaceous chalk deposits exist in western Europe and parts of North America (Key 1960). It has been suggested that unlike other limestones, chalk was originally laid down as a calcite-rich, not an aragonite-rich, mud. Hence, no recrystallization of aragonite to calcite subsequently occurred, and so the rock retained its original porosity and hardness. In Scotland, Antrim, and northeastern England, the chalk is cemented moderately well by secondary calcite that fills the original pores and forms a comparatively hard limestone, whereas in southern England the chalk is much softer and more friable. The density of the Antrim chalk varies between 2.60 and 2.64 gm/cm^3 and that of the soft English chalks between 1.70 and 1.95 gm/cm^3, because of this difference in cementation (Hancock 1963; Harris 1982). The

texture of the chalk varies considerably from bed to bed. A high proportion of Inocera-
mus prisms gives rise to a perceptibly gritty-textured rock, whereas a reduced quantity of
molluscan debris, along with a correspondingly large bulk of coccoliths, results in the
finest varieties of chalk (Black and Barnes 1959). Large Cretaceous chalk deposits of considerable thickness occur in three well-defined belts in the United States. They are
(1) South Dakota-Nebraska-Wyoming-Kansas, (2) Arkansas-Oklahoma-Texas, and
(3) Tennessee-Mississippi-Alabama (Key 1960).

 Marble is a metamorphic rock, and most economic deposits are formed by regional metamorphism. Good examples are found in the Green Mountains, in Vermont
(the United States), at Clifden, in Connemara, Ireland, and at Carrara, in Tuscany, Italy
(Meade 1980; Harben and Kuzvart 1996). Extensive well-bedded lenses of travertine
are mined in New Mexico (Austin 1990). Other travertine-producing states are Montana and Idaho. The finest quality Iceland spar is found only in calcareous veins associated with igneous rock. At Helgustadir, near Eskifjordor, Iceland, Iceland spar is mined
from mud-filled cavities in basaltic flows (Ladon and Meyers 1951).

INDUSTRIAL PROCESSES AND USES

1. Industrial Processes

Quarry and mine operations, including drilling, blasting, and quarrying broken stone,
are the first steps in the production of limestone. These steps are followed by a series of
crushing, screening, classification, and pulverizing processes (Boynton et al. 1983).
Other processes used are washing, drying, heavy media separation, and optical sorting
(Boynton 1980; Carr et al. 1994).

 The large-size limestones are broken down by means of a primary crushing process,
generally either by compression in jaw and gyratory crusher or by impact in impact break-
ers and hammermill crushers. After screening and classification, a secondary crushing
process is usually employed to further reduce the size of small limestone pebbles, granules, and fines. The secondary crushers differ from the primary crushers in that they are
smaller, have less capacity, operate at higher speeds, and their discharge openings can be
more quickly adjusted. Cone crushers are widely used for secondary crushing. Crushed
limestones are generally screened into oversize ($+15$ cm), large ($-15 +5$ cm), medium
($-5 +0.6$ cm), and fine (-0.6 cm). When secondary crushing fails to achieve the required
degree of size reduction, tertiary crushing is usually employed. Double roll crushers, roll
drills, and hammermills are used for this purpose (Brown 1962; Bickle 1976).

 Pulverization is used to produce micron size fines. Both wet and dry grinding operations have been used successfully to pulverize limestone. In wet grinding, limestones
are processed in roll mills, from which particles of from $-$No. 60 to $-$No 100 (0.25 to
0.15 mm) mesh are floated off in countercurrents of water at various velocities. The
fines are sedimented in setting tanks, thickened, dewatered by centrifuges and/or vacuum filters, and dried. Dry grinding relies on air instead of water as the grinding
medium. An air separator is used in connection with the grinding mill to control the air
(Boynton 1980; Severinghaus 1983; O'Driscoll 1990).

 In addition to the above standard procedures for beneficiation of limestone, limestone is washed in order to meet the more stringent quality requirements of both the
chemical and construction industries. Scrubbers, water jets, and log washers are commonly

used. Drying is applied to products used for special industrial needs, such as asphalt, concrete, and others. In some cases, drying is utilized after wet grinding (Boynton 1980).

Heavy-media separation is done by adding medium, finely ground ferrosilicon that contains a small amount of magnetite, to water. With a specific gravity of 2.70 to 2.75, the desired limestone concentrate sinks in the medium, whereas the tailings rise and float to an overflow, and thence to a wastepile. In the optical mineral sorter, pieces of limestone are fed into a highly illuminated optical chamber and inspected by scanning cameras. The pieces that exceed the preset tolerances for light, dark, or specific colors are deflected through a side of the sorter by compressed air ejectors (Boynton 1980).

2. Industrial Uses of Limestone

Limestone is one of the most useful and versatile of all industrial minerals. Its main uses are (1) as construction materials, (2) in chemical and metallurgical industries, (3) for glass and ceramics, (4) in the production of cement, (5) for soil treatment, and (6) as mineral feed in agriculture. Both marble and travertine are used as decorative stone. Iceland spar is restricted to use in the manufacture of optical instruments, such as the Nicol prism.

Construction: Limestone is the most widely used crushed rock, constituting more than twice the combined usage for all other crushed stones, including granite, basalt, sandstone, and others. In the aggregate industries, only sand and gravel outsell limestone. Quality specifications for construction rock are related mainly to its physical properties. Thus, the rock must be clean, strong, dense, durable, free from cracks, and have the required particle size distribution (Harben and Kuzvart 1996).

There are numerous tests recommended by various agencies that are used to determine the type of limestone suitable for construction uses. The ten most important of these tests are (1) Dorry hardness for abrasion resistance, (2) Los Angeles abrasion, (3) Paget impact for toughness and impact resistance, (4) specific gravity, (5) absorption for porosity, (6) compressive strength, (7) cementing value for suitability in concrete, (8) sodium sulfate and magnesium sulfate for soundness, (9) alkali reactivity, and (10) freezing and thawing for soundness under winter conditions (Boynton 1980).

For the principal construction applications, the major aggregate grades of limestone are (1) concrete aggregate, (2) road aggregate, (3) riprap, (4) railroad ballast, (5) fill material, (6) stone sand, (7) roofing granules, (8) asphalt filler, and (9) terrazzo. The physical properties of aggregate used in Portland cement concrete are important since they influence its workability, strength, and durability (Blanks and Kennedy 1955). Top sizes of aggregate and mix proportions vary greatly, depending on the applications, which include dams, highways, walls, foundations, footings, revetments, curbs, sidewalks, reinforced concrete bridges, and concrete products (Anon. 1962). The examples below are given in terms of volume of Portland cement, fine aggregate, course aggregate, and water. The compression strength of the resulting concrete is also given:

Portland cement (ft^3)	1	1	1
Fine aggregate (ft^3)	1–1.5	1.5–2	2–2.5
Coarse aggregate (ft^3)	2–3	3–3.5	3.5–4
Water (gal)	4.5	5.15	6.75
Compression strength (psi)	5,500–6,000	4,500–5,000	3,000–4,000

The boundary line between fine and coarse aggregate is set at No. 4 mesh (4.8 mm). Material that passes the No. 4 mesh is considered fine aggregate; that which is retained on the same screen is coarse.

Road aggregate is a general term that embraces the use of coarse, fine, and graded aggregates in constructing or maintaining road bases and subbases, and as an aggregate in various types of bituminous pavement surfaces (Hewen 1942; Anon. 1976; Barkdale 1991). Railroad ballast, used in the maintenance and construction of railroad beds, has the same general aggregate requirements as roadstone. Additional specifications call for a gradation of 64 to 19 mm and a maximum of 40% wear in the Los Angeles abrasion test.

Finely pulverized limestone is a common mineral filler for asphalt concrete mixes. These fines are usually No. 200 to No. 325 mesh sizes (75 to 45 μm), and are chemically inert in contact with bitumen. Generally, 1 to 10%, with an average of 9%, mineral fillers are added to the aggregate mixture to achieve good density and stability by reducing the percentage of voids in the aggregate to a minimum (Cummins 1960). Limestone sand is used in construction as a fine aggregate in concrete, mortar, and road construction. It is produced in a size distribution of 98 to 100%, passing a No. 18 mesh (920 μm), with only 0 to 2% passing a No. 100 mesh (15 μm) (Kalcheff 1977).

Riprap consists of heavy, irregular stone fragments, some of which may be as large as boulders. It is principally used in river and harbor work. Limestone is used as a fill material in elevating swampy and low-lying land to the desired grade for construction of roadways, industrial parks, and urban developments.

White crystalline limestone or marble chips of 9.5 to 6.4 mm in size are commonly used as the final coat on certain types of flat built-up roofs both for aesthetic considerations and for high reflectivity. They are also used to strengthen the roofing materials by embedded into bituminous bed. Gradation of roofing aggregate is normally 19 to 2.36 mm. Limestone granules for composition roofing have an average gradation of No. 12 to 40 mesh (1.4 mm-350μm). Limestone chips, usually of 9.5 to 3.3 mm size, are frequently used in the finish coats of textural exterior plaster or stucco.

Lime Production: Lime, or quicklime, produced from limestone by calcination, constitutes one of the major industrial uses of limestone. The basic reaction, $CaCO_3 \rightarrow CaO + CO_2$, is a function of both the temperature and partial pressure of carbon dioxide. With the temperature of decomposition of $CaCO_3$ at 895°C under one atmospheric pressure, complete calcination can only be made above 900°C. The release of CO_2 results in a 44% weight loss. The calcination of limestone, like other calcination processes, starts on the exterior surface and proceeds inward. Therefore, the rate of reaction depends upon particle size, porosity, and purity. To expel the CO_2 produced from the interior of a large piece of limestone, sufficient pressure must be generated by means of high temperatures. Such high temperature treatment may overburn the surface layers of the limestone and produce textural changes through excessive shrinkage. Fig. 44a shows time-temperature-particle size relationships, and Fig. 44b, the variation of apparent density with temperature (Oates 1990).

A large variety of kiln systems are used for lime calcination. These include vertical, rotary, rotary with preheaters, and multiple-shift regenerative kilns. The selection of kiln depends upon the burn characteristics of limestone, fuel consumption, and capital equipment costs. In comparison, vertical kilns have the following advantages over

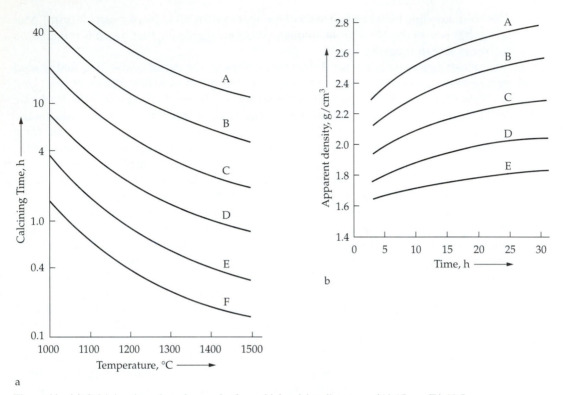

Figure 44 (a) Calcining times for spheres of a dense, high-calcium limestone, (A) 15 cm, (B) 12.5 cm, (C) 10 cm, (D) 7.5 cm, (E) 5.0 cm, and (F) 2.5 cm diameter, and (b) variation of apparent density with temperature and time for a dense, high-calcium limestone at (A) 1400°C, (B) 1300°C, (C) 1200°C, (D) 1100°C, and (E) 1000°, after Oates (1990).

rotary kilns: (1) lower capital investment, (2) more efficient fuel utilization, (3) less refractory wear, (4) less attritive use of limestone, and (5) lower pollution control cost. The advantages of using rotary kilns, on the other hand, include the ability to (1) burn small limestone, (2) produce a wide range of burn from soft to hard, including deadburned, (3) use all types of fuels, (4) have greater capacity, and (5) produce uniform products. The multiple shaft kiln was developed to burn small limestone by utilizing a parallel flow calcining principle in double- and triple-shaft units. Fuel consumption can reach a level of 5.4 to 5.7 Gj/T of lime produced (Boynton et al. 1983).

In most kilns, the lime is exposed to gas temperatures of 1,200° to 1,300°C, just before it enters the cooling zone. Generally, such kilns produce highly reactive lime from high-calcium limestones if the residual $CaCO_3$ content is relatively high. Some designs complete the calcination at a lower temperature of 1,100°C and produce a highly reactive lime with low levels of $CaCO_3$.

When a limestone particle decomposes at temperatures below 900°C, its external dimensions do not change significantly. Limestone with 2% porosity produces lime with a porosity of 54%, corresponding to an apparent density of 1.5 gm/cm³. Prolonged heating of lime above 900°C causes sintering, a reduction of porosity, and an increase in apparent density. The sintering process markedly reduces the reactivity of the lime to water (Boynton et al. 1983).

Mixing lime with water produces slaked or hydrated lime. There are numerous procedures used for slaking lime, ranging from the batch process to continuous slakers, which produce either milk of lime or lime putty. Slaking at 90°C produces the finely divided particles of calcium hydroxide required in most industrial uses.

Lime, or quicklime, has more diverse uses and varied functions than any other commercial material. It is second only to sulfuric acid in industrial consumption. Its major uses are

1. Metallurgical: fluxing steel, copper ore flotation, manufacture of alumina by the Bayer process, and manufacture of magnesium and magnesia by electrolytic and thermal reduction.
2. Chemical: manufacture of sodium carbonate, bicarbonate, and hydroxide by the Solvay process, causticing, bleaching, and pulp digestion in papermaking, production of dead-burned magnesia, and manufacture of most calcium-based organic salts and pharmaceuticals.
3. Environmental: softening and purification of water, sewage and waste treatment, and air pollution abatement.
4. Ceramic: manufacture of glass, refractories, sand-lime brick, and concrete products.
5. Construction: production of masonry mortar, plaster, and stucco, and soil stabilization.
6. Agricultural: soil liming (Boynton 1980).

Glass and Ceramics: Limestone is an important material in glass manufacture. Its fluxing action works on silica to form a chemically fused calcium silicate glass phase. Limestone improves the chemical properties of glass by making it less soluble, the mechanical properties of glass by making it less brittle, and the appearance of glass by providing a more enduring luster. The ranges of variation in composition of different types of soda-lime glass are 68 to 75%SiO_2, 10 to 15%Na_2O, 5 to 14%CaO, and 0 to 10%MgO. Other chemicals or mineral additives are added to the basic ingredients in the glass batch to produce desired opacity, decolorization, heat resistance, and other characteristics (Scholze 1991).

Specifications of limestone for glass uses emphasize uniformity in chemical composition and gradation. Generally, a low iron content is essential, usually less than 0.06%FeO, and for optical glass, a maximum of 0.01 to 0.02%. Very precise gradations are required, such as 100% passing a No. 10 sieve (2 mm), and 96 to 100% retained on a No. 100 sieve (0.15 mm). A British specification established to regulate chemical purity for limestone in colorless glass is as follows (Power 1985):

Chemical Composition (in weight%):
55.2%CaO (min), 98.2%$CaCO_3$ (min),
0.035%Fe_2O_3 & FeO (max),
1.0% acid insoluble including silica (max),
1.0% organic matter (max),
2.0% moisture content (max),
MnO, PbO, P_2O_5, SO_2 must be quantitatively declared by the supplier if these impurities individually exceed 0.1%,
Al_2O_3, MgO must be subjected to agreement upon any limiting values, between supplier and buyer,

Other coloring elements must not be present in such quantities as to stain glass.
Particle Size Distribution:
For use in tank furnace, 100% minus 4.76 mm
    75% plus 0.124 mm
For use in pot furnace, 100% minus 3.175 mm
    95% plus 1.2 mm

Finely pulverized limestone, or whiting, is a necessary component of many formulas for glazes and enamels that are applied to pottery and whiteware.

Paper and Pulp: Limestone is employed in the sulfite process for making paper pulp. In this process it is reacted with SO_2 to form a calcium bisulfite pulp-cooking liquor on the Jennsen Tower system. A high calcium limestone with a minimum $CaCO_3$ content of 95 to 97% is favored over dolomitic limestone in this process. Limestone of large size, up to 8 to 10 in (20.3 to 25.4 cm) is preferred for use in the Tower (Boynton 1980).

Filler and Extender: Limestones, both calcitic and dolomitic, are used as fillers or extenders in well over 100 different industrial products. Some of the industrial fillers could be classed as standard pulverized limestone of nominal No. 100 mesh or No. 200 mesh sizes (150 μm to 75 μm), but a sizable portion involve finer sizes, known commercially as whiting, derived from intensive grinding of limestone, marble, chalk, or oyster shell. The physical properties of whiting are generally analogous to and competitive with precipitated $CaCO_3$. The division line between pulverized limestone and whiting is not well established. A specification of at least 95% to pass a No. 325 sieve (44 μm) for whiting is generally accepted.

Because of the large number of uses for filler, grades tend to be of a more customized nature. Some of the physical and chemical characteristics demanded in many specifications are:

1. good white color and high degree of reflectivity;
2. minute particle size, from 10 microns to submicrons;
3. special particle size distribution for special usage;
4. special particle shape for special usage, and a high specific surface area;
5. known plastic and rheological properties;
6. high absorption of oil, ink, and color pigments;
7. low to zero chemical activity;
8. extreme purity; and
9. particular specific gravity and bulk densities.

Table 13 shows the chemical and physical characteristics of British produced whitings, and typical applications in which they are used (Power 1985; Trivedi and Hagemeyer 1994).

Portland Cement: Portland cement is produced in rotary kilns from a mixture of limestone and clay. As the temperature increases, the water is driven off. CO_2 is liberated around 800° to 900°C, and the CaO from $CaCO_3$ reacts with the aluminosilicates to form a clinker at 1,600°C. This clinker is ground with mineral

Table 13 Chemical and physical characteristics of a British produced whiting and the typical industrial uses, after Power (1985).

Typical Applications	PAINT	RUBBER	PVC	PAPER FILLER	PAPER	COATING	POLISH CHEMICALS ETC
ECC product specification	Queensgate 'Paris White'	Polcarb	G400	Queensfill	NPI00	Carbital (marble)	Queensgate 19
$CaCO_3$	96.15	98.5	98.5	96.15	98.5	98.1	96.15
Water soluble salt content %	0.13	0.10	0.14	0.13	0.14	—	0.13
Moisture content %	0.20	20.0	1.0	15%*	—	—	0.20
Particle size distribution							
plus 53 microns	0.3	0.02	0.1	0.75	0.02	0.01	9.0
plus 10 microns	18	1.0	13	18	1.0	1.0	30
plus 5 microns	33	2.5	35	35	2.5	4.0	53
minus 2 microns	36	80(±3%)	40	40	.80(±3%)	90	28
minus 1 micron	—	42	—	—	42	70	—
Specific surface area m^2/g	2.70	6.0	1.88	2.2	8.0	5-16	1.5
Brightness %	85.5	89.0	84.0	85.5	89.0	96.0	84.7
*slurry form		(±0.7%)					

additives, such as gypsum, to enhance cement performance (Midgley 1964). The reactions may be identified in four stages (Knöfel 1984):

1. In the drying process, water adsorbed by the clay minerals is removed between 100° and 400°C, and the bound water evaporates at temperatures of between 400° and 750°C.

2. In the calcination process, decomposition of clay minerals produces Al_2O_3 and SiO_2, and decalcination of $CaCO_3$ produces CaO and CO_2 in the temperature range of from 600° to 1,000°C. The oxides react to form $CaO \cdot SiO_2$, $CaO \cdot Al_2O_3$, and $2CaO \cdot SiO_2$. The formation of $3CaO \cdot SiO_2$ and $4CaO \cdot Al_2O_3 \cdot Fe_2O_3$ begins at 800°C.

3. In the sintering process, the first melting phases are formed at about 1,260° to 1,310°C, in the system CaO-Al_2O_3-Fe_2O_3-SiO_2. CaO and $2CaO \cdot SiO_2$ dissolve in the presence of melt and produce $3CaO \cdot SiO_3$ by crystallization. At temperatures above 1,400°C, the solid phases in the melt are $3CaO \cdot SiO_2$ and $2CaO \cdot SiO_2$, and the melt has a composition of approximately 56% CaO, 7% SiO_2, 23% Al_2O_3, and 14% Fe_2O_3. Almost all of the Al_2O_3 and Fe_2O_3 are in the melt.

4. In the cooling process, a rapid rate (>40 K/min) leads to a solidification of the melt without dissolving the existing mineral phases.

Portland cement clinker consists mainly of the four crystalline phases listed below, along with their chemical composition and the abbreviations used to represent them. Their relative amounts in the clinker are also given. They are arranged in order of maximum, average, and minimum values (Knöfel, 1984):

Tricalcium silicate(C_3S)	$3CaO \cdot SiO_2$	65–45–38%
Dicalcium silicate (C_2S)	$2CaO \cdot SiO_2$	35–27–15%
Tricalcium aluminate (C_3A)	$3CaO \cdot Al_2O_3$	15–11–7%
Tetracalcium aluminoferrite (C_4AF)	$4CaO \cdot Al_2O_3 \cdot Fe_2O_3$	15–8–4%

There are no specifications for the limestone used in Portland cement manufacture. However, various specifications are used for the finished cement. Typical British specifications for ordinary and rapid-hardening Portland cement are as follows (Power 1985):

1. Lime saturation factor (LSF) should not be greater than 0.66 or less than 1.22, where LSF equals $[(CaO)-0.7(SO_3)]$ divided by $[2.8(SiO_2)+1.2(Al_2O_3)+0.65(Fe_2O_3)]$, with all oxides in wt% in the cement.

2. MgO content should be less than 4.0%, of SO_3 less than 2.5 to 3.0%, weight of acid insoluble residue less than 1.6%, and loss in ignition less than 3.0% in a temperate climate and 4.0% in a tropical climate.

Flue Gas Desulphurization: Pulverized limestone or chalk is used to tackle acid rain at the source of emission by means of flue gas desulphurization (FGD), and acid water at the site of deposition by neutralization. While there exists a strong case for sulfur removal with limestone, problems under evaluation are the design of the scrubbing systems, removal of scale buildup, and sludge disposal. At the present time two scrubbing systems, dry and wet, are in use. The dry process involves the introduction of a finely ground limestone into the fluid bed firing system of a boiler. The resulting CaO produced is used to absorb SO_2, thus forming calcium sulfite and sulfate as a solid assemblage that, along with fly ash, is trapped in an external dust collector. An alternative system of sulfur removal is a wet scrubbing process, in which a wet media of suspended limestone is employed to react with the flue gas as it passes upward toward the stack.

The chemical reactions involved in desulfurization of stack gas with limestone are complex and, in general, may be described as follows: (1) hydrolysis and ionization of SO_2 with H_2O to produce H^+ and SO_3^{2-}, (2) dissolution and ionization of limestone in an acidic medium to produce Ca^{2+} and HCO_3^-, (3) desulfurization of SO_3^{2-} with Ca to produce $CaSO_3$ (sludge), and (4) oxidation of $CaSO_3$ to produce $CaSO_4 \cdot 2H_2O$ (sludge) (Slack 1973).

A high grade limestone of from 85 to 95% $CaCO_3$ is required for the flue gas desulfurization process. Iron content appears to be beneficial to the process, as it catalyses the reaction, but MgO content generally leads to problems in waste disposal, as magnesium sulfate is some 150 times more soluble than the calcium compound. The MgO content should be a maxima of 5%. The limit of acid insolubles should also be 5%. Fine-grained dense limestone shows better reactivity than coarse-grained crystalline limestone. Limestone with a size distribution of 84% passing a 325 mesh (45 μm) has been shown by testing results to be effective (Power 1985).

Pollution Control: Pulverized limestone is used as a neutralizer to treat river and lake water polluted by acid rain and industrial waste. Calcitic limestone of at least 75% passing a No. 200 mesh (75 μm) is the most efficient form to use. In comparison with lime and sodium-based alkalis, limestone has a sluggish rate of reaction and cannot neutralize a strong acid much over pH 6.0. It generally yields a larger volume of sludge for disposal. Yet, limestone is by far the lowest-cost acid neutralizer per unit of basicity, as well as the most widely available (Ford 1974; Boynton 1980).

Flux: Limestone is used in metallurgical processes as a flux. The smelting of pig iron in the blast furnace requires limestone, as lime, to react with impurities in the ore and fuel in order to form a molten slag that can be removed gravimetrically. For nonferrous metals, limestone is also the most widely employed flux for refining copper, lead, zinc, antimony, and manganese. In the process, limestone is charged into the smelters along with the concentrated ores of these metals. Its function is to remove impurities in a way similar to that of pig iron. Limestone is also used to remove silica in the Bayer process for alumina manufacture (Boynton et al. 1983).

Agriculture: Direct application of limestone to soil is by far the largest agricultural use. It reduces the acidity of soil and supplies calcium and magnesium nutrients to the soil. Most crops and plants grow profusely under neutral to slightly acid conditions (pH of 6 to 7). Acid soil of pH lower than 6.0 can be made more fertile and productive for many crops by neutralizing soil acids with limestone. The neutralization is effected by base exchange between the calcium or magnesium cation and the hydrogen ions in the soil (Anon. 1981).

Almost every type of vegetation contains at least traces of calcium and magnesium. Alfalfa, for example, contains 6.0% Ca and 3.0% Mg, and tobacco contains 6.07% Ca and 2.74% Mg. Calcium and magnesium displace hydrogen ions that are present in organic acids through neutralization of Na and K in alkali soils (Adams 1984).

The neutralization efficiency of a liming material, which is designated as its total neutralizing power (TNP), is measured by titration with a standard acid solution. For 100% pure calcium carbonate, a "100" rating is given. Thus, high calcium limestones may have a TNP value ranging between 80 and 98, depending upon the extent of impurities. Pure magnesium carbonate has a high TNP value, since it has a lower molecular weight, but the same neutralizing power per molecular unit as $CaCO_3$. The factor of $MgCO_3$ as calcite equivalent is 1.19 to 1.0. In a pure dolomitic limestone of a 60 to 40 ratio of $CaCO_3$ to $MgCO_3$, the TNP is 108. A liming material with a TNP of 80 would theoretically require a 20% heavier application for equivalent neutralizing power than one with a TNP of 100 (Boynton 1980).

In addition to TNP, fineness and particle gradation constitute other important considerations in evaluating liming materials. It is generally agreed that neutralizing efficiency rises as fineness is increased. Limestone particles of No. 100 mesh (0.15 mm) dissolve and react with soil acids in usually two to four weeks, whereas No. 10 mesh requires 12 months or longer. To correlate the importance of neutralizing power and particle fineness, the state of Iowa contrived the concept of effective calcium carbonate equivalent (ECCE), adopting it as a state specification for evaluating liming materials. The formula is as follows:

$$\text{Fineness factor} = (\%\text{No. 4 mesh} \times 0.1) + (\%\text{No. 8 mesh} \times 0.3) + (\%\text{No. 60 mesh} \times 0.6)$$

$$\text{ECCE} = CaCO_3 \text{ equivalent} \times \text{fineness factor}$$

Assume a limestone with 94.2% $CaCO_3$ and 1.8% $MgCO_3$, total a $CaCO_3$ equivalent of 96.2%. If it has sizes of 100% No. 4 mesh (4.75 mm), 98% No. 8 mesh (2.36 mm), and 65% No. 60 mesh (250 μm), it would have a fineness factor of 78.4 and an ECCE of 7,542. Some states have adopted this evaluation method with modifications, and others have different standards.

Pulverized dolomitic limestone is commonly added to fertilizer as a filler or diluent. It contributes Ca and Mg as plant nutrients and counteracts acidity stemming from acidic nitrogen chemicals. In general, 4 lbs of dolomitic soil counteract the acidity of 1 lb of ammonium sulfate, and 2.5 lbs are needed for ammonium nitrate and urea. Pulverized calcitic limestone is used as a vital ingredient in mineral feed supplements, which fortify organic feed for farm animals and poultry.

Granular limestone is consumed by hatcheries as poultry grits. A uniform size gradation, without fines and coarse aggregate, is required. The predominate sizes are 0.375, 0.25, and 0.125 in (9.5, 6.4, 3.2 mm). Rounded or spherical granules are preferred to flat or elongated ones, and the purity of the limestone is not important (Adams 1984).

REFERENCES

Adams, F., ed. 1984. *Soil acidity and liming.* 2nd ed. Madison, Wisconsin: Amer. Soc. Agronomy, 380p.

Anon. 1976. Guidelines for Construction of Serviceable Crushed Stone Bases. *National Crushed Stone Asso. Bull.,* December, 1976, 21p.

Anon. 1981. Lime and liming. Ministry of Agriculture, Fisheries, and Food, London, Her Majesty's Stationery Office, 44.

Austin, G. S. 1990. Commercial travertine in New Mexico. *New Mexico Geology*, 12: 49–58.

Barkdale, R. D., ed. 1991. *The aggregate handbook.* Washington, D.C.: National Stone Association.

Bickle, W. 1976. Crushing and grinding. *Jour. Amer. Ceramic Soc.* 42:294–304.

Birch, F., J. F. Schariar, and H. Spicer. 1950. *Handbook of physical constants.* 2nd. ed. Geol. Soc. Amer. Special Paper 36. Washington, D.C., 325p.

Black, M., and B. Barnes. 1959. The structure of coccoliths from the English chalk. *Geol. Mag.* 96:321–328.

Blanks, R., and H. Kennedy. 1955. *The technology of cement and concrete.* Vol. 1. New York: John Wiley & Sons, 341p.

Blatt, H. 1982. *Sedimentary petrology.* San Francisco: W. H. Freeman &. Co, 564p.

Boynton, R. S. 1980. *Chemistry and technology of lime and limestone.* New York: John Wiley & Sons, 578p.

Boynton, R. S., K. A. Gutschick, R. C. Freas, and J. L. Thompson. 1983. Lime. In *Industrial minerals and rocks.* 5th ed. Ed. S. J. Lefond, 809–831. New York: Soc. Mining Engineers, AIME.

Brown, G. 1962. High capacity rock crushing plants: Selection, installation and operation. *Jour. Quarry Mag.* 46: 19–24.

Carr, D. D., and L. F. Rooney. 1983. Limestone and dolomite. In *Industrial minerals and rocks.* 5th ed. Ed. S. J. Lefond, 853–868. New York: Soc. Mining Engineers, AIME.

Carr, D. D., L. F. Rooney, and R. C. Freas. 1994. Limestone and dolomite. In *Industrial minerals and rocks.* 6th ed. Ed. D. D.Carr, 605–629. Littleton, Colorado: Soc. Mining, Metal, and Exploration.

Chang, L. L. Y., R. A. Howie, and J. Zussman. 1996. Non-silicates: Sulphates, carbonates, phosphates, halides. Vol. 5B of *Rock-forming minerals.* 2nd. ed. Essex, England: Longman, 383p.

Chou, L., R. M. Garrels, and R. Wollast. 1989. Comparative study of kinetics and mechanisms of dissolution of carbonate minerals. *Chem. Geol.* 78:269–282.

Cummins, A. B. 1960. Mineral fillers. In *Industrial minerals and rocks.* 3rd ed. Ed. J. L. Gillson, 567–584. New York: AIME.

Ford, C. T. 1974. Selection of limestones as neutralizing agents for coal-mine water. *Proc., 10th Forum on Geology of Industrial Minerals, Miscellaneous Rept. 1, Ohio Geol. Survey.* 30–42. Columbus, Ohio.

Hancock, J. M. 1963. The hardness of the Irish chalk. *Irih. Nat. Jour.* 14:157–164.

Harben, P. W., and M. Kuzvart. 1996. *Industrial minerals—A global geology.* London: Industrial Minerals Information Ltd., Metal Bulletin OLC, 462p.

Harris, P. M. 1982. Limestone and dolomite. *Mineral Dossier no. 23, Inst. Geol. Sci.,* Her Majesty's Stationery Office, London, 111p.

Hatch, F., and R. Rastall. 1971. *Petrology of Sedimentary Rocks.* Vol. 2. New York: Hafner, 502p.

Hewen, L. 1942. *American highway practice.* Vols. 1 and 2. New York: John Wiley & Sons, 459p, and 492p.

Kalcheff, I. 1977. Portland Cement Concrete with Stone Sand. *Nat. Crushed Stone Asso. Eng. Rept* (July 1977), 18p.

Key, W. W. 1960. Chalk and whiting. In *Industrial minerals and rocks.* 3rd ed. Ed. J. L. Gillson, 233–242. New York: AIME.

Knöfel, D. 1984. Inorganic binders. In *Process mineralogy of ceramic materials.* Eds. W. Baumgart, A. C. Dunham, and G. C. Amstutz, 52–79. New York: Elsevier.

Ladoo, R. B., and W. M. Myers. 1951. *Non-metallic minerals.* 2nd ed., 122–125. New York: McGraw-Hill.

Leighton, M. W., and C. Pendextor. 1962. Carbonate rock types. In Classification of carbonate rocks. Ed. W. E. Hamm. *Amer. Assoc. Petroleum Geol. Memoir* 1:33–61.

Meade, L. P. 1980. The Vermont marble belt and the economics of current operation, in Proc. 14th. Annual Forum on Geology of Industrial Minerals. Eds. J. R. Dunn et al., Bull 436, New York State Museum, p. 54–56.

Midgley, H. G. 1964. The formation and phases compositions of Portland cement clinker. In *The chemistry of portland cements.* Ed. H. F. W. Taylor, 89–130. London: Academic Press.

National Ready-Mixed Concrete Association. 1962. "Control of Quality for Ready Mixed Concrete" publ. 44, 5th ed. 51 p, Washington: D.C.

Oates, T. 1990. Lime and limestone. In *Ullmann's encyclopedia of industrial chemistry.* Vol. A15, 317–346. VCH Publishing Weinheim: Verlag *GmbH.*

O'Driscoll, M. 1990. Fine carbonate fillers. *Ind. Minerals* (September):21–42.

Pettijohn, F. J. 1957. *Sedimentary rocks.* 2nd ed. New York: Harper & Row, 660p.

Plummer, L. N., and E. Busenberg. 1982. The solubilities of calcite, aragonite and vaterite in CO_2-H_2O solutions between 0° and 90°C and an evaluation of the aqueous model for the system $CaCO_3$-CO_2-H_2O: *Geochim. Cosmochim. Acta* 46:1011–1040.

Power, T. 1985. Limestone specifications—limiting constraints on the market. *Ind. Minerals.* (October):65–91.

Rooney, L. F., and D. D. Carr. 1971. Applied geology of industrial limestone and dolomite. *Indiana Geol. Survey Bull.* 46, 59p.

Scholze, H. 1991. *Glass–nature, structure, and properties.* New York: Springer-Verlag, 454p.

Severinghaus, N., Jr. 1983. Fillers, filters and absorbents. In *Industrial minerals and rocks.* 5th ed. Ed. S. J. Lefond, 243–257. New York: Soc. Mining Engineers, AIME.

Slack, A. 1973. Removing SO_3 from stack gases. *Env. Sci. Tech.* 7:110–119.

Trivedi, N. C., and R. W. Hagemeyer. 1994. Fillers and coatings. In *Industrial minerals and rocks.* 6th ed. Ed. D. D. Carr, 483–495. Littleton, Colorado: Soc. Mining, Metal. & Exploration.

Williams, H., F. J. Turner, and C. M. Gilbert. 1982. *Petrology—an introduction to the study of rocks in thin sections.* San Francisco: W. H. Freeman, 626p.

Lithium Minerals

There are more than one hundred fifty lithium-bearing minerals, but only five of them have industrial importance. They are spodumene, lepidolite, petalite, amblygonite, and eucryptite. Lithium and lithium salts of economic quantities are also found in some brines. Well-known producing areas are Searles Lake and Clayton Valley in the United States, Salar de Uyuni in Bolivia, and Salar de Atacama in Chile. Lithium minerals are mainly used in the production of lithium carbonate and hydroxide, and in the formulation of glass and ceramics. The world's major producers of lithium minerals and salts are the United States, Zimbabwe, and Canada. Other important producers are the former USSR, China, Chile, and Australia.

MINERALOGICAL PROPERTIES

Four of the five major lithium minerals, spodumene, lepidolite, petalite, and eucryptite, are lithium aluminosilicates, and the fifth one, amblygonite, is a lithium aluminum phosphate. The principal salt precipitated from brine at Searles Lake is dilithium sodium phosphate, while at Salar de Atacama and Salar de Uyuni, lithium is precipitated as lithium carbonate. Mineralogical properties of these five lithium minerals are given below.

Spodumene: Spodumene, $LiAl(SiO_3)_2$, is a pyroxene-type mineral consisting of SiO_3 chains linked together by Al^{3+}, with the balance of the positive charge provided by Li^{1+}. Sodium may replace lithium in significant amounts; and iron, chromium, and manganese may also be present (Deer et al. 1992). The theoretical composition of spodumene has a Li_2O content of 8.0%, but in its natural occurrence Li_2O varies from 1.0 to 7.6%. The typical content of the concentrate is 4.0 to 7.5% Li_2O.

Spodumene usually occurs as prismatic crystals in colors varying from white, pale green, and yellow, to emerald green, pink, and lilac. The pure and transparent single crystals of green hiddenite and lilac kunzite are regarded as semiprecious stones. Spodumene has a hardness of 6 to 7, and a specific gravity of 3.1 to 3.2. Its refractive indices range from 1.660 to 1.676. With the replacement of Li by Na and Al by Fe, α decreases in the former and increases in the latter. Birefringence increases in both replacements (Deer et al. 1992).

The naturally occurring spodumene (α-form) undergoes an irreversible phase transition to the β-form at around 1,000°C, accompanied by a 30% volume increase as the specific gravity decreases to 2.4 (Roy et al. 1950; London 1984). The β-form spodumene has a keatite-type structure with a remarkably low thermal expansion. It has an extensive range of solid solution in the system $Li_2O \cdot Al_2O_3 \cdot 2SiO_2 - SiO_2$ (Skinner and Evans 1960; Fishwick 1974).

Lepidolite: Lepidolite, a lithium mica, has a composition of $K_2(Li,Al)_{5-6}$ $(Si_{6-7}Al_{2-1}O_{20})(OH,F)_4$, and a complete series of solid solutions has been proposed to exist between lepidolite and muscovite. Considerable amounts of sodium, rubidium, and caesium may substitute for potassium; and iron, magnesium, and manganese may enter the octahedral sites of aluminium and lithium. The theoretical content of Li_2O is 6.43%, but in economic deposits, the contents are between 3.0 and 4.5% Li_2O (Deer et al. 1992). A lepidolite from Pala, San Diego, California, has the following composition: 50.28% SiO_2, 22.81% Al_2O_3, 0.18% CaO, 0.16% MgO, 4.92% Li_2O, 0.97% Rb_2O, 0.30% Cs_2O, 17.48% K_2O, 1.20% F, 0.41% Mn_2O_3, 2.10% H_2O, and a total of 100.71%.

Lepidolite usually occurs as a compact aggregate of very small flakes that form scaly, granular masses, and occasionally as short prismatic crystals. Its color is commonly pink or lilac, but can also be red, purple, gray, white, or yellow. Lepidolite has a hardness of 2 to 4 and a specific gravity of 2.8 to 2.9. The refractive indices of lepidolite show a wide range of variation depending on the Fe-content, but most lepidolites have $\alpha=1.529$ to 1.537, $\beta=1.552$ to 1.565, and $\gamma=1.555$ to 1.568 (Deer et al. 1992).

Petalite: Petalite has a structure consisting of a framework of SiO_4 and AlO_4 tetrahedra, which are linked by sharing apices. Alternatively the SiO_4 tetrahedra can be regarded as being arranged in sheets parallel to (001) that are joined to one another through AlO_4 (Cerny and London 1983). The chemical formula for petalite is $LiAlSi_4O_{10}$, with a Li_2O content of 4.9%, but in its natural occurrence the range can be between 3.6 and 4.7%. A representative chemical composition of Rhodesian petalite is given as follows (Fishwick 1974): 77.00% SiO_2, 16.80% Al_2O_3, 4.3% Li_2O, 0.39% K_2O, 0.16% Na_2O, 0.06% Fe_2O_3, 0.80% LOI, and a total of 99.51%. Petalite crystallizes in the form of monoclinic prisms, and is white, gray, or reddish in color. It has a specific gravity of 2.4 and a hardness of 6.5. Petalite decomposes at temperatures above 1,000°C, an important temperature for the digestion of its ore, to β-spodumene and SiO_2.

Eucryptite: Eucryptite, $LiAlSiO_4$, is a relatively rare lithium mineral that occurs mainly in association with petalite. It is commonly colorless, with a specific gravity of 2.66 and a hardness of 6.5. There is also an α–β phase transition in eucryptite, and β-eucryptite also shows an extensive range of solid solution in the system $Li_2O \cdot Al_2O_3 \cdot 2SiO_2 - SiO_2$ (Roy et al. 1950; Fishwick 1974).

Amblygonite: Amblygonite is a phosphate of lithium and aluminum with fluorine, and usually contains combined water. Its formula is $LiAlPO_4F$, with a Li_2O content of 10.1%. Sodium often replaces lithium in amblygonite, and in its natural occurrence the Li_2O content ranges from 3.5 to 10%. Amblygonite usually occurs as coarse aggregates or rounded masses. Its hardness is 6, and its specific gravity ranges between 2.85 and 3.09. Amblygonite is normally white, sometimes with brown, yellow, green, or blue tints (Deer et al. 1992).

Of the lithium aluminosilicates, its most unusual physical property is its remarkable thermal expansion coefficient: slightly positive for β-spodumene, almost no variation for β-spodumene-petalite solid solution, and highly negative for β-eucryptite (Hummel 1952; Fishwick 1974). Smoke (1951) identified two regions in the system $Li_2O-Al_2O_3-SiO_2$ in which the thermal expansion is negative. As shown in Fig. 45, the

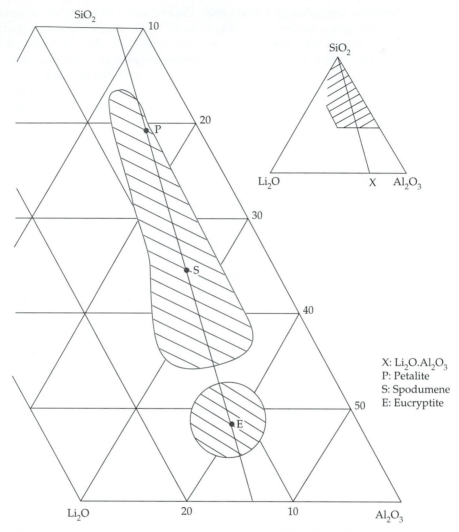

Figure 45   Regions of negative thermal expansion in the system Li_2O-Al_2O_3-SiO_2, after Smoke (1951).

linear thermal expansions range from 0 to -0.38% for the region in which β-eucryptite solid solution is the principal crystalline phase, and from 0 to -0.04% for the region in which β-spodumene solid solution is the principal phase.

GEOLOGICAL OCCURRENCE AND DISTRIBUTION OF DEPOSITS

Lithium minerals occur predominantly in granitic pegmatites that were formed either by the crystallization of postmagmatic fluids or by the metasomatism of residual magmatic fluids. Common minerals present are feldspar, quartz, and mica, along with a group of minerals of rare elements such as beryllium, tantalum, tin, and cesium (Kunasz 1994). Many lithium-bearing pegmatites are zoned, with each zone containing a specific suite of minerals. The lithium minerals are usually found in the intermediate zones. The distribution of lithium content is homogeneous throughout the unzoned pegmatites.

Lithium is present in significant amounts in certain brines of Searles Lake, California (Smith 1976), Clayton Valley, Nevada (Kanasz 1974), Great Salt Lake, Utah (Whelan and Petersen 1976), Salar de Atacama, Chile (Ericksen et al. 1976) and Salar de Uyuni, Bolivia (Eriksen et al. 1977). Unusual amounts of lithium are found in the clay mineral hectorite, a magnesium-rich end member of the smectite group (Kunasz 1994).

The world's major producers of lithium minerals and salts are the United States, Zimbabwe, and Canada. Other important producers are the former USSR, China, Chile, and Australia.

United States: Lithium minerals are mined in North and South Carolina, and lithium-containing brines are exploited in Nevada and California.

The Kings Mountain district, which extends from Lincolnton, North Carolina, to Gaffney, South Carolina, has the world's largest proved reserve of lithium ore. The district is characterized by numerous pegmatites that intrude amphibolites and schists along the eastern margin of the Devonian Cherryville quartz monzonite (Kesler 1961, 1976). The pegmatites, mostly unzoned, commonly contain spodumene that is uniformly distributed throughout. A small amount of amblygonite is present, but not lepidolite or petalite. Pegmatites in the schist are mostly concordant, and those in more competent rocks mostly discordant. On the average, the pegmatites consist of 20% spodumene, 32% quartz, 6% muscovite, 41% albite and microcline, and about 1% trace minerals, including cassiterite. The spodumene contains 7.5% Li_2O (Kesler 1976).

Other known pegmatite sources include the Black Hills of South Dakota, the Pala District of California, and the White Picacho District of Arizona. In addition to the granitic pegmatites, anomalously high concentrations of lithium occur in fluviatile sediments near the McDermitt caldera, Nevada and Oregon (Glanzman and McCarthy, 1978). The sediments, which are vitroclastic in origin, consist mainly of zeolites, clay minerals, feldspar, quartz, calcite, and gypsum. Lithium of an average concentration of 0.01 to 0.1% is hosted by the clay minerals.

In Clayton Valley, Esmeralda, Nevada, lithium-bearing brines occur in an undrained structural depression filled with Quarternary sediments that contains mainly clay minerals, including hectorite, volcanic sands, alluvial gravels, and saline minerals, including gypsum and halite (Kunasz 1974, 1994). The playa lake, 50 km^2 in area, lies at an elevation of about 1,400 meters. The brine which saturates the sediments occurs as a concentrated sodium chloride solution containing subordinate amounts of potassium, and minor amounts of magnesium and calcium. The clay minerals are the hosts for the lithium. The lithium content in the sediments ranges from 260 to 1,713 ppm.

Lithium is produced as a by-product in the treatment of the brines of Searles Lake, California, for the recovery of potassium salts. $Li_2Na(PO_4)$ accumulates with buckeite, a double salt of sodium carbonate and sodium sulfate, which is a residual product from evaporation processes (Whelan and Petersen 1976).

Canada: The production of lithium minerals is based on the zoned Tanco pegmatite, at Bernic Lake, Manitoba (Burt et al. 1988). The Tanco pegmatite is one of a number of subhorizontal pegmatite sheets that makeup the Bernic Lake pegmatite group and are hosted by a synvolcanic metagabbro intrusive. There are eight discrete mineralogical zones within the Tanco pegmatite (Crouse et al. 1979). They are (1) wall zone: k-feldspar, albite, quartz; (2) aplitic albite zone: albite, quartz; (3) lower intermediate zone: k-feldspar, albite, quartz, spodumene; (4) upper intermediate zone: spodumene, quartz, amblygonite; (5) central intermediate zone: k-feldspar, quartz, albite; (6) quartz zone: quartz; (7) pollucite zone: pollucite; and (8) lepidolite unit: lepidolite. Low iron spodumene ($<0.05\%$ Fe_2O_3), with a Li_2O content of 2.76%, is confined to the upper and lower intermediate zones. Lepidolite has a Li_2O content of 2.24%. Other lithium-bearing pegmatites occur in Quebec, Ontario, Saskatchewan, and the Northwest Territories (Flanagan 1979; Lasmani 1979; Williams and Trueman 1979).

Zimbabwe: The lithium minerals occur in the Bikita zoned pegmatite, the largest of a series of pegmatites in the Bikita tin field, some 70 km east of Fort Victoria. All five lithium minerals and bikitaite ($Li_2O \cdot Al_2O_3 \cdot 4SiO_2 \cdot 2H_2O$) were found at Bikita, with lepidolite and petalite as the principal lithium minerals. The Li_2O contents are 4.5% in petalite, 8% in eucryptite, 7.5% in spodumene, 4.1% in lepidolite, 10% in amblygonite, and 6% in bikitaite (Wegener 1981). The pegmatite is divided into three sectors: the Al Hayat and Bikita sectors of the Quarry area, and the Bikita sector of the Southeast area. Only the first two sectors have been exploited. Distinct zonations exist in both sectors. The lithium mineral-bearing zones of the Al Hayat sector consist of a spodumene zone at the center, a petalite-feldspar zone above it, and a feldspathic lepidolite zone below it. Enveloping the Li-zones are mica-feldspar-quartz zones. In the Bikita sector, the core zone consists of massive lepidolite. The intermediate zones above it have petalite and spodumene, and below it have lepidolite. Both extend to the wall zones of feldspar, quartz, and muscovite. Zonations are complex in the Bikita sector (Harben and Bates 1990).

Others: Sizable deposits of spodumene and lepidolite were found in Transbaikal, in the Altai mountains, and on the Kola peninsula of the former USSR (Ginzberg and Lugovskoi 1977). In northwestern China, spodumene deposits were reported in the southern slopes of the Altai mountains, in the Koktohay-Aletai region (Kunasz 1994).

Important reserves of lithium-bearing brines are found at Salar de Atacama, Chile; Salar de Uyuni, Bolivia; Salar del Hombre Muerto, and Salar del Rincon, Argentina; Zha Buye Lake, Tibet; and Dachaidan, Qinghai, China. The lithium contents in these brines are in wt%, respectively, 0.15, 0.024, 0.06, 0.05, 0.12, and 0.02 (Bauer 1990). Ericksen et al. (1977) measured the concentration and distribution of lithium in salars de Uyuni, Coipassa, and Empexa. The concentration of lithium in the brines of the three salars, which are basically saturated chloride solutions, ranges from 80 to 1,150 ppm. The salars of Uyuni, Coipassa, and Empexa are remnants of a former large lake that occupied the southern Bolivian Altiplano in the late Pleistocene age.

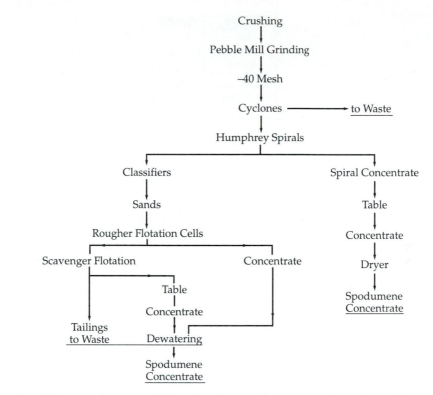

a.

Figure 46   (a) Flow sheet for the beneficiation of spodumene at Foote. Reprinted from Crozier *Flotation: Theory, Reagents and Ore Testing,* 1992. and (b) Flow sheet for the beneficiation of spodumene at Tanco. Source: Industrial Minerals, January 1988.

INDUSTRIAL PROCESSES AND USES

1. Beneficiation and Treatment

In the beneficiation of lithium minerals, the mined ore undergoes several stages of crushing, grinding, and classification, followed by flotation with organic reagents. The production of spodumene concentrate at Foote plants as described by Crozier (1992) involves a combination of gravity separation and flotation to produce a coarse spodumene concentrate from spirals and a fine concentrate from flotation (Fig. 46a). At Tanco plant, a combination of four processes are used, as illustrated in Fig. 46b (Burt et al. 1988): (1) heavy medium separation, with a mixture of ferrosilicon and magnetite to remove K-feldspar; (2) flotation at a pH of 9.2, using starvation quantities of tall oil fatty acid and petroleum sulphonate, to separate amblygonite from spodumene; (3) upgradation of the ore to the 7.25% Li_2O by flotation, to remove quartz and Na-feldspar; and (4) high intensity magnetic separations to remove iron contents. Beneficiation of Bikita lithium minerals by a series of similar processes as described produces petalite concentrates of 4.10% Li_2O, 0.07% Fe_2O_3, 0.30% K_2O, and 0.50% Na_2O; spodumene concentrates of 7.10% Li_2O, 0.08% Fe_2O_3, 0.20% K_2O, and 0.20%

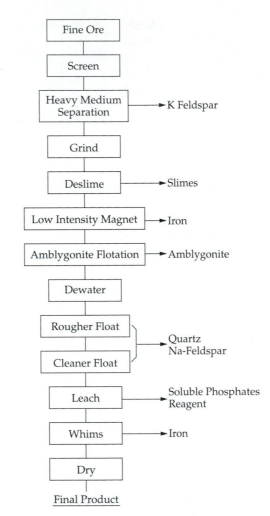

b.

Figure 46 *(Continued)* (a) Flowsheet for the beneficiation of spodumene at Foote, after Crozier (1992), and (b) Flowsheet for the beneficiation of spodumene at Tanco. Source: Industrial Minerals, January 1988.

Na_2O; and lepidolite concentrates of 4.10% Li_2O, 0.06% Fe_2O_3, 8.00% K_2O, and 0.50% Na_2O.

In the recovery of lithium from clay minerals, a lime-gypsum roast extraction procedure was utilized. It consists of feed separation, roasting, leaching, evaporation, product precipitation, and crystallization (Lien 1985).

2. Production of Lithium Compounds

The digestion of lithium concentrate can be carried out with either an acid or alkali process using salts. Depending on the process and the digestion agent, the product obtained will be lithium carbonate, lithium hydroxide, or lithium chloride. Amblygonite and lepidolite can be treated directly with the appropriate digestion agent, but spodumene and petalite require decrepitation to convert α-spodumene to β-spodumene, and petalite to β-spodumene and quartz. Decrepitation, which is usually accomplished in rotary kilns at 1,000 to 1,100°C, results in a 30% increase in the mineral's volume, and a considerable improvement in its grindability (Steinberg et al. 1946; Averill and Olson 1978).

In the acid process, the Li-mineral concentrate is ground and mixed with sulfuric acid to produce a mixture that is heated to 250 to 400°C for the silicates, and 850 to 900°C for the phosphate. The lithium sulfate formed is leached out of the reaction product with hot water, and the filtrate is treated with lime and soda to remove alkaline-earth metals and iron. Lithium carbonate is precipitated by reacting lithium sulfate solution with sodium carbonate solution. Digestion by hydrochloric acid produces lithium chloride.

In the alkali process, lepidolite or spodumene is heated with lime to produce lithium aluminate and calcium orthosilicate, at 900° or 1,040°C, respectively. Lithium hydroxide is formed by leaching the aluminate with water in several stages. The residues are then separated, and the solution is concentrated in vacuum evaporators used to crystallize lithium hydroxide monohydrate.

The lithium carbonate and hydroxide produced in these processes may be further processed into a variety of lithium compounds.

3. Treatment of Brines

The processes used for recovery vary with the character of the brine. At Clayton Valley, it is only economically feasible to rely on solar evaporation to concentrate the brines because of the removal of a large quantity of the water in the original brine. The bittern is treated with lime to remove magnesium, and lithium is then directly precipitated by using sodium carbonate. When the concentrations of calcium and magnesium are high, pretreatment must be done to prevent coprecipitation with the lithium carbonate (Averill and Olson 1978). At Searles Lake, the lithium is recovered by evaporating the brines to produce potash, borax, salt, soda ash, and dilithium sodium phosphate. Froth flotation is used to separate other salts from the dilithium sodium phosphate, which is digested with sulfuric acid, and the lithium is precipitated with sodium carbonate. From the Salar de Atacama brine, recovery of lithium is made from mixed sulfate salts such as $LiKSO_4$, which are crystallized in the summer. K_2SO_4 is produced by reacting the $LiKSO_4$ concentrate with a KCl solution, and Li is recovered in the form of Li_2CO_3 by adding Na_2CO_3 to the solution (Vergara-Edwards et al. 1985).

4. Industrial Uses of Lithium Minerals

Lithia (Li_2O) has four major industrial applications: glasses, glass-ceramics, glazes, and enamels.

Glasses: Acting as a powerful flux, lithia gives better strength, durability, and appearance to soda-lime glasses. Adding lithia to soda-lime glasses results in faster melting and increased productivity, as well as lower melting temperatures and a prolonged furnace life. Consequently, stack emission is reduced for air pollution control. Lithia is the most effective alkali for reducing the viscosity of glasses. The temperature corresponding to a viscosity of 10^{12} poises was found to be 566°C for the base glass, whereas the lithia, potassia, and soda glasses gave 500°, 533°, and 544°C, respectively (Dingwall and Moore 1953). The presence of lithia in the compositions of glass-forming systems produces glasses of low thermal expansion. For fiber glasses, a small amount of lithia added to the glass formulation provides low operating temperatures, high fiber production rates, a low volatilization rate, a low liquidus point, and a slow devitrification rate. Because of the high cross section of thermal neutrons, lithia-containing borosilicate glasses possess good radiation-absorbing properties (Fishwick 1974; Shore and Haigh 1991).

Glass-Ceramics: Glass-ceramics are produced by melting a suitable glass batch containing a nucleating agent, forming the glass into a desired shape, allowing the glass to cool to a temperature at which nucleation occurs, maintaining a temperature suitable for the formation of a sufficient number of nuclei, and increasing the temperature in order to promote crystal growth and complete the recrystallization. Most glass-ceramics are based upon compositions in the system Li_2O-Al_2O_3-SiO_2 because lithium silicate glasses are capable of volume crystallization (Baumgart 1984). The eight initial compositions for the production of glass-ceramics (Stookey 1960) are as follows:

SiO_2	69.8	57.6	61.7	73.1	58.7	56.1	67.4	63.7
Al_2O_3	14.9	15.2	15.3	13.5	13.7	12.1	14.4	12.1
TiO_2	7.0	12.1	10.7	4.5	13.7	13.8	7.0	13.8
Li_2O	4.3	5.2	4.3	4.9	3.9	3.0	4.2	3.1
MgO	—	—	—	—	8.8	3.9	1.2	3.9
CaO	—	—	—	—	—	11.1	1.8	3.4
CaF_2	—	—	—	2.1	—	—	—	—
Na_2O	1.0	1.6	0.9	1.7	1.0	—	1.0	—
K_2O	—	—	0.1	0.2	0.2	—	—	—
B_2O_3	3.0	4.7	3.1	—	—	—	3.0	—
ZrO_2	—	3.6	3.9	—	—	—	—	—

For all of them, β-spodumene or β-eucryptite is the principal crystalline phase, and the resulting glass-ceramics possess a very low thermal expansion. Glass-ceramics of this type are mainly used as oven-to-freezer cookware, range tops, catalyst supports, laboratory bench tops, heat exchangers, radomes, building cladding, and telescope mirror blanks.

Lithium disilicate has a considerably higher thermal expansion than β-spodumene, and glass-ceramics based on this compound are used on capacitors, vacuum gears, transformers, and other devices.

Glazes: Lithium carbonate is used in the formulation of raw alkaline glazes in place of sodium and potassium compounds because it is only slightly soluble in water and provides essential color with a 10 to 12% concentration (Richardson 1939). For high-voltage porcelain glaze, lithium carbonate is used because Li-containing glaze possesses a lower thermal expansion than the porcelain and, therefore, produces tangential compressive stress on the surface of the glazed ware and increases its strength (Maslennikova and Kochetkova 1970). Lithia is an effective substitute for lead oxide in the production of glazes with homogeneous coating and high gloss. Whiteware glazes are produced by a combination of Li_2O and SrO. Li_2O increases brilliance while SrO contributes to fusibility (Orlowski and Marquis 1945). Crystalline glazes that are based on lithia are used for ceramics of low thermal expansion (O'Conor and Eppler 1973).

Enamels: Substitution of lithia, as lithium carbonate, for soda in the composition of a white cover frit, results in improved fusibility, gloss, and opacity (Lewis 1943). Adding lithia has a considerable effect on the recrystallization of TiO_2 in titania-opacified enamels (Houston 1957). Fenton (1963) summarizes the effects of Li_2O in enameling, as follows: (1) increase in fluidity and reduction in viscosity, (2) reduction of maturing temperature or maturing time, (3) increase in application rates, (4) reduction in coating thickness, and (5) improvement in thermal resistance.

Others: Lithium compounds, carbonate, hydroxide, chloride, fluoride, and bromide, each have well-identified industrial uses. In addition to its consumption by the glass and ceramic industries, lithium carbonate is added to aluminum electrolysis cells, resulting in higher production by increasing the conductivity of the molten bath and reducing the operating temperature. Lithium carbonate is also used in the chemotherapeutic treatment of manic depression. By reacting lithium hydroxide with fatty acids, greases are produced that retain their viscosity over a wide range of temperatures and remain stable in the presence of water. Anhydrous lithium hydroxide is an absorber of CO_2. Lithium chloride and lithium bromide solutions are used in absorption refrigeration systems, while lithium fluoride is used as a flux in enamels, glasses, glazes, and in welding and brazing (Kunasz 1994).

Lithium metal, because of its low equivalent weight and high standard electrode potential, is an ideal anode material for high-energy-density battery systems. Ambient-temperature lithium batteries have shown excellent shelf life and performance characteristics over a wide range of temperatures. Some of these battery systems have become alternative power sources for a number of applications where conventional aqueous alkaline-based batteries were once used. Primary lithium systems that operate around 3.0 volts are lithium manganese dioxide, lithium vanadium pentoxide, lithium carbon monofluoride, and lithium silver chromite. Lithium systems that operate around 1.5 volts are lithium copper sulfide, lithium copper oxide, lithium lead bismuthate, and lithium iron sulfide (Venkatasetty 1984).

REFERENCES

Averill, W. A., and D. L. Olson. 1978. A review of extractive processes for lithium from ores and brines. In *Lithium: Needs and resources.* Ed. S. S. Penner, 305–313. Oxford: Pergamon Press.

Bauer, R. J. 1990. Lithium and lithium compounds. In Vol. A15 of *Ullmann's encyclopedia of industrial chemistry,* 393–414. Weinheim: VCH Verlagesgellschaft mbH.

Baumgart, W. Glass. 1984. In *Process mineralogy of ceramic materials.* Eds. W. Baumgart, A. C. Dunham, and G. C. Amstutz 28–50. New York: Elsevier.

Burt, R. O., J. Flemming, R. Simard, and P. J. Vanstone. 1988. Tanco—a new name in low iron spodumene. *Ind. Minerals* (January): 53–59.

Cerny, P., and D. London. 1983. Crystal chemistry and stability of petalite. *Tschermaks Min. Petr. Mitt.* 31:81–96.

Crouse, R. A., P. Cerny, D. L. Trueman, and R. O. Burt. 1979. The Tanco pegmatite, southeastern Manitoba. *Canadian Inst. Min. Metal. Bull.* n.2:1–10.

Crozier, R. D. 1992. *Flotation: Theory, reagents and ore testing.* Oxford: Pergamon Press, 341p.

Deer, W. A., R. A. Howie, and J. Zussman. 1992. *An introduction to the rock-forming minerals.* Essex, England: Longman, 696p.

Dingwall, A. G. F., and H. Moore. 1953. Effects of various oxides on the viscosity of glasses of the soda-lime-silica type. *Jour. Soc. Glass. Tech.* 37:316–321.

Ericksen, G. E., G. Chong D., and Tomas Vila G. 1976. Lithium resources of Salars in the central Andes. *U.S.G.S. Prof. Paper,* 1005, 66–74.

Eriksen, G. E., J. D. Vine, and R. Ballan. 1977. Chemical composition and distribution of lithium-rich brines in Salar de Uyuni and nearby Salars in southwestern Bolivia. In *Lithium: Needs and resources.* Proc. Sym. Corning, New York. Ed. S. S. Penner, 355–363. Oxford: Pergamon Press.

Fenton, W. M. 1963. Lithium as a tool in enameling—a review. *Jour. Canadian Ceramic Soc.* 32:16–21.

Fishwick, J. H. 1974. *Applications of lithium in ceramics.* Boston: Cahners Books, 156p.

Flanagan, J. T. 1979. Lithium deposits and potential of Quebec and Atlantic Provinces, Canada. In *Lithium—needs and resources.* Ed. S. S. Penner 391–398. Oxford: Pergamon Press.

Ginzberg, A. I., and G. P. Lugovskoi. 1977. Deposits of lithium. In Vol. 1 of *Ore deposits of the USSR.* Ed. V. I. Smirmov, V. I. 295–311. London: Pittman Publ.

Glanzman, G. K., and J. H. McCarthy, Jr. 1978. Lithium in the McDermitt caldera, Nevada and Oregon. In *Lithium—needs and resources.* Ed. S. S. Penner, 347–353. Oxford: Pergamon Press.

Harben, P. W., and R. L. Bates. 1990. Industrial Minerals—geology and world deposits. London: Industrial Minerals Division, Metal Bulletin Plc, 312p.

Houston, M. D. 1957. Influence of lithia upon the recrystallization of titania. *Bull. Amer. Ceramic Soc.* 36:139–141.

Hummel, F. A. 1952. Significant aspects of certain ternary compounds and solid solutions. *Jour. Amer. Ceramic Soc.* 35:64–66.

Kesler, T. L. 1961. Exploration of the Kings Mountain pegmatites. *Mining Eng.* 13:1062–1068.

——— 1976. Occurrence, development and long-range outlook of lithium-pegmatite ore in the Carolinas. In Lithium resources and requirements by the year 2000, Ed. J. D. Vine, *U.S.G.S. Prof. Paper* 1005:45–52.

Kunasz, I. A. 1974. Lithium occurrence in the brine of Clayton Valley, Esmeralda County, Nevada. *Proc., 4th symposium on Salt.* Ed. A. H. Coogan, v.1, 57–66, Northern Ohio Geol. Soc., Cleveland, Ohio.

Kunasz, I. A. 1994. Lithium resources. In *Industrial minerals and rocks.* 6th ed. Ed. D. D. Carr, 631–642. Littleton, Colorado: Soc. Mining, Metal. & Exploration.

Lasmani, R. 1979. Lithium resources in the Yellowknife area, Northwest Territories, Canada. In *Lithium—needs and resources.* Ed. S. S. Penner, 399–408. Oxford: Pergamon Press.

Lewis, M. O. 1943. Effects of lithia substitution for soda in vitreous enamel. *Jour. Amer. Ceramic Soc.* 26:77–83.

Lien, R. H. 1985. Recovery of lithium from clay by a roast-leach-precipitation process. In *Lithium: Current applications in science, medicine, and technology.* Ed. R. O. Bach, 61–71. New York: John Wiley.

London, D. 1984. Experimental phase equilibrium in the system $LiAlSiO_4$-SiO_2-H_2O: A petrogenetic grid for lithium-rich pegmatites. *Amer. Mineral.* 69:995–1004.

Maslennikova, G. N., and B. F. Kochetkova. 1970. Effect of additions of certain oxides on the properties of glazes for high voltage porcelain. *Steklo i Keramika,* 28–30.

O'Conor, E. F., and R. A. Eppler. 1973. Semicrystalline glazes for low expansion whiteware bodies. *Bull. Amer. Ceramic Soc.* 52:180–184.

Orlowski, H. T., and J. Marquis. 1945. Lead replacements in dinnerware glazes. *Jour. Amer. Ceramic Soc.* 28:343–357.

Richardson, I. W. 1939. Use of lithium carbonate in raw alkaline glazes. *Jour. Amer. Ceramic Soc.* 22:50–53.

Roy, R., D. M. Roy, and E. F. Osborn. 1950. Compositional and stability relationships among the lithium aluminosilicates: Eucryptite, spodumene, and petalite. *Jour. Amer. Ceramic Soc.* 33:152–159.

Shore, I., and M. Haigh. 1991. Lithium minerals review 1990. *Ind. Minerals* (July): 44–45.

Skinner, B. J., and H. T. Evans, Jr. 1960. Crystal chemistry of beta spodumene solid solution in the system $Li_2O\cdot Al_2O_3$-SiO_2. *Amer. Jour. Sci.* 258A:312–324.

Smith, G. I. 1976. Origin of lithium and other compounds in the Searles lake evaporites, California. *U.S.G.S. Prof. Paper* 1005:92–102.

Smoke, E. J. 1951. Ceramic compositions having negative linear thermal expansion. *Jour. Amer. Ceramic Soc.* 34:87–90.

Steinberg, W. M., E. T. Hayes, and F. P. Williams. 1946. Production of lithium chloride from spodumene by a lime-gypsum roast process. *Rept. Invest., U.S.B.M.,* 3848, 7p.

Stookey, S. D. 1960. Method of making ceramics and products thereof. U.S. Patent 2,920,971, January 12.

Venkatasetty, H. 1984. Primary lithium batteries. In *Lithium battery technology.* Ed. H. V. Venkatasetty, 61–78. New York: John Wiley & Sons.

Vergara-Edwards, L., N. Parada-Frederick, and P. Pavlovic-Zuvic. 1985. Recovery of lithium from crystallized salts in the solar evaporation of salar de Atacama brines. In *Lithium: Current applications in science, medicine, and technology.* Ed. R. O. Bach, 47–59. New York: John Wiley.

Wegener, J. E. 1981. Profile on Bikita. *Ind. Minerals* (June) 1981:50–58.

Whelan, J. A., and C. A. Petersen. 1976. Lithium resources of Utah. *U.S.G.S. Prof. Paper* 1005:75–78.

Williams, C. T., and D. L. Trueman. 1979. An estimation of lithium resources and potential of northwestern Ontario, Manitoba, and Saskatchewan. In *Lithium—needs and resources.* Ed. S. S. Penner, 400–409. Oxford: Pergamon Press.

Magnesite

Magnesite, a common magnesium carbonate, has extensive deposits in Greece, Turkey, Austria, Czechoslovakia, Spain, the former USSR, China, North Korea, and other countries, including the United States. Its major use is in refractories. Various technical terms are commonly used in industry for magnesia produced by magnesite from different thermal treatments. These terms are caustic or caustic-calcined magnesia (or magnesite), dead-burned or sintered magnesia (or magnesite), and fused magnesia.

As the second most abundant metal in seawater, the supply of magnesium is sufficient for the production of magnesium compounds, including magnesium carbonate. In addition, magnesium chlorides are recoverable from some brines. Between 1990 and 1995, annual world production, on the average, was estimated to be 9.5 M tons (Kramer 1996).

MINERALOGICAL PROPERTIES

Magnesite, $MgCO_3$, is a member of the calcite-type carbonates and forms dolomite, $CaMg(CO_3)_2$, with $CaCO_3$ in the system $CaCO_3$-$MgCO_3$. The solid solubility between any pairs of these three carbonates is limited, and this is anticipated from the differences in their cell dimensions (Chang et al. 1996):

	Calcite	Dolomite	Magnesite
a_o,Å	4.989	4.807	4.632
c_o,Å	17.061	16.00	15.012

However, $MgCO_3$ forms extensive solid solutions with both $MnCO_3$ and $NiCO_3$, and a complete series of solid solutions exists between $MgCO_3$ and $FeCO_3$. In its natural occurrence, $FeCO_3$ is constantly present in magnesite. By using the variation of cell dimensions determined in x-ray diffraction analysis, the amount of $FeCO_3$ in $MgCO_3$ can be easily estimated. It can also be determined by a rapid x-ray spectrometric method with a flux of lithium tetraborate (57%) and lithium metaborate (47%) (Jones and Wilson 1991). Magnesite decomposes at 650°C to MgO and CO_2, and for iron-rich magnesite, decomposition occurs at 654°C, followed by an oxidation reaction of $FeCO_3$ in the solid solution at 677°C.

With H_2O and CO_2, magnesium carbonate forms $MgCO_3 \cdot 5H_2O$ (lansfordite), $MgCO_3 \cdot 3H_2O$ (nesquehonite), and $Mg_4(CO_3)_3(OH)_2 \cdot 3H_2O$ (hydromagnesite), a hydrated basic carbonate. Stability relationships between these minerals and magnesite were calculated by Langmuir (1965) using $10^{-5.1}$ as the activity product constant of magnesite at 25°C. Magnesite is not stable in the presence of water, at the unit activity of H_2O, below 60°C. Between 60°C and 10°C, nesquehonite and hydromagnesite are stable, depending on the partial pressure of CO_2; at 10°C nesquehonite transforms to lansfordite. At surface temperatures, nesquehonite is stable for $Pco_2 > 10^{-2}$ bar, and hydromagnesite is stable for $Pco_2 < 10^{-2}$ bar. Solubility studies indicate that for a Pco_2 of approximately 1 bar, the saturated solution is in equilibrium with lansfordite at temperatures up to about 10°C, with nesquehonite up to 55°C, and with hydromagnesite above 55°C. For an activity product constant of $10^{-8.1}$, the only stable mineral under surface or near-surface geological conditions is magnesite.

The crystallization kinetics of magnesite in aqueous solutions were studied utilizing the reaction: $(5MgO \cdot 4CO_2 \cdot 5H_2O) + x\ CO_2 = (4 + x)MgCO_3 + (4 + x)H_2O + (1 - x)Mg(OH)_2$ (Sayles and Fyfe 1973). Results illustrate inhibitions of magnesite crystallization by a species of Mg in solution, and strong positive catalysis by ionic strength and partial pressure of CO_2. The Mg inhibition was demonstrated to be limited by increasing concentration, but no such limit was observed by Sayles and Fyfe for positive catalysis by ionic strength. Therefore, highly saline conditions offer a favorable environment for magnesite crystallization.

Magnesite is usually white, but it may also be light to dark brown, if iron-bearing. With an appreciable amount of nickel, its color appears to be pale to emerald green. Magnesite resembles dolomite in being only slightly soluble in cold dilute HCl, but it dissolves with effervescence in warm acid. Its refractive indices ($\omega=1.700$ and $\varepsilon=1.509$) are higher than calcite ($\omega=1.658$ and $\varepsilon=1.486$) and dolomite ($\omega=1.679$ and $\varepsilon=1.500$), and lower than siderite ($\omega=1.857$ and $\varepsilon=1.633$) and rhodochrosite ($\omega=1.816$ and $\varepsilon=1.597$). Magnesite with a 50% $FeCO_3$ has $\omega=1.788$ and $\varepsilon=1.570$.

Magnesite has a hardness of 4 on the Moh's scale and a specific gravity of 3.00. A staining technique is generally used to differentiate magnesite from calcite and dolomite (Chang et al. 1996).

GEOLOGICAL OCCURRENCE AND DISTRIBUTION OF DEPOSITS

Magnesite deposits are of two general types: massive and crystalline. Massive magnesite, sometimes referred to as cryptocrystalline magnesite, is an alteration product of serpentine that has been subjected to the action of carbonate waters. The alteration reaction between serpentine, and either the surface carbonate waters percolating downward, or waters containing CO_2 rising upward through the fissures, is determined by the temperature and pressure for the formation of magnesite. The conditions at depths are not as favorable for the formation of magnesite as those near the surface. Consequently, deposits of massive magnesite tend to be shallow. The magnesite occurs as massive aggregates, lenticular masses, and veins of varying thickness in serpentine, an alteration product of magnesium-rich rocks such as dunite or peridotite. The magnesite formed by this process usually has limited amounts of impurities of iron, lime, and silica (Spangenberg 1949; Griffin 1972; Barnes et al. 1973; Wetzenstein 1989).

Deposits of crystalline magnesite are usually found in association with dolomite. It is generally thought that the reaction involved is a secondary replacement of magnesite in preexisting dolomite by magnesium-rich fluids on a volume for volume basis (Campbell and Loofbourow 1962). From an investigation of the physical and chemical conditions leading to the formation of magnesite, Johannes (1970) concluded that all magnesite occurrences can be classified into two types: (a) those formed by CO_2 metamorphism, and (b) those formed by Mg metamorphism. In the CO_2 metamorphism, the assemblages of quartz+magnesite and talc+magnesite are stable in contact with both CO_2-rich and CO_2-poor fluid phases, whereas the assemblages of enstatite+magnesite and serpentine+magnesite indicate a low CO_2 content in the fluid phase. In the Mg metamorphism, magnesite is formed by the action of salt-bearing solutions. In equilibrium with chloride solutions, the formation of calcite+dolomite or dolomite+magnesite is a function of temperature, pressure, CO_2 content, NaCl content, and the concentration of $CaCl_2$+$MgCl_2$ in the fluid.

Examples of massive magnesite deposits are found in Greece, Turkey, India, China, North Korea, and Australia, whereas those of crystalline magnesite occur in the United States, Austria, Spain, the former USSR, and Czechoslovakia.

Among deposits of massive or cryptocrystalline magnesite, the best known are those of Euboea island and the Khalkidiki peninsula, Greece. They lie within the Vardar zone consisting of carbonates, shale, graywacke, and volcanoclastics metamorphosed to greenschist facies. Massive magnesite deposits, up to 40 m thick and 960 m long, occur in a 18 km long serpentinite belt; and associated minerals are dolomite, quartz, chalcedony, and calcite. The magnesite in the deposits is of high purity with 97 to 99% $MgCO_3$ (Dabitzias 1980). Turkey has several deposits of massive magnesite, located mainly in the Eskisehir and Kutaya regions. The deposits occur in stockworks, with magnesite filling in weblike cracks in serpentine. In Yugoslavia, massive magnesite occurs in veins and stockworks in peridotite massifs as an extension of the Turkey-Greece ultrabasic belt (Manojlovic 1970). In addition, some sedimentary massive magnesite deposits were found in Miocene basins (Ilic and Popovic 1970). The Bela Stena deposit near Baska has a lenticular form enclosed in marls, shales, and sandstones; and the Beli Kamen deposits near Kosovska Kamenica occur in freshwater marls, sandstones, and tuffs. At Salem, Tamil Nadu, India, veins of white compact massive magnesite deposits occur in association with dunite, peridotite, and pyroxeneite (Harben and Bates 1990).

The largest deposits of crystalline magnesite in the United States are those near Chewelah, Washington (Bennett 1941) and near Gabbs Nevada (Schilling 1966). Extensive deposits of high-grade crystalline magnesite and brucite occur in the Luning district on the northwestern slope of the Paradise Range in Nye County, Nevada. The deposits occur in the upper part of the Luning formation (Triassic) of cherty limestone, dolomite, shale, and volcanics. Large deposits of coarsely crystalline magnesite occur in Stevens County, Washington, which are centered around Chewelah and Valley. Most of the magnesite is colored, varying from white through pink, red, and gray to black. It occurs as replacements of large lenses of dolomite in sedimentary rocks. The most important magnesite deposits in Canada are in the Grenville district, Quebec, and at Mount Brussilof, near Radium Hot Springs, British Columbia. In both deposits, the magnesite is crystalline, and cream white, milky white, or gray in color, and occurs in close association with dolomite (Harben and Kuzvart 1996).

In Austria, a group of about forty deposits extends throughout the "Grauwackenzone" from Semmering to the Tirol-Salzburg border. They are emplaced in a complex of graphitic shales and sandstones with Upper Silurian to Upper Carboniferous carbonates. The magnesite has an iron content up to 6% Fe_2O_3 (Dunning and Evans 1986). The magnesite deposits in Czechoslovkia are situated in a belt of Upper Carboniferous dolomites, graphitic phyllites, and other rocks of the Gemericum, between Podrecany and Ochtina, and near Kosice. Three generations of magnesite and six generations of dolomite were recognized in the Dubravsky massif deposit (Kuzvart 1981).

The magnesite deposits in the former USSR are in the Satka, in the western Ural Mountain district. The deposits of high-grade crystalline magnesite (40.5 to 43.5% MgO) consist of fourteen magnesite bodies in a layer of Proterozoic dolomite and limestone, 300 to 500 m thick and 8,000 m long. The magnesite-dolomite contact is concordant, straight up to lobate (Kuzvart 1981). Other large deposits have been reported near Lake Baikal, East Siberia.

The world's greatest deposits of magnesite are in the Haicheng-Dashiqiao district, Liaoning, China (Zhang 1988; Fountain 1992). The rocks consist of quartzite, sandstone, dolomite, schist, and marble, with minor intrusives of porphyritic greenstone and quartz porphyry. The magnesite is highly crystalline and of a high grade, with MgO content ranging from 44 to 48%. High-grade magnesite deposits extend from China to North Korea. There are a large number of deposits at Tanesengun, South Kankyo Province, and at Gosui in North Kankyo Province (Comstock 1963; Schmid 1984).

INDUSTRIAL PROCESSES AND USES

The major industrial use of magnesite is in the production of magnesia. Magnesia may also be produced from magnesium hydroxide obtained from seawater or magnesium brines. For refractory uses, the magnesia must meet stringent specifications of purity and uniformity. Therefore, beneficiation is usually necessary.

1. Beneficiation of Magnesite

Beneficiation of magnesite varies from plant to plant. Generally, the ore is crushed, scrubbed, and screened into two major fractions; the 1 1/2–1/8 in. (3.8 to 0.32 cm) fraction is sent to the heavy media separation cone (Utley 1954), and the −1/8 in. (0.32 cm) fraction is stockpiled for future treatment by flotation. Since the differences in density are small among the associated minerals (magnesite 3.0, dolomite 2.9, calcite 2.7, layered silicates 2.5 to 3.0, quartz 2.65, and graphite 2.2 g/cm^3), separation depends on exact control of the density. Finely divided ferrosilicon medium is used as the dense medium, and the separation is done in a 6 m diameter cone. A "float" of impurities is discarded, and a magnesite "sink" is crushed, screened, and stored in bins (Wicken and Duncan 1983).

Flotation provides a means for further beneficiation. Sodium silicate is used as a depressor for quartz and serpentine, and oleates or oleic acid used as a collector for the magnesite. The magnesite concentrate is removed as a foam, then thickened, and brought to a residual water content of 8 to 12% in centrifuges or filters. The product is then briquetted and burned (Utley 1954; Cordes 1982).

2. Production of Magnesium Hydroxide

Magnesium hydroxide has a limited occurrence in nature as the mineral brucite. It may be produced by slaking a reactive MgO, but most of the $Mg(OH)_2$ is obtained from seawater or magnesium brines. At the average magnesium concentration of 1.29 g/kg, about 3 grams of magnesium hydroxide can be produced from each liter of seawater (Gilpin and Heasman 1977). The underlying principle in precipitating $Mg(OH)_2$ from seawater involves the addition of hydroxyl ions in the form of calcium hydroxide so that the insoluble magnesium hydroxide will precipitate according to the equation: Mg^{2+} (in seawater) $+ Ca(OH)_2 \rightarrow Mg(OH)_2 + Ca^{2+}$. Dolomite may also be used in the process. The equation is as follows, with the addition of calcination as the first step: $CaMg(CO_3)_2 \rightarrow (Ca,Mg)O + 2CO_2$ and Mg^{2+} (seawater) $+ (Ca,Mg)O + 2H_2O \rightarrow 2Mg(OH)_2 + Ca^{2+}$. The large volume of seawater that must be processed in order to produce significant quantities of $Mg(OH)_2$ requires a large flocculator for mixing, settling tanks for washing, and a rotary vacuum filter for concentrating the slurry of $Mg(OH)_2$ (Comstock 1963).

Magnesium chloride may be precipitated from (1) deep well brines, such as those at Ludington and Midland, Michigan; (2) synthetic brines produced during solution mining, such as those at Veendam, Netherland; and (3) spent liquors obtained in salt production from natural brines, such as those from the Dead Sea. A process for recovery of $MgCl_2$ is illustrated in Fig. 47 (Comstock 1963). $Mg(OH)_2$ is then produced by either pyrohydrolysis of $MgCl_2$ from the brines, or by reacting $MgCl_2$ from the brine with dolomitic slaked lime (Seeger et al. 1985).

3. Production of Magnesium Oxide

Magnesium oxide, or magnesia, is produced by calcining either the magnesite after beneficiation or the magnesium hydroxide precipitated from seawater according to the equations: $MgCO_3 \rightarrow MgO + CO_2$ and $Mg(OH)_2 \rightarrow MgO + H_2O$. The calcination is usually done below 900°C in externally heated rotary kilns, and the product is the reactive form of MgO, caustic-calcined magnesia, which can hydrate to $Mg(OH)_2$ in cold water and is soluble in acids (Seeger et al. 1985).

When calcination temperatures exceed 900°C, magnesium oxide is converted to an unreactive, or dead-burned, form that does not hydrate in cold water and is not soluble in acids. The sintering temperature of pure MgO is above 2,000°C, so impurities such as iron oxides, alumina, silica, and others must be present to achieve sintering at lower temperatures. Natural magnesites contain enough of these impurities to permit good quality, dead-burned MgO to be produced at temperatures as low as 1,400°C. The seawater-derived $Mg(OH)_2$, however, is too pure to sinter at these lower temperatures, and small amounts of ferric oxides are sometimes added in order to achieve sintering at temperatures comparable to those used to calcine natural magnesite. This process produces dead-burned magnesia of lower density and lower purity. A process first converts the seawater-derived $Mg(OH)_2$ to the caustic-calcined MgO, and then makes the reactive MgO into briquettes under pressure, before dead-burning at temperatures of 1,600°C to 2,000°C. The product is dense and lumpy comparable to that produced from natural magnesite and possesses a purity greater than 95%. Chemical analyses of dead-burned magnesia produced from different source materials are given in Table 14 (Seeger et al. 1985).

Fused magnesia is produced by the fusion of high-grade magnesite or magnesia derived from seawater in Higgin's or electric arc tilt furnaces at 2,800° to 3,000°C (Power 1986). Electrical grade fused magnesia is characterized by its high purity, but also by a small, but significant, silica content that helps to enhance the electrical properties of MgO. In comparison, the refractory grade fused magnesia has a lower silica content, higher Ca/Si ratio, and higher density (3.58 gm/cm^3 vs. 2.4 gm/cm^3 for the elec-

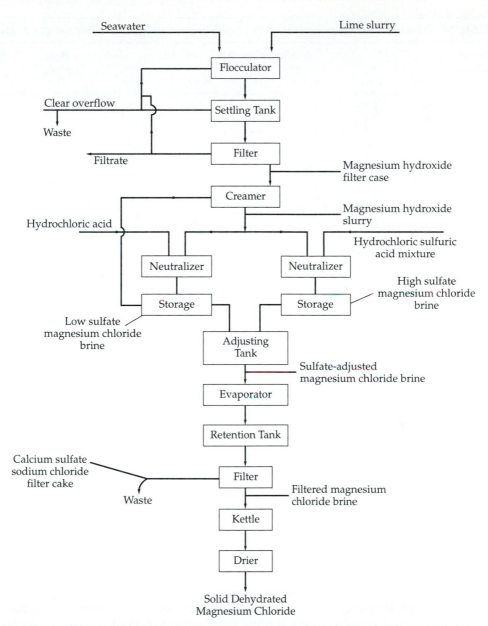

Figure 47 A scheme to illustrate a process for recovery of magnesium chloride, after Comstock (1963).

trical grade). Typical chemical compositions for fused magnesia produced in major industrial countries are shown below (Seeger et al. 1985):

Electric Grade	France	FRG	UK	USA
MgO%	95.5	96.3	96.8	96.0
SiO$_2$%	2.2	2.0	2.3	3.1
CaO%	1.5	1.5	0.8	0.8

Table 14 Typical chemical compositions of dead-burned magnesia, after Seeger et al. (1985).

Type of magnesia*	Composition, wt %						Bulk density g/cm^3	Theoretical density g/cm^3
	MgO	CaO	SiO$_2$	Fe$_2$O$_3$	Al$_2$O$_3$	B$_2$O$_3$		
1	91.5	1.5	4.5	1.1	1.4	0.01	3.27	3.57
2	89.0	2.4	0.5	7.8	0.3		3.26	3.64
3	88.0	2.5	0.6	8.6	0.3	<0.01	3.38	3.65
4	91.3	1.9	0.5	6.0	0.3	<0.01	3.40	3.63
5	98.0	0.7	0.7	0.2	<0.2	0.30	3.28	3.57
6	97.2	1.9	0.5	≤0.2	0.2	0.05	3.43	3.57
7	96.4	2.0	1.1	0.4	<0.1	<0.01	3.43	3.57
8	97.2	2.0	0.4	0.3	≤0.1	<0.01	3.42	3.57
9	98.5	0.7	0.15	0.5	<0.1	0.015	3.45	3.58
10	99.0	0.7	0.1	<0.1	<0.1	<0.01	3.43	3.58

* 1) Lump sintered magnesia with a high silicate (and iron) content
  2) High-iron sintered magnesia produced from flotation concentrate
  3) Lump high-iron sintered magnesia
  4) Lump high-iron sintered magnesia
  5) Sintered magnesia produced from seawater type 11 (C:S=1:1)
  6) Sintered magnesia produced from seawater type 31(C:S=3:1)
  7) Lump low-iron sintered magnesia
  8) Low-iron sintered magnesia with large crystals (LC) produced from flotation concentrate
  9) Sintered magnesia produced from MgCl$_2$ brine
  10) Sintered magnesia produced from pyrohydrolyed MgCl$_2$

Fe$_2$O$_3$%	0.008	0.10	0.10	0.07
CaO/SiO$_2$	0.68	0.75	0.35	0.26

Refractory Grade	France	Canada	Japan
MgO%	98.0	96.7	98.5
SiO$_2$%	0.5	0.2	0.35
CaO%	1.1	2.2	0.9
Fe$_2$O$_3$%	0.15	0.5	0.15
CaO/SiO$_2$	2.2	11	2.6

4. Industrial Uses

The major use of dead-burned magnesite or magnesia is in basic refractory bricks (Coope 1986). These bricks have been used as refractory liners for furnaces since the early days of steelmaking, but the development of the basic oxygen furnace and its higher operating temperatures provided the impetus for a rapid growth of magnesia-based refractories in the middle 1960s. Magnesite-bricks are either tar impregnated or tar-bonded, the pitch residue in the bricks improving their slag resistance (Bakker et al. 1968). Dead-burned magnesite is also mixed with chrome ore to produce magnesite-chrome and chrome-magnesite bricks that are either chemically bonded, unfired, or direct-bonded with treatment above 1,650°C. Bricks with a preponderance of magnesite are known as magnesite-chrome bricks and are used largely in the glass and steel industries, whereas chrome-magnesite bricks that have a greater proportion of chrome ore are used in copper refining. A classification of magnesite brick, magnesite-chrome brick, and chrome-magnesite brick according to MgO content is shown below (Crookston and Fitzpatrick 1983) (also see p. 89):

Class:	Magnesite Brick	Nominal MgO%	Minimum MgO%
	95	95%	91%
	90	90%	86%
	Magnesite-Chrome Brick		
	80	80%	75%
	70	70%	65%
	60	60%	55%
	50	50%	45%
	Chrome-Magnesite Brick		
	40	40%	35%
	30	30%	25%

Dead-burned magnesite bricks have a high heat storage capacity and a high thermal conductivity, and are used in efficient off-peak storage heaters. The best bricks are fired and have the following properties: (1) a high bulk density (up to 3.10 g/cm^3), and thus a high degree of sintering, (2) a high thermal conductivity (up to 9W/m•K at 600°C), and (3) a high specific heat (1.1 kJ/kg•K at 600°C). The influence of impurities on these properties is not critical (Seeger et al. 1985).

The caustic-calcined magnesite has a greater variety of uses than the dead-burned magnesite (Comstock 1963). It is used in the chemical industry for the production of various magnesium compounds and in the preparation of catalysts used in the production of organic chemicals. The reactive MgO in fuels binds sulfur and other compounds and prevents formation of acidic exhaust gases. Caustic-calcined magnesia and magnesite are used in fertilizers and animal feeds to correct magnesium deficiency in plants, cattle, and sheep. Caustic-calcined MgO mixed with concentrated solutions of magnesium salts and sodium phosphate becomes extremely hard and sets in air to form a magnesia binder. It is also an important component in lightweight construction boards for thermal and acoustic insulation. Low-iron caustic magnesia is used as a filler in plastics and rubber, and allows adjustment of viscosity and stiffness. Caustic magnesia and magnesium carbonate are used in cosmetic powders and as antacids in the pharmaceutical industry. In metallurgy, caustic-calcined MgO serves as an absorbent and catalyst in the carbonate leach circuit for uranium ore processing.

Fused magnesia of electrical grade is used as an insulating material in the electrical heating industry because of its combination of high electrical resistance ($2 \times 10^9 - 10^{11}$ $\Omega \cdot$ cm at 600°C and $10^7 - 9 \times 10^7$ $\Omega \cdot$ cm at 1,000°C) and high thermal conductivity (2.9 to 8.4W/m $\cdot$ K for densities of 3.0 to 3.58 g/cm^3). Fused MgO in pulverized form is packed into the space between the heating coil and the outer tube in a heating element for air and liquid. Such elements are used in various domestic appliances such as irons, washing machines, cooking plates, ovens, and coffee machines. In industrial applications it is well represented in water heating elements, kilns, welding machines, and general heating systems (Power 1986).

The refractory graded fused magnesia is used in refractories, especially as a lining for induction furnaces and crucibles. Magnesia-graphite bricks, with an addition of fused magnesia, are used in ultrahigh power furnaces for steel production. Magnesia and graphite act in a synergistic manner in service performance. The low porosity of the fused magnesia and the size of the periclase crystals considerably improve the corrosion resistance and adhesion points for a protective slag coating, thus minimizing oxidation of the graphite. The graphite provides greatly enhanced resistance to thermal shock and slag attack (Coope 1986).

The use of $Mg(OH)_2$ and $MgCl_2$ for the production of magnesium was discussed earlier in this book (Dolomite, p. 101).

REFERENCES

Bakker, W. T., G. D. MacKenzie, G. A. Russel, Jr., and W. S. Treffner. 1968. Refractories. In Vol. 17 of Kirk-Othmer *Encyclopedia of chemical technology.* 2nd. ed. 227–267. New York: Interscience.

Barnes, I., J. R. O'Neil, J. B. Rapp, and D. E. White. 1973. Silica-carbonate alteration of serpentine: wall-rock alteration in mercury deposits of California coast ranges. *Econ. Geol.* 68:388–398.

Bennett, W. A. G. 1941. Preliminary report on magnesite deposits of Stevens County, Washington. *Invest. Rept.,* 5. Washington State Div. Geol, 25p.

Campbell, I., and J. S. Loofbourow. 1962. Geology of magnesite belt of Stevens County. *Bull. U.S.G.S.* 1142F:1–53.

Chang, L. L. Y., R. A. Howie, and J. Zussman. 1996. Rock-forming minerals. In Vol. 5B, *Non-silicates: Sulphates, carbonates, phosphates, halides.* 2nd. ed. Essex: Longman, 383p.

Comstock, H. B. 1963. Magnesium and magnesium compounds, a material survey. *U.S. Bureau Mines Inf. Cir.* 8201, 128p.

Coope, B. 1986. An introduction to refractories. In *Raw materials for the refractories industry.* Ed. E. M. Dickson, 7–13. London: Industrial Minerals Consumer Survey, Metal Bulletin plc.

Cordes, H. 1982. The preparation of magnesite. *Aufbereitungs-Technik* 9:482–489.

Crookston, J. A., and W. D. Fitzpatrick. 1983. Refractories. In *Industrial minerals and rocks.* 5th ed. Ed. S. J. Lefond, 373–385. New York: Soc. Mining Engineers, AIME.

Dabitzias, S. G. 1980. Petrology and genesis of the Vavdos cryptocrystalline magnesite deposits, Chalkidiki Peninsula, northern Greece. *Econ. Geol.* 75:1138–1151.

Dunning, F. W., and A. M. Evans. 1986. Eds. Mineral deposits of Europe. In Vol. 3, *Central Europe.* Mineral. Soc. London, 355p.

Fountain, K. 1992. Liaoning magnesite. *Proc. 10th Ind. Minerals Intern. Congr.,* p. 65–70. Ed. J. B. Griffiths. San Francisco, California.

Gilpin, W. C., and N. Heasman. 1977. Recovery of magnesium compounds from sea water. *Chem. & Ind.* (July): 567–572.

Griffin, R. 1972. Genesis of magnesite deposit, Deloro Twp., Ontario. *Econ. Geol.* 67:63–71.

Harben, P. W., and R. L. Bates. 1990. *Industrial minerals, geology and world deposits.* London: Industrial Minerals Division, Metal Bulletin Plc, 312p.

Harben, P. W., and M. Kuzvart. 1996. Industrial minerals—a global geology. London: Ind. Minerals Inf. Ltd., Metal Bulletin Plc 462p.

Ilic, M., and A. Popovic. 1970. Rezultati novijeg proucavanja magnezitaskog lezista Beli Kamen. *Vesnik Geologija, Zaved za geoloska i geofizicka istrazivanja* 28. Beograd, 25–57.

Johannes, W. 1970. Zur Entstehung von Magnesitvorkommen. *Neues Jahrb. Min., Abh.* 113:274–325.

Jones, M. H., and B. W. Wilson. 1991. Rapid method for the determination of the major components of magnesite, dolomite and related minerals by x-ray spectrometry. *Analyst* 116:449–452.

Kramer, D. 1996. Magnesium compounds. *U.S.G.S. Annual Review,* 9p.

Kuzvart, M. 1981. Industrial minerals and rocks of Czechoslovakia. *Ind. Minerals* (March): 19–35.

Langmuir, D. 1965. Stability of carbonates in the system $MgO-CO_2-H_2O$. *Jour. Geol.* 73:73–754.

Manojlovic, D. 1970. Geoloska karakteristika i praktican znacaj lezista magnezita stokverknog tipa Mramor (okolina Rozane). *Vesnik Geologija, Zavod za geoloska i geofizicka istarazivanja* 28, Beograd, 93–110.

Power, T. 1986. Fused minerals, the high purity, high performance oxides. In *Raw materials for the refractories industry.* Ed. E. M. Dickson, 163–177. London: Industrial Minerals Consumer Survey, Metal Bulletin Plc.

Sayles, F. L., and W. S. Fyfe. 1973. The crystallization of magnesite from aqueous solutions. *Geochim. Cosmochim. Acta* 37:87–99.

Schilling, J. H. 1966. The Gabbs magnesite-brucite deposits, Nye County, Nevada. In *Ore deposits of the United States, 1933–1967.* The Graton-Sales Volume, v. 2, p. 1607–1622, John Ridge, Ed. AIME, New York.

Schmid, H. 1984. China—the magnesite giant. *Ind. Minerals* (August): 27–45.

Seeger, M., W. Otto, W. Flick, F. Bickelhaupt, and O. Akkerman. 1985. Magnesium Compounds. In Vol. A15 of *Ullmann's encyclopedia of industrial chemistry.* 595–628. Weinheim: Wiley-VCH Verlag Gmbit.

Spangenberg, K. 1949. Zur Genesis der Magnesitlagerstätte vom Galenberg bei Zobten (Schlesien). *Neues Jahrb. Min. Mh.* 117–190.

Utley, H. F. 1954. Heavy-media separation supplements flotation in magnesite plant. *Pit and Quarry* 46:90–92.

Wetzenstein, W. 1989. Magnesite. In *Non-metalliferous stratabound ore fields.* Ed. M. K. de Brodtkorb, 255–298. New York: Van Nostrand Reinhold-Chapman and Hall.

Wicken, O. M., and L. R. Duncan. 1983. Magnesite and related minerals. In *Industrial minerals and rocks.* 5th ed. Ed. S. J. Lefond, 881–896. New York: Soc. Mining Engineers, AIME.

Zhang, Q. S. 1988. Origin and metamorphic remobilization of Archean and Early Proterozoic strata-bound ore deposits in China. *Krystalinikum* 19:177–189.

Manganese Minerals

About 95% of manganese ore is used in the production of iron and steel. The other 5% is used by nonmetallurgical industries for batteries, ceramics, and fertilizers. In the years between 1985 and 1990, annual world production was estimated to be 24 M tons (U.S.B.M. Mineral Yearbook 1992). Distribution of manganese deposits is scattered throughout the world. The former USSR, South Africa, Gabon, Brazil, Australia, Morocco, and India have large reserves, and are also the major producers of manganese.

Economically exploitable deposits of manganese occur both on the continents and on the floors of present-day marine and lacustrine basins. Among some one hundred Mn-bearing minerals, pyrolusite, MnO_2, is by far the most important one.

MINERALOGICAL PROPERTIES

In compound formation, manganese occurs in all oxidation states from -3 to $+7$, with $+2$, $+4$, and $+7$ being the most important. The divalent state is generally considered to be the most stable, at least in acid to neutral media, while the tetravalent manganese, as MnO_2, is the most stable under moderately acid and moderately alkaline conditions. The basicity of the oxides decreases with increasing valence. Thus, MnO behaves as a basic anhydride, MnO_2 is amphoteric, and Mn_2O_7 is an acidic anhydride (Kemmit 1973).

Manganese dioxide rarely has the stoichiometric composition of MnO_2. Instead it is generally represented by the formula $MnO_{1.7-2.0}$, because it almost always contains varying percentages of lower valent manganeses. Manganese dioxide also exists in various states of hydration and contains a variety of foreign ions. The structure of manganese dioxide can be described as being composed of MnO_6 octahedra that form chains by sharing edges and cavities. The various chains are cross-linked by sharing corners to form a three-dimensional network (Pisarczyk 1995). MnO_2 and compositions close to MnO_2 have been reported to have six distinct crystalline forms. They are designated as α, β, γ, δ, ε, and ramdellite.

The α-MnO_2 is isostructural to the minerals of the $A_{1-2}Mn_8O_{16} \cdot xH_2O$ series. The structure is distinguished by the corner-shared double chains of MnO_6 octahedra with 2×2 channels. The 2×2 channel structure requires stabilization at the centers by large cations. Hollandite is the Ba-rich member (A:Ba); coronadite the Pb-rich member (A:Pb); cryptomelane the K-rich member (A:K); and manjiroite the Na-rich member (A:Na). In the compositions of this series of minerals, both Mn^{2+} and Mn^{4+} are present (Burns and Burns 1977).

Pyrolusite is the natural equivalent of β-MnO_2. It is the most important manganese mineral in manganese ore. Pyrolusite has single chains of edge-shared MnO_6 octahedra (1×1 channels) with a = 4.39Å and c = 2.87Å. Its color varies from light or dark steel-gray to iron black, colors representative of the tetravalent state, as distinguished from the pink of Mn^{2+}, the red of Mn^{3+}, the blue of Mn^{5+}, the green of Mn^{6+}, and the purple of Mn^{7+}. Usually pyrolusite occurs in massive columnar, fibrous, or divergent forms. The specific gravity ranges from 4.73 to 4.86, and the hardness is between 2 and 2.5 for the massive forms, and between 6 and 6.5 for single crystal forms (Giovanoli 1969; Gutzmer and Beukes 1997). In comparison, pyrolusite is the least reactive and the most highly crystallized of all polymorphic forms of MnO_2. Its composition is the closest to having a stoichiometric MnO_2. Above 650°C, MnO_2 is converted to Mn_2O_3, and above 1,000°C to Mn_3O_4. In the presence of strong alkalis and at elevated temperatures, MnO_2 is readily oxidized by oxygen to Mn^{5+} and Mn^{6+} compounds. Pyrolusite is insoluble in water (Dressel and Kenworthy l961; Reidies 1985).

The γ-MnO_2 was first identified in the Nsuta deposit, Ghana, and named "Nsutite" (Zwicker et al. 1962). Diffused x-ray diffraction data have revealed that Nsutite is poorly crystallized. This feature is related to a structure that consists of randomly alternating layers of double, ramsdellite-type, and single, pyrolusite-type chains of (MnO_6) octahedra (De Wolff 1959). Hydrated MnO_2 from the Yokosuka deposit in Japan was found to also be identical with γ-MnO_2 (Nambu and Okada 1961). The composition and cell dimensions of nsutite are $Mn_{1-x}{}^{4+}Mn_x{}^{2+}O_{2-2x}(OH)_{2x}$ and a = 4.43Å, b=9.36Å, and c = 2.85Å. This nonstoichiometric manganese oxide is known for its good properties for battery use (De Wolff 1959; Faulring 1965).

The δ-MnO_2 is structurally related to minerals in the birnessite group (Jones and Milne 1956; Giovanoli and Stahli 1970; Drits et al. 1997). The structure of δ-MnO_2 consists of both single and double chains of edge-shared MnO_6 octahedra with both 2×1 and 1×1 channels. It has a composition of $(Ca,Na)(Mn^{2+},Mn^{4+})_7O_{14} \cdot 3H_2O$, and is a common Mn-mineral in nodules (Buser et al. 1954; Drits et al. 1997). It is also known as manganous manganite or "7Å manganite" because of its 7Å and 3.5Å reflections. Todorokite, with a composition of $(Na,Mn)Mn_3O_7 \cdot nH_2O$ or $3MnO_2 \cdot Mn(OH)_2 \cdot nH_2O$, is identified as the "10Å manganite" because of its 10Å and 5Å reflections. The ε-MnO_2 has no equivalent that occurs in nature.

Ramsdellite is the orthorhombic form of MnO_2 with a=4.513Å, b=9.294Å, and c = 2.859Å. Its structure is characterized by double chains of edge-shared MnO_6 with 2×1 channels. The distortion of the MnO_6 octahedra is quite similar to that of α-MnO_2 (Miura et al. 1990). Ramsdellite is very rare in its natural occurrence (Fleischer et al. 1962).

Other manganese minerals of some economical values include (1) Psilomelane: a hydrated manganese oxide with variable composition and varying amounts of barium and potassium; (2) Braunite: $3Mn_2O_3 \cdot MnSiO_3$, corresponding to approximately 10% SiO_2; (3) Hausmannite: Mn_3O_4, a member of the Fe_3O_4-Mn_3O_4 spinel series; (4) Wad: hydrous Mn oxides with variable compositions; and (5) Rhodochrosite: $MnCO_3$ with variable amounts of iron, calcium, and magnesium.

Manganese ores exhibit a wide variability in composition, particularly in the balance of the manganese and iron contents. A classification mainly based upon their metallurgical purposes is as follows: (1) manganese ores, containing more than 35% Mn, which are suitable for the manufacture of both high and low grades of ferromanganese; (2) ferruginous manganese ores, containing 10 to 35% Mn, which are used for the manufacture of spiegeleisen; and (3) manganiferous iron ores, containing 5 to 10% Mn, which are used for the manufacture of manganiferous pig iron.

GEOLOGICAL OCCURRENCE AND DISTRIBUTION OF DEPOSITS

1. Manganese Deposits on the Continents

Many classifications of manganese deposits have been made, based upon mode of genesis, tectonic framework, and associated lithology (Varentsov 1964; Roy 1969; Ramkhmanov and Tchaikovsky 1972; Laznicka 1992). It is generally agreed that there are three major types of deposits: (1) hydrothermal, (2) sedimentary with or without volcanic activities, and (3) supergene. Economic deposits of each type are described briefly below.

Hydrothermal Manganese Deposits: In Morocco, a large number of hydrothermal vein deposits occur in the area south of Quarzazate of the Anti Atlas region (Choubert and Faure-Muret 1973). The Precambrian deposits in this area essentially consist of andesitic flows, overlain by rhyolites. The more than 200 Mn-bearing veins occur in swarms, and most of them are restricted to a particular single flow or strata of rhyolite. There are a number of successive generations of vein deposits, with each generation confined and genetically related to an individual unit of the volcanic complex. Choubert and Faure-Muret concluded that the manganese mineralization in hydrothermal veins was connected to pulses of rhyolitic volcanism. The vein deposits consist of a variety of manganese minerals including pyrolusite, psilomelane, todorokite, braunite, cryptomelane, hausmannite, and hollandite. Associated minerals are hematite, goethite, barite, calcite, and dolomite.

In central Kazakhstan, Mn-bearing veins occur in Devonian sandstones and conglomerates at the Dzhezda deposit (Kalimin 1970). Three pulsatory stages of hydrothermal mineralization have been recognized: (1) ores of Pb and Ba-bearing Mn oxides (corondite and hollandite) deposited at relatively low temperatures in an oxygenated condition from an acidic solution, (2) main manganese ore bodies of braunite and manganites deposited from an initial high-temperature solution, and (3) minor mineralization of braunite, rhodochrosite, and manganites near faults and fractures.

In the United States, numerous small hydrothermal Mn deposits are scattered throughout New Mexico, Colorado, Utah, Nevada, Arizona, and California (Hewett et al. 1956). Most of these veins of Mn-oxides fill up simple or complex fractures, and show a steep dip with only a limited extension on the surface and at depths. They are related to barite, fluorite, Pb-Zn-Cu sulfides, and Ag-Au mineralization. A typical example is the Eocene Pb-Zn-Cu deposit at Butte, Montana, that shows three distinct zones, both in plane and in depth. The central mesothermal zone, which contains abundant Cu and Fe-sulfides, is devoid of Zn, Pb, and Mn. The intermediate zone contains major Cu and Fe sulfides and minor amounts of Zn, Pb, and Mn. The peripheral epithermal zone is characterized by abundant Zn-, Pb-, Ag-, and Mn-bearing minerals. Rhodochrosite is the dominant manganese mineral in the deposits (Allsman 1956).

Sedimentary Manganese Deposits: The major manganese deposits of sedimentary origin are those without connection to volcanic activities. In the Ukraine, extensive occurrences of sedimentary manganese deposits occur in terrigenous formations of Oligocene age and are localized in isolated erosion depressions in the basement rocks (Varentsov 1964). The deposits of Nikopol and Bol'shoi Tokmak in the southern

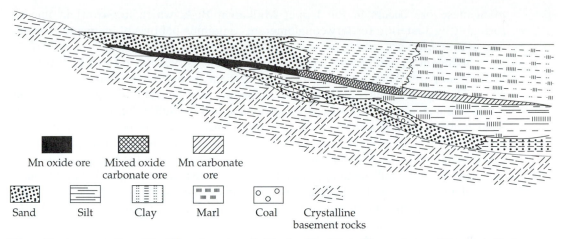

Figure 48   A cross section of the Nikopol ore deposit, Ukraine. Reprinted with permission of Economic Geology, Volume 61, p. 445, D. F. Hewett (1966).

Ukraine are the largest among them and account for about 85% of the reserves of the former USSR. The deposits rest upon a southern edge of the Ukrainian shield that gradually plunges to the south toward the Black Sea. In these deposits, the ore-bearing sequence occurs transgressively on the crystalline basement rocks. Beds of orthoquartzite and glauconitic clay form the basal part of this sequence (Hewett 1966). The Mn-ore beds, showing mineral zones of oxide, oxide-carbonate, and carbonate assemblages, intrude into different stratigraphic horizons in the basal part of the sequence. The zonation corresponds with the basinal configuration, with the oxides occurring nearest to the shore, then the mixed ores, and finally the carbonates toward the deeper parts with a southward plunge of the crystalline rocks (Fig. 48). The oxide zone consists of pyrolusite and psilomelane; the mixed zone of psilomelane, manganite, manganoan calcite, and rhodochrosite; and the carbonate zone of calcian rhodochrosite and manganoan calcite.

In Gabon, the large manganese deposits of Moanda occur in the Precambrian Franceville Series. This series consists of a detrital, marine sedimentary sequence, with only minor components of carbonates, volcanics, and iron formations (Weber 1973). The deposits include important occurrences of ores at the Bangombé, Okouma, and Bafoula plateaus. In the Bangombé plateau, Mn-rich horizons overlie Mn-poor or barren carbonaceous black shales and dolomites. Within the Mn-rich horizon, a thin (0.2 to 0.5 m) basal unit of massive Mn-oxides and hydroxides (pyrolusite, manganite, nsutite) is overlain by a 3 to 9 m thick zone of supergene Mn-oxides interlayered with sandstones and ferruginous red shales. Mn-carbonate bearing carbonaceous black shales represent the protore that, on weathering, produced the Mn-oxide ores. Gabon is the world's leading producer of battery grade manganese ore.

The premier manganese deposit in Australia occurs at Groote Eylandt in the Gulf of Carpentaria (McIntosh et al. 1975). The deposit of Mn-oxides occurs as a single flatbed in the middle of the Lower Cretaceous Mullaman Beds and covers an area of at least 150 km². The Mullaman Beds lie unconformably on the stable platform of the Middle Proterozoic Groote Eylandt Beds of quartzite sandstone and orthoquartzite. The manganese ore occurs in the Upper Mullaman Beds, which are overlain uncomformably by a lateritic Tertiary conglomerate containing pebbles of manganese oxides. The ore minerals are mainly pyrolusite and cryptomelane.

Sedimentary manganese deposits of terrigenous association are widely distributed in different parts of peninsular India, and all of them are of Precambrian age (Roy 1964, 1972). The Sausar Group of Madhya and Maharashtra in central India contains large deposits of manganese ores that occur as an arcuate belt extending for about 200 km, with an average width of 30 km. The sediments of the Sausar Group of pelitic, psammitic, and carbonate rocks represent a miogeosynclinal or shelf-facies of the orthoquartzite-carbonate sequence. Volcanic rocks are totally absent. In most parts, the Sausar sedimentary sequence was regionally metamorphosed to amphibolite facies. Braunite, hollandite, manganite, hausmannite, and bixbyite are the major manganese minerals in the ore deposits.

One of the most extensive sedimentary manganese deposits in the world is located in the northern Cape Province, South Africa (Beukes 1973; von Bezing and Gutzmer 1994). The Kalahari manganese field is represented by stratified manganese deposits conformably enclosed in, and interbedded with, an iron formation. At Hotazel (Fig. 49), the manganese ores consist of only oxides and silicates and are devoid of carbonates. Toward the west (Olive Pan) and south (Smartt), there is a noticeable increase in Mn-carbonate in the ores, and this change corresponds to the facies change from hematite to magnetite in the enclosing iron formation.

Supergene Manganese Deposits: In Brazil, sedimentary manganese deposits of Precambrian age are well distributed in the State of Minas Gerais and the Territory of Amapa, but those of economic importance are supergene enrichment products of metamorphosed manganese carbonate and silicate-carbonate protores (Scarpelli 1973). Both the Rio das Velhas Series in Minas Gerais and the Amapa Series in Amapa Territory contain Mn-carbonate and silicate-carbonate protore. The former is associated with graphitic phyllite, mica schist, quartzite, amphibolite, the carbonate facies of the iron-formation, chert, graywacke, and conglomerate; whereas the latter is associated with graphitic schist and biotite-quartz schist. Rhodochrosite and manganoan calcite are the major minerals in both series (Dorr et al. 1956). The manganese carbonate and silicate-carbonate protores were oxidized in situ to cryptomelane and pyrolusite down to the water table in a continuous process of supergene enrichment. The oxidation extends from 70 to 100 meters.

2. Manganese Deposits In Marine and Lacustrine Basins

Manganese and ferromanganese deposits occur extensively as nodules, crusts, and pavements in present-day oceans, shallow seas, and freshwater lakes (Horn et al. 1972a). Their occurrence covers all ages from the Eocene to the present. Major deposits in the oceans are located in areas of red clay or siliceous ooze where the rate of detrital sedimentation is low. The North Pacific has the most widespread and densest concentration of manganese and ferromanganese nodules, as is shown in Fig. 50.

The manganese nodules are brown to black in color and have a dull luster. With increase in size, the nodules change from spherical to ellipsoidal in shape. Each nodule has one or more nuclei surrounded by concentric layers of oxides of manganese, iron,

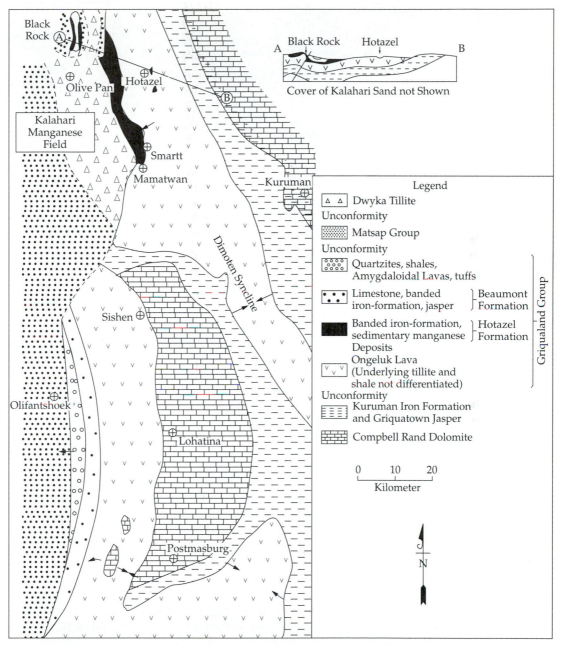

Figure 49 A geologic map showing the distribution of iron-formation and associated manganese deposits in the Griqualand Group. Reprinted with permission of Economic Geology, Volume 68, p. 492, N. J. Beukes (1973).

233

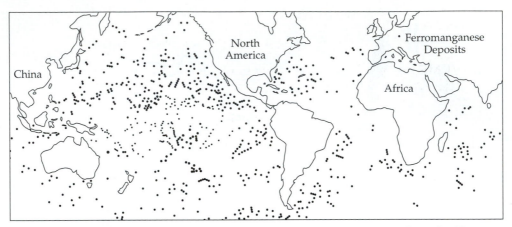

Figure 50   Worldwide distribution of surficial ferromanganese deposits on the ocean floor, after Horn et al. (1972b).

and other metals. Clay layers are often present and illustrate periods of interruption of growth (Hillman and Gosling 1985).

The most common manganese minerals in nodules are todorokite and birnessite. The occurrences of several other manganese oxides and hydroxides have been reported, including nsutite, manganite, cryptomelane, pyrolusite, and psilomelane (Baturin 1991).

INDUSTRIAL PROCESSES AND USES

1. Beneficiation

The process used for beneficiation of manganese ore depends upon the mineralogy of the ore. At Groote Eylandt, Australia, beneficiation is done by heavy medium separation. The ore is processed to be battery active, and a typical composition is as follows: 53.7% Mn, 80.5% MnO_2, 2.23% Fe, 2.44% Al_2O_3, 2.18% SiO_2, 1.56% BaO, 0.08% CaO, 0.66% K_2O, 0.69% MgO, 0.13% TiO_2, 0.02% ZnO, and 5.08% combined H_2O with trace metals in ppm of 121 (Cu), 244 (Ni), 126 (Co), 27 (Cr), 35 (Pb), 30 (As), and 20 (Sb) (Toon 1985). The processes used for the Nikopol manganese ore in the southern Ukraine are more complex. Ore is crushed in a hammer mill to 50 mm and then washed in log washers that separate the ore at 3 mm. Oversize from the washer is crushed in roll crushers and sized on vibrating screens at 3 mm. The primary fines are classified, dewatered on a bowl-type classifier, deslimed at 0.04 mm, and subjected to flotation (Bogdanova et al. 1960).

Flotation with either oil emulsions or conventional soap flotation has been used at Butte, Montana (Jacoby 1983). For beneficiation of low-grade manganese ore, the ammonium carbonate process is considered to be one of the most successful and is used to produce MnO_2 from the Cuyuna iron range ore. The process involves roasting the ore, leaching with ammonium carbonate, and reducing the carbonate to MnO_2 (Jacoby 1983).

2. Synthesis of Manganese Dioxide

Several types of synthetic manganese dioxide are produced for specific end uses. They can be prepared either by chemical process as chemical manganese dioxide (CMD) or by electrochemical methods as electrolytic manganese dioxide (EMD) (Reidies 1985).

The first step in the preparation of CMD is the reduction of MnO_2 ore to MnO using heavy fuel oil at 900°C. The MnO is then treated with H_2SO_4 to form manganese sulfate. After solids are removed by thickening, the $MnSO_4$ solution is then treated with ammonium carbonate to precipitate $MnCO_3$ and $(NH_4)_2SO_4$, a by-product. The $MnCO_3$ is separated and treated at 320°C to form a manganese oxide of the composition $MnO_{1.80-1.85}$. By reacting with $NaClO_3$ in the presence of H_2SO_4, formation of MnO_2 is achieved. CMD is primarily γ-MnO_2 (Pisarczyk 1995).

The preparation of EMD usually starts by using the same raw materials and following the same procedures used to produce the $MnSO_4$ solution. The solution is then adjusted to concentrations of 75 to 160 g/L $MnSO_4$ and mixed with 50 to 100 g/L H_2SO_4. The mixed solution is then subjected to electrolysis at 90 to 98°C. The EMD is deposited as a solid coating on the anode, which is generally made from graphite or titanium (Pisarczyk 1995).

3. Preparation of Potassium Permanganate

One of the most important industrial compounds prepared from MnO_2 is potassium permanganate, $KMnO_4$. It is an oxidant and can be used for a wide variety of inorganic and organic processes (Arndt 1982).

$KMnO_4$ is prepared by mixing MnO_2 with KOH and heating the mixture in air at 300° to 400°C in a rotary kiln, according to the equation: $2MnO_2 + 6KOH + \frac{1}{2}O_2 \rightarrow 2K_3MnO_4 + 3H_2O$. The K_3MnO_4 produced is dried, ground, and subjected to secondary oxidation at 190° to 210°C, also in a rotary kiln, according to the equation: $2K_3MnO_4 + H_2O + \frac{1}{2}O_2 \rightarrow 2K_2MnO_4 + 2KOH$. Oxidation of the manganate to permanganate is accomplished by electrolysis: $K_2MnO_4 + H_2O \rightarrow KMnO_4 + KOH + \frac{1}{2}H_2$ (Reidies 1985).

4. Industrial Uses

The manufacture of batteries is the most important nonmetallurgical use of either naturally occurring manganese dioxide minerals or synthetic CMD and EMD. The function of manganese dioxide in the cell is to act as a depolarizer. Hydrogen released from the electrolyte tends to form around the carbon electrode, thus slowing down the action of the cell. As a component in the mix that comprises the cathode of the cell, manganese dioxide provides oxygen that is available to react with the hydrogen, thus removing it from the carbon electrode. A MnO_2 of high degree of disordering is necessary to promote the reaction (Toon 1985). For battery grade manganese dioxide, it should contain a minimum of 80% MnO_2. As power sources for consumer electronic instruments, Li/MnO_2 cells are rapidly spreading and are universally esteemed for their characteristics (Ikeda 1983).

In ceramic processes, manganese dioxide is used principally in enamels and glazes. Ground coats for sheet steel contain 0.5 to 1.5% MnO_2, which functions to promote adherence of the enamel to the steel (Maskall and White 1986). Manganese, as described, exists in a number of oxidation states and in a number of desired colors. MnO_2 or pyrolusite is used in glazes to give mainly brown colors, and in a high potash glaze an attractive purple color develops (Taylor and Bull 1986). A common type of brown transparent glaze known as Rockingham results from the addition of a combination of manganese and iron oxides to a clear glaze. With the appropriate addition of cobalt, iron, and manganese, a black color can be produced. Like iron oxides, manganese oxides can produce aventurine effects (Taylor and Bull 1986).

Manganese oxides are used to produce manganese-zinc ferrite, a widely used soft ferrite. High purity and specific surface areas are specified: exceptionally low levels of CaO, K_2O, Na_2O, BaO, and SiO_2; and surface areas between 10 to 20 m^2/gm. The ferrite grade EMD has the following composition (Toon 1986): 94.00% MnO_2 (min), 2.50% moisture (max), 0.03% SiO_2 (max), 0.05% CaO (max), 0.03% MgO (max), 1.10% SO_4 (max), 0.05% Na_2O (max), 0.10% $(Fe,Al)_2O_3$ (max) with 99.5% -100 mesh and 95.0% -324 mesh.

Manganese sulfate prepared from manganese ore is used in agriculture as fertilizer. The required manganese can be supplied as an oxide (MnO), sulfate ($MnSO_4$), or tribase manganese sulfate ($3MnO \cdot MnSO_4$). Fertilizer grades require a minimum of 80% $MnSO_4$ (Toon 1985). The $KMnO_4$ prepared from MnO_2 is used, particularly in the United States, for drinking water treatment to remove tastes and odors, waste water treatment to destroy hydrogen sulfide and other toxic and corrosive compounds, including phenols, and air purification to degrade malodorous or toxic constituents in industrial off-gases (Ficek and Sanks 1978; Pisarczyk and Rossi 1984).

REFERENCES

Allsman, P. 1956. Oxidation and enrichment of the manganese deposits of Butte, Montana. *Mining Eng.* 8:1110–1112.

Arndt, D. 1980. *Manganese compounds as oxidizing agents in organic chemistry.* La Salle, Illinois: Open Court Publ. Co. p 183–288.

Baturin, G. N. 1991. Chemical elements in mineral phases of manganese nodules of the ocean. *Akad. Nauk USSR, Dokl. Earth Sci. Sect.* 308:170–172.

Beukes, N. J. 1973. Precambrian iron formations of Southern Africa. *Econ. Geol.* 68:960–1004.

Bogdanova, Z. S., S. I. Gorlovsky, and B. M. Lakota. 1960. Flotation of brown iron ore and slimes from gravity treatment of manganese ores: *5th Intern. Mineral. Proc., I.M.M., London,* 477–490.

Burns, R. G., and V. M. Burns. 1977. Mineralogy. In *Marine manganese deposits.* Ed. G. P. Glasby, 185–248. Amsterdam: Elsevier.

Buser, W., P. Graf, and W. Feitknecht. 1954. Beitrag zur kenntnis der mangan (II)-manganit und des δ-MnO_2. *Helv. Chim. Acta* 37:2322–2333.

Choubert, G., and A. Faure-Muret. 1973. The Precambrian iron and manganese deposits of Anti Atlas. In Genesis of Precambrian iron and manganese deposits. *Unesco, Earth Sciences* 9:115–124.

De Wolff, P. M. 1959. Interpretation of some γ-MnO_2 diffraction patterns. *Acta Cryst.* 12:341–345.

Dorr, J. V. N., J. S. Coelho, and A. Horen. 1956. The manganese deposits of Minas Gerais, Brazil. *20th Int. Geol. Congr., Symposium on Manganese* 3:277–346.

Dressel, W. M., and H. Kenworthy. 1961. Thermal behavior of manganese minerals in controlled atmosphere. *U.S. Bureau Mines, Rept. Invest.,* 5761, 35p.

Drits, V. A., E. Silvester, A. I. Gorshkov, and A. Manceau. 1997. Structure of synthetic monoclinic Na-rich birnessite and hexagonal birnessite: I. results from x-ray diffraction and selected area electron diffraction. *Amer. Mineral.* 82:946–961.

Faulring, G. M. 1965. Unit cell dimensions and thermal transformation of nsutite. *Amer. Mineral.* 50:170–179.

Ficek, K., and R. L. Sanks, eds. 1978. *Water treatment plant design for the practicing engineers.* Ann Arbor Michigan: Ann Arbor Science Publ., Inc., p 461–479.

Fleischer, M., W. E. Richmond, and H. T. Evans. 1962. Studies of the manganese oxides. V. ramsdellite, MnO_2, and orthorhombic dimorph of pyrolusite. *Amer. Mineral.* 47:47–58.

Giovanoli, R. 1969. A simplified scheme for polymorphism in the manganese dioxides. *Chimia* 23:470–472.

Giovanoli, R., and E. Stahli. 1970. Oxide und oxidhydroxide des drei und vierwertigen mangans. *Chimia* 24:49–61.

Gutzmer, J., and N. J. Beukes. 1997. Mineralogy and mineral chemistry of oxide-facies manganese ores of Postmasburg manganese field, South Africa. *Mineral. Mag.* 61:213–231.

Hewett, D. F. 1966. Stratified deposits of the oxides and carbonates of manganese. *Econ. Geol.* 61:431–461.

Hewett, D. F., M. D. Crittenden, L. Pavlides, and G. L. De Huff. 1956. Manganese deposits in the United States. *20th Int. Geol. Congr., Symposium on Manganese* 3:169–230.

Hillman, C. T., and H. B. Gosling. 1985. Mining deep ocean manganese nodules—description and economic analysis of a potential venture. *U.S.B.M. Inform. Circ.,* 9015, 19p.

Horn, D. R., B. M. Horn, and M. N. Delach. 1972a. World-wide distribution and metal content of deep sea manganese deposits, in "Manganese Nodule Deposits in the Pacific," Symposium/Workshop Proc., Honolulu, Hawaii, 16– 17 October, 1972, State Centre Sci. Policy Technol. Assess. Dep. Planning Econ. Develop., State of Hawaii, 46-60.

Horn, D. R., M. Ewing, B. M. Horn, and M. N. Delach. 1972b. Worldwide distribution of manganese nodules. *Ocean Ind.* 7:26–29.

Ikeda, H. 1983. Lithium-manganese dioxide cells. In *Lithium batteries.* Ed. J. P. Gabano, 169–210. London: Academic Press.

Jacoby, C. H. 1983. Manganese. In *Industrial minerals and rocks.* 5th ed. Ed. S. J. Lefond, 897–913. New York: Soc. Mining Engineers, AIME.

Jones, L. H. P., and A. A. Milne. 1956. Birnessite, a new manganese oxide mineral from Aberdeenshire, Scotland. *Mineral. Mag.* 31:283–288.

Kalimin, V. V. 1970. Structural and compositional features of the Karadzhal Fe-Mn ore deposits. In *Manganese deposits of the Soviet Union.* Ed. D. G. Sapozhnikov, 335–340. Jerusalem: Israel Program for Scientific Translations.

Kemmit, R. D. W. 1973. Manganese. In Vol. 3 of *Comprehensive inorganic chemistry.* Eds. J. C. Bailar, H. J. Emeleus, R. Nyholm, and A. F. Trotman-Dickenson, 771–876. Oxford: Pergamon Press.

Laznicka, P. 1992. Manganese deposits in the global lithogenetic system: Quantitative approach. *Ore Geol. Rev.* 7: 279–356.

Maskall, K. A., and D. White. 1986. *Vitreous enamelling.* Oxford: Pergamon, 122p.

McIntosh, J. Z., J. S. Farag, and K. J. Slee. 1975. Groote Eylandt manganese deposits. In Economic geology of Australia and Papua-New Guinea. ed. C. L. Knight. *Austr. Inst. Min. Metall., Monograph Ser.* 5:815–821.

Miura, H., H. Kudou, J. H. Choi, and Y. Hariya. 1990. The crystal structure of ramsdellite from Pirika mine. *Jour. Faculty of Sci., Hokkaido Univ., Ser. IV* 22:611–617.

Nambu, M., and K. Okada. 1961. Re-examination of yokosukalite and ishiganeite. *Chigaku Kenkyu* 12:249–258.

Pisarczyk, K. S. 1995. Manganese compounds. In Vol. 15 of *Kirk-Othmer's encyclopedia of chemical technology.* 4th ed. 991–1055. New York: John Wiley & Sons.

Pisarczyk, K. S., and L. A. Rossi. 1984. Slude odor control and improved dewatering with potassium permanganate, 55th Annual Conf., Water Pollution Control Federation, St. Louis, October 5, 1982. Ed. J. H. Jackson. *Pulp and Paper* 58:147–149.

Ramkhmanov, V. P., and V. K. Tchaikovsky. 1972. Genetic types of sedimentary manganese formations. *Acta Mineral. Petrogr. Szeged* 20:313–323.

Reidies, A. H. 1985. Manganese compounds. In Vol. A16 of *Ullmann's encyclopedia of industrial chemistry,* 123–143, Weinheim: VCH Publ.

Roy, S. 1964. Genesis of the manganese ore deposits of Madhya Pradesh and Maharashtra, India. *22nd. Int. Geol. Congress* 5:199–214.

Roy, S. 1969. Classification of manganese deposits. *Acta Mineral. Petrogr. Szeged* 19:67–83.

——— 1972. Metamorphism of sedimentary manganese deposits. *Acta Mineral. Petrogr., Szeged,* 20:313–324.

Scarpelli, W. 1973. The Sirra do Navio manganese deposit, Brazil. In Genesis of Precambrian iron and manganese deposits. *Unesco, Earth Sciences* 9:217–228.

Taylor, J. R., and A. C. Bull. 1986. *Ceramics glaze technology.* Oxford: Pergamon Press, 263p.

Toon, S. 1985. Manganese—Active batteries attract producers. *Ind. Minerals* (July): 19–37.

Varentsov, I. M. 1964. *Sedimentary manganese ores.* Amsterdam: Elsevier, 119p.

von Bezing, L., and J. Gutzmer. 1994. Das Kalahari manganerzfeld und seine mineralien. *Part 2. Mineralien Welt* 5:41–59.

Weber, F. 1973. Genesis and supergene evolution of the Precambrian sedimentary manganese deposit at Moanda (Gabon). In Genesis of Precambrian iron and manganese deposits. *Unesco, Earth Sciences* 9:307–322.

Zwicker, W. K., W. O. J. G. Meijer, and H. W. Jaffe. 1962 Nsutite—a widespread manganese oxide mineral. *Amer. Mineral.* 47:246–266.

Mica

Mica represents a group of complex hydrous aluminosilicate minerals that are characterized by a layered structure and platy morphology. Two distinct forms of mica are used in industries: sheet mica as insulation material and ground mica as filler and extender.

About 80% of the world's sheet mica is produced in India. The United States is the leading producer of ground mica, representing 55% of world output. Other producing countries are China, Canada, Brazil, the former USSR, and Madagascar. World production of mica in the first half of the 1990s amounted to 0.21 to 0.25 M tons, one-fifth of which was sheet mica (Hedrick 1995).

MINERALOGICAL PROPERTIES

Mica represents a group of minerals that have certain chemical and physical characteristics in common. Chemically, the general formula that can be used to describe the composition of the group is $X_2Y_{4-6}Z_8O_{20}(OH,F)_4$, where "X" is mainly K, Na, or Ca, but also Ba, Rb, Cs, etc.; "Y" is mainly Al, Mg, or Fe, but also Mn, Li, Ti, etc.; and "Z" is mainly Si and Al. Depending on the number of "Y" cations, mica is subdivided into dioctahedral when "Y" is Al_4, and trioctahedral when "Y" is Mg_6. Physically, mica has a platy morphology and perfect basal cleavage, which is a consequence of its layered structure. As shown diagrammatically in Fig. 51, two sheets of

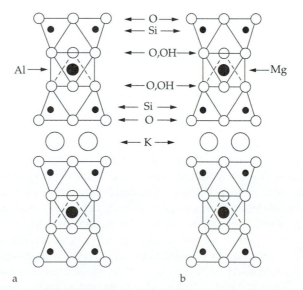

Figure 51 The crystal structures of (a) muscovite and (b) phlogopite.

239

linked $(Si,Al)O_4$ tetrahedra are juxtaposed, with the vertices of the tetrahedra pointing inward. These vertices are cross-linked with either Al^{3+} or Mg^{2+} to form, with additional (OH) groups, the octahedral sheet. The substitution of silicon by aluminum in the tetrahedral sheet is balanced by K^{1+}, Na^{1+}, or others in the inter-layer position, to complete the basic unit of a mica layer. To compensate for dimensional differences between the tetrahedral sheets and the octahedral sheet in the layers, a displacement occurs that reduces the ideal hexagonal to a monoclinic symmetry (Bailey 1984; Deer et al. 1992).

Mica is generally resistant to chemical attack and can withstand elevated temperatures. Mechanically, mica has good flexibility and elasticity, which distinguishes it from other layered silicates. The laminated layered structure enables it to be split into very thin films that are transparent and tough. Mica possesses valuable electrical properties, including high dielectric strength, uniform dielectric constant, capacitance stability, low power loss, and high electrical resistivity (Jahns and Lancaster 1950; Chapman 1983; Tanner 1994).

The most common varieties of mica are biotite, muscovite, phlogopite, and lepidolite. Muscovite and phlogopite are important industrial minerals, whereas the use of biotite is restricted because of its dark color and Fe-content. Lepidotite is a lithium mica and was previously discussed in the section on lithium minerals (p. 206). Muscovite, known as white mica, has an ideal composition of K_2Al_4 $(Si_6Al_2O_{20})(OH,F)_4$. Phlogopite, known as amber mica, with an ideal composition of $K_2Mg_6(Si_6Al_2O_{20})$ $(OH,F)_4$, forms a complete series of solid solutions with biotite, which is known as black mica, and has a complex composition varying within the system $K_2Mg_6(Si_6Al_2)O_{20}(OH)_4 - K_2Fe_6(Si_6Al_2)O_{20}(OH)_4 - K_2Fe_4Al_2(Si_4Al_4)$ $O_{20}(OH)_4 - K_2Mg_4Al_2(Si_4Al_4)O_{20}(OH)_4$ (Bailey 1984; Deer et al. 1992). Typical chemical compositions and selected physical properties of muscovite and phlogopite are listed in Table 15 (Benbow et al. 1985; Deer et al. 1992; Tanner 1994).

GEOLOGICAL OCCURRENCE AND DISTRIBUTION OF DEPOSITS

Mica is a common rock-forming mineral. Muscovite occurs in granites and granitic pegmatites; in phyllites, schists, and gneisses; and in detrital and authigenic sediments. The occurrence of phlogopite, on the other hand, is in peridotites, dunite, and metamorphosed limestones and dolomites.

Although mica is found widely throughout the world, concentrates of sheet mica large enough and sufficiently well crystallized for processing into sheet mica are comparatively scarce. Muscovite of industrial importance occurs predominantly in pegmatites.

The southeastern states are the chief mica-producing region in the United States. Fourteen well-defined mica-mining districts occupy a belt nearly 960 km long and 96 km in average width. The belt extends from east-central Virginia south-westward through the western Carolinas and north Georgia to central Alabama (Fig. 52). The pegmatites in this belt consist chiefly of plagioclase and quartz, with subordinate perthite, muscovite, and biotite. Accessory minerals are apatite, garnet, beryl, columbium-tantalum minerals, uranium minerals, and sulfides. The pegmatites are granodioritic rather than truly

Table 15 Chemical composition and physical properties of muscovite and phlogopite, after Benbow et al. (1985) and Deer et al. Introduction to the Rock-Forming Minerals © 1992, reprinted by permission of Addison Wesley Longman Ltd.

Chemical Composition	Muscovite	Phlogopite
SiO_2	48.42	40.95
Al_2O_3	27.16	17.28
K_2O	11.23	9.80
Na_2O	0.35	0.16
Fe_2O_3	6.57	0.43
FeO	0.81	2.38
MgO	—	22.95
TiO_2	0.87	0.82
H_2O^+	4.31	4.23
H_2O^-	0.19	0.48
Total	99.91	99.48

Physical Property	Muscovite	Phlogopite
density, g/cm^3	2.6–3.2	2.6–3.2
hardness, Mohs	2.0–3.2	2.5–3.0
luster	vitreous	pearly to submetallic
optical axial angle (2 V)	38–47°	0–10°
orientation of optic plane to plane of symmetry	perpendicular	parallel
refractive index, n_D		
α	1.552–1.570	1.54–1.63
β	1.582–1.607	1.57–1.69
γ	1.588–1.611	1.57–1.69
maximum temperature at which no decomposition occurs, °C	400–500	850–100
specific heat at 25°C, J/g	0.049–0.05	0.049–0.05
thermal conductivity perpendicular to cleavage, W/(m · K)	0.6694	0.6694
coefficient of expansion per °C		
perpendicular to cleavage		
20–100°C	$(15–25) \times 10^{-6}$	$1 \times 10^{-6} – 1 \times 10^{-3}$
100–300°C	$(15–25) \times 10^{-6}$	$2 \times 10^{-4} – 2 \times 10^{-2}$
300–600°C	$(16–36) \times 10^{-6}$	$1 \times 10^{-3} – 3 \times 10^{-3}$
parallel to cleavage		
0–200°C	$(8–9) \times 10^{-6}$	$(13–14.5) \times 10^{-6}$
200–500°C	$(10–12) \times 10^{-6}$	$(13–14.5) \times 10^{-6}$
tensile strength, MPa	225–296	255–296
modulus of elasticity, MPa	1.723×10^5	1.723×10^5
water of constitution, %	4.5	3.2
melting point, °C	decomposes	decomposes
dielectric constant	6.9–9	5–6
dielectric strength (highest qualities 25–75 μm thick) at 21°C		
V/μm	235–118	165–83
power factor at 25°C (highest qualities), %		
at 60 Hz	0.08–0.09	
at 1 MHz	0.01–0.02	0.3
resistivity, $\Omega \cdot cm$	$10^{12}–10^{15}$	$10^{10}–10^{13}$

granitic in composition. Muscovite is present in some deposits as disseminated flakes and tiny books, but in others it occurs as very large books, some of which are 0.6 m or more in diameter and weigh several hundred kilograms or more. All variations between these extremes are known. There appears to be little correlation between the size of the pegmatite body and the quantity of mica that it contains (Jahns and Lancaster 1950).

India is the world's premier producer of sheet mica. The most important deposits are situated in the State of Bihar, between Bhagalpur and Gaya, stretching in the form of a belt, 145 km long by 25 to 30 km wide. Muscovite occurs in pegmatites that, along with granite, have intruded the Archean gneiss and schist of the Indian Shield (Murthy 1964; Mishra 1986). Most of the pegmatites are lenticular in form and zoned, whereas others contain a mixture of feldspar, quartz, and muscovite. The muscovite zones in the zoned pegmatites have a thickness of up to 1 meter and occur on either side of the quartz core. Muscovite also occurs at the contact between the pegmatite lenses and the wall rock. Bihar is noted for its production of ruby mica (Rajgarhia 1951). The Andhra Pradesh district, north of Madras, and the Rajasthan district in northwestern India have extensive deposits. Their geologic conditions are identical to those of Bihar (Skow 1962; Clarke 1983). In an occurrence that resembles the Indian deposits, Brazil's muscovite is derived from a belt of pegmatites some 450 km long from north of Rio de Janeiro, running parallel to the Minas Geras coastline (Pecora et al. 1950). The muscovite occurs in zones around the quartz core in the zoned pegmatites, and at the contact between pegmatites and the wall rock. A special feature of the Brazilian pegmatites is their extensive decomposition by weathering, as a result of which hard feldspar has been altered to soft kaolin, permitting open pit mining. Muscovite makes up 5 to 20%, with a maximum of 40% of most of the pegmatites.

The phlogopite deposits of Canada, known as the largest in the world, are located in Suzor, Laviolette, Quebec, some 450 km north of Montreal (Locke et al. 1975; Harben and Kuzvart 1996). Clusters and veins of phlogopite with apatite occur in metamorphosed pyroxenite in association with granites, pegmatites, and marbles. Madagascar is the other major producer of phlogopite. The deposits are situated in the southeastern part of the island around Betroka, some 240 km northwest of Fort Dauphin (Skow 1962; Murdoch 1963). In association with pyroxenite dikes that intrude mica schist, the phlogopite occurs in veins with apatite and in pockets with calcite, diopside, and apatite.

In the former USSR, deposits of muscovite occur in Karelia, where zoned pegmatites with a predominance of plagioclase over K-feldspar constitute veins up to several hundred of meters long and tens of meters thick. Muscovite occurs around the quartz core, and muscovite and biotite in the wall zones. In the East Sayan region some 160 km from Krasnoyarsk, muscovite is produced from pegmatitic bodies in garnetiferous mica schist (Kuzmenko and Nedumov 1965). The Kovdor phlogopite deposit in the Kola Peninsula occurs in geologic conditions similar to those at Suzor, Canada. Mica is produced in a number of regions from pegmatites or as a feldspar by-product. The most important area is Tientsin province China (Harben and Kuzvart 1996).

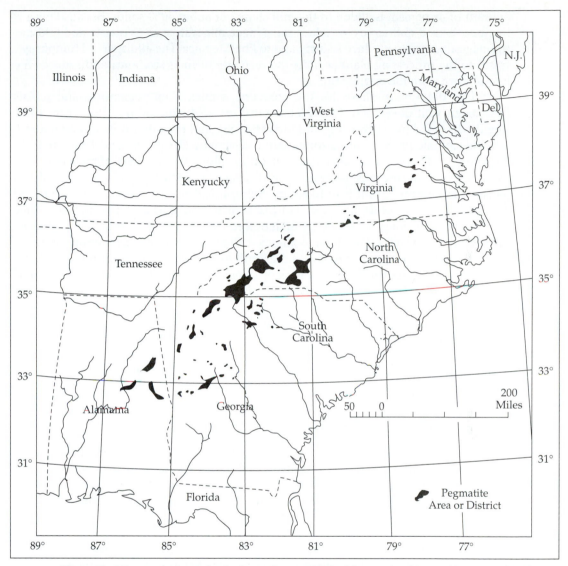

Figure 52   Mica-producing region in the southeastern United States, after Jahns and Lancaster (1950).

INDUSTRIAL PROCESSES AND USES

1. Milling and Beneficiation

The beneficiation of mica is generally done using rod mills, Humphrey spirals, and flotation. The washing or jig method is no longer in use because of environmental and economic reasons (Chapman 1983; Tanner 1994). The mine-run ore is fed into a rod mill, with or without washing, hydraulically from a bin. To facilitate the process, the proper amount of dispersant is added to the mill charge of about 40% solids. Because mica is flexible, the rod mill serves to reduce the gangue minerals such as quartz and feldspar to fine grain sizes, to produce a slurry and to liberate mica. The addition of Humphrey's spiral to the processing plant permits the recovery of finer size mica and reduces the loss of mica from the original ore.

Two flotation methods for the recovery of mica from pegmatites and schists have been well developed. One method utilizes acid-cationic flotation from thoroughly deslimed ore (Browning and McVay 1963), and the other utilizes an alkaline-anionic-cationic process with a minimum of desliming (Browning and Bennett 1965; Browning et al. 1965). The acid cationic flotation allows the recovery of mica as coarse as 14 mesh, but the ore must be completely deslimed at 150 to 200 mesh, which produces a loss of a significant quantity of fine mica. Sulfuric acid is used to condition the ground ore at 40 to 45% solids, and also for pH control and quartz depression. Cationic reagents, such as long-carbon-chain amino acetates, are the most effective collecting agents for floating mica. The alkaline-anionic-cationic process includes conditioning the finely ground ore pulps at 35 to 40% solids with sodium carbonate and sodium silicate, and floating the mica with a combination of anionic and cationic collectors. The cationic amine acetate collecting agents, used in combination with an anionic collecting agent such as oleic acid or a mixture of oleic acid and linoleic acid, are most suitable for floating the mica from gangue minerals. The most effective ratio of anionic to cationic collector for mica flotation is about 4:1. Using this method in a continuous treatment of ore requires grinding, trommel screening, preconcentration, classification, conditioning, and flotation. A process flow sheet is shown in Fig. 53. In addition, a pneumatic process has also been used for mica by using either a zig-zag classifier or an airfloat table.

2. Preparation of Mica Products

Mica is used in two principal forms: sheet and ground. The value of sheet mica depends on the purity, thickness, and useable area of the sheet. The technique used in preparation is to maximize these characteristics, because none of them can be improved by beneficiation. Ground mica is prepared by either dry or wet process (Benbow et al. 1985; Tanner 1994).

Sheet Mica: The preparation of sheet mica requires time- and cost-consuming manual labor. The crude mica books are picked out of the broken rock by hand, and cobbed or cleaned to remove adherent quartz, feldspar, clays, and defective mica. The cobbed mica is then split with knifes into successively thinner sheets, and the principal flaws are removed. Care is taken to provide a final sheet of maximum useable area with minimum waste. Block, thins, films, and splittings are the four categories of sheet mica that are classified according to lamina thickness and defined by the Indian Standards Institute, the British Standards, and the American Society for Testing and Materials. The following classifications refer to thickness: (1) block mica is at least 0.18 mm thick with a minimum useable area of 625 mm^2, (2) thins are defined as knife-dressed sheets with thicknesses of 0.05 to 0.18 mm, (3) films are sheets split from better-quality block mica with thicknesses of 0.02 to 0.18 mm that overlaps with thins in a certain range,

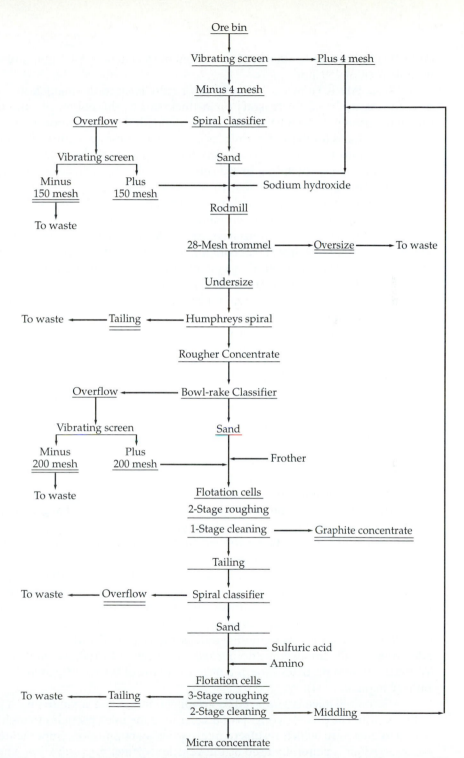

Figure 53 A process flow sheet showing grinding, screening, preconcentration, classification, conditioning, and flotation, after Browning et al. (1965).

and (4) splittings are laminae with a maximum thickness of 0.03 mm and a minimum useable area of 450 mm^2.

Muscovite is known as white mica or ruby mica to distinguish it from phlogopite, the amber mica. In crystals or in thick sheets, the colors of muscovite range through varying shades of red and green, depending mainly upon the iron content. Color is the basis for the industrial classification of muscovite into ruby and nonruby. When split into thin films, both muscovite and phlogopite are nearly colorless and transparent. Stains, which in a broad sense include both inclusions and intergrowths, generally reduce the quality. Clear ruby muscovite is the natural mica of the highest quality, and the most reliable with regard to performance in industrial applications. Green and dark micas are more variable.

In addition to classifications based upon sheet thickness, grading systems also exist for sheet purity as well as the maximum useable rectangular area that can be cut from a single lamina. The Indian grading systems are used as standards in the mica industry. In Table 16, there are sixteen grades listed according to clarity and freedom from stains (thirteen U.S. equivalent grades), and there are thirteen grades listed according to useable rectangles (twelve U.S. equivalent grades).

Small offcuts of sheet mica may be cemented together layer by layer to form "worked" or "manufactured" mica. This material is generally classified as follows (Robbins 1985):

Micanite or build-up mica is produced from thin overlapping splittings arranged in layers that are cemented by a binder (shellac, epoxy resins, or silicon resins), and compressed at 170° to 180°C. Reconstituted mica or mica paper is produced by pressing fine flakes (40 to 2,000 μm) of high-quality scrap mica into sheets in a continuous film that is generally impregnated with a binder. Glass-bonded mica is produced by cementing fine particles with borate or borosilicate glass.

Ground Mica: Ground mica is made from mine scrap as well as factory scrap, by either dry- or wet-grinding methods (Chapman 1984). For dry-grinding, the scrap (>25 mm) is introduced into high-speed hammer mills or pulverizers of the disintegrator type before being topped through vibrating screens. The products are highly delaminated and are referred to as high aspect ratio mica flakes and powders. In the United States, the process of dry grinding is influenced by its use in plasterboard jointing-cements. Using <3.35 mm (6 mesh) scrap as feedstock, production is by means of air micronizers; oversize fractions >150 to 250 μm (10 to 60 mesh) are removed on dry screens.

In the process for wet grinding mica, scrap is ground with up to 20 wt% water in edge-runner mills, followed by drying and classification to 100, 200, and 325 mesh sizes. Wet ground mica particles have a typical thickness of 0.1 to 0.8 μm and a high aspect ratio (Chapman 1984).

Micronized mica is produced by grinding fine-grained scrap to give a particle size below 40 μm. The process consists of accelerating the mica particles to high speeds in a confined enclosure, which results in interparticle comminution. Superheated steam or compressed air is generally used as the vehicle to effect micronization. Particle size is controlled by adjusting the residence time in the mill and removing oversize by air classification.

Table 16 (a) Indian visual classification system for mica and (b) Indian areal classification system for mica, after Benbow et al. (1985).

Designation	Appearance
V-1	ruby clear
V-2	ruby clear and slightly stained
V-3	ruby fair stained
V-4	rudy good stained
V-5	ruby stained "A"
V-6	ruby AQ
V-7	ruby stained "B"
V-8	ruby BQ
V-9	ruby heavy stained
V-10	ruby densely stained
V-11	black dotted
V-12	black spotted
V-13	black/red stained
V-14	green/brown 1st quality
V-15	green/brown 2nd quality
V-16	green/brown stained or BQ

(a)

Designation		Useable	Minimum
Old*	New	rectangle cm^2	dimension, cm
OEE Sp.	630	>645.2	10.2
OEE Sp.	500	516.1–645.2	10.2
EE Sp.	400	387.1–516.1	10.2
E Sp.	315	309.3–387.1	10.2
Sp.	250	232.3–309.7	8.9
1	160	154.8–232.3	7.6
2	100	96.8–154.8	5.1
3	63	64.5–96.8	5.1
4	40	38.7–64.5	3.3
5	20	19.4–38.7	2.5
5.5	16	14.5–19.4	2.2
6	06	6.4–14.5	1.9
7	05	4.8–6.4	1.6

*O= over; E = extra; Sp. = special.

(b)

3. Industrial Uses

Sheet and ground micas are used in two distinct industries with contrasting profit levels. Sheet mica is used principally in electronic and electric applications. These traditional markets have experienced a gradual erosion over the years. Ground mica, on the other hand, is in demand as a filler and extender in plasterboard, paint, and rubber, and as an additive to drilling muds.

Sheet Mica: The various uses of sheet mica take advantage of its unique electrical and thermal insulating properties as well as its mechanical properties, which allow it to be cut, punched, or stamped to close tolerances (Rajgarhia 1951; Montague 1960; Clarke 1983).

The largest use of block mica is in vacuum tubes, where fabricated mica spacers position, insulate, and support the tube elements. Transparent and flat block mica is also used to line the gauge glasses of high-pressure steam boilers. Diaphragms for oxygen equipment are made of high-quality block mica. Other uses include marker dials for navigation compasses, quarter-wave plates for optical instruments, retardation plates in helium-neon lasers, pyrometers, thermal regulators, stove windows, lamp chimneys, and microwave windows. Block mica is fabricated into washers of various sizes that act as insulators in electronic instruments. Disks are punched out of sheets and used as gap separators. Stacked mica, which consists of a multilayered "sandwich" of mica film alternating with silver foil, is used in capacitors. Capacitance is determined by the number of repetitions in the stack and also by the mica thickness. The thinner the mica, the higher the capacitance. Alternative capacitor materials of ceramics or glass, although cheaper than mica, do not possess its range of properties, which include dielectric stability at high frequencies of up to 500 MHz (Benbow et al. 1985). Backing mica used on the top and bottom of capacitors adds rigidity.

Micanite or build-up mica serves as a substitute for natural sheet mica in electrical insulation applications. It is used for segment plate, molding plate, flexible plate, heater plate, and tape (McKetta 1989). Segment plate acts as insulation between the copper commutator segments of d-c universal motors and generators. Molding plate is used to make V-rings for the insulation of copper segments from the steel shaft at the ends of a commutator. It is also fabricated into tubes and rings for insulation in transformers, armatures, and motor starters. Flexible plate is used in electric motor and generator armatures, field-coil insulation, and magnet and commutator core insulation. Heater plate is for uses at high temperatures. Sheets, ribbons, tapes, or other dimensions are made from bonded splittings reinforced with glass cloth, polyester film, Dacron mats, and vanished glass cloth, and have wide industrial applications.

Reconstituted mica or mica paper is used to make a wide range of insulation materials from rigid, cured plate for commutator segments, to reinforced flexible tapes and wrappers, and composite sheets for slot cells and vacuum tubes. Glass-bonded mica is used in a wide variety of molded and sheet electrical insulation in gyro parts, sonar gear, computers, radar, communication equipment, and nuclear devices (McKetta 1989).

Ground Mica: The largest use of ground mica is as a filler and extender in gypsum board cement. Other uses are in paint, rubber, and oil-drilling mud. A summary made by Chapman (1983) and Tanner (1994) is as follows:

1. Wet ground mica is used in wallpaper to provide sheen and in rubber for mould lubrication and dusting.

2. Dry ground mica is used in oil-drilling mud for sealing well walls and in plasterboard to impart crack resistance and provide smooth finishing.

3. Micronized mica is used in paint to facilitate suspension, reduce checking and chalking, prevent shrinking and shearing of the paint film, increase resistance to water penetration and weathering, and brighten the tone of colored pigments.

REFERENCES

Bailey, S. W. 1984. Classification and structures of micas, crystal chemistry of the true micas. *Reviews in Mineralogy* 13:1–60. Washington, D.C.: Mineral. Soc. Amer.

Benbow, R. J., B. H. W. S. De Jong, and J. W. Adams. 1985. Mica. In Vol. A16 of *Ullmann's encyclopedia of industrial chemistry,* 551–562. Weinheim: VCH Publ.

Browning, J. S., and T. L. McVay. 1963. Concentration of fine mica. *U.S.B.M. Invest. Rept.,* 6223, 7p.

Browning, J. S., F. W. Millsaps, and P. E. Bennett. 1965. Anionic-cationic flotation of mica ores from Alabama and North Carolina. *U.S.B.M. Invest. Rept.,* 6589, 10p.

Browning, J. S., and P. E. Bennett. 1965. Flotation of California mica ore. *U.S.B.M. Invest. Rept.,* 6668, 7p.

Chapman, G. P. 1983. Mica. In *Industrial minerals and rocks.* 5th ed. Ed. S. J. Lefond, 915–929. New York: Soc. Mining Engineers, AIME.

Chapman, G. P. 1984. The world mica-grinding industry and its markets. *Proc. 4th Ind. Minerals Intern. Congr, Toronto 1984,* 197–203.

Clarke, G. 1983. Mica—a review of world developments. *Ind. Minerals* (June):29–51.

Deer, W. A., R. A. Howie, and J. Zussman. 1992. An introduction to the rock-forming minerals. Essex: Longman. England, 696p.

Harben, P. W., and M. Kuzvart. 1996. Industrial minerals—a global geology. *Ind. Mineral Inform., Metal Bulletin, London,* 462p.

Hedrick, J. B. 1995. Mica. *U.S.B.M. Annual Review, 1994,* 12p.

Jahns, R. H., and F. W. Lancaster. 1950. Physical characteristics of commercial sheet muscovite in the southeastern United States. *U.S.G.S. Prof. Paper, 225,* 46p.

Kuzmenko, M. V. and I. B. Nedumov, eds. 1965. New data on geology, geochemistry and genesis of pegmatites, Nauka, Moscow, 333p.

Locke, S. R., M. Fenton, and G. C. Hawley. 1975. Suzorite mica flake: A new low-cost reinforcement for structural components. *Proc., 1st Ind. Mineral. Intern. Conference, London.* Ed. R. F. S. Fleming, 41–52.

McKetta, J. J., ed. 1989. Mica supply-demand relationships. In Vol. 30 of *Encyclopedia of chemical processing and design,* 93–123. New York: Marcel Dekker, Inc.

Mishra, H. K. 1986. Mica pegmatites of the Bihar mica belt. *Geol. Surv., India, Spec. Publ. 18.* Ed. K. K. Chakravorty, 107–108.

Montague, S. A. 1960. Mica. In *Industrial minerals and rocks.* 3d ed. Ed. J. L. Gillson, 551–566. New York: AIME.

Murdoch, T. M. 1963. Mineral Resources of the Malagasy Republic. *U.S.B.M. Inform. Circ., 8196,* 121p.

Murthy, M. V. N. 1964. Mica fields of India. *Proc. 22nd. Intern. Geol. Congr., India.* Ed. B. C. Roy, 17p.

Pecora, W. T., M. R. Klepper, D. M. Larrabee, A. L. M. Barbosa, and R. Frayha. 1950. Mica deposits in Minas Gerais, Brazil. *U.S.G.S. Bull., 964C,* 205–305.

Rajgarhia, C. M. 1951. *Mining, processing, and uses of Indian mica.* New York: McGraw-Hill, 176p.

Robbins, J. 1985. Sheet mica—and its changing face. *Ind. Minerals* (February): 33–47.

Skow, M. L. 1962. Mica, a material survey. *U.S.B.M. Inform. Circ., 8125,* 240p.

Tanner, Jr., J. T. 1994. Mica. In *Industrial minerals and rocks.* 6th ed. Ed. D. D. Carr, 693–710. Littleton, Colorado: Soc. Mining, Metal. & Exploration.

Molybdenite

Molybdenite, MoS_2, is the principal source for molybdenum and molybdic oxide, which are used in various chemical and metallurgical applications. As an industrial mineral, molybdenite, along with graphite, ranks at the top as a solid lubricant.

Three-fourths of the world's reserves are in the Western Cordillera of North and South America. They are of porphyry molybdenum or porpyhry copper-molybdenum type deposits. The two most important mines are at Climax and Urad-Henderson in Colorado. World molybdenum mine production ranged from 136,494 tons in 1989 to 95,296 tons in 1993 (*U.S.B.M. Mineral Yearbook 1994*).

MINERALOGICAL PROPERTIES

MoS_2 has three polymorphic forms: (1) molybdenite-2H (hexagonal), (2) molybdenite-3R (rhombohedral), and (3) jordiste (amorphous). The hexagonal and rhombohedral unit cells have identical a-dimensions, with a value of 3.16Å, whereas the c-dimension of the rhombohedral form (18.38Å) is 1.5 times that of the hexagonal form (12.29Å). The only difference between these two crystalline forms of MoS_2 is the layer sequence along the c-axis. Generally, when molybdenite is used with no suffix, it refers to molybdenite-2H (Berry et al. 1983).

In its structure, molybdenite shows the presence of layers of trigonal prismatic coordination polyhedra (Pauling 1952). Each molybdenum atom is surrounded by a trigonal prism of S atoms at distance 2.41Å, and the adjacent layers of S atoms are 3.08Å apart. Thus, the forces between the layers are of the relatively weak van der Waals type (Fig. 54). This distinct layered structure determines the physical properties of molybdenite. Many of those properties resemble the properties of graphite. Both molybdenite and graphite are structural lubricants (Mortier and Orszulik 1992).

Molybdenite is soft, with a hardness of 1 on the Moh's scale. Its specific gravity is 4.62 to 4.73 and it has a metallic luster. Molybdenite is diamagnetic and has an electrical conductivity of 0.790 Ω/cm at 20°C. Chemically, molybdenite is generally unreactive. It resists attack by most acids, except for *aqua regia*, which dissolves MoS_2, and hot concentrated H_2SO_4 and HNO_3, which oxidize MoS_2. Fluorine reacts vigorously with MoS_2, and, on heating, chlorine converts MoS_2 to $MoCl_5$. Molybdenite is insoluble in water, petroleum products, and synthetic lubricants. In dry air, oxidation begins slowly at 400°C and accelerates above 550°C, whereas in moist air, MoS_2 is not stable at room temperature. MoO_3 and SO_2 are the oxidation products of MoS_2 (Killeffer and Linz 1952; Mortier and Orszulik 1992).

The coefficient of friction of MoS_2 varies with its crystallographic orientation and is at a minimum when the cleavage planes are aligned in the plane of sliding. For example, MoS_2 specimens rubbing against a rotating steel disk gave a coefficient of 0.10 when the cleavage planes were parallel to the rubbing surface, and 0.26 when perpendicular to it (Feng 1952). As a consequence of its layered structure, MoS_2 crystals tend to be flat platelets, typically 500 to 1,000Å thick at the edges and 0.05 to 0.1μ thick at

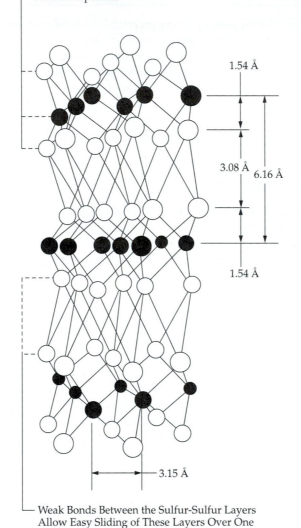

Strong Bonds Within the Sulfur-Molybdenum-
Sulfur Layers Highly Resist Penetration by
Surface Asperities

1.54 Å

3.08 Å   6.16 Å

1.54 Å

3.15 Å

Weak Bonds Between the Sulfur-Sulfur Layers
Allow Easy Sliding of These Layers Over One
Another Resulting in Low Friction Force

Figure 54   Crystal structure of molybdenum disulfide, after Clauss (1972).

the center, with thickness-to-diameter ratios of about 1:20 (Midgley 1956; Lipp 1976). These platelets lie flat on the surfaces on which they are placed. Rubbing and burnishing MoS_2 increases the degree of orientation and lowers the friction coefficient slightly.

Temperature affects the coefficient of friction above 375°C, and this is closely related to the effect of moisture content. The coefficient of friction in dry air is less than that in moist air and does not show as pronounced a decrease with temperature (Peterson and Johnson 1955; Mortier and Orszulik 1992). Unlike graphite, MoS_2 is an effective lubricant in a vacuum. In fact, MoS_2 provides a lower coefficient of friction in a vacuum than in air, with a typical reduction of one-third (Flom et al. 1965).

GEOLOGICAL OCCURRENCE AND DISTRIBUTION OF DEPOSITS

Probably more than 95% of the world's supply of molybdenite has been obtained from porphyry molybdenum or porphyry copper-molybdenum deposits, although the mineral also occurs in pegmatite and aplite dikes, contact-metamorphic zones of limestone, and bedded deposits in sedimentary rocks. In these porphyry deposits, chalcopyrite and molybdenite occur as disseminated grains, in stockworks of quartz veins in fractured or brecciated, hydrothermally altered granitic intrusive rocks, and in intruded igneous or sedimentary country rocks. One or more major faults passing through or close to the ore bodies are common features, and continued fracturing of host intrusives and of enclosing country rocks is a common characteristic (King et al. 1973).

Major molybdenum and molybdenum-copper porphyry deposits are distributed along the entire length of the western Cordillera from Alaska to Argentina (Titley and Hicks 1966; Ridge 1972). Mineralization occurs in calc-alkali intrusions that vary in size from small massifs to large stocks. The host rocks vary from Precambrian crystalline and granitoid rocks, through platform sediments to volcanic rocks. The Climax and Urad-Henderson deposits in Colorado are the most important ones (Fig. 55) (King et al. 1973). At Climax, the deposit consists of a complex dome-shaped mass of fractured, silicified, and mineralized Precambrian granites, gneiss, and schist, which generally overlies a composite porphyry stock of quartz monzonite to granite, of middle Tertiary age. In association with quartz, pyrite, fluorite, topaz, and small amounts of tungsten, scandium, titanium, and tin minerals, molybdenite is dispersed in fracture fillings, in veinlets, and as minute flakes throughout the fractured host rocks, forming a low-grade (0.05 to 0.5% Mo) ore body. In the Urad-Henderson deposit, molybdenite ore body is localized in a zone of fractured and altered granite, around the southern margin of a composite granite porphyry stock of middle Tertiary age. Molybdenite also occurs in veinlets and as disseminations in the altered and fractured granite between the fracture zones and fissures, forming a massive porphyry with 0.3 to 0.5% MoS_2.

The Andean porphyry copper deposits are mostly of Mesozoic and Tertiary age. Among the eleven important deposits now being worked in Chile and Peru, five of them, Toquepala, Cuajone, Chuquicamata, El Salvador, and El Teniente, are accompanied by molybdenite (Sillitoe 1972, 1981). The El Salvador porphyry deposit is the classic type of such Andean occurrence (Gustafson and Hunt 1975). The formation of the porphyry copper and molybdenum deposit culminated from volcanic activity in the Indio Muerto district. Host rocks for the ore are Cretaceous andesitic flows and sedimentary rocks overlain unconformably by lower Tertiary volcanics. Early rhyolite domes, roughly contemporaneous with voluminous rhyolitic and andesitic volcanics, were followed by irregularly shaped subvolcanic intrusions of quartz rhyolite and quartz porphyry. Copper-molybdenum mineralization accompanied this event. Average Cu and Mo contents are 0.5% and 0.02%, respectively.

In Southeast Europe, copper-molybdenum deposits occur in subvolcanic andesite bodies, usually along tectonic faults. Well-known deposits are at Deva, Braza, Arama-Corabia, and Rosia Poieni (Dunning et al. 1982). The porphyry copper ores of the Rosia Poieni

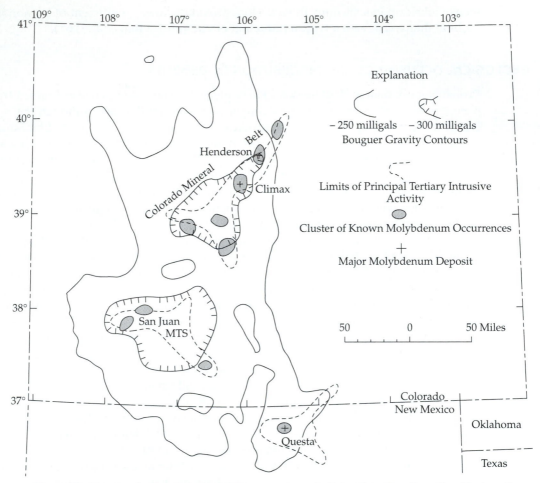

Figure 55   Map showing location of molybdenum resources in Colorado and northern New Mexico, after King et al. (1973).

deposit form an aureole in the marginal zone of the central andesite stock and extend to the enriched breccia zone forming its border.

Porphyry-type copper-molybdenum mineralization belts exist in eastern Queensland, Australia (Horton 1978). The deposits are characterized by weakly developed potassic alteration assemblages. Sulfide and alteration mineralogy are predominantly fracture controlled and usually exhibit rough zonation, commonly about a central core. The deposits are low-grade and the molybdenum content ranges from 0.01 to 0.05%.

INDUSTRIAL PROCESSES AND USES

1. Flotation of Molybdenite

In the porphyry type deposits, molybdenite is always finely disseminated in gangue minerals with low concentrations, and is accompanied by other sulfides such as chalcopyrite, chalcocite, and pyrite. The only beneficiation process for molybdenite that can separate it from all the other minerals is the use of flotation (Pryor 1965; Sutulov 1979; Crozier 1992).

If molybdenite is the major mineral to be separated, grinding can be considerably coarser. As is practiced at Climax, Henderson, and other locations, typical coarseness is around 40% −200 mesh. Molybdenite is a very floatable mineral, and middling with quartz will float readily even if the concentration of molybdenite is low. The collector used in these cases is generally a hydrocarbon, such as kerosene, stove oil, or some kind of light oil. An emulsifier is often added to prevent such hydrocarbons from dispersion and dissolution. Commonly used frothers are pine oil or an alcohol such as methyl isobutyl carbinol (MIBC). The flotation processes are carried out in several stages, with intermediate regrinding, in order to obtain the required grade of final concentrate by elimination of gangue minerals and other impurities that may be present. At Climax and Henderson, such operations are carried out three times to obtain a standard product of 90% MoS_2 (Sutulov 1979).

If the molybdenite is a by-product, the general practice applied is to first float a bulk concentrate and then proceed with selective flotation to separate molybdenite from the major products (Shirley et al. 1967; Wedt 1969). For molybdenite from copper-molybdenum porphyry deposits, the ore is generally ground to about 50 to 70% −200 mesh, and all processes are adjusted for the optimum recovery of copper minerals. Depending on the severity of the pyrite problem, pH may vary from a neutral circuit to as much as 11 or 12 if effective depression of abundant pyrite is sought. The bulk concentrate is generally floated with xanthates or aerofloats. Dithiocarbamates are also used, and light oil is usually added as a complementary collector to promote molybdenite flotation. The frother normally used in these circuits is pine oil or acresylic acid. The bulk concentrate, which assays between 8 to 20% Cu and 0.1 to 0.5% Mo, is reground, and then refloated as many times as needed to obtain a desirable grade of copper concentrate (28 to 35% Mo) for smelter. The molybdenum content in the concentrates simultaneously increases to 0.2 to 2% Mo (Pryor 1965; Sutulov 1979).

There are various methods in use for molybdenite recovery from copper concentrate. They may be grouped as follows:

1. Remove or destroy the collector agent, which adheres preferentially to molybdenite, by steam or dry-heating at 90 to 140°C prior to reflotation.
2. Depress molybdenite with organic colloids such as starch and glue.
3. Depress copper minerals with sodium ferrocyanide in a weakly alkaline pulp, and float molybdenite with fuel oil and an alcohol frother.
4. Depress copper minerals with sodium hypochlorite in an alkaline pulp, and float molybdenite with fuel oil and an alcohol frother.
5. Depress copper minerals with the use of phosphorus, arsenic, or antimony salts (Pryor 1965; Rustagi et al. 1970).

2. Industrial Uses

MoS_2 is classed as a solid lubricant. This is a solid that reduces the mechanical interaction between two moving surfaces and imparts antiwear properties to the system (Braithwaite 1964). The wear life of molybdenum disulfide as a lubricant is limited mainly by oxidation. Oxidation cannot occur in a vacuum, and thus, its wear life is very much extended even at temperatures up to 900°C. Molybdenum disulfide is the best conventional solid lubricant for use in a vacuum. Low friction is an inherent property of its crystal structure, and it is at its best when free of adsorbed substances (Mortier and Orszulik 1992).

MoS_2 is often used in the form of dry powder to facilitate the assembly of parts and to provide long-term lubrication. Particle size must be suited to the applications. The choice of particle size depends upon surface roughness and other factors (Braithwaite 1966). Parts can be tumbled in MoS_2 powder, plus cork, pine cones, or asbestos, to aid in burnishing on a thin, uniform coating. Colloidal particles are used to prepare dispersions that are applied by spraying, dipping, or brushing, and which can be applied from aerosol dispensers. These dispersions are often a more convenient form than dry powders for applying a uniform coating. Common carriers are (1) petroleum oil for general industrial and automotive uses, and as drilling and tapping lubricants; (2) polyglycol for high-temperature lubricants; (3) isopropanol for dry-film lubricants, thread antiseize coatings, and light load mechanisms; (4) trichlorethylene for nonflammable dry-film lubricants and general light load applications; (5) mineral spirits for general high-temperature lubricants; and (6) water for metal-working, such as wire drawing. The solid contents and densities incorporated in various carriers are 10% and 8.4 lb/gal, 10% and 8.9 lb/gal, 20% and 7.9 lb/gal, 20% and 12.2 lb/gal, 25% and 8.0 lb/gal, and 35% and 11.0 lb/gal, respectively (Clauss 1972).

Bonded films of MoS_2 have better friction and wear characteristics than loose powders. They are used in numerous industrial and automotive applications, as well as in aircraft, rockets, missiles, satellites, and nuclear power plants where extreme environments and weight limitations are encountered. Furthermore, the bonded films can prevent adhesion and wear between mating elements. They require little or no maintenance. A comparison of endurance life of MoS_2 used in resin-bonded MoS_2 film, grease containing MoS_2, and MoS_2 powder shows that the numbers of cycles to failure at 35,000 psi are 9,860,000, 1,590,000, and 130,000, respectively (Magie 1960). Some general limitations of bonded solid film lubricants may be summarized as follows (Lipp 1976): (1) solid film lubricants will not last indefinitely, (2) they may promote corrosion and are not corrosion preventive coatings, (3) they have no capacity to remove friction heat, (4) in most cases relubrication is not possible, (5) application techniques and surface preparation must be meticulously controlled or performance is jeopardized, and (6) inspection techniques are not always satisfactory. Bonded composite films containing molybdenum disulfide with graphite and $Sb(SbS_4)$ of optimum formulations were found to have longer wear life than that of each component alone. $Sb(SbS_4)$, which exhibits no lubricating properties whatsoever, improves the tribological behavior of MoS_2 and graphite (Bartz et al. 1986).

MoS_2 films from chemical formation in situ are an alterative to bonded films. They are produced on base metals such as steel and copper, as well as nonmetallic materials, by electroplating molybdenum over the substrate and then reacting in hydrogen sulfide at 200°C (Rowe 1959). The coefficient of friction for sintered molybdenum, steel slider, and 4-kg load, with chemically precipitated MoS_2 film, is 0.06 to 0.10 as compared with 0.4 to 0.55 of untreated sintered molybdenum under the same conditions (Milne 1951).

Plastics, rubber, and powdered metal composites can be impregnated with MoS_2 to impart self-lubricating behavior and to obviate the requirement for externally applied lubrication. These materials are used in cams, thrust washers, ball bearing retainers, compressor piston rings, sheaves, gears, conveyor belting, and dynamic seals (Vukasovich 1985).

The largest consumption of MoS_2 is as an additive in oils and greases used in mining, manufacturing, and transportation. In greases, 1 to 20wt% MoS_2 gives added protection from wear and galling, and provides backup lubrication should the grease deplete or thermally degrade. Paste composed of grease containing 20 to 60wt% MoS_2 is used with open gears, universal joints, spline drives, and in metal forming, press-fitting, and various wear-in operations (Vukasovich 1985). MoS_2 provides increased load-carrying capacity and improves boundary friction properties when added to most greases (Mortier and Orszulik 1992). Fig. 56(a) compares the axial forces and coefficients of friction obtained with various lubricants for press-fitting a 1.001-in. pin into a 1.000-in. bushing. The grease containing MoS_2 has a coefficient of friction of 0.014 under these conditions, as compared with 0.132 for a graphite grease, and 0.214 for zinc oxide grease (Smith 1956). The marked advantage to using MoS_2-filled greases as antigalling compounds is illustrated in Fig. 56(b), which compares the torque required to loosen a 1.25 in. × 7 in. long stud-nut assembly

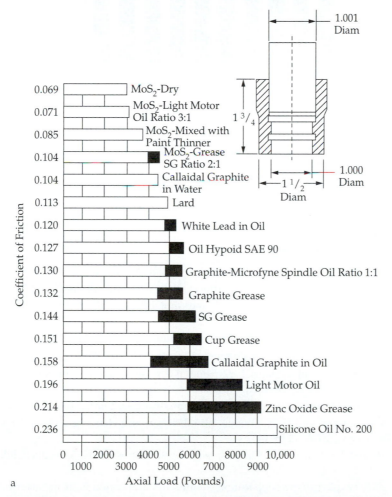

a

Figure 56   (a) Coefficients of friction and axial forces required for interference-fit assembly with MoS_2 and other lubricants, and (b) Torque for loosening stud-nut assembly after use, after Clauss (1972). In (b), 1. no lubricant; 2. graphite powder; 3. colloidal lead; 4. colloidal copper; 5. molybdenum disulfide; 6. aluminum powder; 7. calcium based grease; 8. silicone grease; 9. colloidal zinc; 10. colloidal graphite.

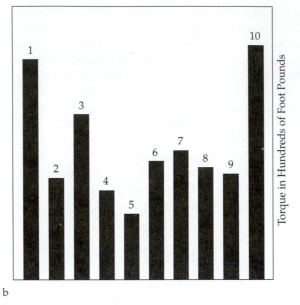

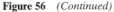

b

Figure 56   *(Continued)*

after 5,000 hours at 540°C (Smith 1956). MoS$_2$ allows the part to be disassembled with the least effort (Clauss 1972).

REFERENCES

Bartz, F. J., R. Holinski, and J. Xu. 1986. Wear life and frictional behaviour of bonded solid lubricants. *Lubr. Eng.* 42:762–769.

Berry, L. G., B. Mason, and R. V. Dietrich. 1983. *Mineralogy.* 2nd. San Francisco: W. H. Freeman, 546p.

Braithwaite, E. R. 1964. *Solid lubricants and surfaces.* New York: MacMillan Co., 286p.

——— 1966. Friction and wear of graphite and molybdenum disulfide. *Sci. Lubrication (London)* 18:17–21.

Clauss, F. J. 1972. *Solid lubricants and self-lubricating solids.* New York: Academic Press, 260p.

Crozier, R. D. 1992. *Flotation-theory, reagents and ore testing.* Oxford: Pergamon Press, 343p.

Dunning, F. W., W. Mykura, and D. Slater. 1982. Mineral deposits of Europe. In Vol. 2 of *Southeast Europe.* London: Inst. Mining Metall., Mineral. Soc., 340p.

Feng, I. M. 1952. Lubricating properties of molybdenum disulfide. *Lubric. Eng.* 8:285–289.

Flom, D. G., A. J. Haltner, and C. A. Gaulin. 1965. Friction and cleavage of lamellar solids in ultravacuum. *Trans. ASLE* 8:133–145.

Gustafson, L. B., and J. P. Hunt. 1975. The porphyry copper deposit at El Salvador, Chile. *Econ. Geol.* 70:857–912.

Horton, D. J. 1978. Porphyry-type copper-molybdenum mineralization belts in eastern Queensland, Australia. *Econ. Geol.* 73:904–921.

Killeffer, D. H., and A. Linz. 1952. *Molybdenum compounds.* New York: InterSciences Publ., 427p.

King, R. U., D. R. Shawe, and E. M. MacKevett, Jr. 1973. Molybdenum. In United States mineral resources. Ed. D. A. Brobst and W. P. Pratt. *U.S.G.S. Prof. Paper 820*: 425–435.

Lipp, L. C. 1976. Solid lubricants—their advantages and limitations. *Lubr. Eng.* 32:574–584.

Magie, P. M. 1960. Moly bonded films. *Electromech. Des.* 4:50–54.

Midgley, J. W. 1956. The friction properties of molybdenum disulphide. *J. Inst. Petr. (London)* 42:316–321.

Milne, A. A. 1951. Lubrication of steel surface with molybdenum disulfide. *Research (London)* 4:93–96.

Mortier, R. M., and S. T. Orszulik. 1992. *Chemistry and Technology of Lubricants*, Blackie, Glasgow and London, and VCH Publ., New York. 302p.

Pauling, L. 1952. The structural chemistry of molybdenum. In *Molybdenum compounds.* Ed. D. Killeffer and A. Linz, 95–109. New York: Interscience Publ.

Peterson, M. B., and R. L. Johnson. 1955. Factors influencing friction and wear with solid lubricants. *Lubric. Eng.* 11:325–330.

Pryor, E. J. 1965. *Mineral processing.* Amsterdam: Elsevier, 844p.

Ridge, J. D., ed. 1972. Annotated bibliographies of mineral deposits in the Western Hemisphere. *Geol. Soc. Amer. Mem.,* 131, 681p.

Rowe, G. W. 1959. Coatings on molybdenum for high-temperature sliding. *Sci. Lubrications (London)* 12:12–15.

Rustagi, M. C., P. Dayal, and A. K. Biswas. 1970. On the molybdenite—chalcopyrite separation. *J. Mines, Metals and Fuels* 18:50–51.

Shirley, J. F., M. L. Campbell, and L. C. De Young. 1967. Recovery of by-product molybdenite at Toquepala. Trans. AIME, 238, 386–393.

Sillitoe, R. H. 1972. A plate tectonic model for the origin of porphyry copper deposits. *Econ. Geol.* 68:799–815.

———— 1981. Regional aspects of the Andean porphyry copper belt in Chile and Argentina. *Trans. Inst. Min. Metall.* 90B:15–36.

Smith, E. E. 1956. Molybdenum disulfide as a grease additive. *J. Nat. Lubric. Grease Inst.* 20:20–36.

Sutulov, A., ed. 1979. Processing and metallurgy. Vol. II of *International molybdenum encyclopedia, 1778-1978.* Santiago, Chile: Internet Publ., 375p.

Titley, S. R., and C. L. Hicks, eds. 1966. Geology of the porphyry copper deposits, southwestern North America. Wilson Volume. Tucson, Arizona: Univ. Arizona Press, 287p.

Vukasovich, M. S. 1985. Uses of molybdenum compounds. In Vol. A17 of *Ullmann's encyclopedia of industrial chemistry,* 682–687. Weinheim: Wiley VCH Verlag.

Wedt, W. 1969. Copper by-product molybdenum recovery in El Salvardor, Chile. *Trans. Inst. Min. Engineer Chile* 24:13–17.

Nitrates

While nitrate minerals occur in many widely scattered deposits around the world, the Chilean deposits are the only ones that have economic significance. The nitrates produced have long been used in fertilizers, but their importance waned as the production of synthetic nitrogen compounds grew.

MINERALOGICAL PROPERTIES

Two principal minerals in nitrate deposits are nitratite ($NaNO_3$) and niter (KNO_3), also known as saltpeter. Physically, they are very much alike (Kapusta and Wendt 1963; Sauchelli 1964; Berry et al. 1983). Both $NaNO_3$ and KNO_3 are white in color and soft, with a hardness close to 2 on the Moh's scale. The specific gravity of $NaNO_3$ is 2.25, and KNO_3 has a value of 2.11. $NaNO_3$ melts at 306°C, and molten $NaNO_3$ remains stable in air up to 500°C, at which point it releases oxygen to produce sodium nitrite. In the same fashion, KNO_3 melts at 334°C, and molten KNO_3 is stable up to 500°C in air, at which point it releases oxygen to produce potassium nitrite. The refractive indices are $\omega = 1.567$ and $\varepsilon = 1.336$ for $NaNO_3$, and $\alpha = 1.335$, $\beta = 1.5056$, and $\gamma = 1.5064$ for KNO_3.

Both nitrates readily dissolve in water and do not form hydrated phases. $NaNO_3$ is soluble in liquid ammonia and forms $NaNO_3 \cdot 4NH_3$ at temperatures below 40°C, whereas KNO_3 is also soluble in liquid ammonia, but no ammoniates are formed. Enthalpies of solution at infinite dilution (ΔH_{298}) are 205 KJ/mol for $NaNO_3$, and 34.9 KJ/mole for KNO_3.

$NaNO_3$ is hexagonal, with a=5.0696Å and c=16.829Å. It is isostructural with calcite, a structure based upon a face-centered cubic NaCl-type lattice. In the structure of KNO_3, the constituent ions are arranged in layers parallel to (001) with alternate layers composed of K^{1+} and a planar $(NO_3)^{1-}$ group. It is isostructural with NH_4NO_3. The structural characteristics of chemical compounds play a key role in determining their behavior in fertilizer usage (Baynham and Raistrick 1960).

GEOLOGICAL OCCURRENCE AND DISTRIBUTION OF DEPOSITS

Deposits of natural nitrates of sodium and potassium occur in many parts of the world including Bengal, China, Iran, Chile, Peru, the West Indies, the United States, Mexico, South Africa, and Egypt. The most important deposits are the well-known Chilean nitrates, which are composed mainly of sodium, with potassium in minor amounts. This is found in the desert region of the Pacific coast, especially in the Provinces of Tarapaca and Antofagasta (Ericksen 1981). The nitrate field is several kilometers inland and rises to an elevation of over 1,200 meters above sea-level.

As shown in Fig. 57(a), it stretches from north to south for about 420 km. Two major types of nitrate ore, collectively known as caliche, can be identified. The alluvial caliche, in which the saline minerals occur as a cement, generally rests on poorly cemented debris, and the bedrock caliche, in which the saline minerals form impregnations,

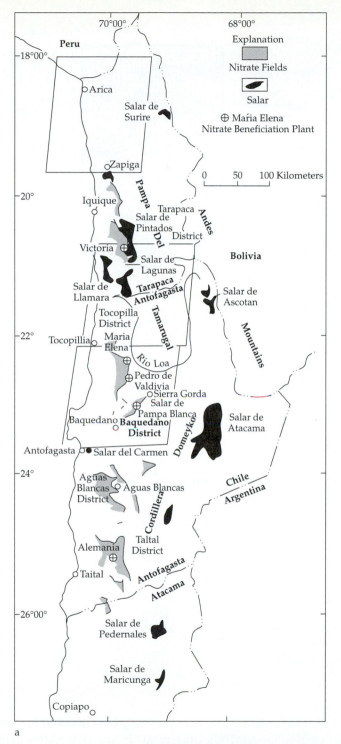

a

Figure 57 (a) Location map of the salars and nitrate fields of northern Chile. In the Figure, "☐" outlines the areas for which landsat images are available, and (b) Characteristic layering of an alluvial-type nitrate deposit, after Ericksen (1981).

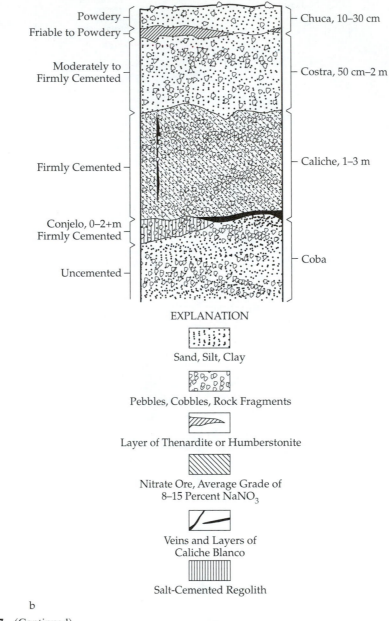

Powdery — Chuca, 10–30 cm

Friable to Powdery —

Moderately to Firmly Cemented — Costra, 50 cm–2 m

Firmly Cemented — Caliche, 1–3 m

Conjelo, 0–2+m Firmly Cemented —

Uncemented — Coba

EXPLANATION

Sand, Silt, Clay

Pebbles, Cobbles, Rock Fragments

Layer of Thenardite or Humberstonite

Nitrate Ore, Average Grade of
8–15 Percent $NaNO_3$

Veins and Layers of
Caliche Blanco

Salt-Cemented Regolith

b

Figure 57   (Continued)

veins, and masses in fractured bedrock, gives way downward to less fractured rock that is relatively free of saline minerals. The typical alluvial-type nitrate deposit consists generally of several layers (Fig. 57b)—chuca, costra, caliche, conjelo, and coba—each having characteristic chemical and physical features (Ericksen 1981):

1. Chuca is the overburden of the nitrate deposit. The powdery to poorly cemented surface layer consists of silt, sand, rock fragments, and saline minerals including abundant gypsum and anhydrite, and small amounts of other saline minerals.

2. Costra consists of either a hard and brittle material similar to the underlying caliche, or poorly cemented friable material that appears to be transitional between chuca and caliche.

3. Caliche is generally 1 to 3 m thick. The bed is as thin as 50 cm and as thick as 5 m exists locally. Veins and layers of high-purity, white nitrate-rich saline material, known as caliche blanco, are widespread in some caliche with thickness ranging from 10 to 50 cm. Caliche may grade downward into conjelo, a saline-cemented regolith containing little $NaNO_3$.

4. Coba is loose uncemented regolith, which is in direct contact with caliche because of the absence of conjelo in the Chilean nitrate deposits. The contact is relatively sharp, generally changing from hard nitrate ore to soft uncemented material within a few centimeters.

The alluvial caliche deposits occur on rocks ranging from granite to limestone and shale, and there is no obvious correlation between deposit and rock type (Tower and Brewer 1964; Ericksen 1981). The ore has a composition averaging 7 to 10% $NaNO_3$, 4 to 10% $NaCl$, 10 to 30% Na_2SO_4, 1 to 2% H_2O, and 41 to 76% gangue, with the remaining percentage balanced by other salts. Compositional variations of the salt assemblages are illustrated by the following analyses (Laue et al. 1991):

Salt	wt%				
$NaNO_3$	34.2	34.4	43.3	28.5	53.5
KNO_3	1.6	—	—	—	17.3
Na_2SO_4	8.4	1.6	25.3	5.4	1.9
$CaSO_4$	6.3	1.6	30.9	2.7	0.5
$MgSO_4$	2.0	5.4	—	3.4	1.4
$NaCl$	32.0	4.0	—	17.2	21.3

The caliche often contains a wide variety of complex salts such as blödite ($Na_2SO_4 \cdot CaSO_4$), polyhalite ($K_2SO_4 \cdot MgSO_4 \cdot 2CaSO_4 \cdot H_2O$), darapskite ($NaNO_3 \cdot Na_2SO_4 \cdot H_2O$), and syngenite ($K_2SO_4 \cdot Na_2SO_4 \cdot H_2O$). In addition, borates, iodates, and perchlorates may be present in small amounts.

INDUSTRIAL PROCESSES AND USES

1. Beneficiation of Caliche Ore

Leaching $NaNO_3$ from caliche is a very complicated process. The ore contains only 7 to 10 wt% $NaNO_3$ and requires multiple-stage crushing and screening before leaching by either Shanks or Guggenheim process, to achieve an acceptable yield (Laue et al. 1991).

Both processes utilize the fact that the solubility of $NaNO_3$ increases with increasing temperature, whereas that of $NaCl$ remains substantially constant. The Shanks process leaches by means of a boiling solution (135° to 140°C), which is thereafter cooled to 10°C. The crude ore must contain at least 13% $NaNO_3$ for profitable operation. The

Guggenheim process uses warm solution (40°C) for leaching and then refrigerates it to about 5°C, utilizing a heat exchanger on the cooling system and power plant for much of the heat input (Tower and Brewer 1964).

A flow sheet illustrating the Guggenheim process is shown in Fig. 58. The crushing and screening process consists of multiple stages, with the jaw crusher as primary, gyratory crusher as secondary, and cone crusher as tertiary crusher. Screening is done between each crushing. The leaching cycle consists of the downward percolation of solutions that are successively advanced through a series of four vats. The leaching solutions increase in concentration as they are advanced through the four vats (Tower and Brewer 1964).

2. Preparation of Nitrates

The simplest and most direct process used to produce synthetic $NaNO_3$ is the neutralization of dilute nitric acid (53%) with caustic soda or soda ash solutions: $HNO_3 + NaOH \rightarrow NaNO_3 + H_2O$ or $HNO_3 + Na_2CO_3 \rightarrow 2NaNO_3 + H_2O + CO_2$. The resulting solution is evaporated upon cooling, and crystals are deposited in the crystallization unit. These crystals are separated in a screw conveyor centrifuge, and then the separated $NaNO_3$ is treated in a dryer to complete the preparation (Stocchi 1990). The principal process used to produce $NaNO_3$, however, is to oxidize ammonia with air to produce nitrogen oxides, which then react with soda ash solution in large reaction towers to produce $NaNO_3$. Sodium nitrate is present along with nitrate during an intermediate stage of the process, but they then react with nitric acid or nitrogen dioxide to produce $NaNO_3$, thus liberating nitric oxide, which is recycled back through the process. An equation can be written as $12NH_3 + 21O_2 + 4Na_2CO_3 \rightarrow 8NaNO_3 + 4NO + 4CO_2 + 18H_2O$. $NaNO_3$ is the only solid product obtained from this process (Miles 1961).

KNO_3 is prepared by conversion of Chilean $NaNO_3$ with potassium chloride: $NaNO_3 + KCl = KNO_3 + NaCl$. The sodium chloride formed crystallizes in the hot liquor due to its relatively low solubility and is then separated by centrifuging or filtration (Seidel 1953). KNO_3 may also be prepared from calcium nitrate and potassium sulfate, and from ammonium nitrate and potassium chloride. In both cases, the KNO_3 prepared can be easily separated from gypsum in the first, and from ammonium chloride in the second.

3. Industrial Uses

Sodium nitrate is mainly used in agriculture as a fertilizer for crops such as cotton, tobacco, and vegetables. As with other nitrates, $NaNO_3$ is prone to leaching in soil, but it has the advantage of possessing a metallic cation. Unlike ammonia and its derivatives, it will not induce cation losses in the soil, which can result in unsuspected soil acidity. However, its usage has declined through the years because of the availability of other nitrogen products (Sauchelli 1964). No $NaNO_3$ is used in mixed fertilizers today. Potassium nitrate also has its main use as a fertilizer. It can be used in the production of liquid fertilizers and is an important constituent of mixed fertilizers as a source of potassium. The absence of chlorine and sulfur is advantageous for many crops (McVickar et al. 1963).

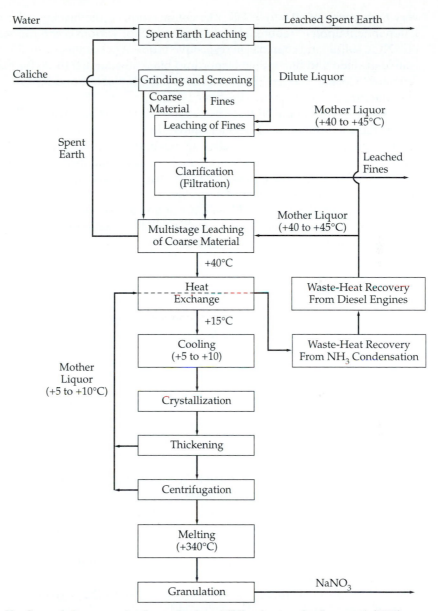

Figure 58 Guggenheim process for the production of Chile saltpeter, after Laue et al. (1991).

Another industrial use of $NaNO_3$ and KNO_3 is in the explosives industry (Sauchelli 1964). Some types of blackpowder used for blasting contain $NaNO_3$ instead of potassium saltpeter, as a cheap substitute. The Chilean saltpeter, known in industry as explosive saltpeter, has an advantage as an oxidizing agent because it contains more oxygen by weight than potassium saltpeter. Chilean saltpeter is also used to produce KNO_3 for the manufacture of blackpowder. KNO_3 used for blackpowder requires a very high purity of 99.0 to 99.8%. The composition of the blackpowder manufactured is dependent upon its required heat of explosion. For various compositions, the amounts of KNO_3, sulfur, and charcoal, in respective order of percentages, are 75, 10, 15 for both coarse-grained and fine-grained (sporting) blackpowder; 62, 16, 20 for blasting; and 80, 2, 18 for "Cocoa" powder (Urbanski, 1967). Both $NaNO_3$ and KNO_3 are used in the manufacture of pyrotechnics (Barbour 1981).

Both $NaNO_3$ and KNO_3 are used in metallurgy as heat transfer baths for quench hardening and tempering of steel and alloys. In the glaze and enamel industry, they are used as oxidizing agents and as active agents in the early stages of frit melting (Taylor and Bull 1986). For example, KNO_3 is a major component in semimatt enamel for sheet steel made from phosphate glass and a silica frit, and both are used to prepare titanium opacified enamels (Maskall and White 1986).

Other uses of $NaNO_3$ and KNO_3 are as an auxiliary in soldering and welding, as a cleaning agent for pipe draining, and as an oxidizing component in an acid-based gas generator system used for the rapid inflation of air bags.

REFERENCES

Barbour, R. T. 1981. *Pyrotechnics in industry.* New York: McGraw-Hill. 190p.

Baynham, J. W., and B. Raistrick. 1960. Structural and x-ray data on chemical compounds found in fertilizers: The influence of structure on behavior. In Chemistry and technology of fertilizers. Ed. V. Sauchelli. *Amer. Chem. Soc. Monograph* 148, 538–575. New York: Reinhold Publ.

Berry, L. G., B. Mason, and R. V. Dietrich. 1983. *Mineralogy.* 2nd. San Francisco: Freeman & Co, 561p.

Ericksen, G. E. 1981. Geology and origin of the Chilean nitrate deposits. *U.S.G.S. Prof. Paper,* 1188, 37p.

Kapusta, E. C., and N. E Wendt. 1963. Advances in fertilizer potash production. In *Fertilizer technology and usage.* Ed. M. H. McVickar, G. L. Bridger, and L. B. Nelson, 189–230. Madison, Wisconsin: Soil Science Soc. Amer.

Laue, W., M. Thiemann, E. Scheibler, and K. W. Wiegend. 1991. Nitrates and nitrites. In Vol. A17 of *Ullmann's encyclopedia of industrial chemistry,* 266–294. Weinheim: VCH Publ.

Maskall, K. A., and D. White. 1986. *Vitreous enamelling.* Oxford: Pergamon Press, 112p.

McVickar, M. H., G. L. Bridger, and L. B. Nelson. 1963. *Fertilizer technology and usage.* Madison, Wisconsin: Soil Science Soc. Amer, 464p.

Miles, F. D. 1961. *Nitric acid—manufacture and uses.* Auspices of Imperial Chem. Inds., Ltd. Oxford Univ. Press: London, 75p.

Sauchelli, V., ed. 1964. Fertilizer nitrogen, its chemistry and technology. *Amer. Chem. Soc. Monograph Series.* New York: Reinhold Publ, 424p.

Seidel, M. 1953. *Solubilities of inorganic and metal compounds.* New York: Van Nostrand, 254p.

Stocchi, E. 1990. *Industrial chemistry,* Vol. 1. New York: Ellis Horwood, 712p.

Taylor, J. R., and A. C. Bull. 1986. *Ceramics glaze technology.* Oxford: Pergamon Press, 263p.

Tower, Jr., H. L., and H. C. Brewer. 1964. Natural Chilean nitrate of soda: Production and use in agriculture. In Fertilier Nitrogen. Ed. V. Sauchelli, 315–330. *Amer. Chem. Soc. Monograph 161.* New York: Reinhold Publ.

Urbanski, T. 1967. *Chemistry and technology of explosives.* Vol. III. Oxford: Pergamon Press, 717p.

Olivine

Olivine represents a series of solid solutions between Mg_2SiO_4 (forsterite) and Fe_2SiO_4 (fayalite). It is mainly used (especially the Mg-rich members) in the production of hot metal in a blast furnace and as foundry sands in the casting of manganese steel. Dunite, a basic igneous rock consisting almost entirely of olivine, is the major resource of this mineral.

Leading producers of olivine are Norway, Austria, and the United States. Other producing countries are Spain, Japan, Mexico, and Italy.

MINERALOGICAL PROPERTIES

The olivine series, $(Mg,Fe)_2SiO_4$, is defined by two end members: forsterite (Mg_2SiO_4) and fayalite (Fe_2SiO_4). Their crystal structures are characterized by isolated (SiO_4) tetrahedra that are crosslinked by chains of edge-sharing (MgO_6) or (FeO_6) octahedra. Each oxygen is bonded to one silicon and three octahedrally coordinated Mg and Fe atoms. Such a structure provides a limited number of nonequivalent cation sites, and, therefore, cation substitutions in olivine are relatively simple, by comparison with other silicates (Birle et al. 1968). Calcium and manganese may represent the two most common cation substitutes for magnesium and ferrous iron in olivine, and they form olivine-type minerals (tephroite, monticellite, glaucochroite, and kirschsteinite). However, their respective amounts in olivine rarely exceed 1 wt% and 2 wt% (Simkin and Smith 1970). Other elements such as nickel, chromium, ferric iron, and alkalis have amounts of olivine below detectability, to about 0.01 wt%.

The olivine minerals are orthorhombic, with cell dimensions of a $=4$ .75Å, b $= 10.20$Å, and c$=5.96$Å for Mg_2SiO_4, and a $= 4.82$Å, b $= 10.48$Å, and c$=6.09$Å for Fe_2SiO_4 (Louisnathan and Smith 1968). The "b" varies linearly with composition in the series, and the variations of "a" and "c" show slightly positive deviations from linearity (Brown 1982). A regression equation relates composition to $d_{(130)}$ (in Å) as follows: mol% fayalite $= 7.522 - 14.9071 (3.0199 - d_{(130)})^{\frac{1}{2}}$. Linear thermal expansion coefficients $(°C^{-1} \times 10^5)$ are a $= 0.87$, b $= 1.54$, and c $= 1.33$ for forsterite, and a $= 0.99$, b $= 0.95$, and c $= 1.19$ for fayalite (Lager and Meagher 1978). At one atmospheric pressure, forsterite melts congruently at 1,890°C, whereas fayalite melts incongruently at 1,205°C to a liquid and iron (Deer et al. 1992).

Olivine is commonly pale olive-green to yellow-green. With oxidation, it becomes brown, reddish brown, or black. As iron content increases in the series, hardness decreases from 7 to 6½, and specific gravity increases from 3.22 to 4.39. Refractive indices increase linearly with Fe: α from 1.635 to 1.827, β from 1.651 to 1.869, and γ from 1.670 to 1.879 (Deer et al. 1962).

GEOLOGICAL OCCURRENCE AND DISTRIBUTION OF DEPOSITS

Olivine occurs mainly in igneous rocks, but it is also found in metamorphic rocks of principally ultramafic composition. Four types of igneous rocks have significant amounts of olivine. In ultramafic intrusions, olivine (Fo_{96}-Fo_{87}) may comprise more than

90% of the rock, known as dunite, or more commonly is present in amounts between 50 and 90%, known as peridotite. The stratiform mafic intrusions have olivines with compositions between Fo_{90} and Fo_{80}, such as at the Stillwater complex in Montana. In gabbros and basalts, olivines in the compositional range Fo_{80} to Fo_{50} are common constituents. Relatively iron-rich olivine occurs in felsic rocks (Brown 1982).

Industrial olivine is recovered principally from dunite. The world's largest producing deposits are located at Aaheim, about 100 km south of Aalesund on the west coast of Norway (Griffiths 1989), and at Norddal, near Tafjord, Norway (Harben and Bates 1990). Both dunite deposits fall in the east-west trending zones within the so-called Basal Gneiss complex and are characterized by their well exposed eclogitic and garnetiferous layers. The Basal gneisses are predominantly banded gneisses, augen gneisses, and potash-poor gneisses of the almandine amphibolite facies. They vary from granitic to granodioritic in composition (Lappin 1967). Olivine is mined from dunite bodies, and the product consists of 91.3% forsterite, 6.8% enstatite, 0.9% serpentine, 0.7% chromite, and 0.4% spinel (Olerud 1995). Chemical analysis of the forsterite gives 49.0% MgO, 42.6% SiO_2, 6.0% FeO, 1.8% other oxides, and 0.6% LOI (Beckius 1970). In Austria, dunite is quarried at St. Stefan near Leoben in Styria (Griffiths 1989). The product has a high serpentine content and must be sintered. The sintered dunite has compositions of 49 to 51% MgO, 39 to 41% SiO_2, 8 to 9% Fe_2O_3, and 0.4% CaO (max.).

In the United States, olivine is produced in the states of Washington and North Carolina. In the middle and northern Washington Cascades there are a number of peridotite and serpentine masses. The Twin Sisters dunite body, elliptical in shape and approximately 90 km^2 in size, is the largest in the Cascade belt, northwestern Washington, 32 km due east of Bellingham (Ragan 1967). The Twin Sisters dunite consists of four minerals: olivine, enstatite, chromite, and clinopyroxene. The olivine has a composition slightly more magnesian than Fo_{90} and has minor compositional variation throughout the dunite mass. The enstatite is also Mg-rich, with a Mg/Fe ratio close to that of olivine. The clinopyroxene is a Cr-diopside (Ross et al. 1954). Serpentine occurs around the entire margin of the dunite mass. Several large dunite masses form a belt extending from south of Boone, North Carolina, southwestward to the state boundary with Georgia, a distance of 280 km (Harben and Bates 1990). The dunite masses occur in the form of either a lenticular intrusion or a ring dike. The core of the bodies consists of unaltered olivine, while in the outer zones the olivine is partially altered to serpentine and talc. Chemical compositions of olivine from North Carolina are 50.5% MgO, 40.1% SiO_2, 6.7% Fe_2O_3, 0.2% CaO, 1.8% other oxide, and 0.7% LOI. In comparison, Washington's olivine has 49.4% MgO, 41.2% SiO_2, 7.1% Fe_2O_3, 0.2% CaO, 1.8% other oxide, and 0.7% LOI (Beckius 1970).

INDUSTRIAL PROCESSES AND USES

1. Milling and Beneficiation

Olivine may be beneficiated by means of either a dry or wet process. Both processes are used in production in North Carolina and Washington (the United States), as well as at Aaheim, Norway (Griffiths 1984). In the dry process, olivine is crushed in jaw crushers, cone crushers, and impact crushers to the required size for special functions, such as, for example, ⅜ to 1¼ in. (0.95 to 3.2 cm) for blast furnace applications, and $-30 +200$ mesh size for nonferrous castings. By adding a washing procedure to the process, the olivine is fed through the primary crusher under wet conditions. Dewatering is done in a screw classifier, which is followed by centrifuging. The moisture content of the final product is reduced to a maximum of 0.5% in a vertical dryer (Anon. 1970).

Because of the relatively high content of impurities, such as serpentine or vermiculite in the North Carolina olivine, it is necessary to process it by flotation or gravity method (Teague 1985). Like most magnesium minerals, olivine is floated by the use of cresylic acid and a hydrocarbon with an alkaline earth soap, in conjunction with a salt of the lead-thallium class and a xanthate.

2. Synthetic Olivine

By using chrysotile (asbestos) tailings, magnesium silicates, including olivine, are manufactured at Thetford Mines in Quebec (Griffiths 1984). The process can be illustrated by the following reaction:

$$2Mg_3Si_2O_5(OH)_4 \rightarrow 3Mg_2SiO_4 + SiO_2 + 4H_2O$$

at $820°$ to $840°C$ to produce olivine and silica, which is stable up to $1,000°C$. Above $1,000°C$ enstatite forms according to the following reaction:

$$3Mg_2SiO_4 + SiO_2 \rightarrow 2Mg_2SiO_4 + 2MgSiO_3.$$

The product has a 85% yield of combined magnesium silicates (olivine and enstatite) and a 15% yield of iron oxide, and other minerals. (Griffiths 1984).

3. Industrial Uses

Olivine is mainly used in blast furnaces for the production of hot metal (75%), and as a foundry sand in mold making for brass, aluminum, magnesium, and manganese steel foundries (15%) (Henning 1994).

In the production of hot metal from the blast furnace process, a slag is formed, which functions as a collector of the unwanted oxides of the ores and coke ash, including sulfur. The efficiency of the slag depends upon two properties, melting temperature and viscosity, both of which contribute to the free-running temperature of the slag. These properties are a function of slag composition or of the ratio of basic oxides to acidic oxides, known as the "basicity" of the slag, which can be defined as:

$$(\% \ CaO + \% \ MgO)/(\% \ SiO_2 + \% \ Al_2O_3).$$

For most blast furnace operations, this ratio varies between 1.10 and 1.25. The addition of olivine to the blast furnace process increases the MgO content of the slag, which in turn lowers its viscosity and enables the lime-plus-magnesia content to be increased, thus improving the desulfurizing power. In the meantime, the addition of olivine to the process does not greatly affect the basicity because of its fixed MgO/SiO_2 ratio (Ross 1964). Olivine is added to the blast furnace process in three size ranges: (1) traditionally, lump feed of 10 to 40 mm is added directly into the furnace, (2) the feed in the sinter stream ranges in size from 0 to 6 mm, and (3) a finer and more reactive feed is used recently in the range of 0 to 3 mm (Griffiths 1989). Olivine can also be used for conditioning electric arc furnace slag, to which olivine is added to provide basicity control and thereby prevent refractory lining from slag attack.

Olivine, along with zircon and chromite, is grouped together as special foundry sands in respect to silica sand, which is the most commonly used foundry sand, especially in large volume and repetitive casting (Taylor et al. 1959). Olivine, zircon, and chromite all have lower thermal expansion and better thermal shock resistance than silica. In relation with binders, zircon is compatible with all common binders and requires relatively low levels to achieve strength. Both chromite and olivine are less compatible with some resin binders. High acid demand values of olivine (at pH 4, 25.70 ml, 0.1m HCl) in comparison with that of chromite (4.30) and zircon (3.20) further restrict its use in acid catalyst systems. As a foundry sand, olivine is specifically used in the casting of austenitic manganese steel because it does not form low melting eutectic points in the system $FeO-MnO-SiO_2$ as silica does (Henning 1994). Other advantages of olivine over silica as foundry sand, summarized by Griffiths (1989), are (1) low, uniform thermal expansion, (2) low required amounts of binder material, (3) higher melting point, (4) high resistance to thermal shock, (5) easier to mould, (6) higher green strength, (7) large grains, and (8) hard ramming without loss of permeability.

Changes in technology and processing have eliminated the open hearth furnaces, and olivine's use in refractories has undergone radical change in recent years. It is now used principally in refractory bricks (Koltermann 1988). Two types are in use. One consists of variable amounts of MgO, and the other of 3 to 6% chromite. Both resin-bonded and inorganically bonded are used (Benbow 1989).

Olivine is used as a sand blasting abrasive because of its hardness, pale color, nontoxic nature, high cleaning rate, and ample supply (Griffiths 1984).

REFERENCES

Anon. 1970. Opportunities for increasing olivine output. *Ind. Minerals* (February): 11–21.

Beckius, K. 1970. Olivine: Its properties and uses. *Ind. Minerals* (February): 22–26.

Benbow, J. 1989. Steel industry minerals. *Ind. Minerals* (October): 27–43.

Birle, J. D., G. V. Gibbs, P. B. Moore, and J. V. Smith. 1968. Crystal structures of natural olivines. *Amer. Mineral.* 58:807–824.

Brown, G. E. 1982. Olivines and silicate spinels. In Vol. 5 of *Reviews in mineralogy.* Ed. P. H. Ribbe, 275–381. Washington, D.C.: Mineral. Soc. Amer.

Deer, W. A., R. A. Howie, and J. Zussman. 1962. Ortho- and Ring-Silicates. Vol. 1 of *Rock forming minerals.* London: Longman, 333p.

Deer, W. A., R. A. Howie, and J. Zussman. 1992. *An introduction to rock-forming minerals.* Essex, England: Longman, 696p.

Griffiths, J. 1984. Olivine—exchanging new uses for old. *Ind. Minerals* (September): 65–79.

Griffiths, J. 1989. Olivine, volume the key to success. *Ind. Minerals* (January): 25–36.

Harben, P. W., and R. L. Bates. 1990. *Industrial minerals, geology and world deposits.* London: Industrial Minerals Division, Metal Bulletin Plc, 312p.

Henning, R. J. 1994. Olivine and dunite. In *Industrial minerals and rocks.* 6th ed. Ed. D. D. Carr, 731–734. Littleton, Colorado: Soc. Mining, Metal. & Exploration.

Koltermann, M. 1988. Olivine $(Mg,Fe)_2SiO_4$ as a refractory material in the steel industry. *Taikabutsu Overseas* 8:3–8.

Lager, G. A., and E. P. Meagher. 1978. High temperature structural study of six olivines. *Amer. Mineral.* 63:365–377.

Lappin, M. A. 1967. Structural and petrological studies of the dunite of Almklovdalen, Nordfjord, Norway. In *Ultramafic and Related Rocks.* Ed. P. L. Wyllie, 183–190. New York: John Wiley.

Louisnathan, S. J., and J. V. Smith. 1968. Cell dimensions of olivine. *Mineral. Mag.* 36:1123–1134.

Olerud, S. 1995. Norway's industrial minerals. *Ind. Minerals* (December): 23–31.

Ragan, D. M. 1967. The Twin Sisters dunite, Washington. In *Ultramafic and related rocks.* Ed. P. J. Wyllie, 100–107. New York: John Wiley.

Ross, C. S., M. D. Foster, and A. T. Myers. 1954. Origin of dunite and of olivine-rich inclusions in basaltic rocks. *Amer. Mineral.* 39:693–737.

Ross, H. U. 1964. Blast furnace technology. In *Aspects of modern ferrous metallurgy.* Ed. J. S. Kirkaldy and R. G. Ward, 65–140. Toronto: Univ. Toronto Press.

Simkin, T., and J. V. Smith. 1970. Minor element distribution in olivine. *J. Geol.* 78:304–325.

Taylor, H. F., M. C. Flemings, and J. Wulff. 1959. *Foundry engineering.* New York: John Wiley & Sons, 607p.

Teague, K. H. 1985. Olivine. In *Industrial minerals and rocks.* 5th ed. Ed. S. J. Lefond, 989–996. New York: Soc. Mining Engineers, AIME.

Perlite and Pumice

Perlite and pumice are similar in chemical composition, physical state, and industrial application. Perlite is a hydrated glass characterized by a perlitic structure, whereas pumice is highly vesiculated and contains less than 1% H_2O. Both are volcanic glasses of rhyolitic composition. Their major uses are as lightweight aggregates in insulation board, plaster, and concrete, as well as in loose-fill insulation. The leading producers of perlite are the United States, Greece, and Japan. Italy and Turkey rank at the top for the production of pumice.

Pumicite, scoria, and cinder are three terms commonly associated with pumice. Pumicite has the same origin, chemical composition, and glassy structure as pumice, but differs in particle size, usually by less than 2 to 3 mm in diameter (Peterson and Mason 1983). Scoria, by petrographic definition (Williams et al. 1982), is the general term applied to a dark, highly-vesicular rock of basaltic composition. Most basaltic lapilli and bombs are composed of scoria and are referred to as cinder.

MINERALOGICAL PROPERTIES

Perlite is a volcanic glass of rhyolitic composition that contains from 2 to 5% combined water. It has a pearly luster and usually exhibits numerous concentric, spheroidal cracks, known as the perlitic structure, which are the result of devitrification (Barker 1983). Perlite generally ranges in color from transparent light gray to glassy black. Its chemical composition varies only within a small range, as shown by representative analyses of perlites from New Mexico and Greece. These are, respectively, 72.1 and 74.2% SiO_2, 13.5 and 12.3% Al_2O_3, 0.8 and 0.95% Fe_2O_3, 0.06 and 0.08% TiO_2, 0.89 and 0.85% CaO, 0.50 and 0.13% MgO, 4.6 and 4.0% Na_2O, 4.4 and 4.4% K_2O, and 3.3 and 2.8% H_2O (Kadey 1983). Perlite is chemically inert and has a pH of between 7 and 8 when placed in solution. It has a low density of approximately 1,750 to 1,875 kg/m^3, and its hardness is between 5.5 and 7 on the Moh's scale (Lin 1989). When perlite is heated, its 2 to 5% combined water vaporizes, forming steam that expands each perlite grain into a mass of glass foam. The original volume of the perlite may be expanded four to twenty times at temperatures between 750° and 1,350°C. The bulk density of expanded perlite varies and usually ranges from 25 to 250 kg/m^3 (Meisinger 1979; Austin and Barker 1995). Some perlites contain crystalline silica at levels close to 0.1%, a level which is beyond the ability of routine quantitative x-ray diffraction analysis to identify. An improved procedure has been developed specifically for this purpose (Hamilton and Peletis 1990). It requires grinding to <50 μm, uniform packing, a long counting time, and the use of accurate standards.

Pumice, like perlite, is also a volcanic glass, usually of rhyolitic composition. It is produced by explosive volcanic eruptions and accumulates as pyroclastics. Its vesicular or cellular structure is due to expansion during the eruption of the steam and gases contained in molten lava. Pumice is usually white or light gray in color. Its hardness is about 6 for individual grains, and its specific gravity varies from a true value of 2.5 for pumicite and ground pumice to an apparent value of <1 for dry block pumice. Depending on the moisture content, particle size distribution, and the specific gravity of the particles, the bulk specific gravity of pumice typically ranges from 500 to 700 kg/m^3, whereas that of cinder ranges from 700 to 900 kg/m^3 (Geitgey 1994). Pumice has a distinct conchoidal fracture resembling that of broken glass. The melting temperature of pumice is approximately 1,343°C, and it undergoes no volume change below 760°C (Schmidt 1956). A representative analysis of pumice from Greece gave 70.55% SiO_2, 12.24% Al_2O_3, 0.89% Fe_2O_3, 2.36% CaO, 0.10% MgO, 3.49% Na_2O, 4.21% K_2O, 0.65% others, and 5.51% LOI (Robbins 1984).

GEOLOGICAL OCCURRENCE AND DISTRIBUTION OF DEPOSITS

1. Perlite

Perlite is formed by hydration of a volcanic glass such as obsidian and occurs in association with lava flows from volcanic domes, glassy zones in welded ash-flow tuffs, and wall zones of felsic intrusive plugs and dikes (Meisinger 1979). Perlite undergoes continuous devitrification and transformation to zeolites and other aluminosilicate minerals in diagenetic environments (Yanez et al. 1993; Klammer and Konrad 1993), and, therefore, the preservation of perlite is very rare in rocks older than Tertiary period. Most economic perlite deposits are associated with the recent lava flows of the Eocene and Oligocene ages.

The United States has the largest known reserves and production of perlite in the world, with more than 85% of its production total originating in New Mexico. The other producing states are Arizona, California, Colorado, Idaho, and Nevada (Jaster 1956; Meisinger 1980; Bolen 1996). In New Mexico, the perlite deposit on the southeastern flank of the Socorro Mountains, 5 km southwest of Socorro, has the form of a glass dome with diameters of approximately 0.5 to 0.8 km, and an exposed vertical continuity of more than 140 meters. The perlite is pale-gray to bluish-gray in color and is flow-banded. Its composition is rhyolitic, and throughout most of the deposit the quality of the perlite is remarkably uniform (Weber and Austin 1982). At No Agua Mountain, Taos County, there exists a cluster of four subconical peaks encircling a breached central valley. The rocks composing these peaks include volcanic glass, massive rhyolitic breccias, and other fragmental volcanics. The basic characteristics of perlite are similar to those of the Socorro deposit (Whitson 1982). A comprehensive investigation of perlite in New Mexico was made by Chamberlin and Barker (1996). In Arizona, large deposits of perlite are in the Superior area, where the perlite rests upon tuff or breccia, and is underlain by glassy rhyolite. Its thickness is approximately 600 meters (Peirce 1969). In California, the Coso perlite deposits are domes that rise to an average height of 100 meters over granitic hills. The perlite is light gray, shows well-developed flow banding, and is found in association with pumice and black obsidian (Austin and Barker 1995).

Perlite from Greece, Europe's leading producer, is derived from deposits on the Islands of Milos and Kos in the Aegean Sea (Harben and Bates 1990; Stamatakis et al. 1996). Numerous volcanic events took place on the islands from the Late Pliocene to the Late Pleistocene, which produced acid tuffs containing pumice and ignimbrite, and lava flows containing andesite, dacite, and rhyolite. The most recent volcanic activity produced perlite (Harben and Bates 1990). Greek perlite is younger than that of other countries in the Carpatho-Balkan region and has a low percentage of phenocrysts.

More than 85% of the Greek perlite groundmass consists of glass and has good expansibility (Koukouzas 1995). The "Armenian volcanic highlands" near the Turkish border represent the largest perlite deposit in the former USSR (Sagatelyan 1973). The volcanic series of the Upper Tertiary to the Quarternary periods has doleritic basalt and trachy-liparitic lava in the lower part, and acid rocks and lava flows such as liparite, obsidian, and perlite in the upper part. Perlite is confined to the uppermost part of the acid lava flows. In Italy, most of the perlite produced is derived from the Island of Sardinia near the town of Uras (Harben and Bates 1990).

2. Pumice

Pumice is formed by the frothing of the gas-rich magma shortly before or at the time of ejection from the vent and occurs in immense heaps around extinct vents, as well as in sheets extending outward from the vents. The finer, lighter pumicite is deposited farther away from the vents than the denser scoria and cinder. On Shinjima island, Kagoshima Bay, southwestern Japan, the rocks comprise subaqueous pumiceous mass-flow deposits, each of which is fine-depleted and shows upward coarsening of water-chilled pumice clasts, a direct consequence of the subaqueous eruption (Kano et al. 1996). Pumicite deposits may be found several kilometers from the source, and they may become interbedded with other sediments (Geitgey 1994).

The Island of Lipari, 45 km off the northeast coast of Sicily, is Italy's major source of pumice. The pumice is highly vesicular and consists almost entirely of glass, with little weathered, altered, or devitrified material. For industrial uses, lipari pumice is produced in two distinct types that differ in major chemical composition. Peerless or black pumice has a composition of 71.76% SiO_2, 12.33% Al_2O_3, 0.11% TiO_2, 1.98% Fe_2O_3, and 0.02% FeO, whereas Lapillo or white pumice has a composition of 70% SiO_2, 12.76% Al_2O_3, 0.14% TiO_2, 1.75% Fe_2O_3, and 0.64% FeO. By comparison, Peerless is higher in Si-, Na-, and K-content and has an apparent density of 500 gm/liter, while Lapillo is higher in Fe, Ca, and Mg, and has a density higher than 500 gm/liter (McMichael 1990). Pumiceous material is prominent in many of the Greek volcanic islands. The principal deposit occurs on the Island of Yali in the Dodecanese group of islands. Deposits of volcanic tuff occur on the Island of Thera in the Cyclades group (Robbins 1984). In Turkey, significant pumice deposits exist in the area around Nevsehir and Kayseri in central Anatolia, and in Bitlis, Agri, and Kars in eastern Anatolia (Harben and Kuzvart 1996). Typical chemical composition is 72.50% SiO_2, 12.59% Al_2O_3, 0.90% Fe_2O_3, 3.62% Na_2O, 4.71% K_2O, and 0.80% CaO.

In the United States, pumice, pumicite, and cinders occur in the western states. In California, most of the pumice occurs in tuffs, tuff-breccias, and pumice-breccias in eastern Siskiyou County, and at Mono Craters, Mono County (Chesterman 1966). In Oregon, block pumice occurs at Newberry Crater and at Rock Mesa, both of which are in Deschutes County (Wagner 1969; Geitgey 1992). The Jemez Mountains in Sandoval and Valencia Counties, New Mexico, have a number of deposits of pumice in association with perlite, and Dona Ana County has deposits of scoria (Austin et al. 1982; Hoffer 1994). In Arizona, deposits of pumice occur in Graham County, east of Phoenix (Keith 1969; Bryan 1987).

INDUSTRIAL PROCESSES AND USES

1. Milling and Processing

Perlite is usually used in an expanded form. The intended use determines the size range into which the crude perlite must be crushed and sized prior to expansion. A general working guide is that the coarser the required product, the coarser the corresponding feed must be. Therefore, laboratory evaluation, especially of expansibility, has to be performed on samples from different deposits, because no two deposits are alike (Kadey 1983; Breese and Barker 1994).

The first milling step is to reduce the crude ore to approximately 1.6 cm (5/8 in) in size in a primary jaw crusher and, if necessary, in a secondary roll crusher. Further grinding is normally accomplished in a closed circuit using vibratory screens, air classifiers, Hammer mills, and rodmills to separate the milled perlites into basic grades for storage, whereupon it is blended to meet the required specifications (Kadey 1983).

The expansion process used for perlite is to soften the glass, volatilize the combined water, and produce a lightweight, cellular aggregate. This is done in either a rotary horizontal furnace or a stationary vertical furnace at temperatures in the range of 750° and 1,160°C (Sharps 1961; Jackson 1986). Rapid heating of perlite to a point of incipient fusion has also been used. In this process, care must be taken to balance the softening of the glass and the volatilization of the combined water. Excessive combined water will cause some perlites to shatter, and insufficient water content will result in incomplete expansion (Kadey 1983). Both the American Society of Testing & Materials (ASTM) and the Perlite Institute have established standards for the determination and specification of the properties of expanded perlite (Breese and Barker 1994).

Pumice is generally used in its natural state and requires minimum milling and treatment (Taggart 1945). The crude ore is sorted, cleaned by screening or water-floating, and graded by size and quality. Standard sizes are lump, 2 in. (5.06 cm) and over; pezzame, 3/4 (1.90 cm) to 2 in.(5.06 cm); chips, 1/4 (0.63 cm) to 3/4 in.(1.90 cm); Lapillo powder, −1/4 in. (−0.63 cm); and waste. For lump pumice, the best grades are light colored, fluffy, and of low apparent density. The lower grades are denser and darker in color and may contain some impurities.

2. Industrial Uses of Perlite

Perlite is used in industries in its expanded state, which has a wide range of applications. Important uses include in lightweight concrete aggregate, fillers, filter aids, plaster and tiles, wallboards, and other formed products, animal feedstuff and crop farming, and low-temperature insulation (Breese and Barker 1994).

In the construction industry, expanded perlite is used in the production of lightweight concrete for a large number of special applications, such as for the docks of long-span bridges, as fire protection for the steelwork of tall buildings, as filling for thick floor or roof slabs, as thermal insulation, and for concrete ships, building blocks, and filler walls (Troxell and Davis 1956). Typical properties of common lightweight aggregate concretes used for both structural and insulation proposes are given in Table 17 (Anon. 1983).

Table 17 Typical properties of some lightweight aggregate concretes (a) generally as used for structural purposes, and (b) generally as used for insulation purposes (*FIP Manual of Lightweight Aggregate Concrete* [1983]).

Aggregate type	Bulk density of aggregate kg/m³			Quantities per m³ of concrete				Air-dry density kg/m³	28-day compressive strength N/mm²	Modulus of elasticity (E) kN/mm²	Drying shrinkage (approx.) %	Thermal conductivity (K) W/m°C
	Fine	Medium	Coarse	Cement kg	Aggregate (m³)							
					Fine	Medium	Coarse					
Expanded clay (made in a rotary kiln)	640	370	320	310	0.65	—	0.65	1050	8.5	3.5	0.055	0.30
				350	0.60	—	0.60	1200	10.5	to	to	to
				440	0.60	0.60	—	1300	14.0	10.5	0.07	0.58
				350	0.40*	—	0.80	1350	15.5		0.045	
				350	0.45*	0.80	—	1500	17.0			
Expanded shale	960	700	590	390	0.75	0.65	—	1500	21.0	14	0.04	0.58
				440	0.75	0.70	—	1550	26.0	to	to	
				480	0.70	0.70	—	1600	31.5	18	0.06	
Expanded slate	860	670	560	450	0.60	—	0.80	1600	27.5	11	0.04	0.68
				500	0.60	—	0.80	1650	34.5	to	to	
				540	0.55	—	0.80	1650	41.5	18	0.045	
Sintered (pulverized fuel-ash)	1040	835 (single size)	770	320	0.55	—	0.70	1550	22.5	12	0.04	0.47
				500	0.40	—	0.80	1600	37.5	to	to	
				550	0.35	—	0.80	1650	45.0	18	0.07	
Foamed slag	920	—	670	480	0.85	—	1.10	1900	21.0	18	0.04	0.92
				540	0.85	—	1.10	2000	27.5	to	to	
				570	0.90	—	0.85	2000	34.5	24	0.06	

*Natural sand

a.

Table 17 (*Continued*)

Aggregate type	Bulk density of aggregate kg/m^3	Mix proportions (by volume) cement:aggregate	Air-dry density kg/m^3	28-day compressive strength N/mm^2	Drying shrinkage (approx.) %	Thermal conductivity (K) W/m°C
Exfoliated vermiculite	60 to 90	1:8 1:6 1:4	400 480 560	0.7 0.9 1.2	0.35 to 0.45	0.09 to 0.15
Expanded perlite	100 to 150	1:6 1:5 1:4	480 560 640	2.1 3.4 4.7	0.14 to 0.20	0.09 to 0.15
Expanded polystyrene	120 to 150	1:37 1:10 1:8	300 600 900	0.3 2.0 6.0	0.10 to 0.20	0.09 to 0.32
Diatomite	400 to 500	1:9 1:6 1:4	690 770 930	2.2 4.5 8.3	0.25 to 0.35	0.17 to 0.26
Graded wood particles	120 to 200	1:4 1:3 1:2	640 880 1200	1.7 4.8 12.1	0.25 to 0.50	0.20 to 0.28
Pumice	500 to 800	1:6 1:4 1:2	1200 1250 1450	14.0 19.0 29.0	0.10 to 0.12	0.29 to 0.55

b.

277

Perlite plaster is made by mixing perlite with gypsum or cement, and is used extensively to fireproof structural steel construction and to reduce the weight of interior walls and ceilings. The grades of plaster, or spreadability, are determined by the aggregate-to-gypsum (or cement) ratio. For a finishing plaster, a low ratio is required to produce a smooth surface, whereas a backing plaster can be made with a higher ratio, which will also impart good insulation characteristics (Power 1986). Expanded perlite is also an important component of roof insulation board, masonry, and floor and wall tiles.

The refractory and insulation properties of perlite enable its expanded form to be used in castable refractories up to temperatures of around 900°C. It is used to slow the rate of solidification of metals by efficient insulation, and thus eliminate casting defects (Dickson 1986).

Expanded perlite is used as a filter aid by breweries to remove extremely fine solids such as algae and bacteria, and by other industries in the processing of juices, syrups, sugar, waxes, lacquers, and drugs. It is also used as a filler to improve properties and add bulk to materials, especially paper and paint (Pettifer 1981).

In agriculture, expanded perlite is used for animal feedstuff and crop farming (Lin 1989). Its functions are identical to those of vermiculite, as a carrier for nutrients, as an absorbent for pesticides and fertilizers, as a digestive to the feed mixtures, and as an additive to feed mixes.

3. Industrial Uses of Pumice, Pumicite, and Cinders

Industrial uses of pumice and perlite overlap in some areas, but have distinct applications in other areas. The main uses of pumice and cinders, like perlite, are in the construction industry as lightweight aggregate, plaster aggregate, concrete building block, and loose-fill aggregate (Robbins 1984). Pumicite is used extensively as a pozzolanic additive to monolithic concrete, where it increases the strength and durability of the concrete and decreases the heat of hydration (Peterson and Mason 1983). Lightweight pumice aggregate is classified as a medium strength aggregate with concrete strengths of 3.5 to 15 N/mm^2 as compared to lightweight perlite aggregate, which is classified as a low strength aggregate with concrete strengths of 0.5 to 3.5 N/mm^2 (Anon. 1983). In the tabulation below, the bulk density (dry, loose) and the particle density of lightweight pumice and perlite aggregates are compared (Anon. 1983):

	Pumice aggregate	Perlite aggregate
Density, kg/m^3		
Individual particles	550–1,650	100–400
In bulk	350–650	40–200
Concrete		
Compressive strength		
N/mm^2–Dry density, kg/m^3	5–1,200	1.2–400
	15–1,600	3–500

Pumice is consumed in large amounts in abrasive applications. It is brittle, continues to break into fresh grains, and has a conchoidal fracture that continues to build up the sharp edges necessary for a good abrasive. As an abrasive, pumice is used for pol-

Table 18 Applications of industrial grade pumice. © Industrial Minerals, May 1990.

Market	Application	Grade
Agriculture	Soil substitute and additive	coarse
Metal detector	Food and chemical processing	coarse/intermediate
Paint manufacture	Nonskid coatings	coarse
	Accoustic insulting ceiling paints	coarse
	Fillers for textured paint	intermediate/coarse
	Flattening agents	extra fine
Chemical industry	Filtration media	coarse
	Chemical carriers	coarse
	Sulphur matches and strikers	intermediate
Metal and plastic	Cleaning and polishing	extra fine
finishing	Vibratory and barrel finishing	extra fine/intermediate
	Pressure blasting	intermediate
	Electroplating	extra fine/fine
	Cleaning lithographic plates	extra fine
Compounders	Powdered hand soaps	intermediate
	Glass cleaners	extra fine
Dental and		
cosmetic	Polishing natural teeth and dentures	fine
	Smoothing rough skin	lump
Rubber	Erasers	intermediate
	Mould release agents	extra fine
Glass and mirror	TV tube processing, glass	
	buffing and polishing	fine
	Bevel finishing	extra fine
	Cut glass finishing	extra fine
Furniture	Hand rubbed satin finishing	extra fine
	Piano keys	extra fine
	Picture frame gold leafing	extra fine
Leather	Buffing	intermediate
Electronics	Cleaning circuit boards	extra fine
Pottery	Filler	extra fine/fine

ishing and cleaning soft metals as well as ceramics and glass. The traditional applications of chrome and nickel plating polishing have declined substantially because these plates are no longer in demand by the automotive industry.

Other industrial uses of pumice, pumicite, and cinders cover a broad range and are listed in Table 18 (McMichael, 1990).

REFERENCES

Anon. 1983. *FIP Manual of Lightweight Aggregate Concrete,* Surrey University Press, Glasgow and London, and Halsted Press, New York and Toronto, 259p.

Austin, G. S., and J. M. Barker. 1995. Production and marketing of perlite in the western United States, Proc., 29th Forum on the Geology of Industrial Minerals. Eds. M. Tabilio and D. L. Dupras. *California Dept. Conservation, Div. Mines & Geol., Spec. Publ.* 110:39–67.

Austin, G. S., F. E. Kottlowski, and W. T. Siemers. 1982. Industrial minerals of New Mexico in 1981. In Proc., 17th Forum on the Geology of Industrial Minerals. Ed. G. S. Austin. *New Mexico Bur. Mines & Min. Res. Circ.* 182:9–16.

Barker, D. S. 1983. *Igneous rocks.* Englewood Cliffs, New Jersey: Prentice-Hall, 417p.

Bolen, W. P. 1996. Perlite. *U.S.G.S. Annual Rev.,* August 1995, 4p.

Breese, R. O. Y., and J. M. Barker. 1994. Perlite. In *Industrial minerals and rocks.* 6th ed. Ed. D. D. Carr, 735–749. Littleton, Colorado: Soc. Mining, Metal. & Exploration.

Bryan, D. P. 1987. Natural lightweight aggregates in the southwest, Proc., 21st Forum on the Geology of Industrial Minerals. Ed. H. W. Peirce. *Arizona Bur. Geol. & Mines, Tech. Spec. Paper* 4:55–63.

Chamberlin, R. M., and J. M. Barker. 1996. Genetic aspects of commercial perlite deposits in New Mexico. *New Mexico Bureau of Mines and Mineral Res., Bull.* 154:171–185.

Chesterman, C. W. 1966. Pumice, pumicite, perlite, and volcanic cinders. In Mineral resources of California. *California Div. Mines and Geol. Bull.* 191:336–341.

Dickson, T. 1986. Insulating refractories. *IM Refractories Survey 1986,* 178–183.

Geitgey, R. P. 1992. Pumice in Oregon. *Oregon Dept. Geol. & Mineral Industries Spec. Paper,* 25.

———— 1994. Pumice and volcanic cinder. In *Industrial minerals and rocks.* 6th ed. Ed. D. D. Carr, 803–813. Littleton, Colorado: Soc. Mining, Metal. and Exploration.

Hamilton, R. D., and N. G. Peletis. 1990. The determination of quartz in perlite by x-ray diffraction. *Adv. in X-ray Analysis,* Eds. C. S. Barrett et al., 32, 493–497.

Harben, P. W., and R. L. Bates. 1990. Industrial minerals, geology and world deposits. *Industrial Minerals Div., Metal Bulletin Plc., London,* 312p.

Harben, P. W., and M. Kuzvart. 1996. Industrial minerals—a global geology. *Industrial Minerals Infor. Ltd., Metal Bulletin Plc., London,* 462p.

Hoffer, J. M. 1994. Pumice and pumicite in New Mexico. *New Mexico Bureau of Mines & Mineral Res. Bull.,* 140, 23p.

Jackson, F. L. 1986. Processing perlite for use in insulation applications. *Mining Eng.* 38:40–45.

Jaster, M. C. 1956. Perlite resource of the United States. *U.S.G.S. Bull.,* 1027I, 28p.

Kadey, Jr., F. L. 1983. Perlite. In *Industrial minerals and rocks.* 5th ed. Ed. S. J. Lefond, 997–1015. New York: Soc. Mining Engineers, AIME.

Kano, K., T. Yamamoto, and K. Ono. 1996. Subaqueous eruption and emplacement of the Shinjima pumice, Shinjima Island, Kagoshima Bay, SW Japan. *Jour. Volcanology and Geothermal Res.* 71:187–206.

Keith, S. R. 1969. Pumice and pumicite. In Mineral and water resources of Arizona. *Arizona Bur. Mines Bull.* 100:407–412.

Klammer, D., and B. Konrad. 1993. Synthetic zeolites formed from expanded perlite: type, formation, conditions and properties. *Mineral. & Petrol.* 48:275–294.

Koukouzas, N. 1995. Greek perlite: Comparison with some deposits of the Carpatho-Balken region. *Geol. Soc. Greece, Spec. Paper* 4:750–753.

Lin, I. J. 1989. Vermiculite and perlite, for animal feedstuff and crop farming. *Ind. Minerals* (July):43–49.

McMichael, B. 1990. Pumice markets—volcanic rise of stone washing. *Ind. Minerals* (May):22–37.

Meisinger, A. C. 1979. Perlite. *U.S.B.M. Commodity Profile.*

———— 1980. Perlite. Mineral Facts and Problems. *U.S.B.M. Bull.* 671:1–12.

Peirce, H. W. 1969. Perlite. In Mineral and water resources of Arizona. *Arizona Bur. Mines Bull.* 180:403–407.

Peterson, N. V., and R. S. Mason. 1983. Pumice, pumicite, and volcanic cinders. In *Industrial minerals and rocks.* 5th ed. Ed. S. J. Lefond, 1079–1084. New York: Soc. Mining Engineers, AIME.

Pettifer, L. 1981. Perlite—diversification the key to overall expansion. *Ind. Minerals* (December):55–75.

Power, T. 1986. Perlite and vermiculite—the market overlap. *Ind. Minerals* (November):39–49.

Robbins, J. 1984. Pumice—the ins and outs of a reinforced market. *Ind. Minerals* (May):31–51.

Sagatelyan, K. M. 1973. Perlites of Armenia. In *Perlite and vermiculite, geology, exploration and production technology.* Translated from Russian and published by the Indian Nat. Science Documentation Center for the U.S. Department of Interior, p. 47–59.

Schmidt, F. S. 1956. Technology of pumice, pumicite, and volcanic cinders. *Calif. Div. Mines Bull.* 174:99–117.

Sharps, T. I. 1961. Perlite in Colorado and other western states. *Mineral Ind. Bull., Colorado Sch. Mines* 4:1–15.

Stamatakis, M. G., U. Lutat, M. Regueiro, and J. P. Calco. 1996. Milos: the mineral island. *Ind. Minerals* (February):57–61.

Taggart, A. F. 1945. *Handbook of mineral dressing,* 3–80. New York: John Wiley & Sons.

Troxell, G. E., and H. E. Davis. 1956. *Composition and properties of concrete.* New York: McGraw-Hill, 434p.

Wagner, N. S. 1969. Perlite, pumice, pumicite and cinders. *Mineral and Water Resources of Oregon, Bull.* 64:222–228. Oregon Dept. Geol. Mineral. Ind.

Weber, R. H., and G. S. Austin. 1982. Perlite in New Mexico. In Industrial rocks and minerals of the southwest. G. S. Austin, Compiler. *New Mexico Bur. Mines and Mineral Res., Circ.* 182:97–101.

Whitson, D. N. 1982. Geology of the perlite deposit at No Aqua peaks, New Mexico, Proc. 7th Forum on the Geology of Industrial Minerals, Ed. G. S. Austin. *New Mexico Bur. Mines and Mineral Res., Circ.* 182:89–95.

Williams, H., F. J. Turner, and C. M. Gilbert. 1982. *Petrography—an introduction to the study of rocks in thin sections.* 2nd ed. San Francisco: Freeman, 626p.

Yanez, Y., P. Ildefonse, and D. Stefanov. 1993. Chemical characterization of clay minerals associated with zeolitization of perlite, Borovitza, Eastern Rhodopes, Bulgaria. *Geolologica Carpathica—Clays* 44:35–42.

Phosphates

Phosphorus is a relatively common element in the earth's crust, and phosphate minerals are present as accessory minerals in virtually every known rock formation. There are more than 250 recognized phosphate minerals. Only the members of the apatite group, however, form economic deposits. This group, which is found primarily in marine sedimentary rocks, contributes approximately 85% of the world's phosphate resources. The sedimentary phosphate deposits are known as phosphorites, with francolite, which is an apatite member, as its main component. The term "phosphate rock" is applied to any rock that contains more than 20% P_2O_5.

Approximately 90% of the phosphate produced (130 M tons in 1993) is consumed by the fertilizer industry, and the remaining 10% is used in the manufacture of detergent, animal feedstuff, and food and drink products. Phosphates are produced in more than thirty countries. The United States, the former USSR, China, and Morocco are the leading producers and contribute about two-thirds of the world's supply of phosphate.

MINERALOGICAL PROPERTIES

Among the more than 250 recognized phosphate minerals (Nriagu 1984), the apatite group deposits are the most important economic deposits. Amblygonite is a lithium mica, whereas monazite and xenotime are rare earth minerals. They are all important industrial minerals, and each has been treated in depth in a separate section of this book (see lithium minerals p. 203 and rare earth minerals, p. 313). The four most common end members of the apatite group are:

Fluorapatite	$Ca_5(PO_4)_3F$
Chlorapatite	$Ca_5(PO_4)_3Cl$
Hydroxylapatite	$Ca_5(PO_4)_3OH$
Carbonate-apatite	$Ca_5[(PO_4)_{3-x}(CO_3,F)_x](F,OH,Cl)$
Francolite	$Ca_5(PO_4,CO_3, OH)_3(F)$
Dahllite	$Ca_5(PO_4,CO_3, OH)_3 (OH)$

Among them, fluorine, chlorine, and hydroxyl ions can mutually replace each other to form solid solutions. Substitutions of Ca^{2+} and $(PO_4)^{3-}$ by other ions are common in apatite. The substituting ions are Sr^{2+}, Mn^{2+}, Mg^{2+}, Ba^{2+}, and Y^{3+}, including rare earths, for Ca^{2+} and $(VO_4)^{3-}$, $(AsO_4)^{3-}$, and $(CO_3)^{2-}$ for $(PO_4)^{3-}$ (Chang et al. 1996).

Fluorapatite is by far the most abundant end member and is the prototype of the group. In its structure, Ca atoms occupy two distinct positions: one seven-fold and one nine-fold coordination. Each F atom is surrounded by three Ca atoms at one level, and, in addition, Ca-O columns are linked with PO_4 groups, thus forming a hexagonal network. This arrangement gives fluorapatite a very stable structure. The structure of hydroxylapatite is slightly expanded compared with that of fluorapatite, and hydroxylapatite is less stable than fluorapatite. The structural configuration and substitution between (PO_4) and (CO_3) groups in carbonate-apatite remain a problem (Chang et al.,

1996). Francolite, a member of the carbonate-apatite group, is the essential component of phosphorite. The cell dimensions and c:a ratios are:

	a,Å	c,Å	c:a
Fluorapatite	9.39	6.88	0.73
Chlorapatite	9.60	6.78	0.71
Hydroxylapatite	9.42	6.87	0.73
Carbonate-apatite	9.32	6.89	0.74
Francolite	9.33	6.87	0.73
Dahllite	9.41	6.89	0.73

Gulbrandsen et al. (1966) demonstrated that not only was P replaced by C, but also that some excess F was found in the P position, whereas there is a balanced substitution of Na for the rare earth ions. Binder and Troll (1989) concluded that $CO_3^{2-} + F^{1-}$ replaces PO_4^{3-}, with C^{4+} substituting for P^{5+}, and an excess halogen replacing one O^{2-} gives $Ca_{10}[(PO_4)_{6-x}(CO_3,F)_x]$ (F, OH, Cl)$_2$. Thus, the fundamental substitution in francolite is $(CO_3)^{2-}$ for $(PO_4)^{3-}$, as confirmed by McClellan and Kauwenberg (1990). They also proposed the presence of excess fluorine in the $(CO_3,F)^{3-}$ group. A general formula for francolite proposed by Nathan (1984) is as follows:

$$Ca_{10-a-b-c}Na_aMg_b(PO_4)_{6-x}(CO_3)_{x-y-z}(CO_3,F)_y(SO_4)_zF_2$$

where $2x = y + a + 2c$, and c is the number of Ca vacancies. Nathan attributes this complexity to the fact that the structure of apatite is an open structure, thus allowing extensive substitution, and the fact that francolite occurs as minute crystallites which produce a nonhomogeneous composition.

Apatite minerals show an extremely wide range of colors, but yellow-green to blue-green are probably the most common. They may also be colorless, however, or yellow, blue, violet, brown, orange, rose-red, and other colors. The hardness of apatite minerals is 5 (Moh's scale), and the specific gravity varies from 3.1 to 3.4. The principal index of refraction (ω) and birefringence vary linearly with composition in the fluorapatite-chlorapatite-hydroxylapatite series (Chang et al. 1996). With an increase in the amount of CO_2 substitution, there is an increase in birefringence and a decrease in the a-axis (Table 19) (Nathan 1984). Gulbrandsen (1970) proposed, by using the peak-pair difference method, to estimate the CO_2 content of francolite. The empirical equation established is $y = 23.641 - 14.7369x$, where "y" is the CO_2 weight percent and $x = \Delta2\theta$ (004 to 410) ($2\theta CuK_\alpha$), with a standard error of estimate of 0.5580. This method cannot be used in samples rich in dolomite because of diffraction peak interference. Upon heating, francolite loses its constituent CO_2 and alters gradually to fluorapatite. This change is accompanied by an increase in crystal size and the appearance of CaO and CaF_2 (Smith and Lehr 1966). The loss of structural CO_2 occurs at two stages between 500° and 700°C, and between 700° and 1,000°C, illustrating two different configurations of CO_2 in francolite: CO_3^{2-} and $(CO_3 \cdot F)^{3-}$.

The sedimentary phosphate deposits, in which francolite is the only phosphate mineral, are known as phosphorites. Associated nonphosphate minerals are quartz, opal, chert, calcite, dolomite, glauconite, illite, montmorillonite, and zeolites (McKelvey 1967; Emigh 1967; Nathan and Soudry 1982). Organic matter is a characteristic con-

Table 19 Changes in unit cell dimensions and birefringence with carbonate substitution, after Nathan (1984).

Sample	CO_2%	F%	a(Å)	Birefringence
Syn Fap	—	3.78	9.371	0.0034
NRM 28259	1.26	3.78	9.361	0.0051
NRM 28260	1.32	3.71	9.360	0.0055
NRM 28258	2.01	3.83	9.352	0.0062
USNM 97339	2.57	3.94	9.350	0.0070
NRM 38648	3.18	3.83	9.349	0.0082
NRM 38745	4.22	4.06	9.342	0.0098
USNM 105532	4.39	4.59	9.334	0.0125

stituent of phosphorites. It appears to be the result of anoxic microbial degradation of marine plankton (Gulbrandsen 1966; Nathan 1990). Sulfur, oxygen, and nitrogen values of organic matter in phosphorites are higher than those in other sediments (Powell et al. 1975). Major elements do not change greatly in phosphorites (Slansky 1980), whereas several trace elements appear to become enriched (Krauskopf 1955). They are Ag, Cd, Mo, Se, Sr, U, Y, Zn, and the rare earths. There is no correlation between geologic age and the chemical composition of phosphorites.

In the weathering of phosphorites, a series of reactions may occur that depend primarily on the intensity of climatic conditions. In order of increasing intensity, the reactions are (1) the oxidation of organic matter, (2) the leaching of carbonate, (3) the evolution of clay and apatite minerals, (4) the appearance of iron hydroxides, and (5) the genesis of phosphates and sulfates rich in iron and aluminum (McArthur 1978, 1980). The carbonate-fluorapatite, francolite, which is the only phosphate in phosphorites, transforms to fluorapatite by decarbonation. The cell dimension "a" increases in this process from that of francolite (9.33Å) to that of fluorapatite (9.39Å). In the presence of clays, the reactions give the following aluminum phosphates:

millisite	$(Na,K)CaAl_6(PO_4)_4(OH)_9 \cdot 3H_2O$,
crandalite	$CaAl_3(PO_4)_2(OH)_5 \cdot H_2O$, and
wavellite	$Al_3(PO_4)_2(OH)_3 \cdot 5H_2O$.

In the presence of sulfates and iron sulfates, the following phosphates form:

Fairfieldite	$Ca_2(Mn,Fe)(PO_4)_2 \cdot 2H_2O$,
Lipcombite	$(Fe^{2+},Mn)Fe^{3+}_2(PO_4)_2(OH)_2$, and
Strengite	$FePO_4 \cdot 2H_2O$.

The occurrence of iron phosphates is less frequent than that of aluminum phosphates, and both only occur in Eocene phosphorites, presently located in the tropical zones.

Guano deposits are formed on rocky islands frequented by sea birds. When freshly formed, guano is a complex mixture of phosphates, nitrates, carbonates, and organic compounds. Diagenetic processes set in after deposition that involve the loss of soluble, volatile, and oxidizable substances, and the concentration of calcium triphosphate. Soluble phosphates, which are leached out in the process, are carried downward and attack the underlying sedimentary or igneous rocks (Cook 1984). Guano, reacting with limestones, forms brushite $CaHPO_4 \cdot 2H_2O$, whitlockite $Ca_3(PO_4)_2$, and hydroxylapatite. In the presence of clays on igneous rocks, the guano gives leucophosphite $K(Fe^{3+}, Al)_2(PO_4)_2 \cdot (OH) \cdot 2H_2O$ and barrandite $(Fe,Al)PO_4 \cdot 2H_2O$.

GEOLOGICAL OCCURRENCE AND DISTRIBUTION OF DEPOSITS

Economically significance deposits of phosphate occur in either carbonatite and alkaline rocks or in marine sedimentary rocks. About 15% of the world's phosphate comes from igneous deposits, whereas marine sedimentary phosphorite contributes 85%. A world map showing the distribution of phosphate deposits of igneous, sedimentary, and three guano types is shown in Fig. 59 (Harben and Bates 1990). A brief description of the major phosphate deposits in the United States, the former USSR, and Morocco is given below.

The United States: The most important phosphorite deposits are concentrated in the Miocene sediments along the Atlantic Coastal Plain, from central Florida to North Carolina (Riggs 1979a, 1979b). As shown in Fig. 60, the Florida peninsula, as a portion of a regional structural platform, extends southward between the Gulf of Mexico Basin, the Atlantic Blake Plateau, and the Florida Straits. It has a central highland

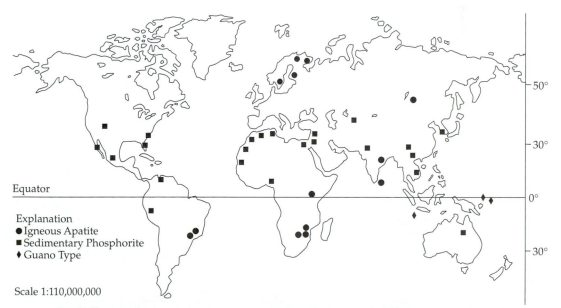

Figure 59   Distribution of phosphate deposits. © P. W. Harben and R. L. Bates. Igneous, sedimentary, and guano types are represented by solid circle, square, and diamond, respectively.

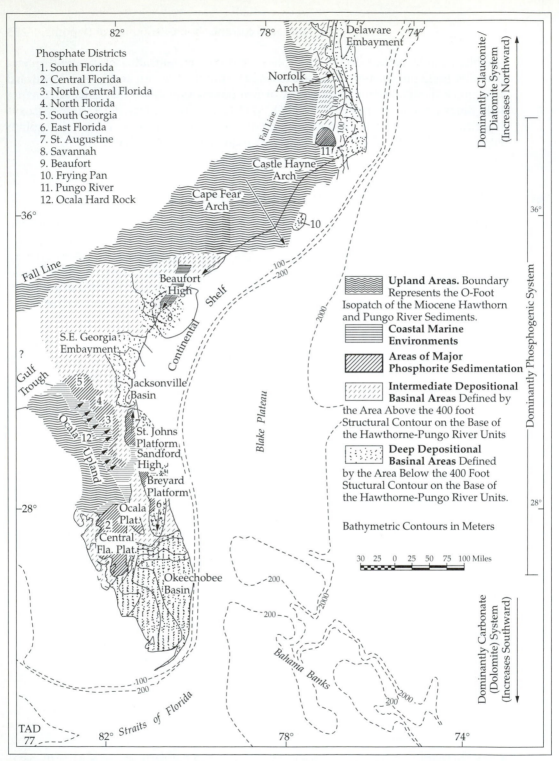

Figure 60 Major phosphate districts in Florida and North Carolina, and their relationship with the structural framework of the southeast Atlantic Coastal Plain-Continental shelf system. Reprinted with permission of Economic Geology, Volume 74, p. 195–220, S. R. Riggs (1979).

consisting of the Ocala Upland that is surrounded by an extensive sequence of coastal lowlands with marine terraces. This Upland affected all subsequent Tertiary sedimentation. The Miocene sediment is characterized by authigenic dolomite and phosphorite mixed with terrigenous sediments. The phosphate-bearing Hawthorn Group can be subdivided into three lithologic formations:

1. The Arcadia Formation (Miocene) consisting predominantly of dolomite mixed with primary phosphorite and subordinate amounts of terrigenous sediments. This Formation, as Riggs (1979a) pointed out, is the phosphate resource of the future.
2. The Noralyn Formation (Miocene) consisting of marine terrigenous sands and clays mixed with primary phosphorite. This Formation constitutes the bulk of the phosphate presently being mined in Florida.
3. The Bone Valley Formation (Pliocene) is a thin unit of fluvial, estuarine, and coastal sediments consisting of terrigenous sands and clays, abundant shell material, and reworked phosphorite. This Formation is famed for its fossil content.

In the Hawthorn Group, the Pliocene Bone Valley Formation unconformably overlies the Miocene Noralyn Formation, which grades down into the Arcadia Formation and rests unconformably upon either the Suwanee Limestone of Oligocene age or the Ocala Limestone of Eocene age (Riggs 1979a). The average P_2O_5 content ranges from 30 to 35%. In North Carolina, phosphate deposits that consist of sand, clay, silt, diatomaceous sediments, and dolomite occur in the Middle Miocene Pungo River Formation (Miller 1982; Riggs and Mallette 1990).

In the western United States, phosphates are distributed extensively in Montana, Idaho, Wyoming, and Utah. This distribution is known as the "Western Phosphate Field." It is of Permian age and contains a classic example of the geosynclinal phosphorite lithofacies (Rhodes 1976; Emigh 1967, 1983). The Field is characterized by black, highgrade phosphorite, phosphatic and carbonaceous mudstones, cherty mudstone, and chert. Calcareous rocks and sandy beds are generally absent. The average ore grade ranges from 18 to 26% P_2O_5.

The Former USSR: The most important phosphate deposit occurs in the Khibiny alkali igneous complex, on Kola Peninsula, just north of Kirovsk, some 16 km south of Murmansk (Notholt 1979). It is a ring complex about 40 km in diameter, consisting essentially of foyaite surrounded by inward and generally steeply dipping sheets of other varieties of nepheline syenite (Fig. 61). Age determinations indicate that the complex is of Lower Permian or Upper Carboniferous age. The apatite-nepheline ore bodies of the Khibiny Complex are restricted to a zone of layered ijolitic rocks and generally contain 15 to 75% apatite (6 to 31% P_2O_5), 10 to 80% nepheline, 1 to 25% aegirine, 5 to 12% sphene, feldspar and titano-magnetite. The phosphate mineral is fluorapatite, and some of the selected specimens have 11.42% SrO and 4.9% RE_2O_3. Other phosphate deposits occur within the Kovdor alkalic and ultrabasic igneous complex, situated in the western part of the Kola Peninsula, and in the upper Proterozoic Oshurkov Complex in the Buryat Republic.

Important sedimentary phosphate deposits also occur in the former USSR. The Karatau Basin, located in the Lesser Karatau Mountain Range, is one of the world's largest sites of phosphorite deposition (Eganov et al. 1986). The productive

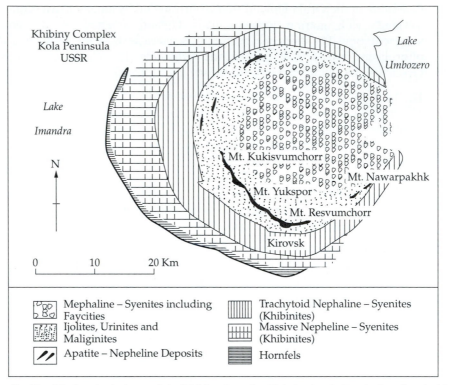

Figure 61 Simplified geological map of the Khibiny Complex, Kola Peninsula, Russia. Reprinted with permission of Economic Geology, Volume 74, p. 339–350, Arthur J. G. Notholt (1979).

phosphorite-bearing series, the Chulaktau Suite of Cambrian age, has three phosphatic members, listed by decreasing percentages: (1) Upper Phosphorite Member (6 to 12 m) of grainstone and mudstone phosphorite, up to 36% P_2O_5, (2) Phosphorite-Chert Member (1.5 to 7 m) of thin, interbedded siliceous phosphorite and chert, 2 to 22% P_2O_5, and (3) Lower Phosphorite Member (1 to 6 m), 26 to 29% P_2O_5 in near surface locations.

Morocco: The major phosphate reserves are located in the marine pelletal phosphorite district (Emigh 1983). The three northern deposits, Khouribga, Youssoufia, and Meskala, are 80 km inland from the Atlantic Ocean along a northeast-southwest trend 250 km long. The Bu-Craa deposit is further south, in the former Spanish Sahara. Among these deposits, the most important one is the Khouribga deposit, which consists of sandy oolitic phosphate beds alternating with marl, clay, phosphatic limestone, and layers of flint. The P_2O_5 contents range from 27 to 33%.

Others: In China, the deposits of phosphate rock are located mainly in the provinces of Yunnan, Jingshan, Hubei, Zhongxiang, Guizhou, and Sichuan. Most of these deposits are of Pre-cambrian or Cambrian age, and are commonly found in association with shale, limestone, and sandstone. The P_2O_5 content is estimated to be between 25 and 30% (Yeh et al. 1986; Harben and Kuzvart 1997).

The Islands of Oceania have the richest guano phosphate resources. The principal resources are found in Australia, the Philippines, Indonesia, Fiji, French Polynesia, Christmas Island, and Nauru Island. Guano phosphate on the Nauru Island contains approximately 40% P_2O_5 (Bartels and Gurr 1994).

INDUSTRIAL PROCESSES AND USES

1. Grades of Phosphate Rock

Phosphate rock is generally graded on the basis of its P_2O_5 content, but may also be graded based on its content of bone phosphate of lime (BPL), or triphosphate of lime (TPL). In some restricted cases, the phosphorus content is used to define the grade. An illustration of this relationship is:

$\% P_2O_5 \times 2.1853 = \% BPL$ or $\% TPL$
$\% PBL$ or $\% TPL \times 0.4576 = \% P_2O_5$
$\% P_2O_5 \times 0.4364 = \% P$
$\% P \times 2.2914 = \% P_2O_5$

and 80% BPL = 80% TPL = 36.6% P_2O_5 = 16% P.

2. Beneficiation of Phosphate Rock

Beneficiation of phosphate rock is commonly processed by disintegrating it with water. The slurry produced is screened, and some coarse sizes are removed for further reduction in size (Habashi 1994). In the Florida and North Carolina phosphate producing mines (Breathitt and Finch 1977; Lawver et al. 1978), ore beneficiation begins at the mine, with high-pressure water disaggregating a matrix placed by the dragline into a sump. The matrix slurry is subjected to washing and screening to disperse the clay and to remove the +14/16−mesh. The washed product usually results in an acceptable grade (60 to 69 BPL) in the central Florida mines. Ore bodies farther south contain fewer pebbles, and the mines of north Florida and North Carolina do not produce pebble products. Flotation has to be used for the production of concentrate (Williams and Zellars 1987). As shown in Fig. 62, the slurry, which has been cleared of pebbles, contains phosphate pellets ranging in size to below 200-mesh, sand grains, disaggregated clays, and other minerals. Desliming at 150/200 mesh eliminates clays, and sizing produces two or three sizes for flotation. Anionic flotation removes the bulk of the sand, and in most cases, the anionic reagents are removed by scrubbing with sulfuric acid. Cationic flotation efficiently removes the remaining sand and yields high-grade phosphate rocks. For rocks high in organics, calcination is required in the beneficiation process.

In the Kola region of Russia, the apatites are beneficiated by crushing the ore, grinding, and sizing to about 60% minus 65 mesh, followed by anionic flotation with three stages of anionic cleaners and two stages of tails scavenging. The product averages about 50% minus 200 mesh and about 39% P_2O_5. Ore beneficiation processes in Morocco vary from mine to mine and generally consist of screening, attrition, washing, dry-

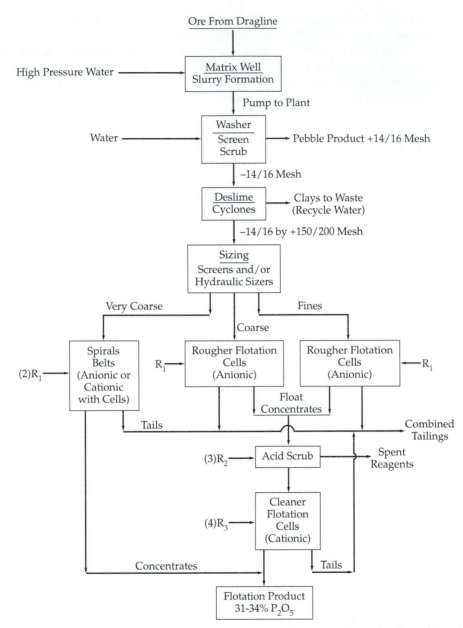

Figure 62 A scheme illustrating beneficiation processes used in Florida and North Carolina mining plants. Reprinted from Manual of Fertilizer Processing, p. 35–78 by Courtesy of Marcel Dekker, Inc. R_1 = reagents, fatty acid, fuel oil, and ammonia; R_2 = sulfuric acid; and R_3 = amine, kerosene, and pH control.

ing, and air classification. Because of the use of sea water in the process, the finished product requires removing the salt on a belt filter with fresh rinse water. The Youssoufia black rocks, which contain 2.5% organic matter, must be calcined (Woodrooffe 1972; Russell 1987).

3. Industrial Uses

By far the largest uses for phosphate rock are in the manufacture of phosphoric acid, of superphosphate, or of other soluble phosphatic salts for fertilizers.

Phosphoric Acid: There are two industrial methods for the preparation of phosphoric acid (Phillips and Boylan 1963; Stocchi 1990). The wet process produces phosphoric acid by the action of sulfuric acid on phosphate rock. The principal reaction is illustrated below:

$$2Ca_5(PO_4)_3F \text{ (fluorapatite)} + 10H_2SO_4 + 20H_2O$$

$$\rightarrow 10CaSO_4 \cdot 2H_2O \text{ (gypsum)} + 2HF + 6H_3PO_4.$$

This is followed by the separation of calcium sulfate and concentration of the acid. A large portion of the weak acid and extraction slurry is recycled to the extraction step in order to minimize blocking of unreacted rock with calcium sulfate. Depending on the temperature at which the reaction is carried out, it is possible to produce semihydrate and even anhydrous $CaSO_4$ at above 125°C. Phosphoric acid produced by this process contains impurities. It requires no purification if it is intended for use in fertilizers, but for chemical use and use in food products, phosphoric acid must be further purified by wet process. Settling and decantation are generally used for solid impurities (Scott et al. 1961; Wilbanks et al. 1961).

   The thermal process of producing phosphoric acid requires the production of elemental phosphorus; therefore, it is a much less impure product, but also a much more expensive commodity to produce than wet process phosphoric acid. In the process, the phosphate, along with coke and silica, is charged into the reaction chamber and heated in an electric furnace to the temperature at which reduction takes place. Phosphorus is volatilized continuously from the top of the furnace and collected in a water-cooled condenser. The slags of calcium silicate and ferrophosphorus are drawn off intermittently from the bottom. The liquid phosphorus is mixed with air as it is injected and burnt in a tower type combustion chamber. The reaction is $P_4 + 5O_2 \rightarrow 2P_2O_5$ with 2,535 kcal/kg(P_4O_{10}), a highly exothermic reaction.

Superphosphates: The most frequently used phosphate fertilizer consists of a mixture of calcium dihydrogen phosphate monohydrate $[Ca(H_2PO_4)_2] \cdot H_2O$ and calcium sulfate dihydrate $[CaSO_4 \cdot 2H_2O]$, which is referred to as simple superphosphate. It is produced by reacting phosphate rock with 65% sulfuric acid:

$$2Ca_5(PO_4)_3F + 7H_2SO_4 + 17H_2O \rightarrow$$

$$3[Ca(H_2PO_4)_2 \cdot H_2O] + 7[CaSO_4 \cdot 2H_2O] + 2HF$$

Other reactions occur that involve impurities in the phosphate rock: a part of the HF reacts with silica to form SiF_4 and a part of the added acid reacts with $CaCO_3$ to yield CO_2. The theoretical content of water-soluble P_2O_5 in simple superphosphate is 22%. By adding H_3PO_4, together with sulfuric acid, in the reaction with phosphate rock, a water-soluble, superphosphate product is obtained that contains 32 to 35% P_2O_5, known as double superphosphate. By using H_3PO_4 alone with phosphate rock according to the reaction:

$$2Ca_5(PO_4)_3F + 14H_3PO_4 + 10H_2O \rightarrow 10[Ca(H_2PO_4)_2 \cdot H_2O] + 2HF,$$

a superphosphate, known as triple superphosphate, is produced, without $CaSO_4 \cdot 2H_2O$. It contains up to 48% P_2O_5 and is water soluble (Inskeep et al. 1956; Hand et al. 1963).

All types of superphosphate, especially the simple and triple types, are used extensively for direct application to the soil and in the production of mixed fertilizers.

Ammonium Phosphates and Polyphosphates: Two salts containing ammoniacal nitrogen and phosphorus are manufactured for use in the food and fertilizer industries (Houston et al. 1955; Getsinger et al. 1962; Young et al. 1962). Mono-ammonium dihydrogen phosphate, $NH_4H_2PO_4$, and diammonium hydrogen phosphate, $(NH_4)_2HPO_4$, are produced in a procedure similar to that leading to the formation of ammonium sulfate from gaseous ammonia and sulfuric acid, by using the appropriate amount of added ammonia to yield the desired salt, and by using a wide range of concentrations of phosphoric acid (30 to 75%).

The grade of a fertilizer is denoted by the use of three numbers such as (18-16-0). The first number defines the percent of elemental nitrogen (N), the second is P_2O_5, and the last is K_2O. In recent years there has been a trend to use numbers referring to the elements only as (N,P,K). Common types of phosphate fertilizers are (1) normal superphosphate, ranging from (0-18-0) to (0-20-0) (solid), (2) concentrated phosphate, from (0-42-0) to (0-50-0) (solid), (3) diammonium phosphate (18-46-0) (solid), (4) monoammonium phosphate (16-48-0), and (5) (11-48-0) (solid). These fertilizers can be granular mixed fertilizers, which are a combination of phosphate fertilizers and K_2O, and liquid mixed fertilizers, which consist of phosphoric acid that has been neutralized by ammonia and nitrogen solution, and to which potash may be added as required.

The other 10% of phosphate rock is used to produce phosphorus and phosphoric acid by a thermal process designed for the production of phosphorus compounds. These compounds are used mainly in the chemical industries that produce detergents, industrial cleaners, food, drinks, animal feedstuff, and flame-retardant additives.

The combination of silicates with P_2O_5 produces a class of glasses with properties analogous to borosilicate glasses. Phosphate glasses have approximately the same dispersion as silicate glasses, but their refractive index is higher. Technical glasses based on P_2O_5 are used because of their transmission properties (Scholze 1990).

REFERENCES

Bartels, J. J., and T. M. Gurr. 1994. Phosphate rock. In *Industrial minerals and rocks*. 6th ed. Ed. D. D. Carr, 751–764. Littleton, Colorado: Soc. Mining, Metal. & Exploration.

Binder, G., and G. Troll. 1989. Coupled anion substitution in natural carbon-bearing apatites. *Contr. Min. Petr.* 101:394–401.

Breathitt, H. W., and E. P. Finch. 1977. Phosphate rock beneficiation. *Preprint 77-H-134, SME-AIME Annual Meeting, Atlanta, Georgia,* 31p.

Chang, L. L. Y., R. A. Howie, and J. Zussman. 1996. Non-silicates: Sulphates, carbonates, phosphates, halides. In Vol. 5B of *Rock-forming minerals.* 2nd ed. Essex, England: Longman, 383p.

Cook, P. J. 1984. Spatial and temporal controls on the formation of phosphate deposits—A review. In *Phosphate minerals.* Eds. J. O. Nriagu and P. B. Moore, 242–274. Berlin: Springer-Verlag.

Eganov, E. A., Yu K. Sovetov, and A. L. Yanshin. 1986. Proterozoic and Cambrian phosphorites deposits: Karatau, southern Kazakhstan, USSR. In Vol. 1 of *Phosphate deposits of the world.* Ed. P. J. Cook and J. H. Shergold, 175–189. Cambridge: Cambridge Univ. Press. Sydney.

Emigh, G. D. 1967. Petrology and origin of phosphorites. Anatomy of the Western Phosphate Field. *Proc., Intermountain Assoc. Geol. 15th Annual Field Conf.,* 103–114.

——— 1983. Phosphate rocks. In *Industrial minerals and rocks.* 5th ed. Ed. S. J. Lefond, 1017–1047. New York: Soc. Mining Engineers, AIME.

Getsinger, J. G., M. R. Siegel, and H. C. Mann, Jr. 1962. Ammonium polyphosphates from superphosphoric acid and ammonia. *Jour. Agr. Food Chem.* 10:341–344.

Gulbrandsen, R. A. 1966. Chemical composition of phosphorites of the Phosphoria Formation. *Geochim. Cosmochim. Acta* 30:769–778.

Gulbrandsen, R. A. 1970. Relation of carbon dioxide content of apatite of the Phosphoria Formation to regional facies. *U.S.G.S. Prof. Paper* 700B:9–13.

Gulbrandsen, R. A., J. R. Kramer, L. B. Beatty, and R. E. Mays. 1966. Carbonate-bearing apatite from Faraday Township, Ontario, Canada. *Amer. Mineral.* 51: 819–824.

Habashi, F. 1994. Phosphate fertilizer industry—processing technology. *Ind. Minerals* (March):65–69.

Hand, L. D., Jr., J. M. Potts, and A. V. Slack. 1963. Use of concentrated sulfuric acid in the production of granular normal superphosphate. *Jour. Agr. Food Chem.* 11:44–47.

Harben, P. W., and R. L. Bates. 1990. *Industrial minerals, geology and world deposits.* London: Industrial Minerals Div., Metal Bull. Plc, 312p.

Harben, P. W., and M. Kuzvart. 1996. *Industrial minerals—a global geology.* London: Ind. Minerals Inform. Ltd., Metal Bull. Plc, 462p.

Houston, E. C., L. D. Yates, and R. L. Haunschild. 1955. Diammonium phosphate fertilizer from wet-process phosphoric acid. *Jour. Agr. Food Chem.* 3:43–48.

Inskeep, G. C., W. B. Fort, and W. C. Weber. 1956. Granulated triple superphosphate. *Indust. Eng. Chem.* 48:1804–1816.

Krauskopf, K. B. 1955. Sedimentary deposits of rare metals. *Econ. Geol.* 50:411–463.

Lawver, J. E. et al. 1978. New Techniques in beneficiation of phosphate rock. *Preprint 78-B-331, SME-AIME Fall meeting, Orlando, Florida,* 36p.

McArthur, J. M. 1978. Systematic variation in the contents of Na, Sr, CO_2 and SO_4 in marine carbonate-fluorapatite and their relationship to weathering. *Chem. Geol.* 21:89–112.

——— 1980. Postdepositional alteration of the carbonate-fluorapatite phase of Moroccan phosphates. *Soc. Econ. Paleo. Mineral. Spec. Pub.* 29:53–60.

McClellan, G. H., and S. J. van Kauwenbergh. 1990. Mineralogy of sedimentary apatites. In Phosphorite research and development. Eds. A. J. G Notholt and I Javis. *Geol. Soc. Spec. Publ.* 52:23–31.

McKelvey, V. E. 1967. Phosphorite deposits. *U.S.G.S. Bull.,* 1252D, 21p.

Miller, J. A. 1982. Stratigraphy, structure and phosphate deposits of the Pungo River Formation of North Carolina. *North Carolina Dept. Nat. Res. & Community Devept. Bull.,* 87, 12p.

Nathan, Y. 1984. The mineralogy and geochemistry of phosphorites. In *Phosphate minerals*. Eds. J. O. Nriagu and P. B. Moore, 275–291. Berlin: Springer-Verlag.

Nathan, Y. 1990. Humic substances in phosphorites: occurrence, characteristics and significance. In Phosphorite research and development. Eds. A. J. G. Nothot and J. Jarvis. *Geol. Soc. Spe. Publ., London* 52:49–58.

Nathan, Y., and D. Soudry. 1982. Authigenic silicate minerals in phosphorites of the Negev, Southern Israel. *Clay Min.* 17:249–254.

Notholt, A. J. G. 1979. The economic geology and development of igneous phosphate deposits in Europe and the USSR. *Econ. Geol.* 74:339–350.

Nriagu, J. O. 1984. Phosphate minerals: Their properties and general modes of occurrence. In *Phosphate minerals*. Eds. J. O. Nriagu and P. B. Moore, 1–136. Berlin: Springer-Verlag.

Phillips, A. B., and D. R. Boylan. 1963. Advances in phosphate manufacture. In *Fertilizer, technology and usage*. Eds. M. McVickar, G. L. Bridger, and L. B. Nelson, 131–154. Madison: Soil Science Soc. Amer.

Powell, T. G., P. J. Cook, and D. M. McKisdy. 1975. Organic geochemistry of phosphorites: Relevance to petroleum genesis. *Amer. Assoc. Petrol. Geol. Bull.* 59:618–633.

Riggs, S. R. 1979a. Petrology of the Tertiary phosphate system of Florida. *Econ. Geol.* 74:195–220.

Riggs, S. R. 1979b. Phosphorite sedimentation in Florida—A model phosphoritic system. *Econ. Geol.* 74:285–314.

Riggs, S. R., and P. M. Mallette. 1990. Carolina continental margin Part 3—Patterns of phosphate deposition and lithofacies relationships within the Miocene Pungo River Formation, North Carolina continental margin. In Vol. 3 of *Phosphate deposits of the world*. Eds. W. G. Burnett and S. R. Riggs, 424–443. Cambridge: Cambridge Univ. Press.

Rhodes, J. A. 1976. Review of the Western Phosphate Field, in Proc., Eleventh Forum on the Geology of Industrial Minerals. *Montana Bur. Mines and Geology, Sp. Publ.* 74:61–67.

Russell, A. 1987. Phosphate rock: Trends in processing and production. *Ind. Minerals* (September):25–59.

Scholze, H. 1990. *Glass, nature, structure and properties*. New York: Springer-Verlag, 454p.

Scott, W. C., G. G. Patterson, and H. W. Elder. 1961. Wet-process phosphoric acid: Pilot-plant production. *Indust. Eng. Chem.* 53:713–716.

Slansky, M. 1980. Géologie des Phosphates Sedimentaires. *Fr. Bur. Res. Geol. Minières Mem.*, 114, 92p.

Smith, J. P., and J. R. Lehr. 1966. An x-ray investigation of carbonate apatite. *Jour. Agri. Food Chem.* 14:342–349.

Stocchi, E. 1990. *Industrial Chemistry*. Vol. 1, 348–381. New York: Ellis Horwood.

Wilbanks, J. A., M. C. Nason, and W. C. Scott. 1961. Liquid fertilizers from wet-process phosphoric acid and superphosphoric acid. *Jour. Agr. Food Chem.* 9:174–178.

Williams, J. M., and M. E. Zellars. 1987. Phosphate rocks. In *Manual of fertilizer processing*. Ed. F. T. Nielsson, 55–78. New York: Marcel Dekker, Inc.

Woodrooffe, H. M. 1972. Phosphate in the Kola Peninsula, USSR. *Mining Engin.* 24:54–56.

Yeh, L., S. Sun, Q. Chen, and S. Guo. 1986. Proterozoic and Cambrian phosphorites deposits—Kunyang, Yunnan, China. In *Phosphate deposits of the world, 1: Proterozoic and Cambrian phosphorites*. Eds. P. J. Cook and J. H. Shergold, 149–154. Cambridge: Cambridge Univ. Press.

Young, R. D., G. C. Hicks, and C. H. Davis. 1962. TVA process for production of granular diammonium phosphate. *Jour. Agr. Food Chem.* 10:442–447.

Potassium Salts

Potassium salts comprise a group of minerals of chloride and sulfate composition that are generally formed by evaporation of seawater or other brines. As industrial minerals, they are mainly consumed by the fertilizer industry, which uses an estimated more than 90% of the total annual production of approximately 30 million tons (1989–1993). The other 10% is used by the chemical industry.

Potash is a generic term used for a variety of potassium minerals, ores, and refined products. The German term "Hartsalz" refers to the greater hardness of sulfate-containing sylvite, whereas the term "Mischsalz" refers to a mixture of hartsalz and carnallite.

Potassium minerals are produced in a dozen countries. The former USSR, Canada, Germany, France, and the United States contribute more than 85% of the world's production.

MINERALOGICAL PROPERTIES

Listed below are the principal and accessory potassium minerals, as well as associated nonpotassium minerals, commonly found in potash deposits (D: specific gravity and H: hardness):

Principal Potassium Minerals:

		K_2O equiv.	D/H
Sylvite	KCl	63.17 wt%	2.00/2.0
Carnallite	$KCl \cdot MgCl_2 \cdot 6H_2O$	16.95	1.60/2.5
Kainite	$4KCl \cdot 4MgSO_4 \cdot 11H_2O$	19.26	2.13/3.0
Langbeinite	$K_2SO_4 \cdot 2MgSO_4$	22.69	2.83/4.0

Accessory Potassium Minerals:

Polyhalite	$K_2SO_4 \cdot MgSO_4$ $\cdot 2CaSO_4 \cdot 2H_2O$	15.62	2.18/3.0
Leonite	$K_2SO_4 \cdot MgSO_4 \cdot 4H_2O$	25.68	2.20/2.5
Schoenite	$K_2SO_4 \cdot MgSO_4 \cdot 6H_2O$	23.39	2.03/2.5
Syngenite	$K_2SO_4 \cdot CaSO_4 \cdot H_2O$	28.68	2.58/2.5
Glaserite	$3K_2SO_4 \cdot Na_2SO_4$	42.51	2.70/2.5

Associated Nonpotassium Minerals:

Anhydrate	$CaSO_4$	0	2.96/4.0
Blödite	$Na_2SO_4 \cdot MgSO_4 \cdot 4H_2O$	0	2.25/2.5
Bischofite	$MgCl_2 \cdot 6H_2O$	0	1.60/1.5
Epsomite	$MgSO_4 \cdot 7H_2O$	0	1.68/2.5
Gypsum	$CaSO_4 \cdot 2H_2O$	0	2.32/2.0
Halite	$NaCl$	0	2.17/2.5
Kieserite	$MgSO_4 \cdot H_2O$	0	2.57/3.5
Löweite	$6Na_2SO_4 \cdot MgSO_4 \cdot 15H_2O$	0	2.34/2.5
Vanthoffite	$3Na_2SO_4 \cdot MgSO_4$	0	2.69/3.5

Sylvite is essentially pure KCl. It is isostructural with halite and is usually intimately grown or mixed with halite, but substitution between Na and K is very limited in the chlorides. Intergrowth of sylvite and halite is common, and an ore consisting of KCl and NaCl is known as sylvinite (Kühn 1959; Phillips and Griffen 1981). Sylvite is generally opalescent or milky white. It often contains hematite inclusions, usually in the form of fine flakes that are strongly concentrated at the outer rim of the sylvite crystal, so that a strong red margin can be observed macroscopically. Its index of refraction (1.490) is lower than that for halite (1.544). Sylvite very rapidly develops a strong violet blue color when exposed to X-rays. In the system $KCl-H_2O$, KCl has a strong positive temperature coefficient of solubility of 0.34 mol K_2Cl_2/1,000 mol H_2O per degree. Solubility is considerably reduced in the presence of NaCl, but the temperature coefficient remains nearly unchanged at 0.31 mol K_2Cl_2/1,000 mol H_2O per degree (Braitsch 1971).

Carnallite, a double salt of KCl and MgCl, is an important primary potash mineral, although secondary carnallite also forms economic deposits. Fiberous carnallite is commonly present in joints of the Hartsalz. Carnallite is orthorhombic and has no distinct cleavage. Its color and refractive indices (1.472 to 1.497) are comparable with those of sylvite. Hematite is more uniformly distributed in carnallite and much flatter in texture than in sylvite (Phillips and Griffen 1981). In the system $KCl-MgCl_2-H_2O$ with saturated NaCl, carnallite precipitates incongruently (Schultz et al. 1993). Fig. 63 shows solubility

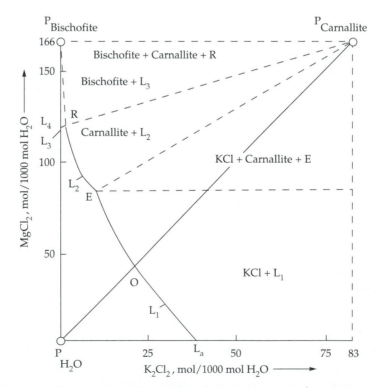

Figure 63 The system $K_2Cl_2-MgCl_2-H_2O$ (not to scale) represented by using van't Hoff coordinates, after Schultz et al. (1993).

relationships by van't Hoff coordinates and points that are indicated by the letter "P" and represent the composition of water, bischofite, and carnallite. All the possible mixtures of water and carnallite are shown on the straight line between P_{water} and $P_{carnallite}$, the molar ratio $K_2Cl_2:MgCl_2$ here always being 1:2, as in carnallite. Point L_0 gives the solubility of sylvite in water. The solutions L_1, L_2, and L_3 are in equilibrium with sylvite, carnallite, and bischofite, respectively. L_4 gives the solubility of bischofite in water. As the solution in the system evaporates, the end products sylvite+carnallite+halite, carnallite+halite, or carnallite+bischofite+halite are produced, depending on the original composition of the solution. It is not possible to produce incongruently saturated solutions by simultaneous dissolution of both solid phases in equilibrium with them.

Kainite, $4KCl \cdot 4MgSO_4 \cdot 11H_2O$, is monoclinic with perfect cleavage on (001). It is also white in color and has refractive indices in the range of 1.494 to 1.516. Langbeinite, $K_2SO_4 \cdot 2MgSO_4$, is cubic in symmetry and has refractive indices between 1.533 and 1.536. It is characterized by a secondary occurrence that forms from the recrystallization of either halite or kainite into langbeinite, sylvite, or kieserite; or of sylvite or kieserite into langbeinite and halite (Rempe 1982).

A portion of the quinary system is shown in Fig. 64, to illustrate the equilibrium relations and crystallization sequence of kainite, langbeinite, and other minerals, including leonite, schönite, glaserite, blödite, and epsomite. This three-dimensional diagram, the so-called NaCl saturation space, represents phase relations at 25°C, and a composition region of 0 to 65 mol $MgCl_2$/1,000 mol H_2O. Each point in the diagram corresponds to a NaCl-saturated solution, in which the concentrations of K_2Cl_2, $MgCl_2$, and $MgSO_4$ are given by the distance of the point from the origin, measured along the three axes. The boundaries of the saturation space shown in the diagram represent the saturation surface of two coexisting salts. Seven of a total twelve stable phases in the system are shown in the diagram because their two-salt surfaces lie in the composition range of Figure 64. In many instances, spontaneous crystallization does not occur when a salt exceeds the concentration indicated on the two-salt surface. As a consequence, supersaturated solutions are formed. In Fig. 64, continuations of the stable schönite+halite two-salt surface to the left and right in the unstable region are shown as broken lines. The KCl-NaCl two-salt surface on the front side of Fig. 64 is of special significance for the potash industry for the treatment of sylvite ores (Schultz et al. 1993).

GEOLOGICAL OCCURRENCE AND DISTRIBUTION OF DEPOSITS

Economic potash deposits are essentially restricted to marine evaporites, which are widespread and occur throughout the geological column. The evaporite deposits are formed in structural basins through the evaporation of sea water or mixtures of sea water and other brines. The salt minerals are precipitated in assemblages and sequences as a function of the brine compositions, according to the solubility relations as described (Schultz et al. 1993). The sequence of precipitation is generally in the order of carbonate, sulfate, and chloride (Adams 1983; Williams-Stroud et al. 1994). The environment of deposition of potash deposits varies with the geological time, from Paleozoic deposits in epicontinental seas, through Mesozoic, and mostly Cenozoic deposits along continental margins, and to Pleistocene and Recent deposits in coastal sabkhas or inland saline lakes (Harben and Bates 1990).

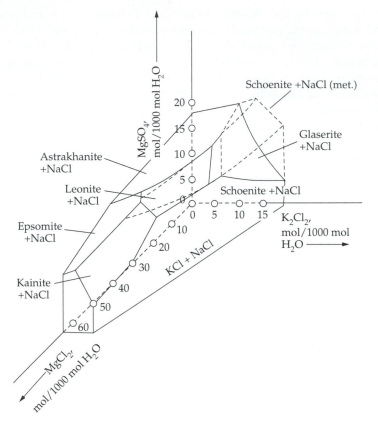

Figure 64   Three-dimensional view of the quinary system Na_2Cl_2-K_2Cl_2-$MgCl_2$-$MgSO_4$-H_2O at 25°C with 0 to 65 mol% $MgCl_2$/1,000 mol H_2O, after Schultz et al. (1993).

Major deposits are distributed in the former USSR, Canada, Germany, France, and the United States. Their total production constitutes 85% of world's annual production, estimated at 30 M tons. Other producing countries are Jordan, Israel, China, Congo, Libya, Brazil, and Thailand.

The Former USSR: As shown in Fig. 65 (Adams, 1983), there are three major potash districts in the former USSR. The former USSR's largest potash deposit lies in the Upper Kama Basin, at Solikamsk-Berezhniki, on the western slope of the Northern Ural mountains. The Upper Kama Basin is a north-trending elliptical basin, 145 km long and 50 km wide, and is part of the 3,000 km long intercontinental trough that formed along the edge of the Urals during the Permian period. Continued uplift cut the trough off from the open sea and produced several isolated basins, including the Upper Kama Basin (Harben and Bates 1990). The deposits are generally thick, high grade, undeformed, and occur at depths of about 250 meters. A cross section through the Permian potash deposit on the upper Kama river in the Perm region has the following sequence (Tatarinov 1969):

Figure 65   Location map of the potash mining districts of Russia, after Adams (1983).

1. Overlying limestone, sandstone, and shale, interbedded with rock salt and gypsum, and with marls at the bottom.
2. Overlying salt layer (1 to 70m).
3. Potash salts (80 to 85 m) over an area of 3,500 km^2, consisting of two parts. The upper part has nine 1 to 15 m layers of sylvinite (35% KCl) and merges into a carnallite layer with anhydrite and magnesite, and the lower part has sylvinite (20 to 22 m) with six layers (0.8 to 6 m) of red sylvinite (10 to 35% KCl), alternating with halite (1 to 5.5 m).
4. Underlying rock salt (240 to 400 m).
5. Clay-anhydrite layer with dolomite.
6. Salt-bearing sandstone and clay.
7. Dolomitized marks.
8. Sandstone and conglomerate.
9. Lower Permian limestone.

The second potash district, which has three sylvinite horizons and one sylvinite-carnallite bed, is in the area near Soligorsk. The deposits are high grade (20 to 25% K$_2$O), and the salt beds are separated by clay, marl, and dolomite/anhydrite. The third potash district, in Stebnik and Kalush, is in the western Ukraine. The Miocene ore consists of hartsalz, sylvinite, kainite, langbeinite, and about 15% insoluble material in at least five potash zones (Adams 1983).

Canada: With developed potash resources in Saskatchewan, Manitoba, and the Maritime Provinces, Canada has the largest reserves of high grade potash ore in the world (Adams 1983). In Saskatchewan and Manitoba provinces, the potash deposits are of Middle Devonian age and occur in beds within the Prairie Formation, an evaporitic sequence composed of halite, potash, and anhydrite (Holter 1969). The evaporites of the Prairie Formation were deposited in the Elk Point Basin, a large trough that extends southeastward from the Northwest Territories and northwestern Alberta, to its southern terminus in northwestern North Dakota and northeastern Montana (Fig. 66) (Pearson 1962; Worsley and Fuzesy 1979; Wittrup and Kyser 1990). A generalized section of Middle Devonian strata in the Elk Point Basin, from the base upwards, consists of (1) dolomitic shale of the Ashern Formation, 10 to 20 m thick; (2) reef limestone of the Winnipegosis Formation, 15 to 90 m thick; (3) salt of halite,

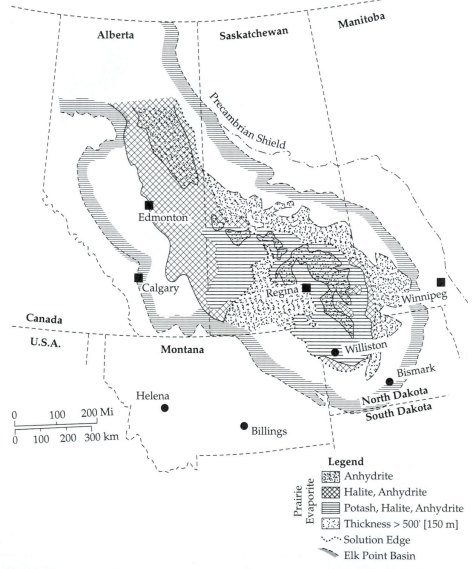

Figure 66   Regional setting of the Prairie Evaporite of Saskatchewan. Reprinted with permission of Economic Geology, Volume 74, p. 377–388, N. Worsley and A. Fuzesy.

anhydrite, and potash with clay of the Prairie Evaporite Formation, up to 200 m in thickness; (4) Dawson Bay Formation, a 20 m sequence of shale, dolomite, carbonate, and evaporite; and (5) Souris River Formation of carbonates and evaporites (Fuzesy 1982). The main potash minerals are sylvite, sylvinite, and carnallite, with K_2O content ranging from 14 to 32% (Boys 1993). Other salt deposits of economic significance in Canada are distributed in eastern New Brunswick, southwestern Newfoundland, and the Madgalen Islands of the Quebec Maritime Provinces (Pearson 1962; Kingston and Dickie 1979; Anderle et al. 1979).

Europe: Potash deposits are distributed in four main regions in Europe, the most important of which is the Zechstein basin (Adams 1983). This basin stretches from northern Britain southeast across the North Sea through the Netherlands, Denmark, Germany, and Poland, to the edge of the Hercynian massifs. The Zechstein sequence was deposited during the Permian period and is made up of four evaporite cycles, each represented by a general series of clastics, carbonates, halite, potassium salts, including sylvite, carnallite, kainite, and hartsalz, regressive halite or anhydrite, and carbonate (Harben and Bates 1990). Other regions that are mined in Europe include the Late Miocene evaporites in Sicily, Italy, with kainite as its ore, and the Lower Oligocene deposits in the Rhine Valley, with potash deposits lying on both sides of the Franco-German border in the area of Mulhouse, France. The fourth deposit is in Spain in the northern part of the Ebro Basin near the border with France (Anon. 1975).

The United States: The principal potash deposits are the Permian evaporites of the Carlsbad District in southeastern New Mexico, on the edge of the Delaware Basin (Cheeseman 1978; Griswold 1982). The evaporites of the Delaware Basin are part of the extensive evaporites of the Ochoa Series, which covers an area about 420 by 350 km. The Ochoa Series, from the lowest to the uppermost levels, contains three evaporite formations (1) the Castille Formation, mainly of anhydrite, (2) the Salado Formation of potash salts, and (3) the Rustler Formation. Potash minerals occur in eleven ore zones. The major ore minerals are sylvite, or sylvinite and langebeinite, with accessory minerals leonite, kainite, carnallite, polyhalite, kieserite, blödite, halite, and anhydrite. The average grade has an 18% K_2O content. Potash salts, found in the Williston Basin in North Dakota and Montana, are extensions of the rich Saskatchewan deposits (Anderson and Swinegart 1979). Potash salts are also exploited in Utah in the paradox Basin, the Great Salt Lake, and the Bonneville Salt Flats, and in California in the Searles Valley region.

Israel and Jordan: The brine of the Dead Sea has an average salinity of 25.7% and is estimated to contain 980 M tons of K_2O. Both Israel and Jordan use a system of solar evaporation ponds along the margins of the Dead Sea to produce carnallite (Williams-Stroud et al. 1994).

China: The principal production of potash in China comes from the Qaidam Basin in the Province of Qinghai (Yang et al. 1997). The Qarhan Salt Lake occupies the central-eastern part of the Basin. It is 168 km long and 20 to 40 km wide, covering an area of

4,700 km^2, and is at about 2,680 meters above sea level. There are ten individual lakes, some yielding halite and carnallite, others with deposits of halite and bischofite. Gypsum, polyhalite, schoenite, kainite, and langbeinite also occur in the lakes. Because of the limited reserves of sylvite in China, carnallite has become the principal alternative source of potash (Liu 1995). The Qarhan Salt Lake has been recognized as one of the largest inland K-forming playas in the world.

Ninety percent of the potash produced worldwide is consumed in the production of fertilizer for the agricultural industry. The remaining 10% is mainly used in the manufacture of carbonate, sulfate, chlorate, nitrate, and hydroxide of potassium by the chemical and processing industries. Annual world production of potash was estimated to be 30 M tons in the years between 1989 and 1993 (U.S.B.M. Mineral Yearbook 1994).

INDUSTRIAL PROCESSES AND USES

Potash chloride (KCl), or muriate of potash, is the major source of fertilizer potash. Currently, it accounts for about 95% of the fertilizer potash (K_2O) used annually in the United States. Sulfate of potash and sulfate of potash-magnesia combined provide about 5% of the total annual supply. The principal fertilizer potash products are classified as follows: (1) mine run salts (manure salts), 16.6 to 20.8% K; (2) muriate of potash (flotation), 49.8 to 51.0% K; (3) muriate of potash (recrystallized) 50.6 to 51.8% K; (4) sulfate of potash, 41.5 to 43.2% K; and (5) sulfate of potash-magnesia, 17.8% K.

Potash ore must be treated to remove insolubles, including clay minerals, before it can be used for the production of potassium chloride and potassium sulfate for agricultural uses (Crocket et al. 1968; Sullivan and Michael 1986). Mine run ore is first sent to a crushing and screening circuit during which potash minerals are liberated from the gangue minerals. Generally, the ore passes through a series of vibrating screens that are designed to classify oversize and undersize particles. The undersize particles are passed on to the next step. The oversize particles must first be returned to the crushing circuit for further size reduction, and then must be rescreened. The crushed ore is then passed through scrubbers where trommels or water jets are used to remove any clay or similar material that is exposed by crushing the ore.

Recovery of brine, in general, uses solar evaporation to concentrate the mineral content of the brine. In a typical system, brine is pumped into shallow solar evaporation ponds ranging in size from 2 to 80 km^2, where the water is evaporated from the brine. The evaporate is harvested every two to three years by a dredge and fed to the mill plant.

1. Production of KCl for Agricultural Use

Potash chloride, or muriate of potash, is produced from sylvinite ore by two different methods: recrystallization and flotation (Sauchelli 1960; Sullivan and Michael 1986; Schultz et al. 1993; Williams-Stroud et al. 1994).

The extraction of KCl from sylvinite ore by recrystallization is based on the fact that the solubilities of NaCl and KCl, as discussed previously, are distinct in hot and cold saturated solutions (p. 292-293). The processed ore is mixed with a brine that is saturated with both KCl and NaCl at 30°C. The mixture is then heated to 100°C, the temperature at which all of the KCl present in the ore dissolves, leaving NaCl as a solid residue. Brine-solid separation is accomplished by centrifuging. The solids (NaCl) are rejected from the process. As the hot saturated brine is cooled to 30°C, essentially all of the KCl entering the system with the ore is recrystallized, typically in a vacuum crystallizer. The KCl is filtered from the brine, dried, and stored.

The production of KCl from carnallite and carnallite-containing mixed salts is based on the phase relations discussed above, in the system $KCl-MgCl_2-NaCl-H_2O$. The carnallite ore is mixed and agitated in water to produce an undersaturated solution. This causes crystallization of KCl in an amount defined by "Point E" on the solubility curve in the system $KCl-MgCl_2-NaCl-H_2O$ as shown in Fig. 63 (Schultz et al. 1993).

Separation of potash from the ore by flotation is widely practiced (Crocket et al. 1968; Williams-Stroud et al. 1994). Although different plants have adopted distinctive circuits, a basic simplified scheme is shown in Fig. 67 (Sullivan and Michael 1986). In the conditioning step shown in Fig. 67, clay depression reagents such as starch are added at a dosage that is dependent on the ore grade, and then the pulp is conditioned with the flotation reagents at 50 to 75% solids. The normal collectors are primary aliphatic amines of about C_6 to C_{24} carbon chain lengths, used at a rate of 250 to 1,000 gm/ton. Polyglycol ether frothers are common, with fuel oil added to control the froth. Among the particle size designations shown in Fig. 67, standard, coarse, and granular have particle size ranges (Tyler sieve) of $-20 +100$, $-10 +35$, and $-6 +14$. All sizes are available in grades containing 49.8% K or 60% K_2O.

Processes for transforming finely divided muriate of potash that is produced by flotation and recrystallization into products of a larger particle size have been developed (Schultz et al. 1993) to meet the demand for muriate of potash products in its coarse and granular form. Two processes of granulation are commonly used: agglomeration of molten or wet material in rotating drums or dishes, or compaction of dry material in roll presses. The latter is the process most often used for potash fertilizers.

2. Production of K_2SO_4 for Agricultural Use

Potasssium sulfate (K_2SO_4) is produced (1) from langbeinite ore by an exchange of ions to remove the magnesia, (2) from burkeite ($Na_2CO_3 \cdot 2Na_2SO_4$) and KCl, (3) by the Hargreaves process from KCl and sulfur, and (4) in Mannheim furnace from KCl and H_2SO_4 (Kapusta and Wendt 1963).

In the langbeinite process, langbeinite ($K_2SO_4 \cdot 2MgSO_4$) ore, recycled mixed salts, principally KCl and leonite, and water are reacted to produce K_2SO_4 as the stable solid phase and as a mother liquor saturated with respect to this salt, KCl and leonite. The process involves complicated reactions of both the liquids and solids in

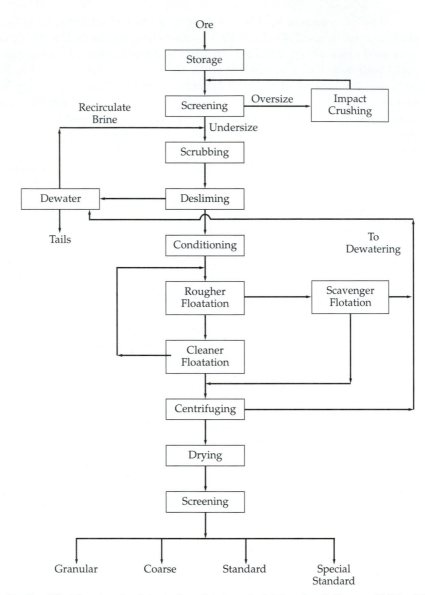

Figure 67 Simplified flotation circuit in the beneficiation of sylvinite, after Sullivan and Michael (1986).

the reciprocal salt pair system. A simplified reaction can be illustrated by $K_2SO_4 \cdot 2MgSO_4 + 4KCl \rightarrow 3K_2SO_4 + 2MgCl_2$.

In the Searles Lake "Trona Process," K_2SO_4 is produced from burkeite and KCl in two stages, with glaserite, $Na_2SO_4 \cdot 3K_2SO_4$, as an intermediate, according to the following reactions:

$$4Na_2SO_4 + 6KCl \rightarrow Na_2SO_4 \cdot 3K_2SO_4 + 6NaCl$$

$$Na_2SO_4 \cdot 3K_2SO_4 + 2KCl \rightarrow 4K_2SO_4 + 2NaCl$$

Potassium chloride and sodium sulfate are reacted at 20° to 50°C in water and recycled brines to form glaserite, which is filtered, and then reacted with more potassium chloride and water to form potassium sulfate. Alternately, sodium sulfate solution may be used to charge an anion exchanger with sulfate, which reacts with potassium chloride solution to give a high yield of potassium sulfate.

In the Hargreaves process (Anon. 1954), K_2SO_4 and HCl are produced from S and KCl. A composite gas of sulfur dioxide, air, and water vapor is passed through a series of chambers of reactors charged with briquettes of KCl. Endothermic reactions take place in the chambers to convert KCl briquettes to K_2SO_4 briquettes, which are cooled, crushed, screened, and stored. The by-product, HCl, is also a marketable product.

Potassium sulfate is produced in the Mannheim Process from potassium chloride and sulfuric acid in two stages:

$$KCl + H_2SO_4 \rightarrow KHSO_4 + HCl$$

$$KCl + KHSO_4 \rightarrow K_2SO_4 + HCl$$

The first stage forms an acid sulfate that is exothermic and proceeds at relatively low temperatures. In the second stage, which is endothemic, the acid sulfate is converted to normal sulfate by further reaction with KCl at higher temperatures. At 260°C, an 80% conversion requires 180 minutes, whereas a total conversion at 550°C takes only 20 minutes. The furnace used has a closed dish-shaped chamber, heated externally, into which KCl and H_2SO_4 are fed, and in which a reaction takes place that produces potassium sulfate and hydrogen chloride gas. The solid is removed from the chamber, neutralized, and cooled. At the same time, the gas is either absorbed in water to form hydrochloric acid is used in its gaseous form.

3. Production of Sulfate of Potash-Magnesia for Agricultural Use

Langbeinite, a double salt of potassium and magnesium sulfate, occurs in the Carlsbad basin, along with halite and sylvite. It is mined and processed for agricultural use under the trade name "Sol-Po-Mag." Langbeinite is soluble, but dissolves rather slowly as compared with the associated halite and sylvite, which dissolve at a much faster rate (Wilson 1966). This solubility distinction is used for the production of sulfate of potash-magnesia in a continuous countercurrent washing process. A flow diagram in Fig. 68 illustrates the process. The end product has 18.26% K or 22.0% K_2O, 10.85% Mg or 18.0% MgO, and a maxima of 2.5% Cl (Harley 1953).

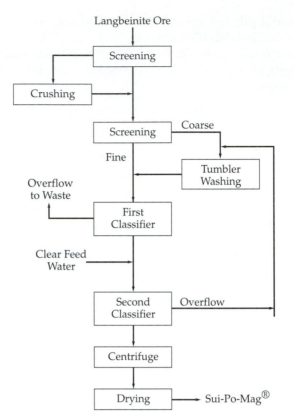

Langbeinite Ore

Screening

Crushing

Screening — Coarse

Fine

Tumbler Washing

Overflow to Waste

First Classifier

Clear Feed Water

Second Classifier — Overflow

Centrifuge

Drying → Sui-Po-Mag®

Figure 68 Simplified flow diagram for langbeinite processing, after Kapusta and Wendt (1963).

REFERENCES

Adams, S. S. 1983. Potash. In *Industrial minerals and rocks.* 5th ed. Ed. by S. J. LeFond, 1049–1077. New York: Soc. Mining Engineers, AIME.

Anderle, J. P., K. S. Crosby, and D. C. E. Waugh. 1979. Potash at Salt Springs, New Brunswick. *Econ. Geol.* 74:389–396.

Anderson, S. B., and R. P. Swinegart. 1979. Potash salts in the Williston Basin, U.S.A. *Econ. Geol.* 74:358–376.

Anon. 1954. For veteran Hargreaves process—a new job in the potash industry. *Chem. Eng.* 61:132–134.

Anon. 1975. *World survey of potash 1975.* London: British Sulphur Corporation Ltd, 143p.

Boys, C. 1993. A geological approach to potash mining problems in Saskatchewan, Canada. *Exploration & Mining Geol.* 2:129–138.

Braitsch, O. 1971. *Salt deposits, their origin and composition.* Berlin: Springer-Verlag, 297p.

Cheeseman, R. J. 1978. Geology and oil/potash resources of Delaware Basin, Eddy and Lea counties, New Mexico. *New Mexico Bur. Mines and Mineral. Res. Circ.* 159:7–14.

Crocket, B. S., J. T. Dew, and R. J. Roach. 1968. Contemporary potassium plant engineering. *Canadian Inst. Metal. Bull.* 737–741.

Fuzesy, A. 1982. Potash in Saskatchewan. *Saskatchewan Energy & Mines Rept.* 181, 44p.

Griswold, G. B. 1982. Geologic overview of the Carlsbad potash-mining district, Proc. 7th Forum on the Geology of Industrial Minerals. Ed. G. S. Austin. *New Mexico Bur. Mines & Mineral. Res. Circ.* 182:17–22.

Harben, P. W., and R. L. Bates. 1990. *Industrial minerals, geology and world deposits.* London: Industrial Minerals Div., Metal Bull. Plc, 312p.

Harley, G. T. 1953. Production and processing of potassium materials. In *Fertilizer technology and resources in the United States.* Ed. K. D. Jacob, 287–322. New York: Academic Press.

Holter, M. E. 1969. The Middle Devonian prairie evaporite of Saskatchewan. *Saskatchewan Dept. Nat. Res. Rept.* 123, 134p.

Kapusta, E. C., and N. E. Wendt. 1963. Advances in fertilizer potash production. In *Fertilizer, technology and usage.* Eds. M. H. McVickar, G. L. Bridger, and L. B. Nelson, 189–230. Madison, Wisconsin: Soil Sci. Soc. Amer.

Kingston, P. W., and C. E. Dickie. 1979. Geology of New Brunswick potash deposits. *Canadian Inst. Mining & Metal. Bull.* 72:134–140.

Kühn, R. 1959. Die Mineralnamen der Kalisalze. *Kali u Steinsalz* 2:331–345.

Liu, X. 1995. Technological process of making potash fertilizer by carnallite. *Jour. Mineral. Petrol.* 15:82–86 (Chinese with English abstract).

Pearson, W. J. 1962. Salt deposits of Canada. *Proc., 1st Sym. on Salt.* Cleveland, Ohio: Northern Ohio Geol. Soc, 197-239.

Phillips, W. R., and D. T. Griffen. 1981. *Optical mineralogy.* San Francisco: W. H. Freeman, 677p.

Rempe, N. T. 1982. Langbeinite in potash deposits, Proc. 7th Forum on the Geology of Industrial Minerals. Ed. G. S. Austin. *New Mexico Bur. Mines Res.* Cir. 182:23–26.

Sauchelli, V. 1960. *Chemistry and Technology of Fertilizers.* New York: Reinhold, 692p.

Schultz, H., G. Bauer, F. Hagedorn, and P. Schmittinger. 1993. Potassium compounds. In Vol. A22 of *Ullmann's encyclopedia of industrial chemistry,* 39–103. Weinheim: VCH Publ.

Sullivan, D. E., and N. Michael. 1986. Potash availability—market economy countries. *U.S. Bur Mines, Infor. Circ.* 9084, 32p.

Tatarinov, P. M., ed. 1969. *Non-metallic mineral deposits* Moscow: Nedra, 692p. (in Russian).

Williams-Stroud, S. C., J. P. Searls, and R. J. Hite. 1994. Potash resources. In *Industrial minerals and rocks.* 6th ed. Ed. by D. D.Carr, 783–802. Littleton, Colorado: Soc. Mining, Metal., & Exploration.

Wilson, W. P. 1966. Leaching potash ores: 4th Sym. on Salt. *Northern Ohio Geol. Soc.* 1:517–525.

Wittrup, M. B., and T. K. Kyser. 1990. The petrogenesis of brines in Devonian potash deposits of western Canada. *Chem. Geol.* 82:103–128.

Worsley, N., and A. Fuzesy. 1979. The potash bearing members of the Devonian Prairie Evaporite of southeastern Saskatchewan, south of the mining area. *Econ. Geol.* 74:377–388.

Yang, Q., B. Wu, S. Wang, K. Cai, and Z. Quian. 1997. Geology of the potash deposit of the Qarhan Salt Lake. *Annual Rept.* Chinese Acad. Geol. Sci., 1995, 98–102.

Pyrophyllite

Pyrophyllite is a layer-type silicate with an ideal composition of $Al_2Si_4O_{10}(OH)_2$. Pyrophyllite is mainly used in ceramics and refractories because of its good thermal shock resistance and high resistance to corrosion. Other uses are found in agriculture as a carrier of insecticides, and in paint, plastics, rubber, and soap as a filler.

As an industrial mineral, pyrophyllite is traditionally treated in association with talc, including in production and consumption data. Except for their crystal structure, pyrophyllite and talc differ in their composition, occurrence, and industrial uses.

The northwestern region of the circum-Pacific provinces including Japan, Korea, and southeastern China contains the bulk of the world's pyrophyllite. The United States and Australia are also important producers.

MINERALOGICAL PROPERTIES

Pyrophyllite has a layered, mica-type structure. The basic feature is sheets of hexagonal rings of silicon tetrahedra (SiO_4) that are stacked along the c-direction. The layer is formed by two tetrahedral sheets that point inward and are cross-linked by octahedrally coordinated aluminum (AlO_6). There is no interlayer cation, such as K^{1+}, in the muscovite or biotite structure, and the layers are bonded together by van der Waals forces. To compensate for the difference in dimensions between the SiO_4 tetrahedron and the AlO_6 octahedron, the SiO_4 tetrahedron has to rotate and tilt from its normal configuration in the tetrahedral sheet. This results in a reduction of symmetry of the pyrophyllite structure from hexagonal to triclinic (Zoltai and Stout 1984).

Pyrophyllite has an ideal composition of 66.0% SiO_2, 28.4% Al_2O_3, and 5.0% H_2O. Chemical data listed in Table 20 illustrate the fact that, in general, two types of pyrophyllite may be identified: high SiO_2 with low Al_2O_3, and low SiO_2 with high Al_2O_3. The associated minerals are diaspore, corundum, kaolinite, alunite, quartz, sericite, and montmorillonite.

Pyrophyllite occurs as a solid mass, commonly flaky and occasionally fibrous, with color varying from green to yellow to white. Observation of pyrophyllite from Hiroshima, Japan by Scanning Electron Microscopy (SEM) shows typical particle morphology, and sizes of 0.1 to 1.0 μm thick and several μm in diameter (Wiewiora and Hida 1996). It has a hardness of 1.5, a specific gravity of 2.8, and a pearly luster. Its refractive indices range between 1.553 and 1.600. At 500° to 800°C, dehydroxylate of pyrophyllite is formed, and this phase has the same structure as pyrophyllite. Decomposition of dehydryoxylate to mullite and cristobalite begins at 1,000°C (Zaremba 1995). Ceramic bodies made of pyrophyllite have high corrosion resistance for molten iron, steel, and slags, good thermal shock resistance, and good hot creep resistance (Cornish 1983). Pyrophyllite has low compressibility and is highly efficient in pressure-transmitting. As compared with other layered silicates, pyrophyllite, along with talc and smectite, shows the best plastic flow properties (Ullrich and Raab 1993).

GEOLOGICAL OCCURRENCE AND DISTRIBUTION OF DEPOSITS

The bulk of the world's pyrophyllite deposits and reserves is located in Japan, Korea, China, the United States, and Australia. Japan's production of pyrophyllite accounted for 50% of the world's annual output (Dickson 1986), which was estimated to be 2 M tons. Most pyrophyllites are formed by a reaction between hydrothermal solutions and acid-volcanic rock complexes, which transforms the country rock from rhyolite or feldspar-quartz to pyrophyllite-quartz. Less commonly, pyrophyllite forms from metamorphosed volcanic ashes (Ciullo and Thompson 1994).

Japan: All the pyrophyllite deposits in Japan were formed by hydrothermal alteration and are related to the volcanic activity that has been active in the islands over a long period of geological time, but especially during the Cretaceous period and the Miocene epoch (Fujii 1983). In general, pyrophyllite deposits are divided into two types, depending upon their relation to (1) acid intermediate intrusives, and (2) exhalative hydrothermal alteration near the surface. Distribution of the main pyrophyllite deposits in Japan is shown in Fig. 69 (Fujii 1983).

The Goto mining district, which is mainly underlain by Eocene sediments and Miocene granite porphyry intrusives, has a number of pyrophyllite deposits occurring in the intrusives and the adjacent sediments. These deposits have various shapes, from massive through bedded to veinlike, and several deposits in the intrusives are characterized by zonal arrangements from the center to the margin of high alumina minerals, pyrophyllite, and quartz + pyrophyllite. Other deposits of this type are located at Shin'yo mine, central Honshu, and in the Abu area of western Honshu (Fujii 1975; Kamitani 1978).

The Mitsuishi district, which contains the largest pyrophyllite deposits in Japan, is mostly underlain by welded acid tuff and tuffaceous shale of Cretaceous age (Fujii 1983). The lower member of the welded tuff is extensively altered and develops zonal arrangements of silica-rich rock at the center and a sericitized zone at the margin. Sandwiched between the silica-rich rock and the sericitized zone is the pyrophyllite-rich zone. The formation of pyrophyllite by the exhalative hydrothermal alteration that took place near the surface is illustrated by the fact that the upper member of the welded tuff series is hardly affected by the hydrothermal alteration (Kinosaki 1963). Pyrophyllite deposits at Bonten'yama (Fujii 1967) and in the Shokozan district (Matsumoto 1968) are also formed by exhalative alteration near the surface. The latter is characterized by the presence of high-temperature mineral assemblages of corundum and andalusite in the core, and the fact that the silica-rich rock seen in the Mitsuishi area is not developed. This fact suggests that the deposits may have been affected by volcanic activity. Chemical data of pyrophyllite are given in Table 20.

South Korea: The pyrophyllite deposits are largely distributed in the southern part, near the coast of the peninsula (Sang 1983; O'Driscoll 1993). The pyrophyllite deposits were mainly formed by hydrothermal alteration of the Cretaceous sequence of andesitic and rhyolitic lavas and tuff, which were intruded by the Bulgugsa Granite Series of granodiorite, granite, and granite porphyry. Deposits are generally small in size and irregular in shape. Typically, the deposits show a zonal arrangement as follows

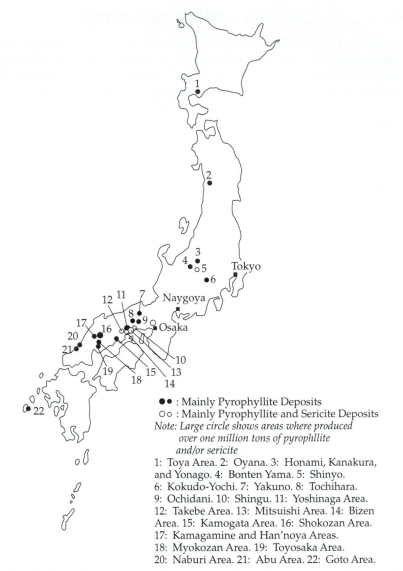

Figure 69   Distribution of the main pyrophyllite deposits in Japan. Source: Industrial Minerals, November 1983.

(in descending order): a siliceous zone, an alunite-quartz zone, a pyrophyllite-kaolin zone, and a sulfide zone. Three mineralogical types of pyrophyllite association are recognized: (1) pyrophyllite with minor quartz, kaolin, and alunite, (2) kaolin-pyrophyllite with some diaspore, and (3) pyrophyllite-sericite with some quartz (Sang 1983). Chemical data from Wando, Sungsan, Milyang, and Dongnae mines are given in Table 20. The first two mines are located at the western end of the peninsula, and the second two are located on the eastern shore.

Table 20 Chemical data for pyrophyllite from world's main deposits: (1) Japan, Industrial Minerals (1983), (2) South Korea, Industrial Minerals (1983), (3) Australia, Raw Materials for the Refractories Industry (2nd edition) (1986), and (4) the United States, after Cornish (1983).

1. Japan

	Ochidani	Hyobu	Ohiro	Shokozan	Gato	Gato*
SiO_2	80.93	75.70	78.32	63.64	76.2	87.38
Al_2O_3	15.21	17.71	16.87	29.64	18.6	10.01
TiO_2	0.15	0.09	0.10	0.22	0.7	0.35
Fe_2O_3	0.36	1.43	0.13	0.02	0.1	0.14
CaO	0.04	tr	0.06	0.03	tr	0.10
MgO	—	0.09	0.10	0.04	0.0	0.05
K_2O	0.05	0.32	0.89	0.04	0.2	0.04
Na_2O	—	0.22	0.12	0.05	0.0	0.04
LOI	3.25	4.31	3.32	6.32	3.3	2.03
Total	99.99	99.90	99.91	99.70	99.1	100.14

*siliceous

2. South Korea

	Wando	Milyang	Dongnae	Sungsan
SiO_2	60.85–63.04	65.31	67.60	63.90
Al_2O_3	31.18–29.50	28.75	26.32	27.68
TiO_2	—	—	—	—
Fe_2O_3 (+ FeO)	0.63–0.58	0.51	0.84	0.56
CaO	0.35–0.15	0.13	0.19	0.67
MgO	0.14–0.10	0.35	0.25	0.27
K_2O	0.13–0.06	0.07	0.25	0.83
Na_2O	0.89–0.58	0.45	0.65	0.76
LOI	6.32–6.02	4.48	3.93	5.35
Total	100.49–100.03	100.05	100.03	100.02

3. Australia

	Chloritic pyrophyllite	Chalcedonic pyrophyllite	Sericitic pyrophyllite
SiO_2	65.70	75.70	79.60
Al_2O_3	27.80	18.80	12.70
TiO_2	0.20	0.39	0.20
Fe_2O_3	0.40	0.38	1.30
CaO	0.01	0.30	0.03
MgO	0.01	0.12	0.19
K_2O	0.16	0.24	2.58
Na_2O	—	0.11	0.88
LOI	5.20	3.90	2.40
Total	99.50	99.80	99.91

(Continued)

311

Table 20 *(Continued)*

4. U.S.A.

	1	2	3	4	5
SiO	76.32	73.50	70.32	69.90	69.38
Al_2O_3	19.80	22.53	24.95	25.13	26.02
TiO_2	—	—	—	—	—
Fe_2O_3	0.18	0.09	0.08	0.07	0.08
CaO	0.14	0.08	0.16	0.16	0.14
MgO	—	—	—	—	—
K_2O	0.27	—	0.13	0.00	0.00
Na_2O	0.07	0.07	0.31	0.08	0.24
LOI	3.44	3.95	4.32	4.67	4.50
Total	100.22	100.21	100.21	100.01	100.36

Samples 1–5, all from North Carolina.

The United States: Major pyrophyllite deposits are concentrated in North Carolina in a belt starting from the Virginia line and extending across the state, through South Carolina to Graves Mountain, Georgia (Stuckey 1967; Dickson 1986). The pyrophyllite deposits occur within the fracture zones in rhyolite volcanic rocks and form oval and lenticular bodies that resulted from metasomatic replacement (Stuckey 1967). In both oval and lenticular bodies, pyrophyllite with quartz and sericite occupy the central part and grade outward into the impure zones. Chemical data of pyrophyllite in the United States are also given in Table 20.

Australia: The main pyrophyllite deposit is located at Pambula near Eden on the southeast coast of New South Wales. It lies in the central part of the ancient Eden Rift Zone, which consists of the Upper Devonian Boyd Volcanic Complex of agglomerate, bassalt, rhyolite, and sedimentary rocks. During the Upper Devonian period, this rift zone was the locus of intense volcanic activity that provided the basis for extensive hydrothermal alteration and burial metamorphism. Mineralization associated with the activity formed the deposit. The pattern of alteration appears to be zoned with a central core of pyrophyllite that is surrounded by chalcedonic pyrophyllite and sericitic pyrophyllite, which grade into unaltered host rocks (Cornish 1983). Chemical data are given in Table 20.

INDUSTRIAL PROCESSES AND USES

1. Processing and Specifications

The processing method used in every processing plant is different, with characteristics suitable for its own ore. The principal methods comprise crushing, washing, picking, and sizing to produce a series of size fractions. Such procedures are used in thirteen out of fourteen processing plants in Japan, with capacities ranging from 1,500 to 15,000 tpm

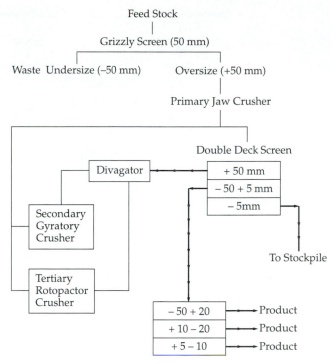

a.

Figure 70 Flow sheet for the crushing and sizing of pyrophyllite from (a) Pambula deposit, Australia, and (b) Standard Mineral, Robbins, North Carolina, after Cornish (1983).

(Dickson 1986). The processing plant at Shokozan has installed both wet and dry systems to process pyrophyllite clay (Fujii 1983).

At Pambula, Australia, and North Carolina (the United States), the processing more or less follows a crushing and screening procedure (Cornish et al. 1980; Nicol 1983; Cornish 1983). Flow sheets are shown in Fig. 70. Ciullo and Thompson (1994) presented a representative flow sheet of processing procedures that consists of the following: vibrating screens → wet ore storage → rotary drier → vibrating screens → dry ore storage → grinding mills → air separator → bulk finished product storage → baggers → shipping. At the end of each of the two vibrating screens, the oversize is crushed and is further screened and mixed with the undersize to go to storage.

2. Industrial Uses

Pyrophyllite has its major industrial uses in refractories and ceramics. It is also used as a filler for paper, rubber, paint, cosmetics, and bleaching soap, and as a carrier in agricultural insecticides. For construction material, pyrophyllite is used in the formulation of white cement clinker and white road aggregate.

Refractories: The use of pyrophyllite has great advantages over clay-based materials, in contact refractories, kiln furniture, and refractory mortars (Nameishi 1976; Cornish et al. 1980; Ciullo and Thompson 1994). By using the quick-action friction process,

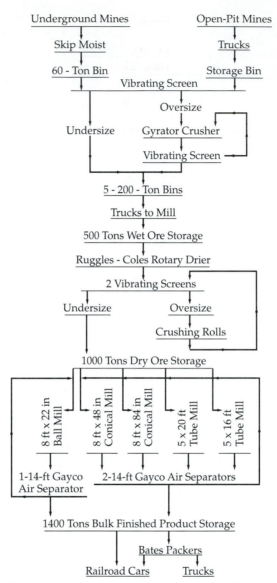

b.

Figure 70 *(Continued)* Flowsheet for the crushing and sizing of pyrophyllite from (a) Pambula deposit, Australia, and (b) Standard Mineral, Robbins, North Carolina, after Cornish (1983).

pyrophyllite and pyrophyllite-zircon bricks are produced for ladle linings. The ladle life increases twofold when lined with pyrophyllite bricks and 4.5 times when lined with pyrophyllite-zircon bricks, compared to the use of clay bricks. For kiln furniture, the use of coarse-sized pyrophyllite in its formulation improves furniture life and lowers maintenance costs. Pyrophyllite-based kiln furniture has excellent thermal shock resistance, a low coefficient of thermal expansion, minimal shrinkage, high resistance to deformation under loads, and a moderate to high PCE (30 to 32).

Formulations made from fine fractions of pyrophyllite are used for the manufacture of mortar of good plasticity and low drying and firing shrinkages (Tauber and Pepplinkhouse 1972). The uses of pyrophyllite in whiteware include wall tile, floor tile, sanitaryware, pottery, chinaware, electric porcelain, and glazes. The addition of pyrophyllite to the mixtures lowers the firing temperature, increases the whiteness of the bodies, lowers firing shrinkage, and improves thermal shock resistance (Tauber 1973; Ciullo and Thompson 1994).

Pyrophyllite is used in the manufacture of white cement because of its low iron content, which improves its whiteness level in excess of 90%. After calcination, pyrophyllite becomes an excellent material to use in the formulation of road surface aggregates. Its benefits include (1) high antiskid properties, (2) high polished stone value, (3) high aggregate impact value, and (4) a high luminance factor.

Pyrophyllite of fine size is used as a filler in paint, plastics, rubber, and both powder and bar soaps, as well as in wallboards (Benbow 1988). Pyrophyllite is an excellent carrier in agricultural insecticides because it is neutral in pH, inert, nonhygroscopic, and has good flowability.

There are industrial grades and specifications for a variety of applications. The refractory grade pyrophyllite must have an average mineralogical composition of 40 to 50% pyrophyllite, 30 to 45% quartz, 5 to 15% kaolinite, and 1 to 3% mica, whereas the composition of the whiteware grade is 20 to 30% pyrophyllite, 50 to 60% quartz, 20 to 25% sericite, and 5 to 10% kaolinite. Refractory pyrophyllite can be further graded by screen analysis (Ciullo and Thompson 1994). The grading system for Australian pyrophyllite is listed below (Cornish 1983):

1. Chloritic pyrophyllite
   Specification: 24 to 28% Al_2O_3, 0.2 to 0.5% K_2O, 0.2% Fe_2O_3
   Application: paint filler, plastics filler, rubber filler
2. Chalcedonic pyrophyllite (refractory grade)
   Specification: 13 to 15% Al_2O_3, 0.2 to 0.5% K_2O, 0.4% Fe_2O_3.
   Application: contact refractories
3. Chalcedonic-sericitic pyrophyllite
   Specification: 15 to 20% Al_2O_3, 0.4 to 0.8% K_2O, 0.5% Fe_2O_3.
   Application: kiln furniture
4. Sericitic pyrophyllite
   Specification: 18 to 22% Al_2O_3, 0.5 to 3.0% K_2O, 0.5% Fe_2O_3.
   Application: whiteware ceramics, floor tiles

REFERENCES

Benbow, J. 1988. Pyrophyllite—far east steels the market: *Ind. Minerals* (June): 37–49.

Ciullo, P. A., and C. S. Thompson. 1994. Pyrophyllite. In *Industrial minerals and rocks*. 6th ed. Ed. D. D. Carr, 815–826. Littleton, Colorado: Soc. Mining, Metal. & Exploration.

Cornish, B. E. 1983. Pyrophyllite. In *Industrial minerals and rocks*. 5th ed. Ed. S. J. Lefond, 1085–1108. New York: Soc. Mining Engineers, AIME.

Cornish, B. E., E. Tauber, and D. Nichol. 1980. Pambula pyrophyllite: Production and applications: *Proc., 9th Australian Ceramic Conf,* 132–134.

Dickson, E. M., Ed. 1986. Pyrophyllite: Asia and American Lead, in Raw Materials for the Refractories Industry, p.131-143, Industrial Minerals, London.

Fujii, N. 1967. Geologic occurrence and formation process of the Bonnten'yama pyrophyllite deposits, Nagano Pref. *Min. Geol. Japan*17:261–271.

Fujii, N. 1975. Pyrophyllite deposits at the Shin'yo mine, Nagano Pref. *Taikabutsu, (Refractories)*27:7–18.

—— 1983. The present position of Japanese pyrophyllite. *Ind. Minerals* (November):21–27.

Kamitani, M. 1978. Genesis of the andalusite-bearing roseki ore deposits in the Abu district, Yamagichi Pref., Japan. *Bull. Geol. Surv. Japan*28:201–264.

Kinosaki, Y. 1963. The pyrophyllite deposits in the Chugoku province, west Japan. *Geol. Rept., Hiroshima Univ., no. 12,* 35p.

Matsumoto, K. 1968. On the geology and pyrophyllite deposits of the Yano-shokozan mine, Hiroshima Pref. *Geol. Rept., Hiroshima Univ., no. 16,* 28p.

Nameishi, N. 1976. Recent status of steel plant refractories in Japan. *XIX Intern. Refractory Colloquium, Aachen, Germany,* 209–271.

Nichol, D. 1983. Pyrophyllite operations at Pambula, Australia. *Ind. Minerals* (November):31–35.

O'Driscoll, M. 1993. South Korea's minerals industry—imports prove Seoul-destroying. *Ind. Minerals* (February):18–37.

Sang, K. N. 1983. Pyrophyllite clay deposits in the Republic of Korea. *Ind. Minerals* (November):30–31.

Stuckey, J. L. 1967. Pyrophyllite deposits in North Carolina. *North Carolina Dept. Conservation and Development, Div. Mineral. Res. Bull.* 80, 38p.

Tauber, E. 1973. Stoneware bodies based on pyrophyllite. *Jour. Australian Ceramic Soc.* 9:47–51.

Tauber, E., and H. J. Pepplinkhouse. 1972. Ceramic properties of pyrophyllite from Pambula, N.S.W. *Jour. Australian Ceramic Soc.* 8:62–64.

Ullrich, B., and S. Raab. 1993. Untersuchungen zum plastisch-mechanischen Verhalten von ausgewählten Schichtsilikaten und Schichtsilikatgesteinen. *Chemie der Erde* 53:79–91.

Wiewiora, A., and T. Hida. 1996. X-ray determination of superstructure of pyrophyllite from Yano-Shokozan mine, Hiroshima, Japan. *Clay Sci.* 10:15–35.

Zaremba, T. 1995. Thermal transformation of synthetic pyrophyllite. *Geologica Carpathica—Clays* 4:67–72.

Zoltai, T., and J. H. Stout. 1984. *Mineralogy, concepts and principles.* Minneapolis, Minnesota: Burgess Publ, 505p.

Rare Earth Minerals

Among more than 150 documented rare earth minerals, only monazite and xenotime, both phosphates, and bastnäsite, a carbonate, are of economic interest. They are the main source of the rare earths that have industrial uses in catalysts, ceramics, glass, metals, and phosphors.

The United States, China, and Australia account for more than 80% of the total production of rare earth minerals, estimated to have been 52,000 tons (RE oxides) in 1993 (U.S.B.M. mineral data). Other productive countries are India, Malaysia, the former USSR, and Brazil.

MINERALOGICAL PROPERTIES

Including La and Y, there are a total of sixteen rare earth (RE) elements: Ce, Pr, Nd, Pm, Sm, Eu, Gd, Tb, Dy, Ho, Er, Tm, Yb, and Lu, which are conveniently classified into two groups. The "Ce-group" or the "light rare earth group" consists of rare earths from La to Eu, and the "Y-group" or the "heavy rare earth group" consists of Y and the rare earths from Gd to Lu. The light rare earths are more abundant than the heavy rare earths, and the rare earths of even atomic number are more abundant than their respective odd atomic numbered neighbors. By comparison with other chemical elements, rare earths are not rare. Cerium is more abundant than tin, yttrium is more abundant than lead, and even the least abundant naturally occurring rare earths are far more abundant than the platinum group metals (Möller 1961; Henderson 1984).

RE-bearing minerals are numerous. Levinson (1966) defined RE minerals as those minerals that contain rare earths as essential constituents. Bayliss and Levinson (1988) further refined the definition of RE minerals as minerals in which the total atomic percentage of RE is greater than any other element within a single set of crystal-structural sites. The International Mineralogical Association (IMA) established a rule for naming a RE mineral: the group name should go first, followed by the dominant element RE in parentheses. Some 150 RE minerals have been documented on the basis of their crystal structure, using the definition above, and may be classified into six groups (Miyawaki and Nakai 1996):

1. RE minerals containing isolated triangular anionic groups—carbonates: bastnäsite $\{CeFCO_3\}$.
2. RE minerals containing isolated tetrahedral anionic groups—phosphates, sulfates, arsenates, and vanadates: monazite $\{(Ce, La, Th, Y) PO_4\}$, xenotime $\{YPO_4\}$.
3. RE minerals containing tetrahedral anionic groups—silicates: cerite $\{(Ca, Mg)_2 (Ce, REE)_8 (SiO_4)_7 (FCO_3)_7\}$, gadolinite $\{Be_2FeY_2Si_2O_{10}\}$.
4. RE minerals containing both tetrahedral and octahedral anionic groups—aluminosilicates and titanosilicates: allanite $\{(Ca,Ce,Th)_2 (Al, Fe, Mg)_3 (SiO_4) (Si_2O_7O (OH)\}$.
5. RE minerals containing octahedral anionic groups—titanates, tantalates, and niobates: brannerite $\{(U, Ca, Fe, Th, Y)_3 Ti_5 O_{16}$, euxenite $\{(Y, Ca, Ce, U, Th) (Nb, Ta, Ti)_2O_6\}$, fergusonite $\{(Y Er, U, th) (Nb, Ta, Ti)O_4\}$.
6. RE minerals containing no anionic groups—oxides and fluorides: fluocerite $\{(Ce, La) F_3\}$, cerianite $\{Ce O_2\}$.

Table 21 Chemical data for monazite, bastnäsite, and xenotime. Reprinted with permission from K. A. Gschneider, Specialty Inorganic Chemicals, The Royal Society of Chemistry Cambridge, 1981, page 411.

Rare Earth%	Bastnäsite California	Bastnäsite Bayun Obo	Monazite Western Australia	Xenotime Malaysia
La	32.0	27.8	20.2	0.5
Ce	49.0	51.5	45.3	5.0
Pr	4.4	4.6	5.4	0.7
Nd	13.5	14.4	18.3	2.2
Sm	0.5	0.6	4.6	1.9
Eu	0.1	0.2	0.1	0.2
Gd	0.3	0.5	2.0	4.0
Tb				1.0
Dy				8.7
Ho				2.1
Er	0.1	0.2	2.0	5.4
Tm				0.9
Yb				6.2
Lu				0.4
Y				60.8
Total	100.0	99.9	100.5	100.0

Of the 150 documented RE minerals, monazite and bastnäsite are the most important by virtue of their abundance and high RE content, and are the main source of the RE used by industry. Xenotime is also used, but to a limited extent. Rare earths have been obtained as a by-product from apatite (Mariano 1989).

Monazite is an anhydrous phosphate, principally of cerium and/or lanthanium, with lesser amounts of the heavy rare earths. Some Th is generally present, commonly to the extent of 4 to 10%; Ca may substitute in a small amount for (Ce,La). Xenotime, another anhydrous phosphate, is essentially YPO_4. Rare earths, particularly Er and Ce, may substitute for Y, and Th, U, Zr, and Ca may also substitute for Y in small amounts. Xenotime is tetragonal with a=6.89Å and c=6.03Å and isostructural with zircon. The yttrium atoms {xenotime-(Y)} form 8-coordinated polyhedra, which connect isolated PO_4 tetrahedra (Kristanovic 1965; Ni et al. 1995). The crystal structure of monazite {monazite-(Ce)} shows a close relationship to that of xenotime, and the coordination number of the cerium atoms is the same as that of the yttrium in xenotime (Ueda 1967; Ni et al. 1995). The difference lies in the shapes and volumes of the RE polyhedra and the arrangement of the PO_4 tetrahedra, which are linked by the RE atoms in both structures. This results in a monoclinic unit cell in monazite with a=6.79Å, b=7.01Å, c=6.46Å, and β = 103°38′. The mean Ce-O distance in monazite is 2.45Å, as compared with the Y-O distance of 2.42Å in xenotime. The larger RE polyhedra in monazite prefer light RE, while the smaller ones in xenotime prefer heavy RE. Selected RE compositions in monazite and xenotime are given in Table 21 (Gschneidner 1980).

The crystal structure of bastnäsite {bastnäsite-(Ce)} consists of (CeF) and (CO₃) layers (Ni et al. 1993). The Ce atom is bonded to three F and six O atoms to form a nine coordinated polyhedron, and both F and Ce form hexagonal sheets parallel to (0001). This characteristic feature stabilizes the structure and results in a hexagonal unit cell of a=7.16Å and c=9.78Å. The large coordinated polyhedra prefers the light rare earths. Selected RE compositions in bastnäsite are also given in Table 21. Bastnäsite has at least three polytypes (4H, 6R, and 3R) (Yang et al. 1992).

The physical properties of monazite, xenotime, and bastnäsite are very similar. They are tabulated as follows:

	Monazite	Xenotime	Bastnäsite
H	5 to 5½	4 to 5	4 to 4½
G	4.6 to 5.4	4.3 to 5.1	4.3 to 5.1
Color	Yellowish brown, reddish brown, brown		
Luster	Vitreous, resinous,waxy		
R.I. α or ω	1.774 to 1.800	1.720 to 1.724	1.712 to 1.723
β	1.777 to 1.801	—	—
γ or ε	1.828 to 1.851	1.816 to 1.827	1.798 to 1.812

All RE elements are metals, as well as strong reducing agents. Their oxidation states vary among 2+, 3+, and 4+, and they react with most of the nonmetals, water, and acids (Möller 1961). The divalent ions are readily oxidized in aqueous solution. Both Sm^{2+} and Yb^{2+} as chlorides react rapidly in the presence or absence of air. Eu^{2+} chloride, however, reacts only very slowly in the absence of air and only slightly more rapidly in the presence of air. All these materials in their anhydrous conditions are more stable, but only the highly insoluble Eu^{2+} compounds such as $EuSO_4$ and $EuCO_3$ are resistant to attack by atmospheric oxygen (Stubblefield and Eyring 1955). The trivalent rare earths are highly ionic in aqueous solution (Spedding and Jaffe 1954a, 1954b; Dye and Spedding 1954; Beaudry and Gschneidner 1991). Common salts of RE of both Y- and Ce-groups, with Cl^{1-}, Br^{1-}, I^{1-}, NO_3^{1-}, ClO_4^{1-}, and SO_4^{2-}, are soluble in H_2O. The only exception is that carbonates of RE of the Y-group are soluble in carbonate solutions, whereas those of the Ce-group are not. The tetravalent rare earths are all strongly oxidizing. Ce^{4+} is best known as an oxide and as double ammonium nitrate. Both PrO_2 and TbO_2 have been prepared by oxidation with oxygen under pressure, but they are apparently unstable with respect to Pr_6O_{11} and Tb_4O_7, respectively, at ordinary oxygen pressure (Guth and Eyring 1954; Beaudry and Gschneidner 1991).

GEOLOGICAL OCCURRENCE AND DISTRIBUTION OF DEPOSITS

RE minerals occur in a variety of geologic environments, but concentrations are restricted to alkalic igneous rocks, especially carbonatite, and unconsolidated secondary deposits such as beach placers and deltaic deposits (Mariano 1989). The two principal deposits of carbonatite are in Mountain Pass, Mariano, California, in the United States, and Bayan Obo, Inner Mongolia, China. Both deposits consist mainly of bastnäsite. Other deposits of alkalic rocks and carbonatite are Kangankunde Hill, Malawi; Mrima, Kenya; Panda Hill, Tanzania; Glenover, South Africa; Araxá and Morro do Ferro, Brazil; Oka, Canada; and the Kola Peninsula, in the former USSR. Placers of monazite with xenotime are widely distributed in Australia, India, Brazil, and Malaysia. Others are located in the Malagasy Republic, Indonesia, India, China, and the United States. Rare earth minerals also occur as vein minerals. Notable deposits are found at Lemi

Pass, in Idaho and Montana; Wet Mountain, Colorado; Llallagua, Bolivia; Karonnge, Burundi; and Steenskampskraal, South Africa (Adams and Staatz 1973).

The Mountain Pass Deposits, the United States: The Mountain Pass district is in a block of metamorphic rocks of Precambrian age, 6.5 to 8.0 km wide and about 30 km long, in a north-northwest direction. This district is bounded on the east and south by the alluvium of Ivanpah Valley, and separated on the west from sedimentary and volcanic rocks of Paleozoic and Mesozoic age by the Clark Mountain normal fault. The northern boundary of the district is a transverse fault (Olson et al. 1954). The metamorphic complex comprises a variety of lithologic types including gneisses, schists, migmatites, pegmatites, and foliated mafic rocks. The RE-bearing carbonate rocks are related to the potash-rich igneous rocks that cut the metamorphic complex. The most abundant site is in and near the southwest side of a large shonkinite-syenite body (Fig. 71). One mass of carbonate rock near the Sulphide Queen mine is 210 meters maximum width and 730 meters long, and is of major economic interest. The mineral assemblages consist mainly of carbonates: calcite, dolomite, ankerite, siderite, and bastnäsite (a fluocarbonate), and the noncarbonates: barite, parisite, and quartz. Other minerals, including monazite, are found in variable small quantities. An estimate of the principal constituents in the carbonate rock of this district is 60% carbonate minerals, 20% barite, 10% RE-bearing fluocarbonates, and 10% quartz and other minerals. In some local concentrations of bastnäsite, the RE oxide content is as high as 40% of the carbonate rock (Olson et al. 1954; Castor 1991).

The Bayan Obo Deposits, China: Situated about 500 km northwest of Beijing in Inner Mongolia, the Bayan Obo deposits are geologically located on the northern margin of the Sino-Korean craton and in the transitional zone between the craton margin and the Mongolian Hercynian foldbelt. The basement rocks are the Wutai group of the lower Precambrian and are made up of schists and gneisses that are intruded by granite, pegmatite, and hornblendite (Lee et al. 1957; Lee 1970; Young et al. 1957; Drew et al. 1990). Overlying the basement unconformably is the mid-Proterozoic Bayan Obo group, which consists of quartzite in the lower part; calcareous quartzite, siliceous and clayey limestone, and dolomite in the middle part; and slate, biotite schist, and metavolcanics in the upper part. The carbonate rocks have a fine- to medium-grained texture, massive and banded structure, and consist mainly of dolomite and calcite. Accessory minerals are feldspar, quartz, hornblende, phlogopite, apatite, fluorite, and barite. Disseminated and aggregated bastnäsite, monazite, columbite, and pyrochlore are abundant. These minerals are very fine grained, 0.02 to 0.1 mm, are associated with magnetite and hematite, and are graded into massive iron ores. The ore bodies are in a syncline that extends 18 km along its axis and varies in width from 1 to 3 km. Within the syncline are two large ore deposits, the Main and East ore bodies, and the sixteen medium ore bodies comprising the West mine. The average ore grades are 6% RE oxide, 0.13% Nb, and 35% Fe (Yuan et al. 1992).

Placer Deposits, Australia: The bulk of the economic resources is located in two areas: Busselton-Bunbury and Eneabba, Western Australia (Fantel et al. 1986; Rattigan and Stitt 1990). As shown in Fig. 72, the placer deposits in the Busselton-Bunbury area

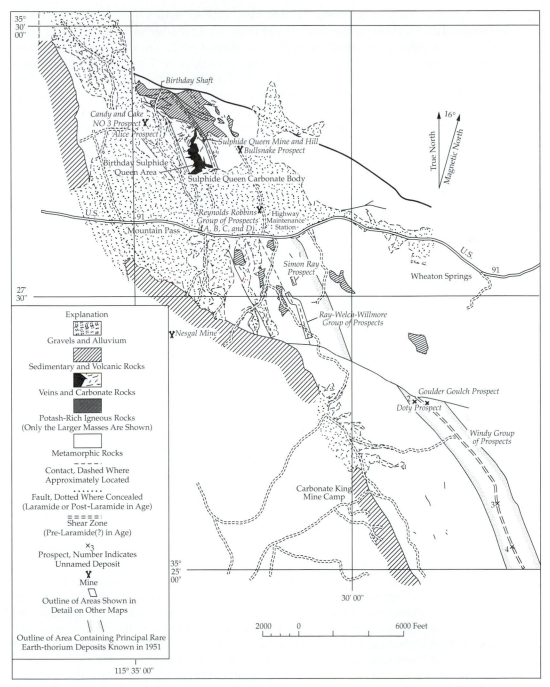

Figure 71 Distribution of veins and carbonates in the Mountain Pass District, after Olson et al. (1954).

are associated with three "fossil" strandlines, located within 15 km of the present shoreline. The Mesozoic sediments, derived from the Archaean hinterland, provided the source of this relatively unconsolidated strata (Baxter 1982). The deposits are lenticular and occur near the surface with an average thickness of 5 meters. Average heavy mineral grades vary from 12 to 20%. The suite contains 75 to 80% ilmenite, 5 to 10% zircon, 5% leucoxene, 0.5% rutile, and 0.4% monazite. The Eneabba deposits are associated with ancient shorelines 70 km inland, and 25 to 130 meters above the present sea level. The deposits have a high clay content and in some areas are interbedded with laterite. The heavy mineral grades have a narrower variation from 10 to 12%. The suite contains 50 to 60% ilmenite, 17 to 22% zircon, 7 to 12% rutile, 4.2% leucoxene, and 0.5% monazite (Lissiman and Oxenford 1975).

A model for heavy mineral deposition has been proposed by Hocking et al. (1982) on the basis of the activity of a barrier system during transgression and regression. During regression, the coastal plain becomes laden with overbank and alluvial fan sediments deposited by streams and rivers. In the transgression that follows, a sand wedge is transported across the coastal plain. Sand and heavy minerals are collected in the barrier systems, and the fine grained silts and clays are sorted toward the ocean basin. Permanent preservation of the deposit normally requires eustatic change shortly after the formation of the deposit.

INDUSTRIAL PROCESSES AND USES

1. Beneficiation of RE Minerals

The crude bastnäsite ore (approximately 9% RE oxides) at Mountain Pass, California, is first crushed in a primary jaw crusher in a series with a cone crusher, from which the ore is passed to a rod mill and a classifier. The slurry produced pours to three heating agitators where it is heated by stages to about 90°C. Following this step, the slurry is cooled down to 60°C and is pumped to rougher flotation, where an oleic-olein collector, Orzan A and silicic acid, are added to the slurry during the conditioning. The barite is depressed to produce a concentrate containing 63% RE oxide. Leaching the flotation concentrate with 10% HCl solution removes calcite and raises the concentrate to 72% RE oxides. Further beneficiation can be done by calcining the leached concentrate (Parker and Baroch 1971; Bautista 1995). At Baiyun Obo, China, the main process used for the flotation of bastnäsite in an alkaline medium includes the following steps: (1) using oxidized paraffin soap as a collector in the floating circuit of the flotation of RE minerals and their intergrowths of calcium and barium bearing minerals, (2) using hydroxamic acid type collectors to separate RE minerals from the bulk froth product, and (3) to separate bastnäsite in a weak acidic medium of pH 5 to 6. In the treatment of bulk froth, sodium carbonate is used as a regulator, water glass as an inhibitor of undesired minerals, and fluosilicate as an activator (Luo and Chen 1986; Meng 1986).

At the heavy mineral plants on the southwestern coast of Australia, the sand slurry is treated first by screening to remove oversize material. The black-sands are concentrated by a series of mechanical treatments, producing a bulk of black-sand concentrate consisting of all heavy minerals present in the sands. Jigs, sluice boxes, shaking tables, and Humphreys spiral concentrators are commonly employed to process this wet

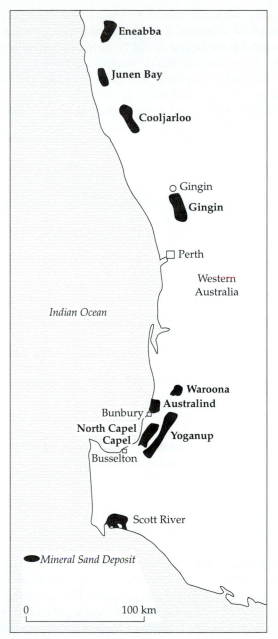

Figure 72 Mineral sand deposits along the west coast of Australia, after Towner (1986).

323

concentration. A low-intensity wet drum magnetic separator is used to remove magnetite and other highly magnetic minerals from the concentrate (Parker and Baroch 1971; Griffiths 1984). The concentrate is then dried, separated from tailings of low specific gravity, and treated by magnetic and electrostatic separators (Svoboda 1987). A high-tension roll separator is used to separate the nonconductors, mainly zircon and monazite, from the conductors, ilmenite and rutile. In an induced magnetic roll separator, the nonconductors are separated into magnetic products, mainly monazite and nonmagnetic zircon, while the conductive product is also separated into magnetic ilmenite and nonmagnetic rutile (MacDonald 1983).

2. Processing of RE Concentrates

A variety of rare earths and rare earth compounds are recovered from the processing of the concentrates, including pure rare earth metals and oxides and mixed rare earth metals and oxides (Parker and Baroch 1971). Both pure and mixed rare earths have wide industrial applications.

Bastnäsite: The 70% RE oxide concentrate, produced from bastnäsite by flotation and leaching with HCl, is heated to oxidize the Ce^{3+} to the Ce^{4+} state. After cooling, the concentrate is leached with HCl to dissolve the trivalent RE, including La, Pr, Nd, Sm, Eu, and Gd. The CeO_2 is thickened and filtered from the solution. The dissolved lanthanides are separated into two groups, La-Pr-Nd and Sm-Eu-Gd, by liquid-liquid solvent extraction. The La-Pr-Nd fraction is further processed for the individual RE oxides, La_2O_3, Pr_6O_{11}, and Nb_2O_3, by use of continuous solvent extraction processes. The Sm-Eu-Gd fraction is treated with a proprietary reducing agent to produce Eu^{2+}, which is through leaching with H_2SO_4, precipitated as $EuSO_4$ and oxidized to Eu_2O_3. The Sm and Gd are separated by a solvent extraction procedure. The four major products obtained from the treatment processing at Mountain Pass are (1) europium oxide, (2) a lanthanum-rich mixture of RE metals, (3) a Sm-Gd rich mixture, and (4) a technical grade CeO_2. A flow sheet illustrating the production of basträsite ore and concentrate from Mountain Pass is shown in Fig. 73 (McGill 1993).

Monazite: A caustic process using NaOH, and an acid process using H_2SO_4, are two common processes used for the treatment of monazite concentrates (Parker and Baroch 1971). A flow sheet illustrating the liquidified extraction is shown in Fig. 74 (McGill 1993). The finely ground monazite concentrate is mixed with a 70% NaOH solution and heated in an autoclave at 140° to 150°C for a period of several hours. Soluble Na_3PO_4 is recovered as a by-product from the insoluble rare earth hydroxide after the addition of H_2O. Two industrial processes are used to remove the Th (1) by dissolving the hydroxide in HCl or HNO_3, and selectively precipitating the $Th(OH)_4$ with NaOH, and (2) by adding HCl slowly to the hydroxide to pH 3 to 4, and removing soluble rare earth chloride from the insoluble $Th(OH)_4$. This separation of the rare earths can be made by either the liquid-liquid solvent extraction process (Fig. 74) or a combination of chemical and solvent extraction processes (Fig. 73). The types of extraction agents used are amines, carboxylic acids, acidic and neutral organophosphorus compounds, and many other agents. Xenotime is treated in the same manner as monazite. Procedures for the separation and

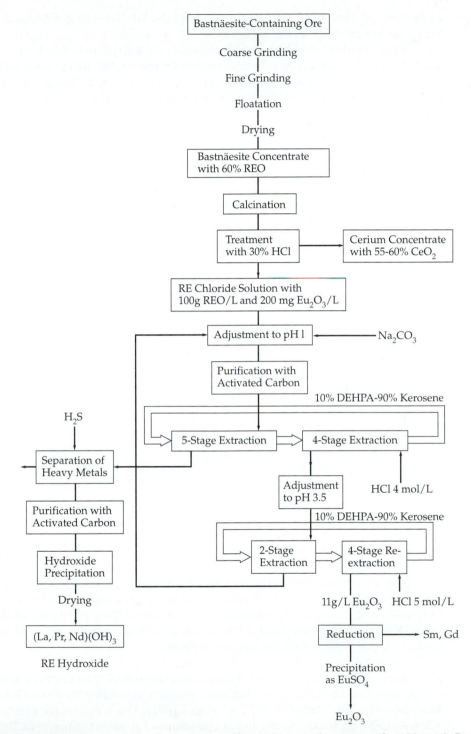

Bastnäesite-Containing Ore

Coarse Grinding

Fine Grinding

Floatation

Drying

Bastnäesite Concentrate
with 60% REO

Calcination

Treatment
with 30% HCl → Cerium Concentrate
with 55-60% CeO_2

RE Chloride Solution with
100g REO/L and 200 mg Eu_2O_3/L

Adjustment to pH 1 ← Na_2CO_3

Purification with
Activated Carbon

10% DEHPA-90% Kerosene

H_2S

5-Stage Extraction ⇒ 4-Stage Extraction

Separation of
Heavy Metals

Adjustment
to pH 3.5   HCl 4 mol/L

Purification with
Activated Carbon

10% DEHPA-90% Kerosene

Hydroxide
Precipitation

2-Stage
Extraction ⇒ 4-Stage Re-
extraction

Drying

11g/L Eu_2O_3 HCl 5 mol/L

(La, Pr, Nd)(OH)$_3$

Reduction → Sm, Gd

RE Hydroxide

Precipitation
as $EuSO_4$

Eu_2O_3

Figure 73 Flow sheet illustrating the production of bastnäsite ore and concentrate from Mountain Pass, after McGill (1993).

extraction of rare earth oxides from monazite at Yao-Lung Chemical plant in Shanghai are given by Bautista (1995).

Both anhydrous and hydrous RE chlorides are also prepared from the 70% oxide concentrate. The anhydrous chloride is made by mixing the concentrate with carbon, briquetting the mixture, and heating it to 1,000° to 1,200°C in a Cl_2 gas atmosphere. The hydrous chloride is prepared by reacting the concentrate, after leaching with HCl, with NaOH in the presence of HCl.

3. Industrial Uses

The rare earths are used in many industrial processes for a wide range of purposes, including in catalysts, ceramics, glass, metals, and phosphors (McGill 1993).

Catalysts: One of the largest single uses for mixed rare earths is to stabilize zeolite cracking catalysts made for the petroleum refining industry (Venuto and Habib 1979; Wallace 1981). The ability of catalysts to crack large molecular weight, high boiling point organic molecules into smaller molecular weight, lower boiling molecules depends to a large extent on their acidic properties. The acid site population and strength of zeolite are several orders of magnitude higher than the older amorphous silica-alumina gel catalysts. The zeolite group consists of a great number of minerals of hydrous aluminosilicates with one or more alkali or alkaline earth elements. The zeolites used in cracking catalysts are synthetic members of the faujasite group, including the X- and Y-types.

The fundamental building block of the X- and Y-type faujasite structures is a truncated octahedron that has hexagonal and square faces, which is generally referred to as a sodalite cage. The addition of rare earth to the zeolite stabilizes the structure and enables it to retain its inherent acidic properties. The thermal stability of zeolite is partially controlled by the exchange level of the rare earth, and zeolite reaches its maximum thermal stability at about 20 wt% rare earth oxide. At this level, a complete exchange between sodium of zeolite and rare earth is nearly obtained. The surface area (m^2/gm) increases from 100 at 5 wt% RE oxide to 700 at 20 wt% RE oxide. Experimental data indicate that 20 wt% RE oxide also produces a maximum hydrothermal stability and a maximum activity, or "microactivity," as used in the petroleum industry, of zeolite. The increase in activity implies a reduced amount of unconverted material recycled to the catalytic cracker.

Rare earths are also widely used as noncracking catalysts. In the synthesis of ammonia (Emmett 1975), a typical catalyst containing mainly iron oxide washed with $Ce(NO_3)_3$, and subsequently reduced, is more active than a conventional catalyst. Mixed RE salts may be used as a substitute for cerium salt. Oxides of Pr, Nd, and Tb exchange oxygen readily with the oxygen molecules of the gas and are active oxidation catalysts for hydrogen (Minachev 1973; McGill 1993).

Rare earths form alloys with transition metals at high temperatures, and these alloys, finely sized, absorb large amounts of hydrogen to produce hydrides (Kuijpers and Van Mal 1971; Van Mal et al. 1976; Morteani 1991). The reaction, for example, $LaNi_5 + 3H_2 \rightarrow LaNi_5H_6$, is exothermic and can be achieved at room temperature. Because hydrogen is chemically bonded in the hydride, its volumetric density is significantly greater

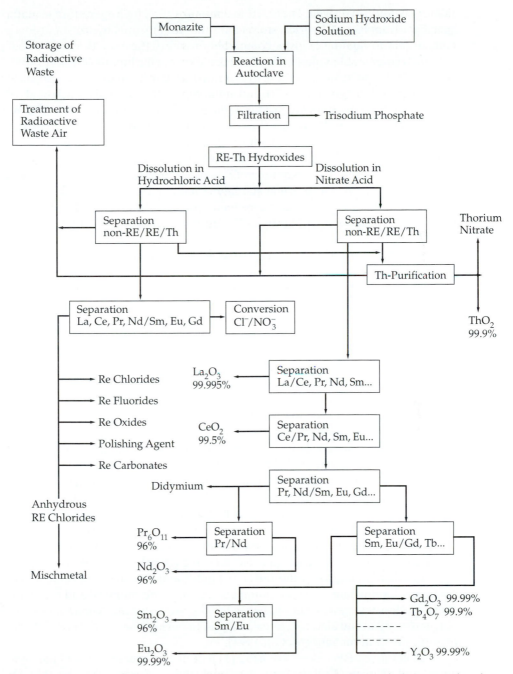

Figure 74 Flow sheet illustrating the liquid-liquid extraction process for rare earths from monazite, after McGill (1993).

than even liquid hydrogen. These RE hydrides are good hydrogen storage materials. The specific advantages of hydride storage are that it (1) stores hydrogen at low pressure, comparable to liquid, (2) gives comparable weight efficiency to compressed gas, and (3) offers comparable volumetric efficiency to cryogenic liquids (Gilbert 1985). In addition, catalytic applications of these hydrides are used in hydrogenation (Soga et al. 1977), methanation (Luengo et al. 1977), and ammonia synthesis (Schlapbach et al. 1979).

Catalytic converters are used to reduce NO_x to N_2, and to oxidize CO and hydrocarbon to CO_2 and H_2O, so that the concentrations of NO_x, CO, and hydrocarbon can reach legally acceptable levels in automobile emission control systems. The catalytic action of RE oxides is similar to that of the oxides of transition metals such as Cu_2O, MnO_2, and MoO_3 (Rosynek 1977). Addition of RE oxides to the catalysts of platinum metals improves their activity and thermal stability (Kummer 1980). Simple RE oxides, such as CeO_2 and La_2O_3, and perovskites-type oxides, such as $LaCoO_3$ and $La_xPb_{1-x}MnO_3$, have all been used (Pedersen and Libby 1972; Voorhoeve et al. 1973; Khittak and Wang 1984).

Glasses and Ceramics: Cerium oxide is used in glass polishing. The polishing compounds have CeO_2 contents varying from 50 to 90%, with the remainder being other light lanthanide oxides. Cerium compounds have replaced cheaper polishing compounds such as Fe_2O_3, SiO_2, and ZrO_2 because they are cleaner, faster, longer lasting, and give a superior finish (Duncan 1970). The mechanism for using CeO_2 in glass polishing, as suggested by Silvernail and Goetzinger (1971), is of a catalytic nature. In the presence of H_2O, a "CeO-Si" activated complex is formed, which breaks the O-Si-O bonds by hydrolysis. The complex then breaks apart, the hydrated silica is swept away from the glass surface, and the polishing repeats. Other uses of cerium oxide include the prevention of chemical decoloration and decoloration by radiation of glass, and the reduction of the UV transparency of glass containers. The decolorizing application depends on the Ce^{4+}/Ce^{3+} oxidation-reduction reaction to maintain iron in the trivalent state, since the ferrous Fe imparts a bluish color to the glass (Schutt and Barlow 1972; Scholze 1991).

RE oxides are used for coloring glasses (Riker 1981). Neodymium is most commonly used to achieve a delicate pink tint with a violet reflection. The color varies from a light pink in thin sections to a beautiful blue-violet in thicker pieces. Praeseodymium gives a green color, which is very similar to the eye, as chromium containing glasses. Erbium gives a pale pink to the glasses that cannot be obtained by any other means. Its use is limited to photochromic glasses and some crystal glasses (Weyl 1959).

Lanthanum improves the refractive index and dispersion in optical glasses used for lenses. Lanthanum and gadolinium also improve the chemical and mechanical properties of glass and enable these glasses to be used under the demanding conditions of magneto-optical and electro-optical systems. In pure glasses, Eu^{2+} and Ce^{4+} give a phototropic effect in sunlight (McGill 1993).

In ceramics, RE oxides are used (1) as yellow pigments (Pr), (2) as opacifiers in glazes and enamels (Ce), (3) for radiation-shielding ceramics (Sm, Eu, Gd, Dy), (4) to improve the light fastness of lead chromate and titanium dioxide pigments, and (5) as stabilizers for cubic ZrO_2 (La, Y) (McGill 1993). A partially stabilized ZrO_2 electrolyte, which uses yttrium oxides as the stabilizer, is widely used as an oxygen sensor in a closed loop emission control system for automotive applications.

RE compounds of the type $Pb_{1-x}La_x(Zr_{1-y}Ti_y)O_3$, known as PLZT, are used as electro-optical ceramics in displays, as electromechanical transducers, and for their piezoelectrical properties (McGill 1993). Compounds of the garnet-type such as $Y_3Al_5O_{12}$ (YAG), $Y_3Fe_5O_{12}$ (YIG), and $YAlO_3$ (YALO), doped with neodymium, are used as solid state lasers (Khittak and Wang 1984).

Phosphors: Stimulation by UV, X-rays, or electron beams causes certain rare earths to emit light of definite wavelengths. The energy is transferred from a host lattice to an activator, with the former determining the efficiency, and the latter determining the emission spectrum. The host lattices are oxides, sulfides, and vanadates of Y, La, Gd, and Lu, and borates, silicates, tungstates, and phosphates of the ruby-type (Stevels 1976). Phosphors containing rare earth oxides are mainly used in color CRT, with other uses including fluorescent lamps, X-ray intensifying screens, optical scanning devices, photocopying equipment, and scintillation crystals.

The red emitting phosphors in CRT are $YVO_4{:}Eu^{3+}$, $Y_2O_3{:}Eu^{3+}$, and $Y_2O_2S{:}Eu^{3+}$ (Levine and Palilla 1964; Wickersheim and Lefever 1964; McGill 1993). By comparison, both $Y_2O_3{:}Eu^{3+}$ and $Y_2O_2S{:}Eu^{3+}$ are brighter than vanadate because of their greater energy conversion efficiency and more orange emissions. The green emitting phosphors of rare earths are $La_2O_2S{:}Tb^{3+}$, $CaS{:}Ce^{3+}$, and $SrGa_2S_4{:}Eu^{2+}$ (Lehmann and Ryan 1971; Peters and Baglio 1972). None of the rare earth phosphors have significantly more favorable emission spectra than the currently most-used green phosphor $(Zn.Cd)S{:}(Cu,Al)$. Blue phosphors of rare earths are $Sr_5Cl(PO_4)_3{:}Eu^{2+}$ and $ZnS{:}Tm^{3+}$.

Metals: One of the most important metallurgical applications of rare earths is their ability to deoxidize and desulfurize (Kippehan and Gschneidner 1970). The rare earths used are in the form of mixed metals principally containing Ce, La, and Nd.

Rare earths are commonly used in alloying to improve mechanical and electrical properties. The addition of rare earth to copper and aluminum alloys, for example, greatly improves their mechanical properties for use as electrical conductors. Yttrium added to a magnesium-aluminum alloy improves tensile strength, heat resistance, and electrical conductivity (McGill 1993).

Rare earth alloys with 3d transition elements have magnetic properties (Gschneidner and Eyring 1988, 1989). Two families of magnet alloys have been developed on the basis of REA_5 and RE_2A_{17}, in which "RE" is mainly Sm or Nd, and "A" is Co or Co with Fe. These alloys give stable magnets, high remanence, and high coercive field strengths.

REFERENCES

Adams, J. W., and M. H. Staatz. 1973. Rare earth elements. In United States mineral resources. Eds. D. A. Brobst and W. P. Pratt. *U.S.G.S. Prof. Paper* 830:547–556.

Bautista, G. 1995. Separation chemistry. In Vol. 21 of *Handbook on the physics and chemistry of rare earths.* Eds. K. A. Gschneidner and L. Eyring, 1–27. Amsterdam: Elsevier.

Baxter, J. L. 1982. History of mineral sands mining in Western Australia. In *Reference paper: Exploitation of mineral sands,* 65–70. West. Aust. Inst. Tech. Perth, Australia.

Bayliss, P., and A. A. Levinson. 1988. A system of nomenclature for rare earth mineral species; revision and extension. *Amer. Mineral.* 73:422–423.

Beaudry, B. J., and K. A. Gschneidner. 1991. Preparation and basic properties of the rare earth metals. In Vol. 1 of *Handbook on the physics and chemistry of rare earths.* Eds. K. A. Gschneidner and L. Eyring, 173–232. Amsterdam: Elsevier.

Castor, S. B. 1991. Rare earth deposits in the southern Great Basin. In *Geology and ore deposits of the Southern Great Basin,* 523–528. Reno, Nevada: Geol. Soc. Nevada.

Drew, L. J., Q. Meng, and W. Sun. 1990. The Bayan Oba iron-rare earth-niobium deposits, Inner Mongolia, China. *Lithos* 26:43–65.

Duncan, L. K. 1970. Cerium oxide for glass polishing. *Glass Ind.* 41:387–393.

Dye, J. L., and F. H. Spedding. 1954. The application of Onsager's theory of conductance to the conductance and transference numbers of unsymmetrical electrolytes. *Jour. Amer. Chem. Soc.* 76:888–892.

Emmett, P. H. 1975. Fifty years of progress in the study of the catalytic synthesis of ammonia. In *Physical basis for heterogeneous catalysis.* Eds. E. Drauglis and R. I. Jaffee. New York: Plenum Press.

Fantel, R. J., D. B. Buckingham, and D. F. Sullivan. 1986. Titanium minerals availability—market economic countries—a mineral availability appraisal. *U. S. B. M. Infor. Circ.* 9061, 48p.

Gilbert, M. 1985. Storage and compression of hydrogen using lanthanum alloys. In Vol. II of New frontiers in rare earth science and applications. *Proc. Intern. Conf. on Rare Earth Development and Applications, Beijing, China, Sept., 10–14, 1985.* Eds. G. Xu and J. Xiao, 1049–1057. New York: Academic Press.

Griffiths, S. J. 1984. Review of placer deposits in Australia. In *Short course—placer exploration and mining.* Eds. C. McKay and P. Mouscet-Jones, 265–307. Reno: Univ. Nevada Div. Continuing Edu.

Gschneidner, K. A. 1980. Rare earth speciality inorganic chemicals. In *Speciality inorganic chemical.* Ed. R. Thompson, 403–443. London: Royal Soc. Chem.

Gschneidner, K. A., and L. Eyring. 1988. Vol. 11 of *Handbook on the physics and chemistry of rare earths,* 293–321. Amsterdam: North Holland Publ.

———. 1989. Vol. 12 of *Handbook on the physics and chemistry of rare earths,* 133–212. Amsterdam: North Holland Publ.

Guth, E. D., and L. Eyring. 1954. Dissociation pressure measurements: X-ray and differential thermal analysis. *Jour. Amer. Chem. Soc.* 76:5242–5244.

Henderson, P. 1984. General geochemical properties and abundances of the rare earth elements. In Rare Earth Geochemistry, Ed. P. Henderson, *Developments in Geochemistry* 2, 1–32. Amsterdam: Elsevier.

Hocking, R. M., J. K. Warren, and J. L. Baxter. 1982. Shoreline geomorphology. In *Reference papers: Exploitation of mineral sands,* 81–91. West Aust. Inst. Tech. Perth, Australia.

Khittak, C. P., and F. F. Y. Wang. 1984. Perovskites and garnets. In Vol. 3 of *Handbook on the physics and chemistry of rare earths.* Eds. K. A. Gschneidner and L. Eyring, 525–608. Amsterdam: Elsevier.

Kippehan, N., and K. A. Gschneidner. 1970. Rare earth metals. In *Steels,* IS-RIC-4, Rare Earth Inform. Center, Iowa State Univ., Ames, Iowa.

Kristanovic, I. 1965. Redetermination of oxygen parameters in xenotime, YPO_4. *Zeit. Krist.* 121:315–316.

Kuijpers, F. A., and H. H. Van Mal. 1971. Sorption hysteresis in the $LaNi_5$-H and $SmCo_5$-H systems. *Jour. Less-Common Metals* 23:395–398.

Kummer, J. T. 1980. Catalysts for automobile emission control. *Prog. Energy Combust. Sci.* 6:177–199.

Lee, K. Y. 1970. Some rare earth element mineral deposits in mainland China. *U.S.G.S. Bull.,* 1312N, 34p.

Lee, Y. Y., L. S. Wang, and Y. C. Ho. 1957. Precambrian stratigraphy of southwestern Inner Mongolia. *Acta Geol. Sinica* 37:241–268.

Lehmann, W., and F. M. Ryan. 1971. Fast cathodoluminescent calcium sulfide phosphor. *Jour. Electrochem. Soc.* 119:275–277.

Levine, A. K., and F. C. Palilla. 1964. New, highly efficient red-emitting cathodoluminescent phosphors, YVO_4:Eu for color television. *Appl. Phys. Lett.* 5:118–120.

Levinson, A. A. 1966. A system of nomenclature for rare earth minerals. *Amer. Mineral.* 51:152–158.

Lissiman, J. C., and R. J. Oxenford. 1975. Eneabba rutile-zircon-ilmenite sand deposit, W.A. In Vol. 1 of Economic geology of Australia and Papua New Guinea. Ed. C. L. Knight. *Metals, Monograph Series* 5: 1062–1070.

Luengo, C. A., A. L. Cabrera, H. B. Mackay, and M. B. Maple. 1977. Catalysis of carbon monoxide and carbon dioxide methanation by cerium-aluminum ($CeAl_2$), cerium cobalt ($CeCo_2$), cerium nickel ($CeNi_2$), cobalt, and nickel. *Jour. Cataly.* 47:1–10.

Luo, J. K., and X. Y. Chen. 1986. Selective flotation of rare earth minerals from fluorite, barite and calcite. In New frontiers in rare earth science and applications. In Vol. 1, *Proc. Intern. Conf. on Rare Earth Development and Applications, Beijing, China, Sept. 10–14, 1985.* Eds. G. X. Xu and J. M. Xiao, 67–70. New York: Academic Press.

MacDonald, E. H. 1983. *The geology, technology and economics of placers.* London: Chapman and Hall, 500p.

Mariano, A. N. 1989. Economic geology of rare earth minerals, in Geochemistry and mineralogy of rare earth elements. Vol. 21 of *Rev. in Mineralogy.* Eds. B. R. Lipin and G. A. McKay, 309–337. Washington, D.C.: Min. Soc. America.

McGill, I. 1993. Rare earth elements. In Vol. A22 of *Ullmann's encyclopedia of industrial chemistry,* 607–647. Weimhein: VCH.

Meng, Y. 1986. Flotation of rare earth minerals from Baiyun-Obo complex iron ore. In New frontiers in rare earth science and applications. In Vol. 1, *Proc. Intern. Conf. on Rare Earth Development and Applications, Beijing, China, Sept., 10–14, 1985.* Eds. G. X. Xu and J. M. Xiao, 79–83. New York: Academic Press.

Minachev, Kh. M. 1973. *Catalytic activity of rare earth oxides,* Proc. 5th Intern. (1972) Congr, on Catalysis. Ed. J. W. Hightower, 219–223. New York: Elsevier.

Miyawaki, R., and I. Nakai. 1996. Crystal chemical aspects of rare earth minerals. In *Rare earth minerals, chemistry, origin and ore deposits.* Eds. A. P. Jones, F. Wall, and C. T. Williams, 21–40. London: Chapman & Hall.

Möller, T. 1961. Chemistry of rare earths. In *The rare earths.* Eds. F. H. Spedding and A. H. Daane, 9–28. New York: John Wiley & Sons.

Morteani, G. 1991. The rare earths: Their minerals, production and technical use. *European Jour. Mineral.* 3:641–650.

Ni, Y., J. M. Hughes, and A. N. Mariano. 1995. The atomic arrangement of bastnäsite-(Ce), $Ce(CO_3)F$, and structural elements of synchysite-(CE), röntgenite-(Ce), and parisite-(Ce). *Amer. Mineral.* 78:415–418.

———— 1995. Crystal chemistry of the monazite and xenotime structures. *Amer. Mineral.* 80:21–26.

Olson, J. C., D. R. Shawe, L. C. Pray, and W. N. Sharp. 1954. Rare-earth mineral deposits of the Mountain Pass District, San Bernardino County, California. *U.S.G.S. Prof. Paper,* 261, 75p.

Parker, J. G., and C. T. Baroch. 1971. The rare earth elements yttrium, and thorium. *U.S.B.M. Inform. Circ.,* 8476, 92p.

Pedersen, L. A., and W. F. Libby. 1972. Unseparated rare earth cobalt oxides as auto exhaust catalysts. *Science* 176:1355–1356.

Peters, T. E., and J. A. Baglio. 1972. Luminescence and structural properties of thiogallate phosphors. *Jour. Electrochem. Soc.* 119:230–236.

Rattigan, J. H., and P. H. Stitt. 1990. Heavy mineral sands. In Geological aspects of the discovery of some important mineral deposits in Australia. Eds. K. R. Glasson and J. H. Rittigan. *Australasian Inst. Mining & Metall., Monograph* 17:369–378.

Riker, L. W. 1981. The use of rare earths in glass compositions. In Industrial applications of rare earth elements. Ed. K. A. Gschneidner. *ACS Sym. Series* 164:81–94, Washington, D.C.: Amer. Chem. Soc.

Rosynek, M. P. 1977. Catalytic properties of rare earth oxides. *Catalytic Rev. Sci. Eng.* 16:111–154.

Schlapbach, L., A. Seiler, and F. Stucki. 1979. A new mechanism for lengthening the lifetime of hydrogenation catalysts. *Mat. Sci. Res. Bull.* 14:785–790.

Scholze, H. 1991. *Glass—nature, structure, and properties.* New York: Springer-Verlag, 453p.

Schutt, T. C., and G. Barlow. 1972. Practical aspects of cerium decoloration of glass. *Bull. Amer. Ceramic Soc.* 51:155–157.

Silvernail, W. L., and N. J. Goetzinger. 1971. The mechanism of glass polishing. *Glass Ind.* 52:130–152.

Soga, K., H. Imamura, and S. Ikeda. 1977. Hydrogenation of ethylene over lanthanium-nickel alloy (LaNi$_5$). *Jour. Phys. Chem.* 81:1762–1766.

Spedding, F. H., and S. Jaffe. 1954a. Conductance, solubilities and ionization constants of some rare earth sulfates in aqueous solutions at 25°C. *Jour. Amer. Chem. Soc.* 76:882–884.

——— 1954b. Conductance, transference and activity coefficients of rare earth perchlorates andnitrates at 25°C. *Jour. Amer. Chem. Soc.* 76:884–888.

Stevels, A. L. N. 1976. Recent developments in the application of phosphors. *Jour. Luminescence* 12/13:97–107.

Stubblefield, C. T., and L. Eyring. 1955. On reaction of europium dichloride with solution of hydrochloric acid. *Jour. Amer. Chem. Soc.* 77:3004–3006.

Svoboda, J. 1987. *Magnetic methods for the treatment of minerals.* Amsterdam: Elsevier, 692p.

Towner, R. B. 1986. Resources of minerals sands—what of their future. In Australia: A world source of ilmenite, rutile, monazite and zircon, 35–50. *Sym. Series, Australasian Inst. Mining & Metall.* (Sept-Oct.) Perth, Australia.

Ueda, T. 1967. Reexamination of crystal structure of monazite. *Jour. Japan. Assoc. Mineral. Petrol. Econ. Geol.* 58:170–179.

Van Mal, H. H., K. H. J. Buschow, and A. R. Miedema. 1976. Hydrogen absorption of rare earth (3rd) transition intermetallic compounds. *Jour. Less-Common Metals* 49:473–475.

Venuto, P. B., and E. T. Habib, Jr. 1979. *Chemical Industries,* v. 1: *Fluid catalytic cracking with zeolite catalysts.* New York: Marcel Dekker, 156p.

Voorhoeve, R. J. H., J. P. Remeika, and D. W. Johnson, Jr. 1973. Rare earth manganites: Catalysts with ammonia yield in the reduction of nitrogen oxide. *Science* 180:62–64.

Wallace, D. N. 1981. The use of rare earth elements in zeolite cracking catalysts. In Industrial applications of rare earth elements. Ed. K. A. Gschneidner, Jr., *ACS Sym Series* 164, 101–116. Washington, D.C.: Amer. Chem. Soc.

Weyl, W. A. 1959. *Coloured glasses.* London: Dawson's Pall Mall, 218-234 and 439-514.

Wickersheim, K. A., and R. A. Lefever. 1964. Luminescent behavior of the rare earths in yttrium oxide and related hosts. *Jour. Electorchem. Soc.* 111:47–51.

Yang, Z., J. Zhang, and Z. Pang. 1992. A study on electron diffraction of bastnäsite. *Scientia Geologica Sinica* n.1:20–25.

Young, C., T. W. Lu, P. K. Liu, and K. C. Hsieh. 1957. Geology of Pai-yun-o-po. *Scientia* 18:572–573.

Yuan, Z., G. Bai, C. Wu, Z. Zhang, and G. Ting. 1992. Geological features and genesis of the Bayen Obo REE ore deposit, Lunar Mongolia, China. *Applied Geochemistry* 7:429–442.

Salt

Salt or halite (NaCl) is one of the most widely distributed minerals. It is found in solution or in rock formation. Deposits of salt have occurred in sedimentary rocks of nearly all geologic ages since the early Cambrian, and salt deposition is still going on at the present time.

Salt is mainly used in chemical manufacturing, consuming about 70% of annual production, which was estimated to be 190 M tons in 1991 (*U.S.B.M. Annual Review*). It has been estimated that there are more than a thousand industrial uses for salt.

The United States, Germany, the former USSR, and China are the principal producers. England, France, Canada, Australia, India, and Romania each had an annual production exceeding 5 M tons in 1991.

MINERALOGICAL PROPERTIES

Halite is generally very pure in its natural occurrence, but in many marine deposits small amounts of admixed clay minerals, iron oxides, gypsum, and other evaporates are present. There does not appear to be much replacement of Na by K, although sylvite (KCl) is isomorphous with halite. Fluid inclusions are relatively common in halite: the trapped liquid phase being brine, the gas phase CO_2, and the solid phases mainly sylvite, anhydrite, or magnesium salts (Chang et al. 1996).

The crystal structure of halite is a simple one, consisting of two penetrating face-centered cubic lattices, one of Na ions and the other of Cl ions. Each Na ion is coordinated by six Cl ions, and each Cl ion by six Na ions, at a distance of a/2 where a = 5.640Å (Chang et al. 1996). Heating of halite crystals at 720°C to 760°C for several hours develops bright lines on their surfaces, which conform with the symmetry of the crystals, and makes rough surfaces smooth. Halite is quite brittle, yet under both compressive and confining pressures, deformation of halite occurs without the development of cracks or losses, due in part to the occurrence of translation gliding. This plastic deformation occurs with greater ease in saturated salt solutions or at moderate temperatures (Ramsdell 1960). Pure halite is colorless and highly transparent to light of wavelengths between the ultraviolet and infrared. The presence of impurities causes coloration. The color gray is due to the presence of clay minerals, and the colors red or brown are due to the presence of iron oxides and hyroxides. The formation of color-center by radiation gives halite blue, violet, and yellow colorations. The refractive index of halite is fairly constant at 1.5443 (Na light) (Chang et al. 1996).

Halite crystallizes from aqueous solutions in well-formed cubes (100), without other modifying faces. In the presence of impurities, octahedra or dodecahedra are sometimes formed. Crystallization from solutions saturated with HCl gives elongated needle-shaped or fibrous crystals. Skeletal crystals, either octahedra or combinations of cubes and octahedra, were reported from sodium chloride solution mixed with alcohol. Crystallization, under the influence of surface tension, often produces small cubes grown together into funnel-shaped, hollow, square-based pyramids, known as hopper crystals (Ramsdell 1960).

The important physical properties of halite are (1) melting point: 800.8°C, (2) density: 2.1680 gm/cm^3 at 0°C and 2.1619 gm/cm^3 at 25°C, (3) hardness (Moh's): 2.0 to 2.5, (4) dielectric constant: 5.90, (5) electrical resistivity: 4.6×10^{16} Ω · cm at 20°C and 1.38×10^{13} Ω · cm at 100°C, (6) thermal conductivity: 0.0610 W/cm · K at O°C, and (7) linear coefficient of expansion: 40.5 μm/m · K. The physical properties of molten NaCl are (1) boiling point 1,465°C at 760 mm, (2) density at 830°C: 1.53 gm/cm^3, (3) electrical conductivity at 1,000°C: 4.17/Ω · cm, (4) surface tension at 1,080°C: 94.0 dyne/cm, and (5) viscosity at 997°C: 0.00708 poises (Kaufmann 1960a).

Halite is readily soluble in water. Solubilities at 0°, 20°, 40°, 60°, 80°, and 100°C are 35.676, 35.92, 36.46, 37.16, 37.99, and 39.12 gm(NaCl)/100 gm(H_2O), respectively. Corresponding densities of the solutions are 1.2093, 1.1999, 1.1914, 1.1830, 1.1745, and 1.1660 gm/cm^3. In the system NaCl-H_2O, only one hydrate, NaCl · $2H_2O$ (hydrohalite), exists. It melts incongruently at 0.15°C to halite and a saturated solution of 26.285% NaCl, and has a eutectic point with H_2O at −21.12°C and 23.31% NaCl. Hydrohalite crystallizes so slowly that, on rapid cooling, the metastable eutectic point reached is 1.3 K (calculated). Its ability to depress the freezing point of water enables sodium chloride to be used in freezing mixtures and as a deicing salt. Pressure has a distinct effect on the melting of hydrohalite. The melting point, 0.15°C at atmospheric pressure, increases to 25.8°C at 9,500 bars and then decreases slightly as the pressure continues to increase. The volume increase on the melting of NaCl·$2H_2O$ to NaCl+saturated solution at 0.15°C is 0.037 cm^3/gm at atmospheric pressure. Increasing pressure lowers this value. Above 9,500 bars, the volume change is negative (Kaufmann 1960b).

GEOLOGICAL OCCURRENCE AND DISTRIBUTION OF DEPOSITS

Geologically, salt is found in solution or in rock formations. In solution, salt occurs in oceans, lakes, and groundwaters. The salt content of seawater is, on the average, 3.5%. This varies slightly with seawater's position on the earth's surface and with its depth. It is quite constant, however, except when diluted by an influx of fresh river water. The figure usually quoted for total oceanic salts is 20 million cubic kilometers, which is 50% greater than the total volume for the entire North American continent above sea level (Landes 1960). Halite begins to crystallize when seawater has been reduced to about 10% of its original volume through evaporation. Halite appears after $CaCO_3$ and $CaSO_4·2H_2O$, but before the very soluble sulfates and chlorides of magnesium and potassium (Chang et al. 1996). Salt crystallizes at the present in areas along the coasts where climate and topographic conditions are ideal for solar evaporation. Marginal salt pans and marine salinas are found along the coast in arid regions. Analyses indicate that NaCl represents 77.6% of the salts formed by evaporation during the present day. Detailed comparisons with similar evaporite deposits in the Stassfürt Permo-Triassic deposits of Germany and the Permian salt deposits of Texas (the United States) reveal no significant variation in this percentage (Chang et al. 1996).

Saline lakes are the result of a climate change that greatly increased the rate of evaporation, causing the lake levels to sink below the elevation of the outlets. The presence of salt within the rocks that lie at or near the surface of the watershed may also cause saline lakes to be formed (Lefond and Jacoby, 1983). Clarke's 1924 classification of saline lakes has withstood the test of time. These classifications are (1) sodium chloride lakes, (2) natural bittern lakes, (3) sulfate water lakes, (4) sulfate-chloride lakes, and (5) alkaline lakes, carbonate lakes, carbonate-chloride lakes, and chloride-sulfate-carbonate lakes (Clarke 1924). Ground waters are classified as either connate water, which is water entrapped in rocks during deposition, or meteoric water, which is fresh water initially, but which increases its soluble material content as it travels through soil and rocks. Therefore the composition of ground water is a function of its geological environment and varies in its degree of salinity from almost pure water to saturated solutions. Clarke (1924) classified ground waters into the following groups: chloride; sulfate; carbonate; mixed (chloride-sulfate; chloride-carbonate; sulfate-carbonate; or chloride-sulfate-carbonate); siliceous; borate; nitrate; phosphate; and acid. The most abundant dissolved solid in ground water is sodium chloride.

In rock formation, halite occurs in playa deposits, as bedded deposits, and as salt domes (Landes 1960). Playa salt deposits, which resulted from the evaporation of mineralized lakes, are fairly common in arid regions. The mineralogy of playa salt deposits varies from lake to lake and is dependent upon the rocks in the area surrounding the playa basin.

Bedded salt deposits are true sedimentary rocks, the same as shale, sandstone, limestone, and coal. They are characterized by chemical precipitation other than physical deposition. Theories of the origin of bedded salt deposits fall into two general types: (1) enrichment of salt content in evaporating interior seas of the terrestrial type, and (2) concentration of ocean water in cut-off seas of the marine type.

Under pressure, salt flows plastically. A salt bed can bulge the overlying sediments to produce a salt anticline, such as those that occur in the Utah-Colorado salt basin and in Europe, or rupture the overlying sediments to produce a salt dome, such as the Gulf Coast Salt Dome and the Zechstein Basin of Germany.

Salt deposits and solar salt pans are distributed in nearly every country of the world (Benbow 1990). With a total production of 190 M tons in 1991, the distributions were as follows: the United States (18.8%), China (12.6%), Germany (7.8%), the former USSR (7.3%), Canada (6.2%), India (5.0%), Australia (4.1%), Mexico (3.9%), England (3.6%), France (3.4%), Brazil (2.5%), Italy (2.1%), Poland (2.0%), the Netherlands (1.8%), Spain (1.8%), and Romania (1.7%).

The United States: The United States has the largest salt deposits in the world. They are located in the Silurian Salina Basin in the northeast, in the Permian-Jurassic Gulf Coast Basin, in the southcentral Permian Basin, and in the Williston Basin in North Dakota, South Dakota, and Montana (Lefond 1969). Of these, the most spectacular salt deposits are the salt domes of the Gulf Coast Basin. More than 300 salt domes, with diameters ranging from 1.5 to 6.5 km, are distributed randomly throughout the Basin. The depth of the domes varies from less than 1,220 meters to more than 3,050 meters. Salt in the salt domes is almost pure sodium chloride, with anhydrite in black bands as the principal impurity (Lefond 1969).

China: China has many widespread and diverse sources of salt. Principal producing regions are Hupeh, Honan, Szechuan, and Yunnan, from brine wells and springs, and the coastal provinces from Liaoning to Guangtung, from seawater by solar evaporation (Lefond 1969).

Germany: The salt resources of Germany are extensive. Oligocene salt deposits exist in the upper Rhine River region. Mesozoic salt deposits include the Jurassic Munder Marl, Triassic Keuper salt-gypsum beds, middle Muschelkalk salt, upper Buntsandstein and Röt salt, and the Werfener Beds. Permian age salt deposits include both Zechstein and Rotlegendes. The Zechstein evaporite sequence is known for the potassium salts found at Stassfürt (Lefond 1969).

The Former USSR: The former USSR contains some of the world's largest salt deposits and sodium chloride brines. The age of the salt deposits varies from Lower Cambrian to Neogene, with the greatest concentration in the Permian. Major deposits are associated with Cambrian rocks of the Angara-Lena-Saian Depression and in the Berezovaia Basin, the Devonian rocks of the Pripiat Depression and the Dnieper-Donets Basin, the Permian beds around Solikamsk, Berezniki, and Sol'Iletsk in the Kama Basin, and the Miocene beds in the Ukraine (Lefond 1969).

INDUSTRIAL PROCESSES AND USES

1. Industrial Processes

Mining: Salt can be mined by three different methods. First, the mining of salt might follow the general principles of other mineral mining and use the room-and-pillar method, also known as dry mining (Von Perbandt 1960). The second, solution mining, is the most modern and economical method of extracting salt from underground deposits. In order for this process to be used, the salt deposits must have an adequate thickness and extent, must be highly pure, and must not contain KCl. As shown in Fig. 75 (Geyer 1993), the design and installation of the brine wells include an outer casing that is cemented to the wall of the borehole and two inner concentric pipes that are inserted into the casing. During operation, fresh water is pumped into the borehole through the annular space between the inner pipes and dissolves salt from the walls of the cavern. As the salt content increases, the fresh water becomes a brine and has a greater specific gravity. It sinks to the bottom, and then rises to the surface in the central pipe due to applied pressure.

The third method used to mine salt is solar evaporation of brine from seawater, salt lakes, and salt springs (Lefond and Jacoby 1983). In operation, salt water is pumped into large concentration ponds. As the NaCl-content of the salt water increases to about 20°Be (20% dissolved solids), it flows by gravity into the "lime" pond, where calcium sulfate ($CaSO_4 \cdot 2H_2O$) and calcium carbonate ($CaCO_3$) precipitate out of the solution. When the salinity reaches 26 to 26.5°Be in the lime pond, salt water is pumped into the crystallizer or harvesting ponds. Pure halite is precipitated when salinity ranges from 26.5 to 30°Be. A layout of a modern salt field in shown in Fig. 76 (Geyer 1993).

To produce solar salt economically, the evaporation of water must exceed precipitation for a long and continuous period. An ideal climate for this can be found in western Australia, where annual evaporation is 3,600 mm and annual rainfall is only 360 mm.

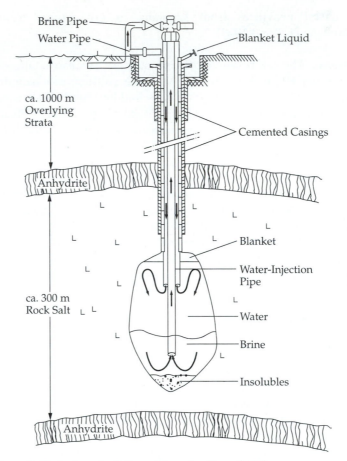

Figure 75   A schematic illustration of solution mining, after Geyer (1993).

Sea salt can be produced all year round. In addition to a favorable climate, the ground for an evaporation pond must be as impermeable and as level as possible. An ideal topographic feature of the ground is for it to be stepped in elevation, which reduces the number of intermediate pumping stations required. In the United States, solar salt production from seawater is restricted to California. On average, five years is required to harvest the salt. Inland solar production is done at the Great Salt Lake in Utah (Kostick 1994).

Milling of Rock Salt: Dry mined rock salt is generally milled by crushing and screening. The crushing is best accomplished in stages in order to hold the amount of −10 to −12 mesh fraction to a minimum. In a three-stage operation, the primary, secondary, and tertiary crushers should be adjusted to pass 22.5 cm (9 in.), 7.5 cm (3 in.), and 2.5 cm (1 in.) materials. A single-roll, spiked-tooth crusher is commonly used in all stages of crushing, and for screening, either vibratory or reciprocating screeners are preferred. Based upon screen sizes, three types of rock salt are normally produced in

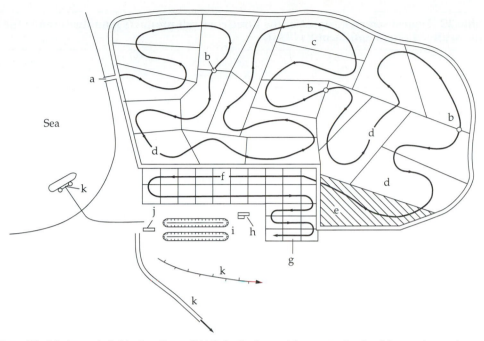

Figure 76   Modern salt field, after Geyer (1993). In the figure, (a) seawater intake; (b) pumping station; (c) dikes; (d) condensers; (e) brine storage tanks; (f) crystallizers; (g) bitterns area; (h) washing plants; (i) salt storage areas; (j) packaging plant; (k) dispatch.

the United States. They are (1) fine crystal (FC), (2) coarse crystal (CC), and (3) grade No. 1. Typical screen analyses are tabulated in Table 22. The NaCl content is 98.193% in FC, 98.272% in CC, and 97.710% in No. 1 rock salts (Lefond and Jacoby 1983).

Processing of Solar Salt: Solar salt is generally processed by feeding the salt into a vibrating feeder or screw conveyor, where washing is done with saturated salt brine. The washed salt is brought to a classifier and then is dewatered and dried before being conveyed to a storage site (Kostick 1994).

Brine Purification: The most common impurities of crude brine are calcium, magnesium, and sulfate ions. They are commonly removed by chemical treatment, evaporation, or recrystallization, as well as other methods (Kaufmann 1960c; Geyer 1993):

In chemical treatment, magnesium ions are precipitated from the crude brine as magnesium hydroxide by adding calcium or sodium hydroxide according to the reactions: $MgCl_2 + 2Na(OH)$ or $Ca(OH)_2 \rightarrow Mg(OH)_2 + 2NaCl$ or $CaCl_2$. Ferric iron can be removed by this process, if it is present in the brine. Calcium ions are precipitated as calcium carbonate by adding soda ash: $CaSO_4 + Na_2CO_3 \rightarrow CaCO_3 + Na_2SO_4$ and $CaCl_2 + Na_2CO_3 \rightarrow CaCO_3 + 2NaCl$, or by pumping the brine with carbon dioxide. Salt is obtained from the purified brine by evaporation, which is stopped prior to the beginning of sodium sulfate crystallization.

The solubility of sodium chloride increases only slightly with temperature, and it can be crystallized out in an evaporative crystallizer at 50° to 150°C. The most common

Table 22 Typical screen analyses of the three rock salt products manufactured in the United States, after Lefond and Jacoby (1983).

Sieve No.	Cumulative % Retained on	Component % of Aggregate
Screen Analysis: Fine Crystal (FC)		
4	0.0	0.0
8	0.0	0.0
12	12.2	12.2
20	52.9	40.7
40	72.1	19.2
70	87.1	15.0
80	89.3	2.2
Pass		
Screen Analysis: Coarse Crystal (CC)		
3/8 in.	0.0	0.0
4	24.0	24.0
8	82.0	58.0
10	89.7	7.7
12	94.7	5.0
16	98.2	1.8
Pass		
Screen Analysis: No. 1		
3/4 in.	0.0	0.0
3/8 in.	5.0	5.0
4	90.3	85.3
8	98.7	8.4
Pass		1.3

practice used currently is the multistage evaporation process. In this process, the vapors from the boiling brine in the first stage are used to boil the brine in the second stage, which is at a lower pressure than the first. The vapors from each succeeding stage are circulated through the following stage, which is in turn at a lower pressure and enables it to boil at a correspondingly lower temperature. Three general types of crystallizers are used. Type 1 is an evaporator with forced circulation and external heating, Type 2 is an evaporator with internal heaters and a circulating pump in the central pipe that produces forced circulation, and Type 3, known as the Oslo crystallizer, in which the recirculating brine and fresh brine are heated in a heat exchanger and evaporated in an evaporator. The supersaturated crystal-free brine passes down the central pipe into the crystallizer, and from there passes upward through a bed of crystals. Type 1 crystallizers generally produce cubic crystals with 50% greater than 400 μm by sieve analysis. Type 2 crystallizers produce cubic crystals with rounded corners with 50% greater than 650 μm. Type 3 crystallizers produce a granular product (Geyer 1993).

The recrystallization process is based on the fact that the solubility of calcium sulfate in brine decreases with increasing temperature, while that of sodium chloride increases slightly. The process consists of three stages (Besticker 1963): (1) steam is introduced into a dissolving vessel, where fine salt and mother liquor are heated to about

105°C. Sodium chloride dissolves, but calcium sulfate remains in the mother liquor as a solid; (2) calcium sulfate is removed from the brine by filtration; and (3) salt is precipitated during cooling when the brine is pumped to an expansion evaporator.

2. Industrial Uses

Salt is mainly used in chemical manufacturing, which consumes more than two-thirds of annual production. The principal applications are chloralkali electrolysis, which is used to produce chlorine and caustic soda, and the ammonia-soda process, which is used to produce soda ash (see p. 390). Sodium, hydrogen, sodium sulfate, and hydrochloric acid are other chemicals that are derived from salt. They are all major industrial chemicals, and each has numerous applications, as summarized by Lefond and Jacoby (1983).

Salt is used as a nutrient (table salt, salted food, canning, baking), as a preservative (meat and fish curing, sausage casings, vegetable salting), and in food processing (blanching seafood and vegetables, chicken deboning, crabmeat pickling, oyster shucking, yeast processing) (Lefond and Jacoby 1983; Kostick 1994).

In metallurgical processing, salt is used as a flux, molten metal cover, metal refining, sink and float baths, heat treating baths, and for chloride roasting.

Two other major uses of salt are (1) to deice road surfaces to give safe driving conditions in winter, and (2) the regeneration of water softening in the treatment of hard water. To this date, there is no alternative to the established method, which is based upon cation exchange. Salt is also used in tile glazing, textile dyeing, leather tanning, and dye processing, as well as a freezing point depressant, soil stabilizer, rubber coagulant, and dehydration agent (Kostick 1994).

REFERENCES

Benbow, J. 1990. Salt, A status report. *Ind. Minerals* (April): 19–29.

Besticker, A. C. (ed.) 1963. *Proc. first symposium on salt.* Cleveland, Ohio: Northern Ohio Geol. Soc, 600p.

Chang, L. L. Y., R. A. Howie, and J. Zussman. 1996. Non-silicates: Sulphates, Carbonates, Phosphates, Halides. In Vol. 5B of *Rock-forming minerals.* Essex, England: Longman, 383p.

Clarke, F. W. 1924. Data of geochemistry. *U.S.G.S. Bull.* 770: 181–217.

Geyer, H. 1993. Controlled solution mining, sodium chloride. In Vol. A24 of *Ullmann's encyclopedia of industrial chemistry.* Weinheim: VCH Verlagsgesellschaft, mbH.

Kaufmann, D. W. 1960a. Physical properties of sodium chloride in crystal, liquid, gas, and aqueous solution states. In Sodium chloride: The production and properties of salt and brine. Ed. D. W. Kaufmann, 587–626. *Amer. Chem. Monograph Soc. Series.* New York: Reinhold Publ.

———— 1960b. Low temperature properties and uses of salt and brine. In Sodium chloride: The production and properties of salt and brine. Ed. D. W. Kaufmann, 547–568. *Amer. Chem. Soc. Monograph Series.* New York: Reinhold Publ.

———— 1960c. Brine purification. In Sodium chloride: The production and properties of salt and brine. Ed. D. W. Kaufmann, 186–204. *z* New York: Reinhold Publ.

Kostick, D. S. 1994. Salt. In *Industrial minerals and rocks.* 6th ed. Ed. D. D. Carr, 851–868. Littleton, Colorado: Soc. Mining, Metal. & Exploration.

Landes, K. K. 1960. The geology of salt deposits. In Sodium chloride: The production and properties of salt and brine. Ed. D. W. Kaufmann. *Amer. Chem. Soc. Monograph Series.* New York:
Reinhold.

Lefond, S. J. 1969. *Handbook of world salt resources.* New York: Plenum Press, 384p.

Lefond, S. J., and C. H. Jacoby. 1983. Salt. In *Industrial minerals and rocks.* 5th ed. Ed. S. J. Lefond,
1119–1149. New York: Soc. Mining Engineers, AIME.

von Perbandt, L. K. 1960. Salt mining. In Sodium chloride: The production and properties of salt
and brine. Ed. D. W. Kaufmann. *Amer. Chem. Soc. Monograph Series.* New York: Reinhold.

Ramsdell, L. S. 1960. Mineralogy of salt. In Sodium chloride: The production and properties of
salt and brine. Ed. D. W. Kaufmann. *Amer. Chem. Soc. Monograph Series.* New York: Reinhold Publ.

Sepiolite and Attapulgite

Sepiolite and attapulgite are clay minerals, but are distinguished from common clays by their chain-like structure. Both are magnesium-rich in composition, and both have great sorptive capacity, catalytic action, and rheological behavior. The use of sepiolite for pipe making first started in Budapest some two hundred and fifty years ago, while attapulgite served various industries long before it was identified.

Palygorskite, another reported chain-type clay mineral, has a structure identical to that of attapulgite and a composition varying within the same range as that of attapulgite. Attapulgite is used here in this text to represent both minerals.

Spain dominates the world with its sepiolite resources and production, while the United States is the principal producer of attapulgite. Both India and China have extensive deposits.

MINERALOGICAL PROPERTIES

Sepiolite and attapulgite (or palygorskite) are distinguished from common clay minerals by their chain-like structure. As shown in Fig. 77a, the crystal structure of sepiolite consists of three pyroxene-type chains that are linked to form two amphibole-type chains (Brauner and Preisinger 1956). The two sheets of (SiO_4) tetrahedra are continuous, but the direction of the spical extremes of the tetrahedral sheet of silica is inverted after every six tetrahedral units. This fibrous structure has channels oriented in the longitudinal direction of the fiber. The section of these channels, where water and other fluids can penetrate, is 3.6×10.6Å. There are eight octahedral sites and four hydroxyls per unit cell. In addition, two types of H_2O are present: (1) the crystallized water that completes the coordination of the octahedral cations, and (2) the zeolitic water that is bonded by hydrogen at the external surface or in the channels. The structural formula is $Mg_8(OH)_4(H_2O)_4Si_{12}O_{30} \cdot 8H_2O$ with a = 13.4Å, b = 26.8Å, and c = 5.28Å. Al^{3+} substitutes for Si^{4+} or Mg^{2+} in sepiolite, and a sepiolite with 19% of the octahedral sites filled by Al^{3+} has been reported (Weaver and Pollard 1973). Sepiolite is expandable when it is exposed to ethylene glycol vapor (Argast 1989). The $d_{(011)}$ increases from 12.4 to 12.8Å, an expansion along the c-axis, and the $d_{(130)}$ decreases from 4.53 to 4.45Å, a contraction along the b-axis. This structural distortion is not permanent, and the sepiolite returns to its original state six to twelve hours after it is removed from the ethylene glycol saturated atmosphere.

The crystal structure of attapulgite is shown in Fig. 77b. It has a general feature that is similar to the crystal structure of sepiolite, which consists of layers of SiO_4 tetrahedra that have unshared corners that point up or down (Bradley 1940; Christ et al. 1969).

They alternate after two chains in attapulgite, while in sepiolite they alternate after three chains. The formula is $Mg_5(OH)_2(H_2O)_4 Si_8O_{20} \cdot 4H_2O$ with a = 12.65Å, b = 17.9Å, and c = 5.26Å. Monoclinic attapulgite, with a β-angle varying between 106° and 108°, has also been identified, and many attapulgite crystals consist of both orthorhombic and monoclinic forms (Chisholm 1992). The octahedral coordinated cations in attapulgite are Mg and Al with a Mg:Al ratio varying between 2:3 and 3:2. A recent study

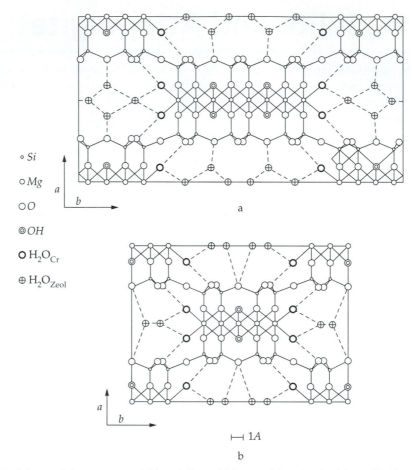

° Si
○ Mg
○○ O
◎ OH
○ H₂O_{Cr}
⊕ H₂O_{Zeol}

Figure 77   Scheme of the structures of (a) sepiolite and (b) attapulgite, after Preisinger (1963).

by Galan and Carretero (1999) showed that sepiolite is a true trioctahedral mineral, and attapulgite has a structure intermediate between di- and tri-octahedral. The theoretical formula of attapulgite should be $(Mg_2R^{3+}{}_2\blacksquare_1)(Si_{8-x}Al_x)O_{20}(OH)_2(H_2O)_4 \cdot R^{2+}{}_{1/2}(H_2O)_4$, where x=0.5. Hence, sepiolite and attapulgite are more compositionally limited than previously proposed.

The water content of both sepiolite and attapulgite depends upon humidity. The zeolitic water lies in the channels, its loss is reversible, and there is no change in the cell dimension up to 380°C for sepiolite and 350°C for attapulgite (Preisinger 1959). Crystallized water is lost above these temperatures. Sepiolite and attapulgite are transformed into what is known as "anhydride," in which the silicate layers become tilted (Preisinger 1963). Both sepiolite anhydride and attapulgite anhydride are stable up to 650° to 680°C.

Sepiolite and attapulgite possess a great surface area, with zeolitic channels throughout the structure into which a large quantity of water or polar molecules can be absorbed. The surface area of sepiolite, computed on the basis of section size of 3.6 × 10.6Å for the channels, is approximately 900 m²/gm, of which 500 m²/gm is internal surface and 400 m^2/gm is external surface (Serna and Van Scoyoc 1978). The surface area increases as the temperature increases and as the absorbed water and zeolitic water are removed. At temperatures above 300°C, a sharp decrease in the accessible surface area occurs because of the folding of the structure (Preisinger 1959). Sepiolites treated with HCl of various strength (1.25, 2.5, 5.0, 10.0, and 20.0 wt%) at 25°C for 2, 6, 24, and 48 hours show changes in surface areas. A maximum value of 549 m^2/gm was obtained after treatment with 1.25 wt% HCl for 48 hours (Rodriquez et al. 1994).

The needle-shaped particles of sepiolite and attapulgite appear agglomerated, forming large bundles of fibers similar to a haystack or brush-heap. A typical attapulgite needle is about 1μ in length and approximately 0.01μ across (Haden 1963). These agglomerated fibers are easily dispersed in water or other solvents. The resultant individual fibers are randomly distributed, are capable of more or less independent motion relative to one another, and can entrap the liquid. The rheological behavior in high or medium polarity media is related to several factors, including the concentration of clay, shear stress, pH, and electrolytes. In nonpolar solvents, sepiolite and attapulgite can form stable suspensions if the hydrophilic surface is treated with a surface active agent (Alvarez 1984). Koga (1992) proposed a procedure for the determination of the rheological properties of sepiolite suspensions, which includes

1. Observation of the micro-structure of sepiolite by rapid freeze-drying of sepiolite suspension.
2. Quantitative estimation of opening degree for sepiolite bundles.
3. Determination of viscoelasticity of sepiolite suspension.
4. Comparison between viscosity of sepiolite suspension and that of other inorganic fibrous materials.
5. Adjustment of the equation for viscosity of sepiolite suspension to its lower concentration and estimation of rheological properties of suspensions with multi-compositions.

Sepiolite and attapulgite are normally white to grayish with earthy to waxy luster. Their hardness is 2 to 2.5, and specific gravity ranges from 1.0 to 2.6, depending upon the state of dehydration and porosity. Refractive indices are α = 1.490 to 1.522 for sepiolite and 1.500 to 1.520 for attapulgite and γ = 1.505 to 1.530 for sepiolite and 1.540 to 1.555 for attapulgite. Minerals commonly associated with sepiolite and attapulgite are clays and nonclays such as carbonates, quartz, feldspars, and phosphates.

GEOLOGICAL OCCURRENCE AND DISTRIBUTION OF DEPOSITS

Sepiolite and attapulgite are formed in three distinct geological environments: (1) in epicontinental and inland seas and lakes as chemical sediments, (2) in oceans by hydrothermal alteration of basaltic glass and volcanic sediments, and (3) in calcareous soils by crystallization (Callen 1984). The geological conditions specified for the formation of sepiolite and attapulgite are high pH (alkaline), high Si-, high Mg-, and low Al-activity, and arid or semiarid climate (Singer 1979, 1980). Spain dominates the world with its sepiolite resources and production, while the United States is the principal producer of attapulgite. Both India and China have extensive deposits.

Spain: Tertiary deposits of sepiolite and attapulgite in Spain are formed in lacustrine or perimarine environments under semiarid or seasonably arid climatic conditions during periods of tectonic calm. Among the deposits, the most important is the Tajo Basin sepiolite deposit, which accounts for 90% of the world's known reserves (Galan and Castillo 1984).

The Tajo deposit is located in the central region of the Peninsula and northeast of the city of Madrid. It has an area of approximately 6.6 km^2 and is bordered by the Guadarrama and Gredos Sierras on the north and northwest, the Toledo Mountains on the south, and the Iberian Range on the east. It was formed by the fragmentation of the Iberian Massif during the Alpine orogenesis and was filled by a continental sedimentation during the Tertiary. The thickness of the sediments varies from more than 1,600 meters to as much as 3,000 meters (Carames et al. 1973).

The Basin is divided into three sub-basins: western, eastern, and central, the last one of which carries the sepiolite deposits. The stratigraphic sequence of the central sub-basin is as follows, with increasing age from Tertiary to Mesozoic: Pediment, Pontian limestone, transition facies of marly subfacies and detrital subfacies, detrital facies, evaporitic facies, Mesozoic sediments, basement, and granite. In the detrital subfacies, a pinkish sepiolite layer occurs in the upper portion. It is continuous, with a thickness of 4 meters, and is in association with chert and calcite. A discontinuous layer occurs in the lower portion, with a thickness of 2 meters, and is in association with dolomite, illite, and smectite. Among the deposits in the Tajo basin, the Vallecas-Vicalvaro deposit is the largest and most important one, with fairly pure sepiolite (65 to 95%). A representative analysis is 63.10% SiO_2, 1.08% Al_2O_3, 0.27% Fe_2O_3, 23.80% MgO, 0.49% CaO, 0.09% Na_2O, 0.21% K_2O, and 10.88% H_2O. The marly subfacies consists of limestones, marly limestones, and smectitic clays. The top of this subfacies is made up of a brown layer of chert and massive sepiolite that occasionally contains attapulgite as a major component (Megías et al. 1982).

The United States: The Miocene sediments of the southeastern United States contain economic deposits of attapulgite-sepiolite. Overall, montmorillonite is the dominant clay mineral in the Tertiary of the Atlantic and Gulf Coast Plains. In the Upper Oligocene and Lower and Middle Miocene deposits of northern Florida, Georgia, and southern South Carolina, attapulgite and sepiolite are commonly dominant (Weaver 1984). They were deposited in shallow, brackish to schizohaline coastal waters. The distribution of attapulgite and sepiolite in the Lower Miocene is illustrated in Fig. 78, and Fig. 79 shows the core lithology and clay mineral composition (Weaver 1984). Chemical analyses of montmorillonite and attapulgite in association show that Al_2O_3 and MgO are inversely related.

Sepiolite deposits exist in Pleistocene Lake Tecopa, located in the Mojave Desert, southeastern Inyo County, California (Sheppard and Gude 1968). Sepiolite occurs in mudstones that contain quartz, mica, saponite, chlorite, and feldspars. It was formed when silica, which was taken into solution from the volcanic ash, became available to the magnesium-rich, high-pH lake waters. In the ash beds where alumina is present, zeolite is formed instead.

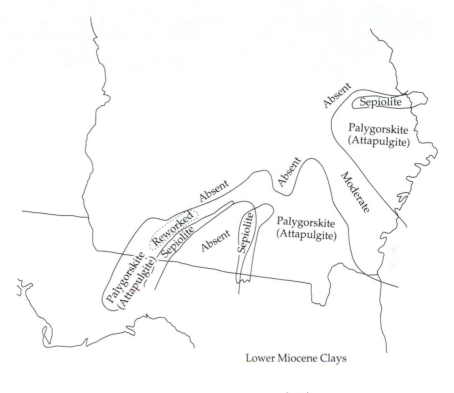

Lower Miocene Clays

20 Km

Figure 78 Distribution of attapulgite and sepiolite in the Lower Miocene clays in Florida, Georgia, and South Carolina. Reprinted from Weaver/*The Chemistry of Clay Minerals, Development in Sedimetolgy,* Copyright 1973, 213p, with permission from Elsevier Science. Montmorillonite is the dominant clay in unlabeled areas. Dotted line indicates location of concentration of detrital attapulgite in Middle Miocene sediments..

India: Extensive deposits of high grade attapulgite occur in the Hyderabad district of Andhra Pradesh. The clay is derived from the weathering of basic igneous rocks of Late Cretaceous to Eocene age. A representative chemical analysis gives 53.70% SiO_2, 7.78% Al_2O_3, 1.23% TiO_2, 7.96% Fe_2O_3, 0.92% CaO, 8.45% MgO, 0.14% Na_2O, 1.57% K_2O, and 18.13% H_2O (Siddiqui 1984).

China: Attapulgite deposits are found in Miocene basaltic pyroclastic sediments in Liuhe, Jiangsu, and the marine sepiolite deposits of early Permian age are found in association with limestone, dolomitic limestone, and cherty limestone in Jingdezhen, Fuliang and Leping, Jiangxi (Zhang 1984).

INDUSTRIAL PROCESSES AND USES

1. Milling and Beneficiation

Sepiolite and attapulgite are initially air-dried to reduce their moisture content before size reduction to approximately 2.5 cm in roller crushers. The clay is then treated in a

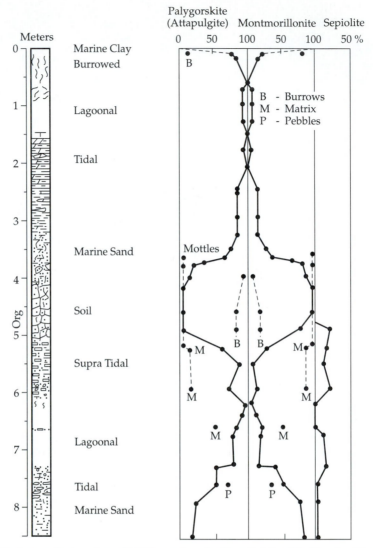

Figure 79 Lithology and mineralogy of MC-1 core from La Camelia Mine, Florida, after Weaver (1984).

rotary dryer where the moisture content is reduced to about 6 to 10% (Siddiqui 1968). The product, known as regular volatile material (RVM), is screened to the specified size for industrial uses. Further treatment is made at a temperature of 850°C to produce low volatile material (LVM). The difference between the two materials is that RVM tends to slake easily in water (Russell 1991).

For the preparation of colloidal grades, the clay, after the removal of moisture, is further treated in roller crushers to about 0.5 cm size. The granules are dispersed in a liquid medium to the extent that the individual fibers or needles are capable of more or less independent motion relative to one another. In the noncolloidal grade, the fibers or needles are attached to one another to give rigid particles, each of which is made up of many discrete fibers or needles. Treatment at high temperatures tends to remove most of the bound water, and to produce a colloidal dispersion cannot be achieved.

2. Industrial Uses

Sepiolite and attapulgite are used in various industries because of their sorptive, rheological, and catalytic properties (Galan 1996).

Oil Well Drilling Fluids: In drilling operations by the rotary method, a fluid is circulated continuously through a well and serves the primary function of removing bit cutting from the hole. In addition, it lubricates the bit, prevents hole sloughing, and forms an impervious coating on the wall of the drilled hole in order to prevent loss of fluid to porous rock formations (Lundie 1986).

The drilling fluids must be thixotropic with certain gel strengths when the fluids are not in motion, in order to avoid setting of the drilling cuttings. The fluids should not be affected by large variations in the concentrations of electrolytes encountered in the drilling process. Bentonite has been widely used in oil fields, but it cannot be used without chemical treatment in situations where contaminations such as sodium chloride, calcium sulfate, or magnesium sulfate are encountered. These salts prevent the swelling of bentonite, and thus reduce its ability to maintain effective viscosity in their presence. Attapulgite and sepiolite, on the other hand, do not depend on swelling to yield viscosity and are not affected by the presence of these salts. Both attapulgite and sepiolite have excellent stability under the high temperatures encountered in deep drilling wells (Lummus and Azar 1986). Specifications for sepiolite and attapulgite for use in drilling have been established by the American Petroleum Institute (API RP 13b, 2nd edition, 1969, 19p and API STD 13a, 5th edition, 1969, 11p).

Agriculture: Granular sepiolite and attapulgite are used to condition the soil by maintaining its required porosity and improving drainage and aeration. In the production of liquid fertilizers, a complete solution of the components is required in order to be useful. This requirement limits the amount of plant food available and the use of insoluble components (Wilbanks et al. 1961). These limitations can be eliminated by using a suspending agent. Attapulgite and sepiolite provide the necessary requirements for fertilizer suspensions because they have highly stable colloidal properties in high concentrations of salt.

Sepiolite and attapulgite are used as fluid carriers for pregerminated seeds and in seed capsulation (Alvarez 1984).

Paints and Polyesters: Sepiolite and attapulgite are added to paints as suspension agents to prevent pigment from setting during storage, as thickening agents to provide suitable viscosity, and as thixotropic agents to provide easy application (Haden 1963). Sepiolite and attapulgite are also used as thickening and thixotropic agents in liquid polyester resins to prevent the pigments from settling and to sag the polyester resins after application. In many cases, the hydrophilic surfaces of the clay minerals must be modified with a surface active agent in order to be compatible with the polyester.

Pharmaceuticals: Sepiolite and attapulgite are incorporated into pharmaceutical preparations and have been found to be superior to kaolin in the adsorption of alkaloids, diphtheria toxin, and bacteria (Barr 1957). The gel-forming properties of both sepiolite and attapulgite allow them to be used for the protection of intestinal mucous membranes and for the formation of a coating on intestinal walls. The capacity of sepiolite and attapulgite to neutralize considerable amounts of strong acids is of interest to the pharmaceutical industry (Anon. 1982).

Animal Nutrition: Sepiolite and attapulgite possess some distinct properties that are widely used in animal feed. Clay minerals, in concentrations from 0.5 to 3.0%, improve the feed efficiency up to 7% in pigs. This is related to an increase in the digestibility of the protein, which is produced by the slow flow of feed through the intestines due to the formation of a gel (Alvarez and Perez Castells 1983). Using clay minerals as binders for feed also improves the feed efficiency as compared with non-clay binders.

Sepiolite and attapulgite are used as carriers of nutritional supplements such as vitamins, minerals, and antibiotics. A particle size of 60 to 120 mesh provides an efficient carrier, with good homogenization of the micro-ingredients.

Decolorizing Agent: Sepiolite and attapulgite are used widely for the decolorization of greases and oils. The distinct processes can be categorized as (1) retention during filtration or percolation of colored particles, (2) absorption of colored compounds, and (3) catalytic conversion of colored compounds into colorless compounds or easily adsorbed compounds (Chambers 1959). In the decolorization of a petroleum oil, granular clay (30 to 60 mesh, for example) is first heat-activated at 250° to 420°C and charged to the filter shell. The oil is percolated throughout the bed of clay until the adsorptive capacity is reduced to the point where the effluent oil reaches a predetermined quality level. After draining the bed, washing with naphtha, and steaming, the clay is sent to a regenerating kiln. The absorbed organic matter is burnt off at 600°C, and the clay is reused. Clay with a high decolorant capacity should naturally have low retention of oils and very good filtration characteristics. An addition of small quantities of sepiolite and attapulgite to normal bleaching earths improves filtration capacity (Chambers 1959). Attempts have also been made to use sepiolite for decolorization of sugar juice (Ünal and Erdogan 1998). A fully developed fiber structure, smaller particle size when dispersed in water, and larger surface area promote a successful decolorization. The American Oil Chemists Society has recommended procedures for oil bleaching tests of attapulgite Fuller's earth for use in bleaching (AOCS Official Methods Cc 8a 52, 1958).

The ability of both clay minerals to decolorize oils and greases depends on the nature of the colored compounds. Sepiolite and attapulgite have a greater decolorant activity with respect to mineral than to vegetable oils. The colored compounds in vegetable oils are commonly molecules of considerable size that cannot penetrate into the channels and pores of the chain-type structures of clay minerals. On the other hand, the colored compounds in mineral oils are simple molecules, which have no difficulty penetrating into the pores and channels (Chambers 1959).

Absorbent Agent: The sorptive capacities of sepiolite and attapulgite are greater than those of any other clay mineral. In granular form, they are used as an adsorbent of water and oil, as well as of benzene and methyl alcohol (Robertson 1957). In addition, sepiolite has a very high capacity to absorb molecules responsible for foul odors. It is especially effective for compounds that have ammonium groups or various types of nitrogen in their compositions. Sepiolite in an environmental application can reduce the concentration of gaseous NH_3 from an initial concentration of 100 ppm to 18 ppm at a rate of 40 gm/m^3.

Catalytic Carrier: Sepiolite and attapulgite have the capacity to be catalytic carriers, since they have good thermal and mechanical stability and large surface areas. They can be used in supporting Zn, Cu, Mo, W, Fe, Co, Ni, and rare earths in hydrogenation, desulfuration, denitrogenation, or demetallization (Alvarez 1984).

Other Uses: Sepiolite and attapulgite are also used as filter aids, as anticaking and free flow agents, as fillers in rubber, and in cosmetics.

REFERENCES

Alvarez, A. 1984. Sepiolite: properties and uses. In Palygorskite-sepiolite: Occurrences, genesis and uses. Eds. A. Singer and E. Galan. *Development in Sedimentology* 37:253–288. Amsterdam: Elsevier.

Alvarez, A., and R. Perez Castells. 1983. Sepiolite in the field of animal nutrition. *5th Intern. Congress, Industrial Minerals,* 37–45.

Anon. 1982. Absorcion de bacterias por sepiolita. *Tolsa Research Rept.* 21.

Argast, S. 1989. Expandable sepiolite from Ninetyeast ridge, Indian Ocean. *Clays and Clay Minerals* 37:371–376.

Barr, M. 1957. Adsorption studies on clays. *Jour. Amer. Pharm. Assoc.* 46:486–497.

Bradley, W. F. 1940. The structural scheme of attapulgite. *Amer. Mineral.* 25:405–410.

Brauner, K., and A. Preisinger. 1956. Struktur und entstehung des sepioliths. *Tschermarks Min. Petr. Mitt.* 6:120–140.

Callen, R. A. 1984. Clays of the palygorskite-sepiolite group: Depositional environment, age and distribution. In Palygorskite-sepiolite: Occurrence, genesis and uses. Eds. A. Singer and E. Galan. *Developments in Sedimentology* 37:1–38. Amsterdam: Elsevier.

Carames, M., F. Lopez Aguayo, and J. L. Martin Vivaldi. 1973. Note sobre la mineralogia del sondeo de Tielmes en el Terciario de la Cuenca del Tajo. *Estudios Geol.* 29:307–313.

Chambers, C. P. C. 1959. Some industrial applications of the clay mineral sepiolite. *Silicates Inds.* (April): 181–189.

Chisholm, J. E. 1992. Powder-diffraction patterns and structural models for palygorskite. *Can. Mineral.* 30:61–73.

Christ, C. L., J. C. Hathaway, P. B. Hostetler, and A. O. Shephard. 1969. Palygorskite: New X-ray data. *Amer. Mineral.* 54:198–205.

Galan, E. 1996. Properties and applications of palygorskite-sepiolite clays. *Clay Minerals* 31:443–453.

Galan, E., and A. Castillo. 1984. Sepiolite-palygorskite in Spain Tertiary basins: Genetic patterns in continental environments. In Palygorskite-sepiolite: Occurrences, genesis and uses. Eds. A. Singer and E. Galan. *Developments in Sedimentology* 37:87–124. Amsterdam: Elsevier.

Galan, E., and M. J. Carretero. 1999. A new approach to compositional limits for sepiolite and palygorskite. *Clays and Clay Minerals* 47:399–409.

Haden, W. L. 1963. Attapulgite: Properties and uses. *10th Nat. Conf. on Clays and Clay Minerals,* 284–290. New York: Plenum Press.

Koga, M. 1992. Rheological properties of sepiolite suspensions. *Jour. Clay Sci. Soc., Japan* 32:190–198.

Lummus, J. L., and J. J. Azar. 1986. *Drilling fluids optimization, a practical field approach.* Tulsa, Oklahoma: Penn Well Books, 283p.

Lundie, P. 1986. Standardisation of minerals for drilling fluids. *Ind. Minerals* (March): 113–117.

Megías, A., G. Ordoñez, J. P. Calvo, and M. A. García del Cura. 1982. Sedimentos de flujo gravitacional yesiferos y facies asociadas en la cuenca neogena de Madrid,España. *5° Congreso tinoamericano de Geología, Argentina.* Actas II: 311–328.

Preisinger, A. 1959. X-ray study of the structure of sepiolite. *6th National Conf. on Clays and Clay Minerals,* 61–67. New York: Plenum Press.

——— 1963. Sepiolite and related compounds. Its stability and application. *10th National Conf. on Clays and Clay Minerals,* 365–371. New York: Plenum Press.

Robertson, R. A. S. 1957. Sepiolite: A versatile raw material. *Chem. Ind. (N.Y.)* 1492–1495.

Rodriquez, Vincente, M. A., J. de Lopez Gonzalez, and M. A. Bañares Muñez. 1994. Acid activition of a Spain sepiolite: physiochemical characterization, free silica content and surface area of products obtained. *Clay Minerals* 29:361–367.

Russell, A. 1991. Speciality clays, Market niches taken by unique properties. *Ind. Mineral.* (June): 49–59.

Serna, C., and G. E. Van Scoyoc. 1978. Infared study of sepiolite and palygorskite surfaces. *Proc. 1978 Intern. Clay Conference,* 197–206. Oxford: Elsevier.

Sheppard, R. A., and A. J. Gude, 3rd. 1968. Distribution and genesis of authigenic silicate minerals in tuffs of Pleistocene Lake Tecopa, Inyo County, California. *U.S.G.S. Prof. Paper* 597.

Hasnuddin Siddiqui, M. K. 1968. *Bleaching earth.* Oxford: Pergamon Press, 86p.

——— 1984. Occurrence of palygorskite in the Deccan Trap Mountain in India. In Palygorskite-sepiolite: Occurrence, genesis and uses. Ed. A. Singer and E. Galan. *Developments in Sedimentology* 37:243–250. Amsterdam: Elsevier. Oxford, New York, Tokyo.

Singer, A. 1979. Palygorskite in sediments: Detrital, diagenetic or neoformed—A critical review. *Geol. Rdsch* 68:996–1008.

——— 1980. The paleoclimatic interpretation of clay minerals in soils and weathering profiles. *Earth Sci. Rev.* 15:303–326.

Ünal, H. I., and B. Erdogan. 1998. The use of sepiolite for decolorization of sugar juice. *Appl. Clay Sci.* 12:419–429.

Weaver, C. E. 1984. Origin and geologic implications of the palygorskite deposits of S.E. United States. In Palygorskite-sepiolite: Occurrences, genesis and uses. Eds. A. Singer and E. Galan. *Developments in Sedimentology* 37:39–58. Amsterdam: Elsevier.

Weaver, C. E., and L. D. Pollard. 1973. *The chemistry of clay minerals, development in sedimentology 15.* Amsterdam: Elsevier, 213p.

Wilbanks, J. A., M. C. Nason, and W. C. Scott. 1961. Liquid fertilizers from wet-process phosphoric acid and superphosphoric acid. *Jour. Agr. Food. Chem.* 9:174–178.

Zhang, R. J. 1984. Sepiolite clay deposits in south China. In Palygorskite-sepiolite: Occurrences, genesis and uses. Eds. A. Singer and E. Galan. *Developments in Sedimentology* 37:251–252. Amsterdam: Elsevier.

Silica

Silica is a compound composed of the two most abundant elements, silicon and oxygen. In the form of quartz, it is an essential mineral in most igneous and metamorphic rocks, and because of its high resistance to weathering attack, it becomes concentrated in sedimentation to form sand and sandstone. In addition to quartz, four silica minerals, (1) tridymite, (2) cristobalite, (3) coesite, and (4) stishovite, are well established. Other minerals and mineral aggregates consisting predominantly of silica, are chert, chalcedony, tripoli, novaculite, opal, and diatomite (p. 94).

The major industrial uses of silica are glassmaking and the production of ceramics. It is also used as foundry sand, as an abrasive, in construction, as fillers, and in well fracturing. Natural and cultured quartz crystals have applications in optical and oscillator devices. For a large group of manufactured silicas (silica gel, pyrogenic silica, and precipitated silica), naturally occurring silica is used as a raw material.

Long-term exposure to crystalline and amorphous SiO_2 dust has significant effects on the human respiratory tract. A strong association between silicosis and lung cancer has been reported (Goldsmith et al. 1982).

MINERALOGICAL PROPERTIES

Silica is polymorphic, and the phase relations of stable forms are shown in Fig. 80 (Zoltai and Stout 1984). Except for stishovite, all silica phases have the SiO_4 tetrahedron as their fundamental structural unit. Each tetrahedron is linked to the other four tetrahedra by shared apical oxygens to form a three-dimensional network of SiO_4. Various polymorphic forms result from different linkages of the tetrahedron (Heaney 1994; Hemley et al. 1994). Stishovite has a rutile-type structure with Si^{4+} in an octahedrally coordinated site. The coesite structure is composed of four-membered rings of SiO_4 tetrahedra linked at the corners, to form chains parallel to the c-axis.

The structure of tridymite is based on a stacking module, in which sheets of SiO_4 tetrahedra are arranged in hexagonal rings that point alternately upward and downward. The sheets are stacked in such a way that the hexagonal rings directly overlie one another, producing the continuous tunnels normal to the sheets. The stacking sequence may be represented by the double periodicity of the module . . . ABABAB . . . along the direction of the stacking, where A refers to a sheet and B is its reflected equivalent. This gives a hexagonal symmetry to the structure. As temperature decreases to below 400°C, this arrangement of SiO_4 tetrahedra undergoes several changes caused by lattice distortion, with the transformation of symmetry from hexagonal to orthorhombic or monoclinic. These forms are known as the low-temperature forms of tridymite or simply low tridymite. By analogy with the . . . ABABAB . . . sequence of high-temperature tridymite, the crystal structure of high-temperature cristobalite can be described as having a stacking sequence of . . . ABCABCABC. . . . The structural relationship is identical to that between the hexagonal (HCP) and cubic (CCP) closest packing. The hexagonal rings do not superimpose directly upon each other, and cristobalite, therefore, does

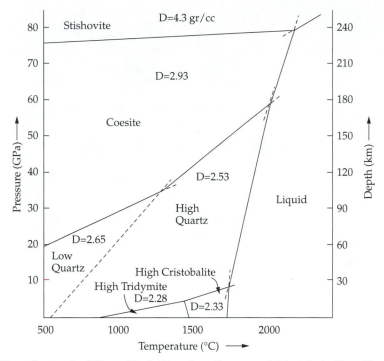

Figure 80 Phase diagram for SiO_2 modified. *Mineralogy: Concept and Principles* by Zottai/Stout, ©1986. Reprinted by permission of Prentice-Hall, Inc., Upper Saddle River, New Jersey.

not have the continuous tunnels normal to the sheets. Like tridymite, the structure of cristobalite also undergoes distortional changes at low temperatures and produces several low-temperature modifications.

In high-temperature or β-quartz, the distinct structural features are helical chains of SiO_4 tetrahedra along the c-axis, which have a repeat distance of three tetrahedra (Fig. 81a). The structure of the low-temperature modification or α-quartz can be described as a collapsed framework of β-quartz with a symmetrical change from hexagonal to rhombohedral (Fig. 81b). The winding of the helices can be left- or right-handed, which results in the enantiomorphism of quartz crystals (Donnay and Le Page 1978). A polarity exists along the a-axes that accounts for the piezoelectricity of low quartz and its use in oscillator and voltage-pressure transducers. The low sensitivity, however, requires high-voltage amplification (Newnham 1975). Quartz has low thermal conductivity (κ_{11}=6.2 W/mK, κ_{33}=10.4 W/mK) and high thermal expansion (α_{11}=13.3x10^{-6}/K, α_{33}=7.1x10^{-6}/K). A combination of these properties gives quartz low thermal shock resistance (Newnham 1975).

The transformation between α- and β-quartz is displacive which involves rotatory dislocation of the tetrahedra, leaving the Si-O bonding unaffected. The transformation is reversible and occurs without hysteresis. Transformations between other polymorphic forms are reconstructive in nature with sluggish rates of reaction. Therefore, the high-temperature modifications of both tridymite and cristobalite may be metastably present or may transform to their metastable low-temperature modifications, in the stability p-T region of quartz (Megaw 1973; Heaney 1994).

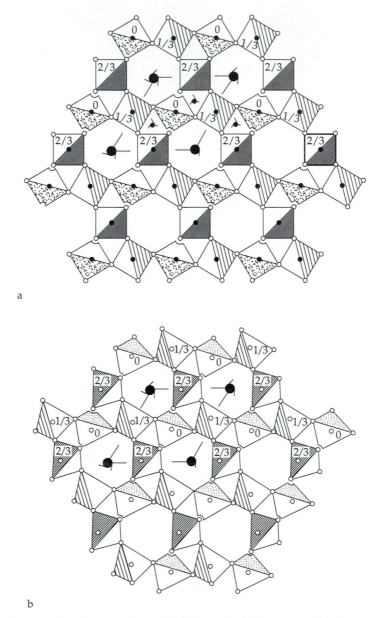

a

b

Figure 81   Projections of quartz along the c-axis (a) β-quartz and (b) α-quartz, after Heaney (1994).

Quartz is essentially pure SiO_2 without major or common impurities. The substitution of Al^{3+} for Si^{4+} ion, which is very limited in concentration, appears to be accompanied by the introduction of the alkali ions Li^{1+} or Na^{1+}. Both tridymite and cristobalite have more open structures compared with quartz and may contain appreciable amounts of alkalis, balanced by the replacement of the Si^{4+} ion by Al^{3+} in the tetrahedrally coordinated site (Deer et al. 1992).

Stability range, specific gravity, hardness, refractive indices, and the cell dimensions of silica minerals are listed below:

Low Quartz (α): stable up to 573°C under atmospheric pressure, H=7 and G=2.65, ω=1.544 and ε=1.533, rhombohedral with a=4.913Å and c=5.405Å.

High Quartz (β): stable from 573°C to 870°C, H=7 and G=2.53, ω=1.540 and ε=1.533, hexagonal with a=4.999Å and c=5.457Å.

Low Tridymite (α): present metastably between 190° and 380°C, H=6–7 and G=2.28, α=1.478, β= 1.479, and γ=1.481Å, orthorhombic with a=8.73Å, b=5.04Å, and c=8.28Å. Other forms of low tridymite present metastably between 150° and 190°C, between 110°C and 150°C, and below 110°C.

High Tridymite (β): stable between 870°C to 1,470°C, present metastably from 380°C to 879°C, metastably melting at 1,670°C, hexagonal with a=5.03Å and c=8.22Å.

Low Cristobalite (α): present metastably up to 200° to 275°C, H=6–7 and G=2.33, ω=1.489 and ε=1.482, tetragonal with a=4.97° A and c=6.93Å.

High Cristobalite (β): stable from 1,470°C to the melting point of 1,713°C, present metastably above 200 to 275°C to 1,470°C, cubic with a=7.13Å.

Coesite: stable above 10 GPa at 25°C (by extrapolation), a triple point with β-quartz and liquid at 4.0 to 4.5 GPa and 2,000° to 2,200°C (Jackson 1976; Hemley et al. 1994), H=7–8 and G=2.93, α=1.59, β=1.60, and γ=1.60, monoclinic with a=7.17Å, b=12.33Å, c=7.17Å, and β=120.0°.

Stishovite: stable above 75 GPa, an inferred triple point with coesite and liquid at 12.5 Gpa and 2,220°C (Davies 1972; Hemley et al. 1994), G=4.30, ω=1.799, and ε=1.826, tetragonal with a=4.18Å and c=2.66Å.

Normally, all forms of SiO_2 are colorless and transparent. Other colors do occur including yellow (citrine), gray-brown (smoky quartz), pink (rose quartz), and violet (amethyst). The cause of coloration in most cases is due to inclusions of iron and titanium minerals in either tiny crystalline or colloidal states (Deer et al. 1992).

All polymorphic forms of SiO_2 dissolve readily in HF, except for coesite, and especially for stishovite, both of which dissolve with difficulty (Deer et al. 1992). In H_2O, the solubility of quartz at ambient temperatures is calculated to be 6 ppm (Morey et al. 1962) and increases strongly with increasing temperature and pressure, as well as with the addition of NaOH or NaCl solutions (Anderson and Burnham 1965). As shown in Fig. 82 (Kennedy 1950), the solubility passes through a retrograde region. Fluid inclusions are common in quartz with dimensions varying from less than 1 μm to several millimeters. These inclusions are mainly saline solution, CO_2, and CH_4, and under ambient conditions halite may be partly precipitated from the saline solution.

In addition to quartz and other polymorphic forms of SiO_2, there are several minerals and mineral aggregates consisting predominantly of silica. They are opal, chalcedony, chert, flint, jasper, tripoli, novaculite, and diatomite, which is an organic product (p. 94).

Opal is a hydrous silica formed by evaporation from silica-containing solutions, hydrothermal alteration of silicates, or the transformation of biogenic amorphous silica. The composition of opal is $SiO_2 \cdot nH_2O$, with water content usually varying between 5 and 10%. There are three well-defined structural groups: (1) opal-A is highly disordered and nearly amorphous, (2) opal-C is well-ordered and consists predominantly of low cristobalite, and (3) opal-CT is disordered and consists predominantly of low cristobalite and low tridymite. Opal has a hardness of 5½ to 6½ and a specific gravity of 1.99

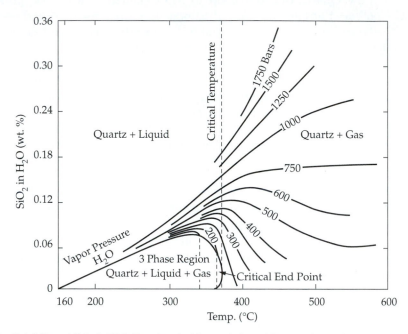

Figure 82   Solubility of SiO_2 in H_2O. Reprinted with permission of Economic Geology, Volume 45, p. 629-653, G.C. Kennedy (1950).

to 2.25. The color of opal is highly variable, usually white or a pale shade of yellow, red, or brown. Refractive index varies from 1.441 to 1.459. Diagenetic process converts opal to quartz (Deer et al. 1992).

Chalcedony is a microfibrous variety of quartz with refractive indices (1.53 to 1.54) somewhat lower than those of normal quartz. Chert is a siliceous rock, nonporous and hard, authigenic in origin, and composed of microcrystalline quartz, chalcedony, and opal. When the constituent mineral is predominantly microcrystalline quartz, the name novaculite is used, and when opal is the predominant mineral, the name porcellanite is used. Flint and jasper are varieties of chert that are, respectively, of a dark-gray color with a typical occurrence as nodules in chalk, and of a red-brown color that is caused by the presence of ferric oxides. Tripoli is a microcrystalline, finely particulate, and friable form of silica that appears to be the leaching product from calcareous chert or siliceous limestone (Blatt 1982; Williams et al. 1982).

GEOLOGICAL OCCURRENCE AND DISTRIBUTION OF DEPOSITS

Silica minerals with their distinct physical and chemical properties form under conditions that prevail in all geological environments. Although industrial silica may come from igneous rock (such as quartz vein), metamorphic rock (such as quartzite), and sed-

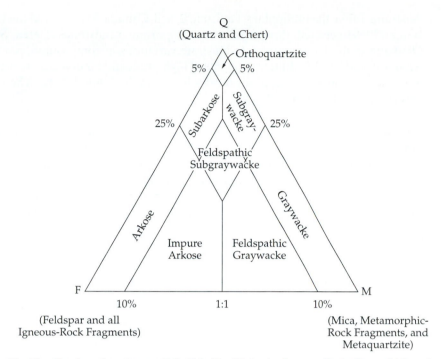

Figure 83 Classification of sandstones. Folk, R.L. *The Distinction between Grain Size and Mineral Composition in Sedimentary Rock* Nomenclature, 1954, p. 344-359. Reprinted with permission of The University of Chicago Press.

imentary rock (such as chert and flint), the main sources of silica are high-silica sand and sandstone because of their abundance and distribution.

Sands accumulate in environments characterized by rapidly moving water, including beaches, sandbars, river channels, desert dunes, and alluvial fans. The areal extent and depth of the deposits depends upon the amount of tectonic displacement over time. A beach sand deposit may have a large increase in areal extension, and only a slight increase in thickness, when the beach migrates inland during a marine transgression. It is also possible for a beach sand deposit to exist at the deposition site for a long period of time. This leads to a tectonic setting of a slowly subsiding basin, which results in pure sand deposits hundreds of meters thick, with a relatively restricted areal extent (Blatt 1982).

The mineral composition of sands and sandstones is directly determined by the tectonic setting in which they were formed. Cratonic sands are rich in quartz and chert because of repeated reworking in shallow water environments. Along both divergent and convergent continental margins, as well as in intracontinental rift zones, the sandstones formed are less silica rich, and lithic fragments tend to be abundant (Dickinson and Suczek 1979). One of the commonly used mineralogical classifications for sandstone (Folk 1954) is given in Fig. 83. Silica-rich sandstones are classified as those which have 90 to 95% silica content.

Silica-rich sand and sandstone are distributed widely throughout the world (Murphy and Henderson 1983; Benbow 1989), including England, Belgium, Germany, France, Austria, Denmark, Finland, Italy, Norway, Turkey, Saudi Arabia, South Africa,

Australia, India, the Philippines, Pakistan, Brazil, Canada, Mexico, and the United States. Bates (1969) selected three industrially important sandstones: Berea, St. Peter, and Oriskany, in the United States, to illustrate variations in composition, cementation, distribution, and age. In addition, deposits of high-grade quartz crystals are well established in Brazil, and significant deposits of tripoli exist in the United States. These deposits are described briefly below:

Berea Sandstone: Berea, one of the most used dimension sandstones in the United States, is quarried in Lorain County, northern Ohio, about 40 km west of the city of Cleveland. It is a medium- to fine-grained sandstone, light gray in color, and is cemented by silica and clay. It contains more than 90% quartz and small percentages of chert, feldspar, rock fragments, and heavy minerals (Pepper et al. 1954). The Berea sandstone, and the genetically related underlying Bedford shale, both of Early Mississippian age, form a wedge of sediments lying between two black shales, the Ohio below and the Sundbury above. The Bedford shale was deposited, in part, subaerially on a delta and, in part, subsequently as offshore beds along the delta front. The Berea sandstone was deposited above the Bedford shale first as a delta and later as a marine pavement that formed as the sea inundated the delta. The Bedford shale and Berea sandstone, therefore, represent cycles of deposition during oscillations of the land and sea between two periods of quiescence. Berea sandstone may be divided into two units. The upper unit, 6 to 12 meters in thickness, is thinly bedded, with oscillation ripple marks, or massive and cross bedded, whereas the lower unit is massive uniform with thicknesses up to 70 meters and is free from contaminating materials (Pepper et al. 1954). Dimension stone is quarried from the lower unit.

St. Peter Sandstone: The St. Peter sandstone of early Middle Ordovician age is an orthoquartzite of perhaps the highest possible purity. It has an extensive areal distribution, cropping out along the south margin of the Wisconsin dome in Minnesota, Wisconsin, Iowa, and Illinois, and on the flanks of the Ozark dome in Missouri. It exists beneath younger rocks in Iowa, Illinois, Indiana, and Missouri. It has an average thickness of 20 meters, over an area of 60×10^4 km^2 (Dapples 1955; Zdunczyk and Linkous 1994). Major quarrying districts are centered at the Ottawa district in northern Illinois, about 129 km southwest of Chicago, and in the Klondike and Pacific deposits of east-central Missouri, about 50 km west of St. Louis (Lamar 1927; Dapples 1955).

St. Peter sandstone commonly contains more than 97% quartz, with clay minerals constituting most of the nonquartz fraction. Limonite averages about 0.35%, and feldspar and heavy minerals are present in trace amounts. Fluid inclusions in quartz grains are very rare. The sandstone is sorted well, with 80% of the grains falling into coarse and medium grades in the upper part, which is restricted to the Ottawa district, and 80% of the grains falling into the medium and fine grades in the lower part. St. Peter sandstone is light gray to white in color. It is soft and poorly cemented either with clay minerals or by compaction (Willman and Payne 1942).

The origin of the St. Peter sandstone according to a regional analysis by Dapples (1955) shows that it is a transgressive sheet sand representing reworked beach deposits with sources from the Canadian Shield, and with prevailing minor subsidence.

Oriskany Sandstone: The Oriskany sandstone of Early Devonian age is an important source of high silica sand in the eastern United States (Harris 1972). It extends from central New York southward to Pennsylvania, Virginia, Maryland, and Ohio. Major quarries are located in the folded belt of the central Appalachians, including Mifflin and Huntingdon, Pennsylvania; Cumberland, Maryland; and Hancock and Berkeley Springs, West Virginia (Zdunczyk and Linkous 1994).

Oriskany sandstone is highly pure and white to bluish-gray in color. It is loose or partly disaggregated, and with the removal of the cement, the normally hard Oriskany becomes friable sandstone or loose sand. In the quarry districts, the sandstone reaches a thickness of 60 to 100 meters. The Oriskany sandstone is of near-shore, shallow-water origin and was deposited under tectonically stable environments (Stow 1938).

Quartz Crystal Deposits, Brazil: Brazil is known to be the world's principal quartz-producing country, but there are producing deposits of comparatively smaller scale that exist in other countries, including the former USSR, Canada, Madagascar, India, and Australia.

Brazilian quartz producing districts are in the states of Minas Gerais, Santa Catarina, São Paulo, Goiaz, and Bahia (Campbell 1946; Knouse 1947). The deposits occur in the veins and pockets of sandstone and quartzite of late Precambrian age and consist of milky or gray quartz of high quality and good size. The crystals have grown inward from the vein and pocket walls to fill, either partially or completely, the void space, illustrating a simple fracture filling process by aqueous solution. The bulk of the economic quartz occurs as clear terminations on pyramids of bull-quartz crystals (Johnston and Butler 1946). Quartz crystals are also produced from the residual and alluvial deposits that overlie the primary deposits.

The Brazilian crystals are classified into three quality classes: (1) class 1 of more than 200 grams, and with 60% usability; (2) class 2 of more than 200 grams, and with 60 to 30% usability, and (3) class 3 of less than 200 grams, and with 60% usability, or of more than 200 grams, and with 30% usability (Waesche 1983). The pure quartz chips and small crystals not meeting the dimensional requirements for electronic grade are known by the Brazilian term "lasca." They are used as source material for crystal growth and for production of fused quartz.

Tripoli Deposits, the United States: Tripoli is known to occur in many regions of the world, but the United States is the only nation with significant production. Tripoli deposits are distributed in Missouri, Oklahoma, Illinois, Arkansas, Tennessee, and Georgia. Principal producing deposits are centered in Ottawa County, Oklahoma, and Newton County, Missouri, all within a 16 km radius of Seneca, Missouri (Fellows 1967), and in Alexander and Union Counties in southwestern Illinois (Lamar 1953).

At Seneca, the tripoli occurs in flat-lying deposits of up to 6 meters thick, with an overburden consisting of about 3 meters of "rotten" tripoli, and grades upward to a red, cherty clay at the surface. It rests on the Mississippian Boone Fromation, which is composed of limestone, cherty limestone, and chert. The regional tripoli deposits are a product of the leaching of cherty limestone or calcareous chert (Fellows 1967). In the southwestern Illinois district, the tripoli deposits were derived from the chert and cherty limestone of the Devonian Clear Creek Formation and the underlying Grassy Knob

Formation (Weller 1944; Lamar 1953). Tripoli is normally white, but iron oxide-staining is prevalent at the top and bottom of the deposits, defining the limit of the economic grades.

INDUSTRIAL PROCESSES AND USES

1. Beneficiation of Silica Sand

The common method used to beneficiate silica sand includes coarse and fine crushing, stage washing, and high-density attrition to remove loosely adhering impurities. A surface treatment in a very dilute solution of hydrofluoric acid and in the presence of the reducing agent sodium hyposulfate may be used to remove surface contamination (Segrove 1956; Zdunczyk and Linkous 1994). The remaining impurities such as mica, hematite, magnetite, rutile, and feldspar are generally reduced by the use of a froth flotation, low or high intensity magnetic separation, or high-tension electrostatic separation. The treated product is then dried and screened into industrial sizes. A flow sheet proposed by Pryor (1965) has the following steps: Drier → comminution → 20-mesh screen → air classifier → attrition mills → wet cyclone → flotation → drier → magnetic separator → sand. Zdunczyk and Linkous (1994) outlined nine steps for the production of high quality silica sand. They are

1. drying and screening;
2. washing to remove clay, drying and screening;
3. washing, scrubbing, flotation, drying, and screening;
4. washing, scrubbing (high pulp density attrition), drying, screening, and sizing in an air separator;
5. iron removal (magnetic, wet, or dry);
6. grinding (rod mill or ball mill);
7. acid leaching for high-purity products;
8. bagging facilities for some or all products; and
9. cooling facilities for foundry sand.

The processing of vein quartz is more or less similar to that described above, although different plants have adopted various modifications. A general scheme is as follows: lump quartz → washing drum (fine fraction and sludge) → sorting belt (minerals, accumulation of impurities) → calcination (700°C/6 hr) → sorting belt (iron pigmentation) → crushing and grinding (jaw crusher, ball mill in circuit classifier) → screening (fine sludge, 0.08 mm) → wet magnetic separation (magnetic particles) → dewatering → chemical refining (20% HF, 50° to 65°C/2 hr) → washing, dewatering drying → packing (Fanderlik 1991).

The processing of quartz lascas to granular quartz for special industrial uses is made by (1) washing with aqueous 2% HF solution; (2) heating to a temperature > 573°C to produce transformation from α to β quartz; (3) quenching in cold H_2O, which causes fracturing of the lascas by thermal expansion shock; (4) sorting out iron-containing grains and grains with rutile inclusions; (5) grinding; and (6) sizing. For tripoli, the crude ore is stored in drying sheds for about six months, crushed, and then ground. The product is sized for market use by screening or air classification (Bradbury and Ehrlinger 1983).

2. Industrial Uses of Silica Sand

Glass: By far the largest use of silica sand is in the glass manufacturing industry, which accounts for approximately 42% of the total consumption of silica sand. In each of the four common industrial glasses, soda-lime, borosilicate (Pyrex), flint, and silica (Vycor) glasses, SiO_2 is the principal component and the glass-forming oxide. Silica content reaches a value of 96.6% in silica glass (Baumgart 1984a; Scholze 1991). (See Borax and Borates, p. 71.)

To the glass industry, iron oxide is the most undesirable impurity in silica sand. In order to produce a clear, colorless, ordinary container glass, the Fe_2O_3 content of the sand must be less than 0.06%, although 0.25% can be tolerated for amber glass. The amount of oxides, such as Al_2O_3, CaO, MgO, and K_2O, in the sand are not too critical, as they are normal constituents of the glass batch; however, the content of these oxides should be constant. The silica sand for container glass should be at least 99.5% SiO_2. The size of the sand is required to be 90% between 40- and 140-mesh, as the coarser grains are difficult to melt and the finer ones tend to produce seeds in the glass. Specifications for silica sand used in glasses other than container glass are more stringent, especially where silica content must reach a minimum value of 99.7 to 99.8% (Segrove 1956; Murphy and Henderson 1983; Scholze 1991).

By blowing, pressing, drawing, rolling, and casting, glass can be shaped into different forms that can be used for various purposes, such as container glass, optical glass, glass insulator for electrical applications, window glass, polished plate glass, tempered plate glass, colored structural glass, laminated safety glass, insulating glass, glass building blocks, glass foam, glass fabrics, and fiberglass, which has three major forms: (1) glass wool, (2) continuous filament glass fiber, and (3) optical glass fiber (Anon. 1987a; Benbow 1989).

Special applications of silica glass include composite fibers for telecommunication, mercury discharge lamps, tubing for the production of semiconductors, tubing for gas lasers, optical instruments and apparatus, thermally stressed light bulbs, electric valves, melting crucibles and pots, reacting vessels and equipment for chemical processing of corrosive substances, high-quality electrical insulators, and structural materials for space technology (Fanderlik 1991).

Ceramics: Silica is used in the formulation of batches for stoneware, porcelain, fireclay, and aluminosilicate refractories (Anon. 1987b). It is added to the ceramic bodies in order to reduce their drying shrinkage and plasticity. During thermal treatment, silica reacts with the other constituents in the batch, thus precipitating the development of the desired structure of the ceramic body. The type of silica used in porcelain influences the development of its microstructure. The higher the transformation rate to cristobalite during thermal treatment, the faster it dissolves in the melting phase of a ceramic body, thus creating a more translucent porcelain (Schüller 1984).

Silica bricks containing 96 to 98% of SiO_2 are manufactured from mixtures of silica sandstone and quartzite in a mass ratio in order to balance the rate of phase transformation during heat treatment. Both sandstone and quartzites consist of α-quartz, whereas cristobalite is the predominant silica phase in heat-treated silica bricks. The α-quartz in quartzite changes easily to cristobalite, whereas the rate of transformation from α-quartz to cristobalite in sandstone is very sluggish. The transformation from α-quartz to cristobalite is bound to a volume expansion of about 15%. In the production of silica bricks, 1 to 3% CaO, or $Ca(OH)_2$, is added to the batch, acting as an activator for the phase transformation and as a sintering aid (Baumgart 1984b).

By controlled devitrification of glass, a class of material known as "glass-ceramics" is produced, in which silica is a major component (Beall 1986). The foundation of controlled crystallization lies in efficient internal nucleation. In the process, glass is melted, fabricated to shape, and then converted by heat treatment to a predominantly crystalline state. The silica concentration must be at a well-defined range to allow the appropriate viscosity for high speed formation. Therefore, the viscosity at the liquidus temperature is critical. Cristobalite, as the most common form of silica precipitating from glass, has viscosity characteristics that are high enough for many forming processes. Solid solutions based upon β-quartz are also used for the preparation of glass-ceramics.

Foundry Sand: Silica sand is used more widely as foundry sand than any other mineral sand, including olivine, zircon, spinel, sillimanite, and chromite (Taylor et al. 1959; Wilborg and Henderson 1983). Clay is used for bonding molding sand; and linseed oil, resins, and plastics are used for making cores. The bonding action of clay and sand is electrostatic in nature. Bonding functions to develop the strength and plasticity of the clay and to make the aggregate suitable for molding. Only relatively pure clays are suitable for bonding purposes. The three general types of clays for bonding molding sands that are used in the foundry industry are montmorillonite, kaolinite, and illite. For cores, they are baked by a series of treatments including drying, condensation, polymerization, or oxidation, depending upon the type of binder used.

Abrasives: Silica sand and powder are used for grinding, scouring, and polishing. The hardness (H = 7), fracture pattern, and brittleness provide and maintain grains with sharp cutting edges during use. Tripoli is friable, with a grain size of 0.1 to 5 μm and a grain shape with no sharp edges, and is used for polishing (Harben 1983).

Fillers and Extenders: Silica powder, silica flour (micronized silica), and tripoli act as hard, inert fillers and extenders in paints, plastics, and rubber (Bradbury and Ehrlinger 1983). They improve chemical resistance to acids and durability in paints, flexural strength and resistance to thermal shock in plastics, and adhesion and tear strength in rubber tires.

Filters: Silica sand and pebbles are used as bedding materials for water filtration. Because of their hardness and toughness, they more effectively resist physical and chemical breakdown during filtering and backwashing (Tien 1989).

Enamels and Glazes: Most enamels are made from borosilicate glasses containing alkalis. Silica, the major constituent, can be added to the batch in many different ways, ranging from the addition of pure silica sand with 99%$^{+}$ purity, to the addition of one of the many silica-containing minerals, combined with other oxides (Maskall and White 1986). Since silica is highly resistant to most mineral and organic acids, acid resisting enamels are always enriched with silica. However, such enrichment in the batch composition also increases viscosity and decreases expansion. To compensate, the composition of the batch is modified in many cases by adding small amounts of titania in place of silica, and by substituting lithium for sodium or potassium.

As it is in enamels, silica is the major and, in many cases, the only glass former used in glazes. The higher the proportion of silica in the glaze, the higher the glost temperature. This disadvantage is outweighed, however, by the benefits of higher resistance to mechanical damage and chemical attack. The silica to flux ratio (in mol%) is 2:1 for low-temperature glazes, whereas high-temperature glazes require a ratio of 10:1 (Taylor and Bull 1986).

Well Fracturing: High silica sand is utilized in fracturing the reservoir rocks of oil and gas wells (Griffiths 1987). A fluid containing suspended quartz sand is pumped at high pressures into the wells, thus enlarging the existing openings and creating new voids. After the fluid is withdrawn, the sand remains and holds the fractures open. For well fracturing silica sand must have 98% SiO_2, with well-rounded grains, in order to provide good permeability.

Metallurgical Use: Quartz lumps are used to produce metallurgical grade silicon and ferrosilicon by reduction with coke in electric-arc furnaces. Source iron, in the form of turnings or shredded iron, is added to the charge during the production of ferrosilicon (Büchner et al. 1989).

Construction: Silica sand is an integral part of mortar, and silica sand and gravel are additives for Portland cement and asphalt (Blanks and Kennedy 1955; Kalcheff 1977). Tripoli is used as an inert filler and extender in asphalt (Bradbury and Ehrlinger 1983). Silica sandstones with various cementing materials and colors have long been used as dimension stones. The quality of the stone is measured on the basis of the following parameters: (1) water absorption, (2) bulk specific gravity, (3) modulus of rupture, (4) compressive strength, (5) abrasion resistance, and (6) flexural strength (Harben and Purdy 1991).

3. Growth and Use of Quartz Crystals

The limited availability of natural quartz crystals of high quality stimulated the experimental growth of quartz crystals. Among all the known crystal growth methods, hydrothermal growth produces quartz crystals, which are also known as cultured quartz crystals, of good size, high quality, and well-developed morphology. Hydrothermal crystal growth is defined as growth at elevated pressures and temperatures in aqueous solution, both in sub- and super-critical states (Laudise 1970).

Growth is generally carried out in a steel or special alloy vessel of the "modified Bridgman type," a simplified model of which is shown in Fig. 84 (Laudise 1970). Natural quartz chips, which serve as nutrients, are added into the lower part of the vessel, which

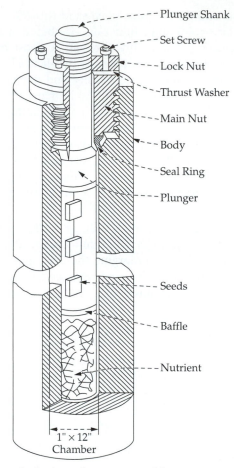

Plunger Shank

Set Screw

Lock Nut

Thrust Washer

Main Nut

Body

Seal Ring

Plunger

Seeds

Baffle

Nutrient

1" × 12"
Chamber

Figure 84   Modified Bridgman hydrothermal pressure vessel for quartz growth. Laudise, R.A. The Growth of Single Crystals, 1970, p278. Reprinted by permission of Prentice-Hall, Upper Saddle River, New Jersey.

is separated from the upper part by a baffle with a defined opening. Seed crystals are cut by defined crystallographic orientation and are selected from natural or cultured quartz crystals that are free from twinning and structural defects. They are placed in a rack in the upper part of the vessel. Preferred seeds are usually cut parallel to the basal plane (0001) and perpendicular to the optic axis, or parallel to (1120) and (0001) and perpendicular to the a-axis.

After the nutrient chips and seed plates or rods are positioned, an aqueous solution with NaOH or Na_2CO_3, which is used as a mineralizer, is introduced to fill the vessel to 70 to 80% of its capacity. The vessel is sealed and heated electrically to 400°C at the lower part, while the temperature of the upper part is kept a constant 30 to 40°C lower. The pressure required for growth at a given temperature is usually determined by the p-T curves for water (Holser and Kennedy 1959). As shown in Fig. 82, the solubility of SiO_2 increases with temperature and pressure above 750 bars. In NaOH (0.5 to 2.0 M) solutions, the solubility of SiO_2 can reach a level of about 2 to 3% at 400°C and 1,000 bars. The duration of growth runs varies between one and four months, and the

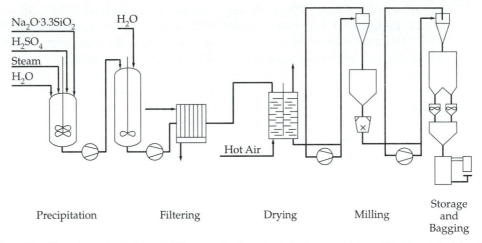

Na₂O·3.3SiO₂
H₂SO₄
Steam
H₂O
H₂O
Hot Air

Precipitation   Filtering   Drying   Milling   Storage and Bagging

Figure 85   Flow sheet of precipitated silicas production. Reprinted by permission of the Royal Sociaty of Chemistry from Kleinschmit, *Silicas and Related materials,* (1981).

yield is directly related to the size of the vessel. For a vessel with an inner diameter, length, and volume of 40 cm, 4.5 m, and 560 liters, respectively, the average yield per run is 250 kg. With an increase in inner volume to 4,500 liters, the yield per run is also increased by eight-fold to 2,000 kg (Ettlinger 1993).

Cultured quartz crystal has excellent ultraviolet and infrared transmission, better than natural quartz and silica sand. It is used as a low-loss light transmitter. In electronics, it is principally used as filters, oscillators, timers, and sensors (Dodd and Frazer 1965; McLaren and Phakey 1969). In recent years, the use of quartz for piezoelectric devices has been challenged by the development of polyvinylidene fluoride, but the demand for quartz crystal continues to increase in the applications of high-sensitivity thermometers, in integrated circuits, microprocessors and synthesizers, and in color television, video, and other communication systems.

4. Production and Use of Precipitated Silicas

Precipitated silicas are the amorphous form of SiO_2. They are prepared by reacting silica sand with Na_2CO_3 or NaOH to form sodium silicate (water glass), which in turn is reacted with H_2SO_4 or CO_2 + HCl to produce precipitated silicas according to the reaction: $Na_2O \cdot 3.3SiO_2 + H_2SO_4 \rightarrow 3.3SiO_2 + Na_2SO_4 + H_2O$. The precipitation is carried out in a neutral or alkaline media, and at elevated temperatures. The reaction mixture is stirred to avoid gel formation. Discrete silica particles are formed that grow by further precipitation of silica on the surface of the primarily formed seed particles. The production process consists of precipitation, filtration, drying, milling, and bagging, as shown in Fig. 85 (Kleinschmit 1981). In some cases, additional compaction and granulation is required.

Precipitated silicas consist mainly of spherical particles, generally intergrown in three-dimensional aggregates. They are in a fluffy white powder form and are hydrophilic. The specific surface area ranges from 25 to 800 m^2/gm, and the weight of the powder, which is determined as the tapped density, lies in the range of 50 to 500 kg/m^2. The dibutyl phthalate (DBP) absorption measurement for precipitated silicas gives a range of 175 to 320gm/100 gm. A representative composition has 98% SiO_2, 1% Na_2O, 0.03% Fe_2O_3, 0.8% SO_3, and a residue of 0.5% (Kerner and Kleinschmit 1993).

The most important use of precipitated silicas is the reinforcement of elastomer products such as shoe soles, rubber articles, and tire compounds (Kraus 1965). Precipitated silicas are used as free-flow agents for powder formulations, particularly of hygroscopic and adhesive substances. Their high absorptive capacity enables precipitated silicas to absorb liquids or solutions to yield powder concentrates that contain up to 70% of the liquid. The liquid material is converted to a dry and free flowing form, which can be used in mixtures with other solids. Precipitated silicas are added to toothpaste as cleaning agents, and to paper to ensure high depth of color and good contrast (Kerner and Kleinschmit 1993).

5. Production and Use of Silica Gels

Silica gels have the nominal formula $SiO_2 \cdot xH_2O$. They are amorphous and are distinguished by their microporosity and hydroxylated surfaces. Silica sand and Na_2CO_3 or NaOH are used to produce alkali silicates, and by the neutralization of alkali silicates with acid silica gels are formed. Such a process initiates a polymerization of $(SiO_2)_4$ tetrahedra in a random manner to form small spheroids, with the ultimate particles being known as micelles. The micelles consist of SiO_4 in their interior and Si-OH on their surface (Iler 1979; Scott 1993). Gel formation occurs when the interaction of micelles through hydrogen bonding and interparticle condensation becomes significant and continues until the hydrosol, a liquid containing micelles, has solidified into a hydrogel. The resulting transparent rigid mass is then broken, washed with water to remove dissolved salts, dried, and classified. In some cases, a milling step is added (Kleinschmit 1981; Scott 1993). The drying process removes water from the washed hydrogel, which causes a collapse of the gel structure. Fast drying can minimize shrinkage. Hydrogels that can be dried with a negligible loss of pore volume are known as aerogels. If the drying process is slow, a structural state is obtained in which the water is continuously evaporated, and in which there is no more shrinkage of the gel structure, thus producing a xerogel (Welch 1993; Scott 1993).

Silica gels have distinct physical properties, including surface area: 750 to 800 m^2/gm, pore volume: 0.43 cm^3/gm, average pore diameter: 22Å, apparent bulk density: 0.72 gm/cm^3, skeletal density: 2.19 gm/cm^3, specific heat: 920 J/kg·K, thermal conductivity: 522 J/m·h·K, and refractive index: 1.45. Pure silica gel has a hydroxylated surface covered with silanol groups and is hydrophilic. Thermal treatment at 600° to 800°C is required to dehydroxylate silica gels to 1 $OH/100Å^2$, and the surface becomes hydrophobic. A representative composition of silica gels contains 99.71% SiO_2, 0.10% Al_2O_3, 0.09% TiO_2, Fe_2O_3, and trace oxides of 0.07% (Welch 1993; Scott 1993).

Silica gels are used as absorbents for water and polar substances and in food processing (McDonald 1975), as a component in dentifrice formulations, and as matting, free flow, and antiblocking agents.

6. Production and Use of Pyrogenic Silicas

Pyrogenic silicas are highly dispersed silicas formed from the gas phase at high temperatures. In the production process, silica sand is reduced in an electric arc furnace to

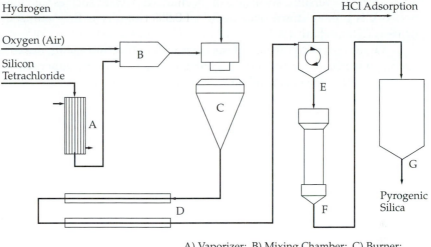

Hydrogen

Oxygen (Air)

Silicon
Tetrachloride

HCl Adsorption

A) Vaporizer; B) Mixing Chamber; C) Burner;
D) Cooling Section; E) Separation; F) Deacidification;
G) Hopper

Figure 86   Flow sheet of flame process for the production of pyrogenic silicas, after Ettlinger (1993).

silicon metal, which is then chlorinated with chlorine or hydrogen chloride to form silicon tetrachloride. Pyrogenic silicas are produced from silicon tetrachloride by flame hydrolysis according to the reaction: $SiCl_4 + 2H_2 + O_2 \rightarrow SiO_2$ (gas) $+ 4HCl$ (gas). After cooling, the silica is separated from the HCl-containing off-gas by cyclones or filters. The hydrogen chloride is washed from the off-gases in absorption columns to give hydrochloric acid for reuse. A general scheme of flame hydrolysis is shown in Fig. 86 (Ettlinger 1993).

Pyrogenic silicas are amorphous, extremely fine, mostly spherical particles, which form aggregates by intergrowth, and agglomeration through cohesion forces. Both the siloxane group, which is hydrophobic, and the silanol group, which is hydrophilic, occur on the surface of the silica particles. As a result of this combination, pyrogenic silicas are wettable. Like precipitated silicas and silica gels, pyrogenic silicas have distinct properties, including surface area (BET): 50–400 m^2/gm, density: 2.2 gm/cm^3, average particle size: 70 to 100Å, and refractive index: 1.45. Pyrogenic silicas are thermally quite stable. Treatment at 1,000°C for one week does not result in any change of morphology nor does it result in crystallization. A representative chemical composition has >99.8% SiO_2, <0.05% Al_2O_3, <0.003% E_2O_3, <0.03% TiO_2, <0.025% HCl, and <0.05% of residue (Kleinschmit 1981).

Pyrogenic silicas are used in large quantities as active fillers in silicone rubber. Ten percent of both hydrophilic and hydrophobolic pyrogenic silicas is required to modify the mechanical properties of vulcanizing silicon rubber, whereas 30% is needed for high temperature vulcanizing silicon rubber (Vondracek and Schätz 1977; Berrod et al. 1981).

As a thickening and thixotropic agent, pyrogenic silica is used to adjust the rheological properties of liquid systems such as paints, thermosetting resins, and printing inks. This is brought about by a reaction between the silanol groups of adjacent particles through hydrogen bonding, which produces a three-dimensional network that extends throughout the liquid and causes an increase in viscosity, a thickening process.

This effect is reversible. By means of mechanical stresses such as shaking or stirring, the network is broken down once again and the viscosity is thus reduced, a thixotropic process (Kleinschmit 1981).

Other uses of pyrogenic silicas include as antisetting agents, dispersants, free-flow agents, adsorbents, antiblocking agents, coatings, catalyst supports, matting agents, grinding agents, thermal insulation, and additive carriers.

REFERENCES

Anderson, G. M., and C. W. Burnham. 1965. The solubility of quartz in supercritical water. *Amer. Jour. Sci.* 263:495–511.

Anon. 1987a. The world's glass industries. In Raw materials for the glass and ceramics industries. Eds. G. M. Clarke and J. B. Griffiths. *IM Glass and Ceramics Survey,* 7–14. London: Industrial Minerals.

Anon. 1987b. The world's ceramics industries. From construction to advanced engineering. In Raw materials for the glass and ceramics industries. Eds. G. M. Clarke and J. B. Griffiths. *IM Glass and Ceramic Survey,* 15–26. London: Industrial Minerals.

Bates, R. L. 1969. *Geology of the industrial rocks and minerals.* New York: Dover Publ., 459p.

Baumgart, W. 1984a. Glass. In *Process mineralogy of ceramic materials.* Ed. W. Baumgart, A. C. Dunham, and G. C. Amstutz, 28–50. New York: Elsevier.

——— 1984b. Refractories. In *Process mineralogy of ceramic materials.* Eds. W. Baumgart, A. C. Dunham, and G. C. Amstutz, 80–103. New York: Elsevier.

Beall, G. H. 1986. Glass ceramics. In Commercial glasses. Eds. D. C. Boyd and J. P. MacDowell. *Adv. Ceramics* 18:157–173.

Benbow, J. 1989. Industrial silica sand, an operational review. *Ind. Minerals* (July):19–39.

Berrod, G., A. Vidal, E. Papirer, and J. B. Donnet. 1981. Reinforcement of siloxanes elastomers by silica. Interactions between an oligomer of poly(dimethyl siloxanes) and a fumed silica. *Jour. Appl. Polym. Sci.* 23:2579–2590.

Blanks, R., and H. Kennedy. 1955. Vol. 1 of *The technology of cement and concrete.* New York: John Wiley, 341p.

Blatt, H. 1982. *Sedimentary petrology.* San Francisco: W. H. Freeman, 564p.

Bradbury, J. C., and H. P. Ehrlinger. 1983. Tripoli. In *Industrial minerals and rocks.* 5th ed. Ed. S. J. Lefond, 1363–1374. New York: Soc. Mining Engineers, AIME.

Büchner, W., R. Schliebs, G. Winter, and K. H. Büchel. 1989. *Industrial inorganic chemistry.* Weinheim: VCH Verlagsgesellschaft mbH, 614p.

Campbell, D. F. 1946. Quartz crystal deposits in the state of Goiaz, Brazil. *Econ. Geol.* 41:773–799.

Dapples, E. C. 1955. General lithofacies relationship of St. Peter sandstone and Simpson group. *Bull. Amer. Asso. Petroleum Geol.* 39:444–467.

Davies, G. F. 1972. Equation of state and phase equilibria of stishovite and a coesitelike phase from shock-wave and other data. *Jour. Geophys. Res.* 77:4920–4933.

Deer, W. A., R. A. Howie, and J. Zussman. 1992. *An Introduction to the Rock-Forming Minerals.* Essex, England: Longman, 698p.

Dickinson, W. R., and C. A. Suczek. 1979. Plate tectonics and sandstone compositions. *Bull. Amer. Assoc. Petroleum Geol.* 63:2164–2182.

Dodd, D. M., and D. B. Frazer. 1965. The 3000–3900 cm^{-1} absorption bands and aneleacticity in crystalline α-quartz. *J. Phys. Chem. Solids* 26:673–686.

Donney, J. D. H., and Y. Le Page. 1978. The vicissitudes of the low-quartz crystal setting or the pitfalls of enantiomorphism. *Acta Cryst.* A34:584–594.

Ettlinger, M. 1993. Silica. In Vol. A23 of *Ullmann's encyclopedia of industrial chemistry,* 635–642. Weinheim: VCH Verlagsgesellschaft, mbH.

Fanderlik, I. 1991. *Silica glass and its applications.* Amsterdam: Elsevier, 304p.

Fellows, L. D. 1967. Tripoli. In Vol. 43, 2nd Series *Mineral and water resources of Missouri,* 220–223. Missouri Div. Geol. Surv. and Water Resources.

Folk, R. L. 1954. The distinction between grain size and mineral composition in sedimentary rock nomenclature. *Jour. Geol.* 62:344–359.

Goldsmith, F. D., T. L. Guidotti, and D. R. Johnston. 1982. Does occupational exposure to silica dust cause lung cancer? *Amer. Jour. Industrial Med.* 3:423–440.

Griffiths, J. 1987. Silica: Is the choice crystal clear? *Ind. Minerals* (April):25–43.

Harben, P. 1983. Tripoli and novaculite—the little known relations. *Ind. Minerals* (January): 28–32.

Harben, P., and J. Purdy. 1991. Dimension stone evaluation. *Ind. Minerals* (February): 47–61.

Harris, W. B. 1972. High-silica resources of Clarke, Frederick, Page, Rockingham, Shenandoah, and Warren Counties, Virginia. *Vir. Div. Mineral. Res. Rept.* 11, 42p Charlottesville, Virginia.

Heaney, P. J. 1994. Structural and chemistry of the low-pressure silica polymorphs. In Vol. 29 of *Reviews in mineralogy.* Ed. P. H. Ribbe, 1–40. Washington, D.C.: Mineral. Soc. Amer.

Hemley, R. J., C. T. Prewitt, and K. J. Kingma. 1994. High-pressure behavior of silica. In Vol. 29 of *Reviews in mineralogy.* Ed. P. H. Ribbe, 41–82. Washington, D.C.: Mineral. Soc. Amer.

Holser, W. T., and G. C. Kennedy. 1959. Properties of water V. pressure-volume-temperature relations of water in the range 400°–1000°C and 100–1400 bars. *Amer. J. Sci.* 257:71–77.

Iler, R. K. 1979. *The chemistry of silica.* New York: John Wiley & Sons, 866p.

Jackson, I. 1976. Melting of silica isotypes SiO_2, BeF_2 and GeO_2 at elevated pressures. *Phys. Earth Planet. Int.* 13:218–231.

Johnston, Jr., W. D., and R. D. Butler. 1946. Quartz crystal in Brazil. *Bull. Geol. Soc. Amer.* 57:601–649.

Kalcheff, K. 1977. Portland cement concrete with stone sand. *Nat. Crushed Stone Assoc. Eng. Rept.* (July 1977), 18p.

Kennedy, G. C. 1950. A portion of the system silica-water. *Econ. Geol.* 45:629–653.

Kerner, D., and P. Kleinschmit. 1993. Silica. In Vol. A23 of *Ullmann's encyclopedia of industrial chemistry,* 642–646. Weinheim: VCH Verlagsgesellschaft, mbH.

Kleinschmit, P. 1981. Silicas and related materials. In *Speciality inorganic chemicals,* Ed. R. Thompson. London: The Royal Soc. of Chemistry.

Knouse, F. L. 1947. Deposits of quartz crystal in Espirito Santo and eastern Minas Gerais, Brazil. *Trans. Amer. Inst. Min. Met. Eng.* 173:173–184.

Kraus, G. 1965. *Reinforcement of elastomers.* New York: Interscience, 611p.

Lamar, J. E. 1927. Geology and economic resources of the St. Peter Sandstone of Illinois. *Bull., Ill. Geol. Surv.,* 53.

——— 1953. Siliceous materials of extreme southern Illinois. *Rept. Invest. 166, Illinois Geol. Surv.,* 30p.

Laudise, R. A. 1970. *The growth of single crystals.* Englewood Cliffs, New Jersey: Prentice-Hall, 352p.

Maskall, K. A., and D. White. 1986. *Vitreous enamelling.* Oxford: Pergamon Press, 112p.

McBride, E. F. 1963. A classification of common sandstones. *Jour. Sedi. Petr.* 33:664–669.

McDonald, R. S. 1975. Study of the interaction between hydroxyl groups of aerosil silica and adsorbed non-polar molecules by infrared spectrometry. *Jour. Amer. Chem. Soc.* 79:850–854.

McLaren, A. C., and P. P. Phakey. 1969. Diffraction contrast from Dauphine twin boundaries in quartz. *Phys. Stat. Solidi* 31:723–737.

Megaw, H. D. 1973. *Crystal structures: A working approach.* Philadelphia: W. B. Saunders, 563p.

Morey, G. M., R. O. Fournier, and J. J. Rowe. 1962. The solubility of quartz in water in the temperature interval from 25° to 300°C. *Geochim. Cosmochi. Acta* 26:1029–1043.

Murphy, T. D., and G. V. Henderson. 1983. Silica and silicon. In *Industrial minerals and rocks.* 5th ed. Ed. S. J. Lefond, 1167–1185. New York: Soc. Mining Engineers, AIME.

Newnham, R. E. 1975. *Structure-property relations.* Berlin: Springer, 324p.

Pepper, J. F., W. De Witt, Jr., and D. F. Demarest. 1954. Geology of the Bedford shale and Berea sandstone in the Appalachian Basin. *U.S.G.S. Prof. Paper* 259, 112p.

Pryor, E. J. 1965. *Mineral processing.* 3rd ed. Amsterdam: Elsevier, 844p.

Scholze, H. 1991. *Glass: Nature, structure, and properties.* New York: Springer-Verlag, 454p.

Schüller, K. H. 1984. Silicate ceramics. In *Process mineralogy of ceramic materials.* Eds. W. Baumgart, A. C. Dunham, and G. C. Amstutz, 1–27. New York: Elsevier.

Scott, R. P. W. 1993. *Silica gel and bonded phases: Their production, properties and use in LC.* Chichester John Wiley & Sons, 261p.

Segrove, H. D. 1956. The production of sand for making colorless glasses. *Jour. Soc. Glass Tech.* 40:363T–375T.

Stow, M. H. 1938. Conditions of sedimentation and sources of the Oriskany sandstone as indicated by petrology. *Bull. Amer. Asso. Petroleum Geol.* 22:541–564.

Taylor, H. F., M. C. Flemings, and J. Wulff. 1959. *Foundry engineering.* New York: John Wiley, 407p.

Taylor, J. R., and A. C. Bull. 1986. *Ceramics glaze technology.* Oxford: Pergamon Press, 265p.

Tien, C. 1989. *Granular filtration of aerosols and hydrosols.* Boston: Butterworth, 365p.

Vondracek, P., and M. Schätz. 1977. Bound rubber and "crepe hardening" in silicone rubber. *Jour. Appl. Polym. Sci.* 21:3211–3222.

Waesche, H. H. 1983. Quartz crystal and optical calcite. In *Industrial minerals and rocks.* 5th ed. Ed. S. J. Lefond, 607–698. New York: Soc. Mining Engineers, AIME.

Welch, W. A. 1993. Silica. In Vol. A23 of *Ullmann's encyclopedia of industrial chemistry,* 629–635. Weinheim: VCH Verlagsgesellschaft, mbH.

Weller, J. M. 1944. Devonian System in southern Illinois. *Bull. Illinois Geol. Surv.* 68:89–102.

Wilborg, H. E., and G. V. Henderson. 1983. Foundry sand. In *Industrial minerals and rocks.* 5th ed. Ed. S. J. Lefond, 271–278. New York: Soc. Mining Engineers, AIME.

Williams, H., F. J. Turner, and C. M. Gilbert. 1982. *Petrography.* 2nd ed. San Francisco: W. H. Freeman, 626p.

Willman, H. B., and J. N. Payne. 1942. Geology and mineral resources of the Marseilles, Ottawa, and Streator quadrangles. *Bull. Illinois Geol. Surv.* 66:71–80.

Zdunczyk, M. J., and M. A. Linkous. 1994. Industrial sand and sandstone. In *Industrial minerals and rocks.* 6th ed. Ed. D. D. Carr, 879–891. Littleton, Colorado: Soc. Mining, Metal. & Exploration.

Zoltai, T., and J. H. Stout. 1984. *Mineralogy, concepts and principles.* Minneapolis: Burgess Publ. Co., 605p.

Sillimanite, Andalusite, and Kyanite

Sillimanite, andalusite, and kyanite are three of the polymorphic forms of Al_2SiO_5 that are used in refractories. Their uses rely upon their ability to produce mullite at high temperatures. Mullite, with a high melting point and excellent thermal properties, is the main material sought after by industry. Mullite is rare in nature.

The major world producers of Al_2SiO_5 minerals are South Africa (andalusite), the United States (kyanite), the former USSR (kyanite), and India (kyanite and sillimanite). France, Brazil, Spain, Australia, Canada, China, and Malawi are also important producers. More than 90% of the annual production of Al_2SiO_5 minerals (about .5 M tons) is used by the steel, glass, and ceramic industries for refractory applications. England, Germany, the United States, Japan, and Brazil are the only producers of synthetic mullite.

MINERALOGICAL PROPERTIES

1. Sillimanite, Andalusite, and Kyanite

Sillimanite, andalusite, and kyanite are polymorphic forms of Al_2SiO_5, and their equilibrium relations are shown in Fig. 87 (Ribbe 1982). Two generally accepted sets of data for the triple point, as given in Figure 87, are 5.5 kb and 622°C (Richardson et al. 1969), and 3.76 kb and 510°C (Holdaway 1971). By calculating specific heat from measured phonon spectra associated with lattice vibration, Salje and Werneke (1982) obtained a triple point at 3.0 to 3.2 kb and 420° to 440°C. The free energies of the three polymorphic forms differ by less than 100 cal/mole.

The structural differences among polymorphic forms are small, but distinct. In sillimanite, the AlO_6 octahedra form chains by edge-sharing. The chains originate at the corners and from the center of an orthorhombic cell and extend parallel to the c-axis. Lateral linkage between chains is made by a double chain of both AlO_4 and SiO_4 tetrahedra. Al coordination in sillimanite is thus half octahedral coordination, and half tetrahedral coordination (Burnham 1963a). Both andalusite and kyanite have similar sets of AlO_6 octahedra chains, but the chains are linked laterally by SiO_4 tetrahedra and AlO_5 polyhedra in alternating positions in andalusite, and by SiO_4 tetrahedra and AlO_6 octahedra in kyanite (Burnham 1963b; Burnham and Buerger 1961). Therefore, Si^{4+} is tetrahedrally coordinated, and one-half of Al^{3+} is octahedrally coordinated in all polymorphic forms. The other one-half of Al^{3+} is in four-, five-, and six-coordinated positions in sillimanite, andalusite, and kyanite, respectively. Andalusite has the largest molar volume among the three polymorphic forms and prefers the lowest pressure environment. On the other hand, kyanite has the smallest molar volume and a preference for the highest pressure environment.

The chemical compositions for all three forms are quite close to the formula, Al_2SiO_5. Ferric iron is the most common cation to enter the structures as a substitute for Al^{3+}, but the amount is very limited. The crystallographic, physical, and optical properties for sillimanite, andalusite, and kyanite are summarized in Table 23 (Ribbe 1982;

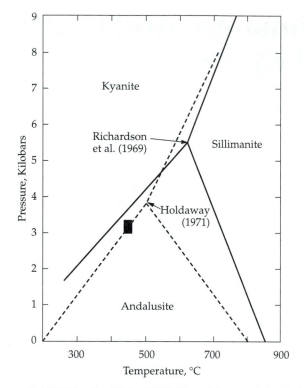

Figure 87 Phase diagram for Al_2SiO_5, after Ribbe (1982). The solid rectangle in the figure represents the triple point of Salje and Werneke (1982).

Deer et al. 1992). The hardness of kyanite varies between 5½ and 7, and depends upon crystallographical orientation: H=5½ parallel to c on {100} and H=7 perpendicular to c on {110}. The well-defined structural positions for Al^{3+} and Si^{4+} in the chemical formula are given in Table 23. Volume expansion as a function of temperature has been studied by means of X-ray diffraction (Winter and Ghose 1979). From 25° to 800°C, the molar volume (in $Å^3$) increases from 331 to 335.92 in sillimanite, from 342 to 348.82 in andalusite, and from 293 to 299.29 in kyanite. All three polymorphic forms decompose to mullite and silica at high temperatures under atmospheric pressure, and none of them is a stable phase in the system Al_2O_3-SiO_2.

2. Mullite

Mullite is a stable phase in the system Al_2O_3-SiO_2, with a composition close to $3Al_2O_3 \cdot 2SiO_2$. The variation in composition resulting from the excess of Al^{3+} in the tetrahedral sites leads to the existence of a composition-dependent number of oxygen vacancies: $2Al^{3+} + \square = 2Si^{4+} + O^{2-}$. The chemical formula for mullite may therefore be expressed as $Al(Al_{1+2x}Si_{1-2x})O_{5-x}$ with $0.085 < x < 0.295$ (Cameron 1977a, 1977b). When x=0, the composition that results is that of sillimanite, andalusite, or kyanite, and when x=0.5, the composition that results is that of corundum. A form of mullite with 89 mole% Al_2O_3, which represents the most alumina-rich form that has been synthesized, has an

Table 23 Crystallographic, physical, and optical properties of sillimanite, andalusite, kyanite, and mullite.

	Sillimanite	Andalusite	Kyanite	Mullite
Formula	$Al^{VI}Al^{IV}Al^{IV}O_5$	$Al^{VI}Al^{V}Si^{IV}O_5$	$Al^{VI}Al^{VI}Si^{IV}O_5$	$Al^{VI}(Al_{1+2x}Si_{1-2x})^{IV}O_5$
Symmetry	Orthorhombic	Orthorhombic	Triclinic	Orthorhombic
Unit Cell				
a (Å)	7.48	7.79	7.12	7.54
b	7.67	7.90	7.85	7.69
c	5.77	5.55	5.57	2.89
α	90°	90°	89.98°	90°
β	90°	90°	101.12°	90°
γ	90°	90°	106.01°	90°
D	3.23–3.27	3.14–3.16	3.53–3.56	3.11–3.26
H	6½–7½	6½–7½	5½–7	6–7
Color	colorless, white	usually pink, also white, rose	blue to white	colorless, white
R.I. α	1.653–1.661	1.633–1.642	1.710–1.718	1.630–1.670
β	1.657–1.662	1.639–±.644	1.719–1.724	1.636–1.675
γ	1.672–1.682	1.644–1.650	1.724–1.734	1.640–1.690

orthorhombic symmetry of a=7.7391Å, b=7,6108Å, and c=2.9180Å (Fischer et al. 1994) (*see* Table 23 for mullite). The general features of the structure of mullite are very similar to the structure of sillimanite, with straight chains of edge-sharing AlO_6 octahedra parallel to the c-axis. They are cross-linked by AlO_4 and SiO_4 tetrahedra that form double chains and run parallel to the c-axis (Burnham 1963c).

The melting relations of mullite are in dispute. Incongruent melting was first proposed by Bowen and Greig in 1924, but a subsequent study by Aramaki and Roy (1962) established congruent melting for mullite at 1,850°C. Experimental data obtained by both Aksay and Pask (1975) and Klug et al. (1987) support incongruent melting, but the melting point is 1,828°C in the former and 1,890°C in the latter. The positions of the α-Al_2O_3 liquid are also different. In addition, a displacement of mullite solid solution toward Al_2O_3 with increasing temperature in the system $Al_2O_3 - SiO_2$ was shown by Klug et al. (1987). These studies suggest that the Al_2O_3 content of the liquidus peritectic, relative to that of mullite at the melting temperature, appears to determine the nature of melting. When Al_2O_3 in mullite is higher than that of the liquidus peritectic, it melts incongruently. Mullite melts congruently when the Al_2O_3 content is lower.

Mullite has good thermal properties. Its thermal conductivity is low and is nearly independent of temperatures up to 800°C (about 4 W/m · K), and then decreases to close to 1 W/m · K at 1,000°C. A further decrease may be made by adding a second phase (Ismail et al. 1990). The coefficients of thermal expansion along three crystallographic axes, a, b, and c, are (x10^{-6}/°C) 3.9, 7.0, and 5.8 for sintered mullite and 4.1, 5.6, and 6.1 for fused mullite (Schneider and Eberhard 1990). The corresponding values of the same set of axes for sillimanite are 2.3, 7.6, and 4.8 (Winter and Ghose 1979).

Mullite also has excellent mechanical properties. Although its properties vary with methods of preparation, a set of representative values is as follows: bending strength of 405 Mpa at room temperature and 350 MPa at 1,300°C; fracture toughness,

2.73 MPa · m$^{0.5}$; Young's modulus, 246 GPa; and Vickers hardness, 11.96 GPa (Ismail et al. 1987). Mullite is characterized by a combination of good thermal and mechanical properties. The following properties make mullite useful in refractories:

1. high melting point,
2. high thermal shock resistance,
3. good hot strength under load,
4. volume stability at high temperatures,
5. low shrinkage,
6. high resilience to aggressive melts and slag,
7. good abrasion resistance,
8. high creep resistance, and
9. low thermal expansion (Metcalfe and Sant 1975; Benbow 1991).

The crystallographic, physical, and optical properties of mullite are listed in Table 23 along with those of sillimanite, andalusite, and kyanite, for the purpose of comparison.

3. Transformation

Sillimanite, andalusite, and kyanite transform to mullite and silica on heating at high temperatures and at oxidizing conditions according to the following reaction:

$$3Al_2SiO_5 \rightarrow 3Al_2O_3 \cdot 2SiO_2 + SiO_2.$$

For kyanite, andalusite, and sillimanite, the temperatures (°C) at which the transformation begins are 1,150°, 1,250°, and 1,300°, respectively. The temperatures at which the transformation completes are 1,300°, 1,500°, and 1,700°. The widths of the transformation interval are 150°, 250°, and 400° (Schneider and Majdic 1979, 1980, 1981). Both the temperature of transformation and the width of the transformation interval increase from kyanite to andalusite and to sillimanite.

Transformations from kyanite to mullite and from andalusite to mullite are topotactic (Schneider and Majdic 1979, 1980), whereas the transformation from sillimanite to mullite is a multiple step reaction, with an intermediate formation of disordered sillimanite (Guse et al. 1979; Winter and Ghose 1979). The kinetics of transformation for each of the reactions, plotted as a function of reaction time and temperature, is given in Fig. 88 (Schneider and Majdic 1979, 1980, 1981).

MODE OF OCCURRENCE AND DISTRIBUTION OF DEPOSITS

Sillimanite, andalusite, and kyanite all occur in metamorphosed rocks, but each forms under certain distinctive ranges of temperature and pressure (Espenshade 1973). Sillimanite is normally the product of regional and contact metamorphism of argillaceous rocks at relatively high temperatures, as in sillimanite-cordierite gneiss, biotite-sillimanite hornfels, micaceous sillimanite schist, and coarse-grain quartz-sillimanite gneiss. Andalusite is typically found in the argillaceous rocks of contact aureoles around igneous intrusives, most commonly in hornfels, in association with biotite and cordierite. Kyanite forms in quartzite, schist, and gneiss under conditions of moderate

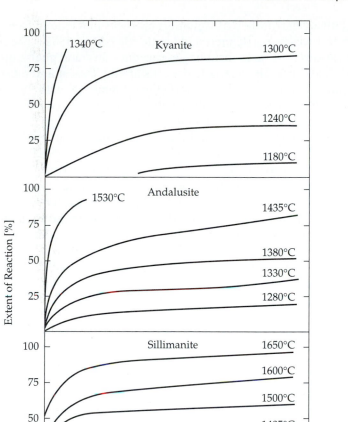

Figure 88 Transformation of sillimanite, andalusite, and kyanite to mullite, as a function of reaction time and temperature modified, after Schneider and Majdic (1979, 1980, 1981).

regional metamorphism. It is commonly found in association with quartz, garnet, corundum, and muscovite. All three polymorphic forms occur in pegmatites and quartz veins as well as in sediments as detrital minerals (Hyndman 1985).

Economic deposits of Al_2SiO_5 minerals have been identified in at least twenty-five to thirty countries. The principal producers are South Africa (andalusite), the United States (kyanite), the former USSR (kyanite), and India (kyanite and sillimanite). Other producers are France (andalusite), Australia (sillimanite), Spain (kyanite), Brazil (kyanite), Canada (sillimanite), China (sillimanite and andalusite), and Malawi (kyanite).

The United States: The United States is an important producer of Al_2SiO_5 minerals, mainly kyanite. Its deposits occur in the Piedmont provinces and Blue Ridge, in a relatively narrow belt of metamorphic rocks that extend from southeastern Virginia to central Alabama (Bennett and Castle 1983; Marr 1992). The deposits of kyanite occur in various host rocks including micaceous schist and gneiss, quartz veins, quartzite, and residual sands. The quartzite deposits are the largest and most numerous of the group. Kyanite and quartz are the predominant minerals, with kyanite generally constituting 10 to 30% of the rocks. Muscovite is a common associated mineral, and rutile and pyrite are widespread minor constituents. The rock is extremely resistant to weathering and erosion, and the larger deposits form ridges or monadnocks several hundred meters high and one kilometer or more long. The deposits have a great range of sizes. Principal deposits are in the Farmville district, Virginia, the Kings Mountain district of North Carolina and South Carolina, and the Graves Mountain district, Georgia (Espenshade and Potter 1960). In the kyanite schist belt of Hebersham and Rabun Counties, Georgia, the Al_2SiO_5 mineral has accumulated in residual soil and in stream placers. Kyanite is among the heavy minerals in the sandy formations of the Atlantic Coastal Plain (Espenshade and Potter 1960).

Sillimanite is also distributed in quartzite, micaceous schist, and gneissin deposits, but in small quantities. Sillimanite is also present in the heavy mineral assemblages of the sandy formations of the Atlantic Coastal Plain.

South Africa: South Africa is the world's leading producer of andalusite (Batha 1989; Lange and McCracken 1992). Deposits occur in the andalusite-bearing slates of the Pretoria Group, which exists within the metamorphic aureole associated with the intrusion of the Bushveld igneous complex (Power 1986). In the Transvaal region, there are three principal mining districts: (1) the Groot Marico-Zeerust in the west, (2) the Thabazimbi in the northwest, and (3) Northern Lydenburg in the east (Blain and Coetzee, 1976). The quality of the andalusite produced in the three mining districts varies from 53 to 54% Al_2O_3 and 1.5 to 2.0% Fe_2O_3 at Groot Marico-Zeerust, to 59 to 60% Al_2O_3 and less than 0.6% Fe_2O_3 at Thabazimbi, and to 59 to 60% Al_2O_3 and 0.9% Fe_2O_3 at Lydenburg (Power 1986). Sillimanite is also mined in South Africa in the Namaqualand region close to the border with Namibia. Alluvial andalusite rich sand, which contains 10 to 50% andalusite, is found in the drainage system of the Klein Marico River and its tributaries (Van Rooyen 1951; Grobbelaar 1994).

The Former USSR: Kyanite deposits occur on the Kola peninsula as pockets and lenses in micaceous schist, gneiss, and quartzite and are the result of the metamorphism of kaolinitic clays. The principal mines are at Kiev, where kyanite and kyanite-staurolite mica schist are traceable over a distance of 140 km at a thickness of 80 to 150 meters. Kyanite occurs in either a fibrous form or as an acicular aggregate, with an Al_2O_3 content of 57 to 58% (Kuzvart 1984).

India: Massive kyanite-quartz rock and kyanite schist deposits occur at various areas in a metamorphic belt that is about 110 km long in Singhbhum, with principal deposits in the Lapsa Buru district at the western end of the belt (Klinefelter and Cooper 1961). The deposits exist either as rock ridges outcropping at the surface, or as boulders that are generally found at depths of 1.2 to 1.5 meters and have sizes varying from 6 mm in diameter to several tons in weight. The average kyanite content of the deposits is approximately 35%, with 50% quartz, 10% mica, and 5% of associated rutile, garnet, and amphibole. The Al_2O_3 content reaches a level of 58% (Power 1986).

Large deposits of massive sillimanite and sillimanite-corundum occur in sillimanite schists in two districts: the Khasi Hills plateau of Assam, and the Sidhi District of Madhya Pradesh State (Kleinfelter and Cooper 1961). At Khasi Hills, sillimanite occurs in the form of large boulders that have been exposed by soil removal and quarried by blasting. The composition of the ore has 61% Al_2O_3, with a maximum of 1.0% Fe_2O_3 and 1.5% TiO_2. In the Sidhi District, the ore occurs as clusters of massive boulders in association with pelitic schists. Chemical analysis gives an alumina content of 57.5%, with an average of 0.7% Fe_2O_3 and 2.16% TiO_2 (Power 1986). Alluvial sillimanite is mined in Kerala, Tamil Nadu, and Orissa States.

INDUSTRIAL PROCESSES AND USES

1. Beneficiation of Sillimanite, Andalusite, and Kyanite

When Al_2SiO_5 minerals are marketed as raw materials, they require minimal treatment, except for cobbing and hand-sorting of the crushed lumps to remove schist, quartz, muscovite, rutile, and iron oxide minerals. At Assam, India, sillimanite boulders are sliced into blocks by using circular saws with diamond teeth or frame saws with steel shot as an abrasive (Foster 1960). In India, a two-stage froth flotation is used for the beneficiation of Lapso Buru kyanite, which is characterized by a high mica content: the cationic flotation of mica in a neutral/acidic medium, and the anionic flotation of kyanite in an acidic medium, with boramine C and sodium petroleum sulphonate as collectors, respectively. After the rougher kyanite flotation, the process requires three cleaner kyanite flotations in order to reject the gangue minerals to the maximum extent possible, without effecting kyanite recovery. The final stage of beneficiation is the wet high intensity magnetic separation stage, which is used to reject the feebly magnetic minor gangue minerals such as tourmaline, garnet, iron oxide, and others (Amanullah et al. 1978).

Various beneficiation procedures have been used in different plants for South Africa andalusite. Generally, the procedure consists of from one to three stages of crushing, screening and scrubbing, two stages of heavy medium separation and drying, and multistage dry magnetic separation (Grobbelaar 1994).

The beneficiation processes used in various mines in the United States are similar in operation. The treatment for kyanite ore, which is used at Henry Knob, South Carolina, is illustrated in Fig. 89. The broken ore is crushed in a jaw crusher to yield a 20 cm product. This step is followed by a series of secondary crushing steps performed in a cone crusher to yield a product of 5 cm. The product is then screened, crushed in a cone crusher again, ground in a rodmill, and then screened once again. The preliminary flotation, which is followed by flotations in six roughing and six cleaning cells, removes sulfides. After drying in a rotary dryer, the concentrates are further treated in a magnetic separator. Dry grinding in a rodmill or a pebble mill completes the process (Klinefelter and Cooper 1961; Sweet 1994).

Specifications for sillimanite, andalusite, and kyanite have not been well established. A set of general requirements for kyanite in the United States is as follows: a minimum of 59% Al_2O_3, a maximum of 39% SiO_2, a maximum of 0.75% Fe_2O_3, and a

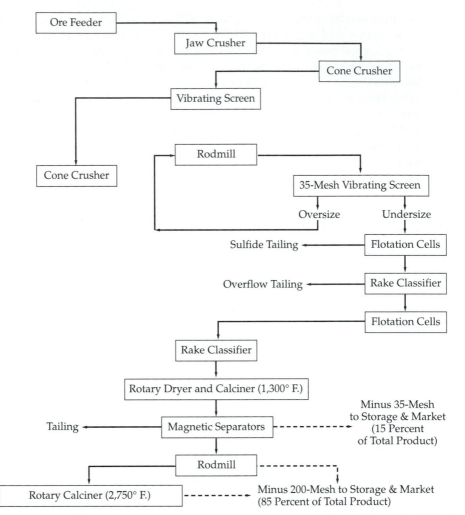

Figure 89 Flow sheet illustrating the beneficiation of kyanite at Henry Knob, near Clover, South Carolina, after Klinefelter and Cooper (1961).

maximum of 1.0% (CaO+MgO+ Na$_2$O+K$_2$O). A pyrometric cone equivalent of 37 or better is required.

2. Industrial Uses of Sillimanite, Andalusite, and Kyanite

The Al$_2$SiO$_5$ minerals are all used in refractories. Andalusite and sillimanite have lower volume expansion with increasing temperatures than kyanite, as illustrated by their mineralogical characteristics. They may be used without calcination, and consequently the cost of energy consumption is much reduced. A volume expansion of 6% occurs in sillimanite, 4% in andalusite, and 16% in kyanite, during conversion to mullite. However, the industrial uses of the Al$_2$SiO$_5$ minerals are, in many cases, dependent on availability. Kyanite is the most common mineral used because of its abundance and the higher quality of minable deposits.

Developments in iron and steel production have led to increasingly severe furnace conditions. The current tendency is to move away from fire-clay materials, toward materials with a higher alumina content. Four classes of silica-alumina refractories are generally recognized. They are (1) low-medium alumina (25 to 50% Al_2O_3), including fireclays, aluminous fireclays, and pyrophyllite; (2) medium-high alumina (30 to 70% Al_2O_3), including sillimanite, andalusite, kyanite, and mullite; (3) high alumina (75 to 90% Al_2O_3), including bauxite-based aluminosilicates; and (4) high alumina (>90% Al_2O_3), including alumina, high alumina-carbon, and alumina-chrome (Benbow 1989).

Kyanite is used mostly for monolithic forms, but also for whitewares, acoustic tiles, high tension insulators, ceramic fibers and blankets, and kiln furniture. Andalusite is, on the other hand, used for both shaped and monolithic forms. Andalusite that contains 54 to 60% Al_2O_3 has become the standard general purpose refractory, replacing the more traditional firebricks. In the manufacture of refractory bricks for heavy duty, materials consisting of as much as 60% coarse-grained Al_2SiO_5 minerals are required. Fine-grained material is suitable for the production of monolithic refractories (McMichael 1990).

The main uses of Al_2SiO_5 bricks and shapes in the steel industry are as (1) checker bricks, domes, and combustion shafts in hot blast stoves; (2) linings for stack and bosh areas, bustle pipes, tilting troughs, swivelling troughs, and iron and slag runners in blast furnaces; and (3) large capacity ladles, torpedo ladles, steelworks ladles, and continuous casting tundishes in ladles (McMichael 1990). In addition, Al_2SiO_5 bricks and shapes are also used in cement kilns, copper roasting, and glass melting furnaces.

3. Preparation and Processing of Mullite

Mullite is formed with silica during the treatment of sillimanite, andalusite, and kyanite, at temperatures above 1,200°C. This conversion process, which is used to produce mullite, represents a major use of Al_2SiO_5 minerals in refractories. Mullite that is formed without silica is prepared by means of many processes and is generally grouped into three main classes: (1) sintered mullite, (2) fused mullite, and (3) chemical-mullite.

Sintered Mullite: Sintered mullite is mullite that is prepared by mixing powdered raw materials through solid-solid or transient liquid-phase reactions. Sintered mullite can be prepared by the direct union of alumina and silica, but sintering temperatures required for mullitization by this process are too high. Mixtures of kaolinite, pyrophyllite, the Al_2SiO_5 minerals, boehmite, and diaspore are generally used to facilitate a reaction at lower temperatures (Sacks and Pask 1982a, 1982b; Schneider et al. 1987).

Mixtures of Bayer alumina, kaolinite (or other aluminosilicates), and refractory bauxite are used as starting materials in the preparation of sintered mullite. These minerals are crushed, dried, and milled. The blended mixtures are ground to the required grain size, pressed into briquets, and treated in a tunnel dryer to reduce the moisture content from 20% to about 2%. The process that produces mullite by calcination takes place in a rotary kiln at temperatures above 1,750°C. The hard briquets produced are crushed in a jaw crusher, passed through a pan mill to produce rounded grains, and treated with a magnetic separator to remove iron-containing materials (Hawkes 1962).

Fused Mullite: Fused mullite is produced by melting a mixture of Bayer alumina and silica sand in an electric arc furnace above a temperature of about 2,000°C. The melt obtained is then cast into ingot molds and cooled to room temperature. The chemical composition of fused mullite depends largely on the crystallization temperature and the cooling rate. Fused mullite usually has the composition $2Al_2O_3 \cdot SiO_2$, with a 66.6 mole% Al_2O_3 content, as compared to the 60 mole% Al_2O_3 content of sintered mullite. Fused mullite with 76 mole% Al_2O_3 has been reported (Kriven and Pask 1983).

Zirconia mullite is produced by fusion of alumina and zirconium silicate or zircon with silica. It is highly resistant to the leaching effects of molten glass and certain types of slag. It also has a low coefficient of thermal expansion.

Chemical-Mullite: Mullite, which is prepared by the sol-gel method, hydrolysis method, and spray pyrolysis and other chemical methods, is designated as chemical-mullite. Although many different routes have been chosen for the preparation of mullite in these methods, the basic features of each method are more or less correlated with one another. A brief description of sol-gel, hydrolysis, and spray pyrolysis methods is given below:

In the Sol-gel method, mullite is prepared by mixing sols of silica and alumina in a neutral pH environment (Ghate et al. 1973). Alumina sol is prepared by dispersing ultrafine particles of γ-Al_2O_3 in HCl and silica sol, by using fumed or colloidal silica. At pH 7, the surfaces of the alumina particles are positively charged, whereas those of the silica particles are negatively charged. Intimate mixing of the two sols particles is induced by heteroflocculation. The temperature of mullitization is at about 1,200°C, much lower than that for the preparation of sintered mullite.

In the hydrolysis method, xerogels are first prepared by dissolving TEOS and aluminum nitrate ($Al(NO_3)_3 \cdot 9H_2O$) in absolute ethanol. The solution is then gelled in a water bath by heating at 60°C. The gels are dried at 60°C to produce powders, which are then treated at high temperatures. As shown in DTA, the formation of mullite, which is poorly crystallized, starts at 1,015°C (Hoffman et al. 1984).

The spray pyrolysis method uses droplets formed from a solution that is sprayed into a furnace and heated at high temperatures. This process combines the evaporation of solvents, the precipitation of solids, and thermal decomposition, instantaneously. For example, TEOS and aluminum nitrate are dissolved in a water-methanol solution (1:1 by volume). Droplets are formed by using a borosilicate glass atomizer with compressed air that is sprayed into a furnace that is heated at 350° to 650°C (Kanzaki et al. 1985). DTA shows that the product, as sprayed, is amorphous, and mullite appears at 1,000°C.

4. Industrial Uses of Mullite

Whether mullite is sintered, fused, or chemically prepared, its main uses are in refractories of both shaped and monolithic forms. In the production of glass and steel, mullite is the major component of furnace linings, as well as for any structural parts required for service under severe thermal conditions. Mullite fibers are used as heat-insulating materials for furnaces (Benbow 1991).

Mullite ceramics are suitable for high-strength applications at high temperatures. Mullite ceramics are used for shelf, saggers, and props, because these forms cannot tolerate any creep deformation. Because mullite ceramics are impermeable to gases, they are in demand for protection tubes, thermocouple tubes, and crucibles. The use of stainless steel for belt conveyors in tunnel kilns has been replaced by mullite ceramics. In addition, mullite is considered a good candidate for use in heat exchangers (Schneider et al. 1994).

Mullite that has a dielectric constant of 6.7 results in about 17% lower signal transmission delay time than alumina (dielectric constant: 9.8), and its low thermal expansion coefficient matches well with that of silicon. Furthermore, when mullite is prepared by the sol-gel and hydrolysis methods, it can be sintered at temperatures as low as 1,250°C. Mullite glass-ceramics that contain cordierite or spodumene further improve these properties, thus allowing mullite to become a candidate for high performance packaging applications (Ismail et al. 1990; Aksay et al. 1991).

Mullite and mullite glass ceramics have the potential to be used as an infrared transparent window material for infrared wavelength ranges of 3 to 5 μm and in the visible light range under high-temperature, chemically harsh, and mechanically stressed environments (Aksay et al. 1991).

REFERENCES

Aksay, I. A., and J. A. Pask. 1975. Stable and metastable equilibria in the system SiO_2-Al_2O_3. *Jour. Amer. Ceramic Soc.* 58:507–512.

Aksay, I. A., D. M. Dabbs, and M. Sarikaya. 1991. Mullite for structural, electronic, and optical applications. *Jour. Amer. Ceramic Soc.* 74:2343–2358.

Amanullah, S., G. M. Rao, and K. Satyanarayana. 1978. Beneficiation of mica-quartz-bearing kyanite. *Ind. Minerals Process. Suppl.*, 24–29.

Aramaki, S., and R. Roy. 1962. Revised phase diagram for the system Al_2O_3-SiO_2. *Jour. Amer. Ceramic Soc.* 45:229–242.

Batha, J. C. 1989. Andalusite—trends in the supply and demand. *Mineral. Bur. South Africa Bull.* B11/89, 7p.

Benbow, J. 1989. Steel industry minerals. *Ind. Minerals* (October): 27–43.

——— 1991. Synthetic mullite: Refractories' shock absorber. *Ind. Minerals* (February): 39–45.

Bennett, P. J., and J. E. Castle. 1983. Kyanite and related minerals. In *Industrial minerals and rocks.* 5th ed. Ed. S. J. Lefond, 799–807. New York: Soc. Mining Engineers, AIME.

Blain, M. R., and C. B. Coetzee. 1976. Andalusite, sillimanite, and kyanite. In *Mineral resources of the Republic of South Africa.* 5th ed. Ed. C. B. Coetzee. *Dept. Mines Geol. Surv. Handbook* 7:255–260.

Bowen, N. L., and J. W. Greig. 1924. The system Al_2O_3-SiO_2. *Amer. Jour. Sci.* 7:238–254.

Burnham, C. W. 1963a. Refinement of the crystal structure of sillimanite. *Zeit. Krist.* 118:127–148.

——— 1963b. Refinement of crystal structure of kyanite. *Zeit. Krist.* 118:337–360.

——— 1963c. Crystal structure of mullite. *Carnegie Inst. Washington, Yearbook* 62:223–227.

Burnham, C. W., and M. J. Buerger. 1961. Refinement of crystal structure of andalusite. *Zeit. Krist.* 115:269–290.

Cameron, W. E. 1977a. Composition and cell dimensions of mullite. *Bull. Amer. Ceramic Soc.* 56:1003–1007.

———— 1977b. Mullite: A substituted alumina. *Amer. Mineral.* 62:747–755.

Deer, W. A., R. A. Howie, and J. Zussman. 1992. *An introduction to the rock-forming minerals.* Essex, England: Longman, 698p.

Espenshade, G. H. 1973. Kyanite and related minerals. *U.S.G.S. Prof. Paper* 820:307–312.

Espenshade, G. H., and D. B. Potter. 1960. Kyanite, sillimanite, and andalusite deposits of the southeastern states. *U.S.G.S. Prof. Paper* 336, 121p.

Fischer, R. N., H. Schneider, and M. Schmücker. 1994. Crystal chemistry of Al-rich mullite. *Amer. Mineral.* 79:983–990.

Foster, W. R. 1960. The sillimanite group—kyanite, andalusite, sillimanite, dumortierite, topaz. In *Industrial minerals and rocks.* 3rd ed. Ed. J. L. Gillson, 773–789. New York: AIME.

Ghate, B. B., D. P. H. Hasselman, and R. M. Spriggs. 1973. Synthesis and characterization of high purity, fine-grained mullite. *Bull. Amer. Ceramic Soc.* 52:670–672.

Grobbelaar, A. P. 1994. Andalusite. In *Industrial minerals and rocks.* 6th ed. Ed. D. D. Carr, 913–920. Littleton, Colorado: Soc. Mining, Metal. & Exploration.

Guse, W., H. Saalfeld, and J. Tjandra. 1979. Thermal transformation of sillimanite single crystals. *Neues Jahrb. Mineral. Monatsh,* Abt. A, 175–185.

Hawkes, W. H. 1962. The production of synthetic mullite. *Trans. Br. Ceramic Soc.* 61:689–703.

Hoffman, D. W., R. Roy, and S. Komarneni. 1984. Diphasic xerogels, a new class of materials: Phases in the system Al_2O_3-SiO_2. *Jour. Amer. Ceramic Soc.* 67:468–471.

Holdaway, M. J. 1971. Stability of andalusite and the aluminum silicate phase diagram. *Amer. Jour. Sci.* 271:97–131.

Hyndman, D. W. 1985. *Petrology of igneous and metamorphic rocks.* New York: John-Wiley & Sons, 786p.

Ismail, M. G. M. U., Z. Nakai, and Z. Somiya. 1987. Microstructure and mechanical properties of mullite prepared by the sol-gel method. *Jour. Amer. Ceramic Soc.* 70: C7–C8.

Ismail, M. G. M. U., H. Tsumatori, and Z. Nakai. 1990. Preparation of mullite-cordierite composite powder by the sol-gel method: Its characteristic and sintering. *Jour. Amer. Ceramic Soc.* 73: 537–543.

Kanzaki, S., H. Tabata, T. Kumazawa, and S. Ohta. 1985. Sintering and mechanical properties of stoichiometric mullite. *Jour. Amer. Ceramic Soc.* 68:C6–C7.

Klinefelter, T. A., and J. D. Cooper. 1961. Kyanite, a material survey. *U.S.B.M. Infor. Circ.,* 8040, 55p.

Klug, F. J., S. Prochazka, and R. H. Doremus. 1987. Alumina-silica phase diagram in the mullite region. *Jour. Amer. Ceramic Soc.* 70:750–759.

Kriven, W. M., and J. A. Pask. 1983. Solid solution range and microstructure of melt-grown mullite. *Jour. Amer. Ceramic Soc.* 66:649–654.

Kuzvart, M. 1984. *Industrial minerals and rocks.* Amsterdam: Elsevier, 545p.

Lange, S. P., and M. H. McCracken. 1992. Potential and reliability of supply for high-grade andalusite from RSA and alternative sources. *Proc. 10th Industrial Minerals Intern. Congr., San Francisco* Ed. J. B. Griffiths, 82–88.

Marr, J. D. 1992. Geology of the kyanite deposits at Willis Mountain, Virginia, Proc. 26th Forum on the Geology of Industrial Minerals. ed. P. C. Sweet. *Virginia Mineral Res. Publ.* 119:129–134.

McMichael, B. 1990. Alumino-silicate minerals. *Ind. Minerals* (October): 27–43.

Metcalfe, B. L., and J. H. Sant. 1975. The synthesis, microstructure and physical properties of high purity mullite. Trans. Brit. Ceramic Soc., 74, 193–201.

Power, T. 1986. Sillimanite minerals, Europe places demands on andalusite, in "Raw Materials for the Refractories Industry," *IM Refractories Survey,* 2nd edition, Ed., E. M. Dickson, Ind. Minerals, p. 77–91.

Ribbe, P. B. 1982. Kyanite, andalusite, sillimanite, and other aluminum silicates. In Vol. 5 of *Reviews of mineralogy.* Ed. P. B. Ribbe, 189–214. Washington, D.C.: Mineral. Soc. Amer.

Richardson, S. W., M. C. Gilbert, and P. M. Bell. 1969. Experimental determination of kyanite-andalusite and andalusite-sillimanite equilibria: The aluminum silicate triple point. *Amer. Jour. Sci.* 167:259–272.

Sacks, M. D., and J. A. Pask. 1982a. Sintering of mullite-containing materials: I, Effect of composition. *Jour. Amer. Ceramic Soc.* 65:65–70.

——— 1982b. Sintering of mullite-containing materials: II, Effect of agglomeration. *Jour. Amer. Ceramic Soc.* 65:70–77.

Salje, E., and C. Werneke. 1982. The phase equilibrium between sillimanite and andalusite as determined from lattice vibration. *Acta Cryst.* 15:65–68.

Schneider, H., and E. Eberhard. 1990. Thermal expansion of mullite. *Jour. Amer. Ceramic Soc.* 73:2073–2076.

Schneider, H., and A. Majdic. 1979. Kinetics and mechanism of the solid-state high-temperature transformation of andalusite (Al_2SiO_5) into 3/2-mullite ($3Al_2O_3 \cdot 2SiO_2$) and silica (SiO_2). *Ceramurgia Int.* 5:31–36.

——— 1980. Kinetics of the thermal decomposition of kyanite. *Ceramurgia Int.* 6:32–37.

——— 1981. Preliminary investigation on the kinetics of the high-temperature transformation of sillimanite to 3/2-mullite plus silica and comparison with the behaviour of andalusite and kyanite. *Sci. Ceram.* 11:191–196.

Schneider, H., K. Okada, and J. Pask. 1994. *Mullite and mullite ceramics.* Chichester: John Wiley & Sons, 251p.

Schneider, H., J. Wang, and A. Majdic. 1987. Firing of refractory-grade Chinese bauxites under oxidizing and reducing atmospheres. *Ceramic Forum Intern., Ber. Dtsch. Keram. Ges* 64:28–31.

Sweet, P. C. 1994. Kyanite and related minerals in *Industrial minerals and rocks.* 6th ed. Ed. D. D. Carr, 921–927. Littleton, Colorado: Soc. Mining Metal. & Exploration.

Van Rooyen, D. P. 1951. Fluvial andalusite: Deposits in the Marico District. *Union of South Africa Dept. Mines, Geol. Ser. Bull.,* 19.

Winter, J. K., and S. Ghose. 1979. Thermal expansion and high-temperature crystal chemistry of the Al_2SiO_5 polymorphs. *Amer. Mineral.* 64:573–586.

Soda Ash

Soda ash, or sodium carbonate, is one of the most important raw materials used in the chemical and glass industries. Although the bulk of the world's soda ash is manufactured by the Solvay process in more than fifty countries, in the United States it has been increasingly produced from trona and brine since World War II. The annual world production of soda ash was estimated to be from 25 to 35 M tons during the first half of the 1990s. Seventy-five percent of this total was manufactured and 25% was derived from natural sources.

MINERALOGICAL PROPERTIES

There are more than a dozen Na_2CO_3-bearing carbonate minerals, but only trona, $Na_2CO_3 \cdot NaHCO_3 \cdot 2H_2O$, which contains 70.4% Na_2CO_3, is of major economic importance. Other carbonate minerals include natron ($Na_2CO_3 \cdot 10H_2O$, 37.1% Na_2CO_3), thermonantrite ($Na_2CO_3 \cdot H_2O$, 85.5% Na_2CO_3), nahcolite ($NaHCO_3$, 63.1% Na_2CO_3), shortite ($Na_2CO_3 \cdot 2CaCO_3$, 34.6% Na_2CO_3), Gaylussite ($Na_2CO_3 \cdot CaCO_3 \cdot 5H_2O$, 35.8% Na_2CO_3), pirssonite ($Na_2CO_3 \cdot CaCO_3 \cdot 2H_2O$, 48.8% Na_2CO_3), and hanksite ($2Na_2CO_3 \cdot 9Na_2SO_4 \cdot KCl$, 13.6% Na_2CO_3). Trona is a glassy or translucent salt, and crystallizes with a monoclinic symmetry. It is normally white or gray, but when stained, a yellowish or brownish color becomes dominant. The mineral has a hardness of 2.5 and a specific gravity of 2.15. Its refractive indices are 1.412 (α), 1.492 (β), and 1.540 (γ). Trona is commonly fibrous, columnar, bladed, or massive. Trona is very soluble in water and effervesces with acids.

Both thermonantrite ($Na_2CO_3 \cdot H_2O$) and natron ($Na_2CO_3 \cdot 10H_2O$) are sodium carbonate hydrates, and both are stable in the system Na_2CO_3-H_2O. With the additional hydrate phase $Na_2CO_3 \cdot 7H_2O$, they form equilibrium assemblages in the system below 105°C. All hydrates melt incongruently. Their melting temperatures are 105°C for monohydrate, 35.37°C for heptahydrate, and 32.0°C for decahydrate. Above 105°C, phase relations in the system Na_2O-H_2O illustrate a simple solubility relation of Na_2CO_3, which melts congruently at 851°C. It decomposes at approximately 1,000°C to Na_2O and CO_2 (Newkirk and Aliferis 1958; Thieme 1993).

The basic properties of sodium carbonate solutions including density, pH, viscosity, and vapor pressure have been well established. The density (gm/cm^3) increases with Na_2CO_3 concentration and decreases with temperature in a linear relation (Goncalves and Kestin 1981). Selected data are given below:

Na_2CO_3 wt%	20°C	30°C	40°C	50°C	60°C
3.168	0.99204	0.995647	0.992211	0.988036	0.983199
8.368	1.06411	1.06039	1.05609	1.05125	1.04600
13.28	1.11439	1.11226	1.10520	1.09998	1.09441
15.185	1.15343	1.14865	1.14350	1.13799	1.13223
17.20	1.16018	1.15534	1.15012	1.14458	1.13879
23.92	1.2240	1.2149	1.2136	—	—
30.30	1.2946	1.2917	1.2836	—	—

The pH values of Na_2CO_3 solutions at 25°C increase with an increase in concentration (gm Na_2CO_3 in 100 gm solution), as illustrated by the following selected data (Lortie and Demers 1940; Peiper and Pitzer 1982):

wt%	pH	wt%	pH
0.0055	10.274	2.076	11.63
0.0217	10.612	4.241	11.453
0.0887	10.919	5.979	11.79
0.2246	11.063	7.816	11.82
0.4376	11.171	9.583	11.84
0.8609	11.254	12.034	11.617
1.806	11.357	18.555	11.699

The viscosity (centipoise) of sodium carbonate solutions increases with concentration and decreases with temperature (Correia et al. 1980; Goncalves and Kestin 1981). Selected data are given below:

Na_2CO_3 wt%	20°C	40°C	60°C	80°C
0	1.0038	0.65782	0.47498	0.335
3.17	1.1778	0.7669	0.5576	—
5.66	—	0.872	0.627	0.456
9.16	—	1.060	0.745	0.555
11.02	1.8974	1.1937	0.8314	—
14.58	2.4787	1.5096	1.0278	—
16	2.76	1.636	—	0.78
20	—	2.240	—	0.97
24	—	3.182	—	1.10
27.10	8.695	4.1772	—	1.28
30.30	13.228	5.760	—	—

The vapor pressure (mm Hg) of sodium carbonate solutions decreases with concentration and increases with temperature, as illustrated by the selected data given below (Taylor 1995):

Temp. °C	Na_2CO_3 wt%		
	5	15	25
20	17.2	16.3	—
40	54.2	51.6	48.4
60	146.5	139.9	131.6
80	345.0	335.2	315.0
100	746.0	715.0	676.0

Na_2CO_3 is hygroscopic, and when it absorbs moisture and carbon dioxide from the atmosphere, its alkalinity decreases with the formation of $NaHCO_3$ according to the following reaction:

$$Na_2CO_3 + H_2O + CO_2 \rightarrow 2NaHCO_3.$$

By reacting with H_2O vapor above 400°C, Na_2CO_3 produces sodium hydroxide and carbon dioxide, according to the following reaction (Thieme 1993):

$$Na_2CO_3 + H_2O \rightarrow 2NaOH + CO_2.$$

GEOLOGICAL OCCURRENCE AND DISTRIBUTION OF DEPOSITS

Na_2CO_3-containing minerals commonly occur as precipitates from alkaline lakes and marshes at shallow depths, on the bottom, or around the shore, as buried beds of variable thicknesses, and as constituents of brine.

There are many known occurrences of sodium carbonate in the world. Brief descriptions of twenty occurrences in the United States and thirty-seven occurrences in other parts of the world have been given by Mannion (1983). From a list of the largest soda ash occurrences compiled by Garrett (1992), estimated reserves of more than 80 million metric tons and with high concentration are given below:

> Green River, Wyoming: Trona, 90%, 100,000 M tons.
> Piceance Creek, Colorado: Nahcolite, 40%, 29,000 M tons.
> Lake Magadi, Kenya: Trona, 94%, 4,000 M tons and brine.
> Searles Lake, California: Trona, 25%, 2,000 M tons and brine.
> Lake Natron, Tanzania: Trona, 80%, 1,000 M tons and brine.
> Lake Shalla, Ethiopia: Brine, 1.1% Na_2CO_3, 700 M tons.
> Denison Trough, Australia: Brine, 2.0% $NaHCO_3$, 500 M tons.
> Lake Van, Turkey: Brine, 1.6% Na_2CO_3, 340 M tons.
> Beypazari, Turkey: Trona, 83%, 235 M tons.
> Anpeng, China: Trona, nahcolite, 85%, 130 M tons.
> Wucheng Basin, China: Trona, nahcolite, 70%, 130 M tons.
> Sua Pan, Botswana: Brine, 2.5% Na_2CO_3, 230 M tons.
> Texcoco, Mexico: Brine, 3.7% Na_2CO_3, 80 M tons.

The soda ash resources of Green River, Searles Lake, and Lake Magadi have been selected for a brief description because they are the largest and the richest resources, and because each has distinctive characteristics:

Green River Basin, Wyoming: Numerous and extensive beds of trona were formed in a vast Eocene lake that is now represented by the deposits of the Green River Formation. In the Green River basin, southwest Wyoming, some 3,100 to 3,600 km^2 are underlain by nearly flat trona beds at depths of 100 to 1,000 meters (Mannion 1983; Kostick 1994). The Green River Formation is of lacustrine origin and has the form of a great lens, or pile of lenses, within an enormous volume of flutiatile sediments, which have been divided into the Wasatch and Bridger Formations. The former underlies the Green River Formation, whereas the latter overlies it in most parts of the basin, except around the margins, where the Bridger rests directly upon the Wasatch. Volcanic activity began early in the Green River epoch and increased in frequency and intensity up to the end of the Eocene epoch. Most tuff beds in the lower parts of the Green River Formation are thin, but stratigraphically upward they increase in number and thickness. The upper part of the Bridger Formation is dominantly tuffaceous (Bradley 1964). A playa-lake complex model has been proposed for the deposition of the Green River Formation (Surdam and Wolfbauer 1975).

The Gosiute Lake, in which the Green River Formation was deposited, changed in characteristics considerably and repeatedly during the Green River epoch. Three major stages have been recognized, each of which corresponds to a Member of the Green River Formation. During the first and third stages, the Lake was a large freshwater lake and had an outlet. The climate became very arid during the second stage. The Gosiute Lake shrank, occupied only part of the Green River Basin, and had no outlet. The Wilkins Peak Member was deposited during the middle of this stage. At least 25 beds of trona have been identified in this Member. They range from 1 meter to 11 meters in thickness and from a few hundred to more than 1,800 km^2 in areal extent (Kostick 1994).

Trona of the Green River Formation is light to brown due to the contamination of organic matter. It has a bladed habit ranging in size from needles of less than 0.5 mm long to laths 100 mm long that commonly occur in radiating aggregates. The associated minerals are halite, shortite, dolomite, calcite, and analcite, the most abundant silicate mineral. Most of the above listed Na_2CO_3-containing minerals, although found in small amounts, are all present in the saline mineral assemblages of the Green River Formation. Economic beds normally contain about 90% trona (Mannion 1983).

Searles Lake, California: Searles Lake is a large playa about 16 km long and 10 km wide, which occupies the lowest part of Searles Valley, about 226 km northeast of Los Angeles. The Searles Valley is a block-faulted basin filled with about 1,000 meters of alluvium and nonmarine evaporites lying on a quartz monzonite basement. The Searles Lake may represent the principal terminal sump, that is, a sump that received drainage from a chain of lakes until a brine was developed and salt deposits were formed. Two-thirds of the surface of the lake is covered by mud and one-third by halite. From the surface downward, there are a series of permeable crystalline saline lenses that contain interstitial brine and interbedded saline muds. The deposit extends from the surface to a depth of about 180 meters and has an areal extent of about 100 km^2 (Smith 1979).

A summary of the stratigraphic units in the Searles Lake evaporite sequence is shown in Fig. 90. The upper salt is a complex lens of salts of halite, trona, hanksite, and borax, whereas the lower salt consists of a series of seven salt lenses and six mud layers. These two young salts, which crystallized during the most recent Ice Ages, about 40,000 years B.P., are of major economic importance (Smith 1979). The mixed layer is older, about 120,000 years B.P., and is classified into six zones. Trona is concentrated in the top two zones (Garrett 1992).

At Searles Lake, the brine contained in the salt beds, like the salts in the upper and lower salt layers, is used for the production of sodium salts. The brine density ranges from 1.25 to 1.3, and its salt contents are 4.5% Na_2CO_3, 16% NaCl, 7% Na_2SO_4, 3.5% KCl, and 1.5% $Na_2B_4O_7$ (Mannion 1983).

Lake Magadi, Kenya: Lake Magadi in southern Kenya appears to be the largest, and also the richest, soda ash deposit in the east African Rift Valley (Vincens and Casanova 1987). Approximately 75 km^2 of the lake's 164 km^2 total area are covered by trona deposits that are as much as 20 to 40 meters thick. The lake is essentially a trona salt pan. Brine is present at or within a meter of the surface. Surrounding the trona pan are alkaline lagoons, which are fed by numerous warm spring waters, with a high

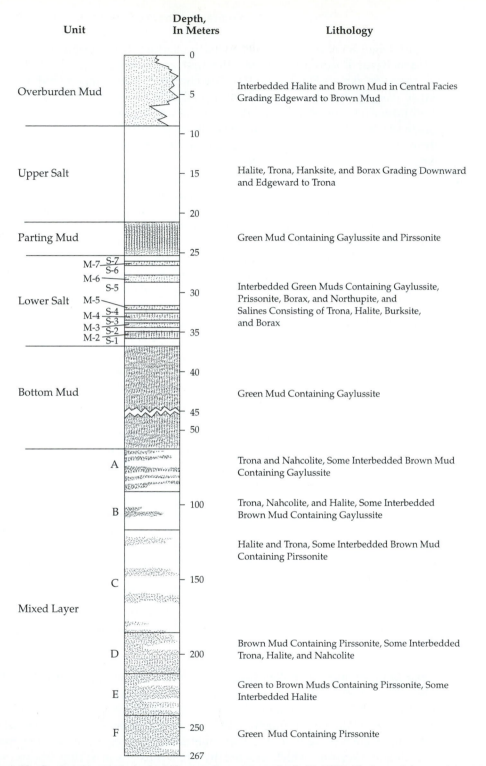

Figure 90 Stratigraphic units in Searles Lake evaporite sequence, after Smith (1979). The Lower Salt was subdivided into thirteen units (S-1 to S-7 and M-2 to M-7), and the Mixed Layer was subdivided into six units (A to F).

Na_2CO_3 content as a result of the weathering and leaching of igneous silicates (Baker 1958; Jones et al. 1977; Eugster 1980).

The distribution of trona and surface brine in Lake Magadi is shown in Fig. 91 (Smith 1979). The chemical compositions for the spring waters from Little Magadi Lake in the north and Spring 16, which is located at the south end, are, respectively (in ppm), 93,000 and 15,480 Na_2CO_3, 32,000 and 6,300 $NaHCO_3$, and 62,000 and 10,250 NaCl. The composition of the brine from borehole, 6 meters in depth, is 200,000 Na_2CO_3, 1,200 $NaHCO_3$, and 109,000 NaCl (Baker 1958).

INDUSTRIAL PROCESSES AND USES

1. Preparation and Beneficiation

The sodium carbonate-containing minerals, like most other ores, must be treated through beneficiation processes to improve their quality. Generally, raw trona is dredged from the salt beds, crushed, washed, and dewatered. Flotation is a commonly used method of beneficiation for soda ash minerals. Trona, nahcolite, thermonatrite, and natron can be separated from salt and sulfate minerals by a cresylic acid or a related compound float (Garrett 1992). Trona can be separated from salt and sodium sulfate (Farag et al. 1975) by using sulfonation of caster oil and sodium laurate, with lead nitrate as an activator. For mixtures of oil shale and trona, the usual procedure is to float the oil shale away from the trona. The system is both self-frothing and collecting, so it can function without reagents (Garrett 1992).

2. The Wyoming Trona Process

Two variations of the beneficiation process, the trona process and the monohydrate process, are used by the Wyoming trona industry (Frint 1971). The trona process begins by leaching trona ore at 95° to 98°C. Next, the hot solution is cooled to precipitate trona, which is then calcined to produce soda ash. In the monohydrate process, the trona ore is first calcined at 200°C, followed by leaching. The hot solution is evaporated to produce monohydrate, which in turn is heated to produce soda ash. For both processes, the ore is ground to a fairly fine size, and leaching is done in leaching tanks. After leaching, the hot slurry is sent to a clarifier, and flocculating agents are added to speed the process. Carbon is removed by adding activated carbon, and iron and other metals are removed by adding Na_2S to the hot solution, which is then filtered. Before the hot solution is sent to the cooling crystallizers, an antifoaming agent and a crystal habit modifier are added to improve the size and form of the crystals to be produced. Crystallization is done in vacuum crystallizers. The crystallized product is then centrifuged, washed, dried, and calcined to produce soda ash and shipped to storage (Anon. 1953; Muraoka 1985). A calcination temperature of 200°C during the monohydrate process is the preferred temperature because it helps maintain a balance between the sluggish rate of decomposition of trona and the removal of organics at lower temperatures, and the formation of harmful and soluble silicates at higher temperatures (Anon. 1989). A two-stage leaching procedure is carried out during the monohydrate process in order to minimize the effect of scaling and to prevent the shortite from dissolving. During the

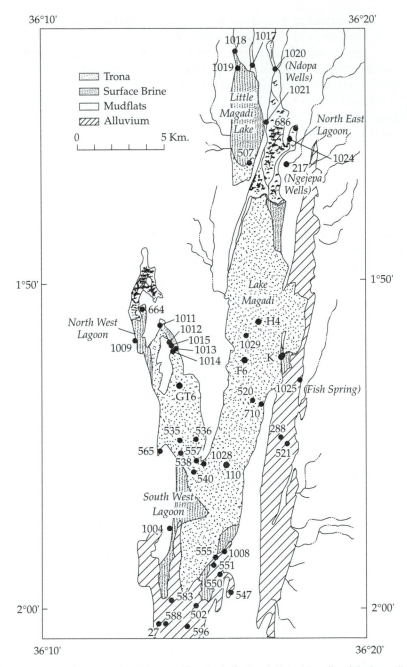

Figure 91 Distribution of trona, surface brine, mudflats, and alluvium in Lake Magadi and the immediate vicinity, after Jones et al. (1977). All numbers with or without alphabets denote sampling locations.

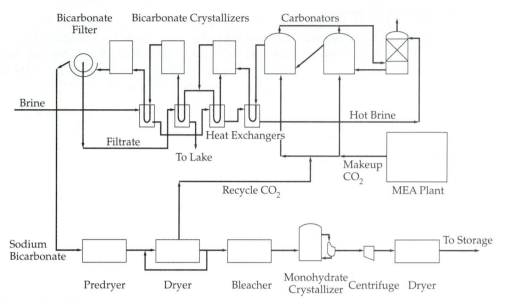

Figure 92   A flow sheet showing the plant process at Searles Lake, after Parkinson (1977).

first stage, the calcined ore is washed with riverwater that has sufficient carbonate to precipitate the excess calcium and magnesium, and to leave the desired Na_2CO_3 content at greater than 6.2%. Shortite will not dissolve appreciably when the Na_2CO_3 content is above 6%. During the second stage, the calcined ore is washed with softened water in order to minimize scaling and to dissolve much of the remaining sodium carbonate (Garrett 1992).

3. The Searles Lake Process

At Searles Lake, soda ash is produced from brine, with a high sodium carbonate content, by carbonation precipitation. The brine has a complex composition, which varies from well to well, and at different depths. The largest operation at Searles Lake is the Argus Plant, which draws brine, produced by solution mining, from the Mixed Layer of the trona-nahcolite-halite formation. A flow sheet that illustrates the Argus Plant process is given in Fig. 92 (Parkinson 1977). The brine, along with some recycled plant liquors used previously for solution mining, is first heat exchanged with hot carbonate brine in bicarbonate crystallizers. The brine is then pumped to the carbonating towers. The CO_2 source for the towers is derived from the conversion of sodium bicarbonate back to soda ash, and from the monoethanolamine (MEA) towers that are used to purify power plant flue gas into a pure carbon dioxide steam. The brine is sent from the carbonating towers to cooling-crystallizers to produce additional sodium bicarbonate slurry. The wet sodium bicarbonate filter cake is dried and calcined at about 260°C. The crude product from the calciners, which is known as "light ash," is treated in rotary oxidizing or bleaching units that are operated at 427°C to burn off the organic matter. A monohydrate crystallizer is used for the conversion of light ash to sodium carbonate monohydrate. The final processing steps include washing, centrifuging, and drying before storage (Garrett 1992).

4. The Solvay Process

Virtually all manufactured soda ash is produced by the Solvay process (Hou 1969; Kostick 1994), which involves the formation of sodium bicarbonate by reacting sodium chloride with ammonium bicarbonate in aqueous solution: $NaCl + NH_4HCO_3 \rightarrow NaHCO_3 + NH_4Cl$. The Solvay process has seven stages (Thieme 1993):

1. Production of a saturated salt solution, $NaCl + H_2O$: The sodium chloride from brine is used to prepare the saturated solution, and purification is made by the lime-soda process to remove Ca^{2+} and Mg^{2+}.
2. Burning of limestone, $CaCO_3 \rightarrow CaO + CO_2$: The CO_2 produced is used for carbonation, and the CaO is used for the distillation recovery of ammonia.
3. Saturation of the salt solution with ammonia: NH_3 and CO_2 are absorbed by the purified brine. The reactions can be represented by $NH_3 + H_2O \rightarrow NH_4OH$ and $2NH_4OH + CO_2 \rightarrow (NH_4)_2CO_3 + H_2O$. These reactions are made in absorption towers that are equipped with effective internal or external coolers to remove heat of reaction. Both of the above reactions are exothermic.
4. Precipitation of bicarbonate: The ammoniacal brine is reacted with CO_2 from the lime kiln to precipitate sodium bicarbonate, $NaCl + H_2O + NH_3 + CO_2 \rightarrow NH_4Cl + NaHCO_3$. Carbonation is achieved by means of a counter current flow in Double-tray columns that are equipped with tubular coolers at the base of the columns. The maximum temperature for the process is 50 to 60°C, and the suspension of bicarbonate in the mother liquor leaves the carbonation stage at 30°C.
5. Filtering and washing: The precipitated bicarbonate is filtered out from the mother liquor by rotary vacuum filters or centrifuges, and then washed with softened water. The crude bicarbonate has an approximate composition of 75.6% $NaHCO_3$, 6.9% Na_2CO_3, 3.4% NH_4HCO_3, 0.4% $NaCl + NH_4Cl$, and 13.7% H_2O.
6. Calcination of bicarbonate: Thermal decomposition of bicarbonate is carried out in rotary calciners, heated either externally by oil, gas, or coal, or internally by steam. The product is technical grade soda ash from the reaction, $2NaHCO_3 \rightarrow Na_2CO_3 + CO_2 + H_2O$. Two side reactions are $NH_4HCO_3 \rightarrow NH_3 + CO_2 + H_2O$ and $NaHCO_3 + NH_4Cl \rightarrow NH_3 + CO_2 + H_2O + NaCl$. The gases that evolve from the reactions, after purification, are used in the carbonation stage.
7. Recovery of ammonia: Both $(NH_4)_2CO_3$ and NH_4HCO_3 completely decompose at 85°C and produce CO_2 and NH_3. Liberation of ammonia from NH_4Cl requires $Ca(OH)_2$. The light ash can be converted to dense ash by dissolution, reprecipitation as a monohydrate, and drying to remove H_2O.

The light soda ash and the dense soda ash differ in bulk density and particle size, and the quality of both is generally assessed by the loss on heating at 250°C, the amount of insolubles, total alkalinity, and the percentages of $NaHCO_3$, Na_2CO_3, $NaCl$, Na_2SO_4, Fe_2O_3, and H_2O. A typical composition for the soda ash produced is 99.6% Na_2CO_3, 0.15% NaCl, 0.02% Na_2SO_4, 0.002% Fe_2O_3, 0.01% CaO, and 0.02% MgO. The bulk density of light soda ash is 0.5 to 0.6 kg/L and that of dense soda ash is 1.05 to 1.15 kg/L. Most of the particle sizes in light soda ash are distributed within the range of between 0.25 and <0.063mm, and the particle sizes in dense soda ash are distributed within a range of between 1 to 0.125mm (Thieme 1993).

5. Industrial Uses

Many industries, including the chemical, glass and ceramics, detergent, pulp and paper, metal, iron and steel, food, petrochemical, and environmental industries, use soda ash for a variety of purposes (Kostick 1994).

Chemical: Soda ash is used in the production of a great number of chemicals. The important ones include:

1. All chromium compounds are produced from sodium chromate and dichromate, both of which are made by reacting chrome ore with soda ash with or without lime (Büchner et al. 1989).
2. Sodium orthophosphates are produced from phosphoric acid and soda ash. A great number of polyphosphates are made from the orthophosphates and are used for food, fertilizers, detergents, water treatments, and other purposes (Hurst 1961).
3. Caustic soda is produced by reacting soda ash with lime. This is used in the Bayer process for aluminum (see Bauxite).
4. A group of soluble silicates is produced by reacting soda ash with silica. These soluble silicates are known as water glass and have many industrial uses (see Silica p. 367).

Glass and Ceramics: The largest single use of soda ash is in the production of glass. Although a great number of chemical compositions can produce glass, about 90% of all glass contains soda ash including soda-lime glasses, borosilicate glasses, aluminosilicate glasses, and lead glass (see Borax and Borates p. 71 and Silica p. 363). Soda ash is used in ceramics for the production of glazes, enamels, and ultramarine pigments.

Detergent: The production of detergents, soaps, and cleaning compounds is one of the major uses of soda ash. It is added directly to the detergent formulations, where it functions as a builder, an agglomerating aid, a carrier for surfactants, and as a source of alkalinity (Greek 1990). In addition, phosphates and silicates used in detergents are also manufactured from soda ash (Hunter 1991).

Pulp and Paper: Soda ash is the starting chemical and makeup for the neutral sulfite semichemical pulping process (Keller 1969) and is used for delignification in the alkali-sulfite-anthraquinone-methanol process (Samdani 1991). Soda ash plays a dual role in the kraft process as a minor pulping agent and as an intermediate compound formed in the regeneration process. An adhesive can be made by digesting wood bark in Na_2CO_3-Na_2SO_3 solution (Jenkin 1984).

Metal: Soda ash is widely used in the digestion and beneficiation of the ores of antimony, lead, chrome, cobalt, nickel, bismuth, and tin (Gilchrist 1989).

Water Treatment: Soda ash has broad general use in the treatment of water for pH adjustment and carbonate precipitation. The most commonly used process, known as the lime-soda process, reduces calcium and magnesium hardness, and removes various contaminants from the water. Depending upon the initial pH of the water, the exchange reaction can reach an equilibrium after about 25 minutes (Alexander and

McClananhan 1975). Soda ash has the ability to reduce calcium content to about 35 ppm $CaCO_3$, and magnesium content to 50 ppm equivalent of $CaCO_3$ (Bellow 1978). In municipal water treatment, soda ash is generally added to various processes in order to increase the pH after prechlorination, flocculation, activated carbon treatment, and filtration. The sludge produced from the lime-soda process is voluminous. It is essentially inert and is customarily disposed of in lagoons or landfills.

Purification of Flue Gases: After fine grinding, sodium bicarbonate or sodium sesquicarbonate is injected into the flue gas stream just ahead of the baghouse. Soda ash with a large and reactive surface area is produced by thermal decomposition and has high absorption efficiency (Muzio et al. 1980; Yeh et al. 1982).

REFERENCES

Alexander, H. J., and M. A. McClananhan. 1975. Kinetics of calcium carbonate precipitation in lime-soda ash softening. *Jour. Amer. Water Works Assoc.* 67:618–620.

Anon. 1953. Trona to soda ash. *Chem. Eng.* (May): 118–120.

Anon. 1989. Soda ash from General Chemical. General Chemical, Parsippany, New Jersey, 10p.

Baker, B. H. 1958. Geology of the Magadi area. *Rept. 42, Kenya Geol. Surv,* 82p.

Bellow, E. F. 1978. Comparing chemical precipitation methods for water treatment. *Chem. Eng.* 85:85–91.

Bradley, W. H., 1964. Geology of Green River Formation and associated Eocene rocks in southwestern Wyoming and adjacent parts of Colorado and Utah. *U.S.G.S. Prof. Paper,* 496A, 86p.

Büchner, W., R. Schliebs, G. Winter, and K. H. Büchel. 1989. *Industrial inorganic chemistry*. Weinheim: VCH Verlagsgesellschaft mbH, 614p.

Correia, R. J., J. Kestin, and H. E. Khalifa. 1980. Viscosity and density of aqueous sodium carbonate and potassium carbonate solutions in the temperature range 20-90°C and the pressure range 0–30 Mpa. *Jour. Chem. Eng. Data* 25:201–206.

Eugster, H. 1980. Lake Magadi, Kenya, and its precursors. In *Hypersaline brines and evaporitic environments.* Ed. E. Nissenbaum, 213–235. Amsterdam: Elsevier.

Farag, A. A., T. H. Mulla, and J. L. Skander. 1975. Concentration of sodium carbonate (Trona) by flotation. *Bull. Fac. Eng. Univ. Alexandria,* Chem. Eng. Ser. 14:31–41.

Frint, W. R. 1971. Processing of Wyoming trona. *Contr. to Geol.,* Univ. Wyoming 10:25–27.

Garrett, D. E. 1992. *Natural soda ash, occurrences, processing, and use.* New York: Van Nostrand Reinhold, 630p.

Gilchrist, J. D. 1989. *Extractive metallurgy*. Oxford: Pergamon Press, 431p.

Goncalves, F. A., and J. Kestin. 1981. The viscosity of sodium carbonate and potassium carbonate aqueous solutions in the range 20–60°C. *Int. Jour. Thermophys.* 2:315–322.

Greek, B. F. 1990. Detergent industry ponders products for new decade. *Chem. Eng. News* (January):37–59.

Hou, J. P. 1969. *Manufacture of soda.* New York: Hafner Publ, 590p.

Hunter, D. 1991. Soaps and detergents. *Chem. Week* (January):38–54.

Hurst, T. L. 1961. Manufacture of phosphate salts. In Vol. 2 of *Phosphorus and its compounds.* Ed. J. R. Van Wazer, 1214–1219 New York: Interscience.

Jenkin, D. J .1984. Adhesives from Pinus Radiata bark extractives. *Jour. Adhes.* 16:299–310.

Jones, B. F., S. A. Rettig, and H. R. Eugster. 1977. Hydrochemistry of the Lake Magadi Basin, Kenya. *Geochim. Cosmochim. Acta* 42:53–72.

Keller, E. L. 1969. Is caustic soda suitable for buffering NSSC digestions? *Soc. Pulp Paper Mfr.* 32:32–36.

Kostick, D. S. 1994. Soda ash. In *Industrial minerals and rocks.* 6th ed. Ed. D. D. Carr, 929–958. Littleton, Colorado: Soc. Mining, Metal. & Exploration.

Lortie, L., and P. Demers. 1940. Physicochemical studies of the alkali carbonates 1. pH. *Can. Jour. Res.* 18:B160–167.

Mannion, L. E. 1983. Sodium carbonate deposits. In *Industrial minerals and rocks.* 5th ed. Ed. S. J. Lefond, 1187–1206. New York: Soc. Mining Engineers, AIME.

Muraoka, D. 1985. Monohydrate process for soda ash from Wyoming trona. In Minerals and metallurgical processes. *Trans. AIME* 278:102–103.

Muzio, L. J., J. K. Arand, and N. D. Shah. 1980. Bench scale study of dry sulfur dioxide removal with nahcolite and trona. In Vol. 1 of *Proc. Conf. Air Qual. Mgt. Elect. Power Ind. 2nd,* Austin, Texas. Ed. H. B. Cooper, 566–599.

Newkirk, A. F., and J. Aliferis. 1958. Drying and decomposition of sodium carbonate. *Anal. Chem.,* 30:982–984.

Parkinson, G. 1977. Soda ash plant exploits mineral-laden brine. *Chem. Eng.* 84:62–63.

Peiper, J. C., and K. S. Pitzer. 1982. Thermodynamics of aqueous carbonate solutions. *Jour. Chem. Thermo.* 14:613–638.

Samdani, G. 1991. Pulp bleaching. *Chem. Eng.* 98:37–43.

Smith, G. I. 1979. Subsurface stratigraphy and geochemistry of Late Quaternary evaporites, Searles Lake, California. *U.S.G.S. Prof. Paper,* 1043, 130p.

Surdam, R. C., and C. A. Wolfbauer. 1975. Green River Formation, Wyoming, A playa lake complex. *Bull. G.S.A.* 86:335–345.

Taylor, C. E. 1955. Thermodynamics of sodium carbonate in solution. *Jour. Phys. Chem.* 59:653–657.

Thieme, C. 1993. Sodium carbonate. In Vol. A24 of *Ullmann's encyclopedia of industrial chemistry,* 299–316, Weinheim: VCH Verlagsgesellschaft mbH.

Vincens, A., and J. Casanova. 1987. Modern background of Natron-Magadi Basin (Tanzania-Kenya), physiography, climate, hydrology and vegetation. *Sci. Geol. Bull.* 40:9–21.

Yeh, J. T., R. J. Denski, and J. I. Joubert. 1982. Control of sulfur dioxide emissions by dry sorbent injection. In Flue gas desulfurization. *ACS Sym. Ser.* 188:349–368.

Talc

Talc is a soft, hydrous magnesium silicate mineral with an ideal composition of $Mg_3Si_4O_{10}(OH)_2$. Its highly pure, massive variety is known as steatite. Commonly, the talc used in industry is a mixture of magnesium-rich silicates, with compositions ranging from pure talc to mineral assemblages that have physical properties in common, but contain very little actual talc.

Ground talc is mainly used for ceramics, paint, rubber, roofing, plastics, cosmetics, and insecticides. Soapstone is a talcose rock that may be used in block form and sawed into various shapes.

The United States, China, India, Finland, France, Austria, Italy, and the former USSR contribute more than 50% of the world's total production of talc.

MINERALOGICAL PROPERTIES

Talc is a layered, mica-type silicate with a crystal structure similar to that of pyrophyllite (p. 308) except that the octahedral sites are occupied by magnesium instead of aluminum, and no site is vacant. Like pyrophyllite, the SiO_4 tetrahedron has to rotate and tilt from its normal configuration in the tetrahedral sheet in order to accommodate the dimensionally different octahedral sheet. This gives the structure of talc a triclinic symmetry. Talc has an ideal composition of $Mg_3Si_4O_{10}(OH)_2$, from which there appears to be little variation. Small amounts of Fe^{2+} and Fe^{3+} may replace Mg^{2+}, and a very minor amount of Al^{3+} may replace Si^{4+}. Small numbers of Na^{1+} or K^{1+} may enter the interlayers of the structure in order to balance the charge differences. Talc is white to silvery white or pale to dark green in color. It has a hardness of one, a specific gravity of about 2.75, a pearly luster, and a greasy feel. It usually occurs in massive foliated or fibrous aggregates. The indices of refraction for pure talc are $\alpha = 1.539$ to 1.550, $\beta = 1.589$ to 1.594, and $\gamma = 1.589$ to 1.600 (Deer et al. 1992).

Talc undergoes thermal decomposition at about 800°C to yield enstatite and cristobalite. Equilibrium curves calculated from thermodynamic data for reactions involving talc in the system MgO-SiO_2-H_2O are shown in Fig. 93. The common association of talc with forsterite is illustrated. This is replaced by enstatite at high temperatures and by serpentine at high pressures. The presence of FeO in talc generally lowers its thermal stability (Evans and Guggenheim 1988). Intensive grinding of talc in oscillating or ball mills produces marked changes in physicochemical properties, which are accompanied by structural deterioration. The main effects are the agglomeration and cementation of individual particles and dissolution of Mg^{2+} in water because of the weakening of the Mg-OH linkage (Aglietti 1994).

Commonly, the talc used in industry, known as commercial or industrial talc, is a mixture of magnesium-rich silicates with compositions ranging from pure talc to mineral assemblages that have similar physical properties, but contain very little actual talc. The most common associated mineral is tremolite. Other minerals associated with industrial talc are chlorite, serpentine, anthophyllite, diopside, magnesite, chrysotile, and pyrophyllite. Quartz, dolomite, and calcite are also present in many commercial talcs

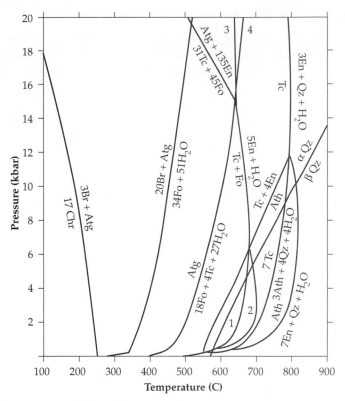

Figure 93 Pressure-temperature curves for equilibria involving talc in the system $MgO-SiO_2-H_2O$, after Evans and Guggenheim (1988). Symbols used are Tc: talc, Atg: antigorite, Chr: chrysotile, Fo: forsterite, Br: brucite, En: enstatite, Ath: anthophyllite, Qz: quartz. 1: $9Tc + 4Fo = 5Atg + 4H_2O$; 2: $Atg + Fo = 9En + H_2O$; 3: $Atg + 14Tc = 90En + H_2O$; 4: $Atg = 14Fo + 20En + 31H_2O$.

(Brown 1973). The mineralogical composition of commercial talc varies from deposit to deposit. Talc from Montana has a talc content of up to 90 to 95% with minor amounts of chlorite (2 to 4%) and dolomite (1 to 3%). The composition of talc from New York is characterized by a high content of tremolite (30 to 55%), with an amount of talc in the range of between 35 and 60%. Others are anthophyllite (3 to 10%) and serpentine (2 to 5%), and minor amounts (1 to 3%) of magnesite, quartz, calcite, and dolomite. Magnesite is a major associated mineral in Vermont's talc, which has 20 to 30% magnesite and 40 to 60% talc (Grexa 1987). Results from an analysis of nine California talc samples give a range of talc content from a high of 78.4% to a low of only a trace amount. The associated tremolite has a range from a low of 0.87% to a high of 55.4% (Heystek and Planz 1984). Commercial talc from Trimouns, France, is a mixture of talc and chlorite, with the latter making up between 5 and 60% of the ore (Russell 1988). Talc from some deposits in Andhra Pradesh and Madhya Pradesh, India, considered to be the world's best in purity, has a talc content of 98 to 99% (Russell 1991).

Health problems related to talc exposure have been a controversial subject in many discussions. Whether talc pneumoconiosis or talcosis is caused by talc or by its associated asbestiform minerals is in dispute (Spiegel 1974).

GEOLOGICAL OCCURRENCE AND DISTRIBUTION OF DEPOSITS

Talc is usually a secondary mineral that results from the alteration of other magnesium containing silicates or carbonates. Talc deposits of economic importance occur mainly in metamorphosed dolomite and altered ultramafic igneous rocks. These rocks, which contain talc deposits, are present in metamorphic terranes of various ages throughout the world (Brown 1973). Talc deposits derived from carbonate rocks are generally considered to have been produced by contact metamorphism of dolomite or dolomitic limestone where it is intruded by granitic rocks. Additional silica and sufficient water is required for the reaction, which releases $CaCO_3$ and CO_2, and which results in a considerable decrease in volume. Talc deposits related to ultramafic rocks were formed by metamorphism and hydrothermal alteration.

The United States and China are the top producers of talc, with a combined contribution of 25% of the total world production (8 M tons), as estimated by O'Driscoll (1992). Other important producers are France, Finland, Austria, Italy, India, and the former USSR.

The United States: About 90% of the talc is produced in Vermont, New York, Montana, Texas, and California (Chidester et al. 1964). In Vermont, the main belt of ultramafic rocks approximately follows the central meridian of the State from Massachusetts to Canada. Except for a few fresh dunite bodies in the northern and southern parts of the belt, the rest is extensively serpentinized and has appreciable amounts of talc associated with it. Most of the talc deposits form concentric shells of talc-carbonate rock and talc rock around central cores of serpentinite. The talc and talc-carbonate rocks range from light gray to dark gray and from massive to schistose. All the talc is high in iron (3.36 to 6.88% combined FeO and Fe_2O_3), and most contains appreciable carbonate (Chidester et al. 1951). The average SiO_2 content ranges from 62.24% to 35.98%, and MgO content from 33.16% to 27.89%. The major mines in Vermont are at Johnson, in Johnson, and at Waterbury, in Moretown. There are two minor mines located at Hammondsville and Windham, near Chester. The Waterbury mine is the sole producer of talc crayons in Vermont.

The New York talc deposits of economic importance are located in the northwest Adirondack Mountains, with the largest deposit located in St. Lawrence County near Gouverneur. Talcose rock occurs in elongated zones within a northeast-trending belt of impure marble of the Precambrian Grenville series. These zones have a composite strike length of more than 8 km, an extensive down dip of about 600 meters, and a width about 120 meters (Brown and Engel 1956). The talcose rock mined is a "fibrous talc" that is composed largely of tremolite, anthophyllite, serpentine, and talc. Harben and Bates (1990) showed that a typical mineralogical composition for New York talc is 50 to 70% tremolite, 10% anthophyllite, 20 to 30% talc, and 20 to 30% serpentine, in the form of antigorite.

In Montana, the talc deposits are in the Dillon-Ennis district, within an area of about 300 m^2 that embraces parts of the Gravelly and Ruby Ranges between Dillon and the Madison River. The talc is a replacement of dolomite of early Precambrian age, and much of it is of steatite grade and is the principal resource for steatite in the United States. The talc ranges from nearly white through various shades of gray, green, and brown, but is mostly pale to medium green. Most of the talc is very fine grained and blocky. Except for a minor amount of chlorite, no magnesium silicate minerals other than talc have been observed. The main mines are the Yellowstone mine at Johnny Gulch, the Smith-Dillon mine in Sweetwater Creek, and the Stone Creek mine in Stone Creek (Berg 1979; Harben and Bates 1990).

Talc deposits in Texas are associated with ultramafic igneous rocks and with metamorphosed sedimentary and volcanic rocks. In the Allamoore district in western Texas, about 160 km southeast of El Paso, the talc deposits are in an area about 30 km long from east to west, and about 10 km wide. They are in the Precambrian Allamoore Formation, which consists of interbedded highly deformed metamorphosed limestone and volcanic rocks. The latter contain thin bodies of phyllite that are commonly talcose. The principal talc deposits developed are enclosed in cherty limestone and may be an alteration of dolomitic marl or magnesium-rich tuff (King and Flawn 1953; Roe and Olson 1983).

The major talc deposits in California lie in a narrow belt about 320 km long, along the eastern state line, which extends into western Nevada. All deposits have formed through metamorphism and hydrothermal replacement of siliceous magnesium marbles and limestones. There are three principal districts: (1) The most southerly district is east and north of the Silver Lake playa in San Bernardino County, and the deposits, which are of Precambrian age, are medium-grained tremolite schist that is in part altered to serpentine and talc (Wright 1954). (2) In southern Death Valley, deposits of fine-grained tremolite and talc are found in both foliated and massive varieties. They are also of Precambrian age (Wright 1952). (3) In Inyo Mountain, there are talc deposits of steatite grade, which were the product of replacement that occurred during the Cretaceous or Tertiary periods, mostly of dolomite and quartzite (Page 1951; Roe and Olson 1983).

China: Extensive talc deposits are distributed in at least five provinces including, in order of importance, Liaoning, Shantung, Guangxi, Chekiang, and Fukien (Huang 1983). In Liaoning, the talc deposits were formed by metasomatic replacements of magnesite and dolomite within the carbonate and calc-silicate rocks that had undergone metamorphism, ranging from greenschist facies to amphibolite facies. Talc veins are generally podiform and sheared, and have widths ranging from 2 to 15 meters and strike lengths ranging from tens to hundreds of meters. The talc is of high purity. The talc deposits of Shantung, which consist of schists, gneisses, magnesite, and dolomite, belong to the Late Proterozoic Jutuen-Zhanggezhuang Group. Like the deposits in Liaoning, they were also formed by metasomatic replacement of carbonates. The talc veins generally have minable widths of from 7 to 30 meters, and extents of hundreds of meters of downdip and along the strike. The largest talc deposits are located in Guangxi Province and are of high purity. The host rock is dolomitic marble, and the talc veins have an average thickness of 5 meters, and strike lengths can reach hundreds of meters (Piniazkiewicz et al. 1994).

India: Talc deposits are widely distributed in India, and some of them contain talc as pure as 98 to 99%. The largest and the most productive deposits are in the state of Rajasthan, which contributes 85% of total production (about 300,000 tons/year)

(Russell 1991). The deposits in Rajasthan resulted from contact metamorphism of dolomite by intruded basic dikes. Other deposits are located in Andhar Pradesh, Madhya Pradesh, and Bihar.

The Former USSR: The Onot deposits in the Eastern Sayan Mountains at the margin of the Siberian platform are in a belt form and are 500 meters long and 40 meters thick. They are situated in magnesite that was formed by hydrothermal replacement of lower Proterozoic dolomite from solution along a regional fault zone. The deposits show the following zonation: amphibolite, chloritized amphibolite, talquite, magnesite with talc, and magnesite. At the foothills of the Yeniseie Mountains in the Krasnoyarsk region, deposits of lenticular bodies of talc occur in dolomite at its point of contact with schist. The deposits are 550 to 800 meters long and have thicknesses ranging from 20 to 100 meters (Tatarinov 1969).

Europe: Europe is a major producer of talc. Large deposits occur in France, Italy, Austria, and Finland, with smaller deposits located in other nations (Engel and Wright 1960). In France, on the north slopes of the Pyrenees in Ariège, Luzenac's Trimouns deposits consist of numerous lenses of talc at the point of contact between mica schist, slate, and limestone. Some of the lenses have thicknesses of up to 300 meters and are of high grade. In Italy, deposits of high grade steatite occur in interlayers of dolomitic limestone in a mica schist complex located in the Chisone valley near Roussa in the Cottian Alps. Near Leoban, Austria, the Mautern deposits are situated in a narrow band of Carboniferous graphitic shales with interlayered dolomitic limestone, partly altered into magnesite and dolomite. Elongated talc lenses occur beneath the dolomitic limestone. They were formed by metamorphism with metasomatism of underlying limestones. The thickness of talc lenses is up to 4 meters (Kuzvart 1984).

Deposits of talc occur in Sotkamo, near Kajaaani, in central Finland and in Polvijarvi, near Outokumpu in eastern Finland. The large lenticular bodies of talc are approximately 450 meters long and 180 meters wide, and are enveloped in mica gneiss. The ore is estimated to be 50% talc, and the associated minerals are breunnerite, magnetite, and some nickel-containing minerals (Isokangas 1978).

INDUSTRIAL PROCESSES AND USES

1. Processing and Beneficiation

Talc milling is largely a dry processing operation. Ore taken from the mine is crushed in primary and then secondary crushers to a size of 0.15 to 1.25 cm. Further fine grinding is performed in Raymond mills or by other types of air separation. In some plants, vertical-shaft pulverizing mills are used for ultrafine grinding (1μ to 5μ range). In India, large high-quality lump talc is manually sorted before the ore is sent to be ground (Russell 1991). In France, the processing of talc includes drying, grinding, and micronising. Optical sorting by laser has been used to optimize the recovery of high grade talc (Russell 1988).

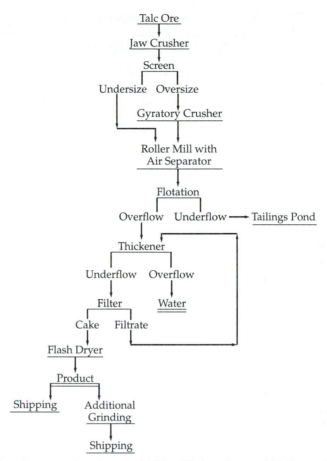

Figure 94   Flow sheet for processing talc ore at (a) West Windsor plant, and (b) Eastern Magnesia plant in Vermont, after Roe and Olson (1983).

The milling procedures employed at the Gouverneur district in New York (Roe and Olson 1983) are as follows: primary crushing by jaw crushers, secondary crushing by gyratory crushers, conveyor to plant, wet ore storage, drying to minus three-eighths of an inch (or 0.95 cm) product, tertiary crushing, dry ore storage, grinding of coarser product grades in pebble mills in a closed circuit with Raymond and Sturtevant separators, grinding of finer grades with fluid energy mills, screening, and conveying to storage bins.

At the West Windsor processing plant in Vermont, flotation is used for the production of several high-grade talcs (Trauffer 1964). A flow sheet illustrating the process is shown in Fig. 94a. A mill flow sheet for the Eastern Magnesia Talc plant in Vermont is shown in Fig. 94b (Roe and Olson 1983). Concentrating tables are installed to remove high-gravity products containing nickel, cobalt, and iron minerals. A combination of froth flotation and high-intensity magnetic separation has been studied for the removal of iron-bearing minerals (Andrews 1986).

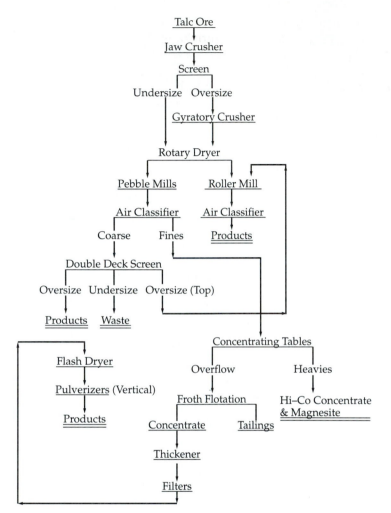

Figure 94   *(Continued)* Flow sheet for processing talc ore at (a) West Windsor plant, and (b) Eastern Magnesia plant in Vermont, after Roe and Olson (1983).

2. Industrial Uses

Talc is used as a ceramic base material for wall and floor tiles; a filler for plastics, paper, roofing, and rubber; a paint extender; and an insecticide diluent and carrier. In the United States, of the approximately 1 M ton used in 1989, the industry breakdown was 32% ceramics, 15% paint, 14% paper, 10% roofing, 10% plastics, 5% cosmetics, and 14% insecticides, rubber, and other products (Burger 1991). There are specifications for each type of use. In addition, soapstone is cut into table tops and sinks, steel-making crayons, and blocks for art carvings. High-frequency ceramic insulator shapes are machined from steatite.

Ceramics: Large amounts of talc are used in the preparation of ceramic bodies, including glazed wall and floor tiles, semivitreous tablewares, electrical porcelain, and high-frequency insulators. Uniform chemical and physical properties are required that will ensure constancy in both shrinkage and color in the whiteware body when fired. Color specifications for fired, talc-bearing bodies depend upon the intended use. Wall tile, for example, must be nearly white, whereas a buff color for insulators is not objectionable. The talcs that are to be used in ceramic bodies are ordinarily ground sufficiently to enable them to pass through a 200-mesh screen. Grinding to finer sizes is common but not always necessary. In the steatite grade talcs that are to be used in the manufacture of high-frequency insulators, the contents of CaO, Al_2O_3, and iron oxides must stay below 1.5%, 4%, and 1.5%, respectively (Roe and Olson 1983).

Thermal decomposition of talc at high temperatures produces a phase assemblage of silica and enstatite, which will reduce moisture expansion but will not greatly affect porosity. Some of the quartz produced is converted to cristobalite. The presence of cristobalite leads to an increase in thermal expansion of the ceramic body, and consequently to an increase in the degree of compression in the glaze, thus adding to its craze resistance. Talc is also used in glazes as a source of MgO (Taylor and Bull 1986).

Cordierite is produced by mixing talc with clay in the required proportions, and then firing the mix at high temperatures. This silicate has very low thermal expansion, an important and favorable property in the compounding of ceramic bodies.

Paints: For use as an extender in paint, the important properties of talc are color, particle shape and size, packing index, and oil absorption, all of which are evaluated according to the specifications for the end product required. A higher grade of white talc is required when it is necessary not to alter the tints of pigments. Fibrous talcs appear to act as bonding agents in paint films, and flaky talcs seem to act as laminal pigments. In order to yield the desired smooth paint film, a relatively fine particle size, at least 98.5% through a 325-mesh screen, is commonly required for paints. Grinding to a finer size increases oil absorption, which results in a low gloss paint. For any given particle size, commercial talc predominantly consists of talc that has higher oil absorption properties than commercial talc that is high in tremolite, quartz, and carbonates (Roe and Olson 1983). Talc used for extender pigments must have a refractive index below 1.75.

Paper: Talc is used in the filling and coating of paper. Its distinct properties are chemical inertness, softness, freedom from grit, satisfactory ink absorption, good adhesive suspension, brightness, dispersibility in water, and rheological properties that allow ready flow and application (Burger 1991). Talc filler should not contain more than 2 to 5% $CaCO_3$. Particle size should be within the range of $.5\mu$ to 5μ, with surface areas from 4 to 25 m^2/gm (Huuakonen and Ahonen 1984). When talc is used in place of a titanium dioxide extender in paper, it has a distinct weight advantage, with a specific gravity of 2.75 versus 4.2 for TiO_2. Ultrafine talc can be manufactured to exacting specifications, which facilitates its use in paper coating. It can be used with calcium carbonate and clay to control rheological properties, gloss, opacity, and brightness.

Various methods have been used for pitch control in the paper-making process. One of them is to use pitch absorbing agents that are capable of adsorbing pitch onto its surface, thus preventing agglomeration and adhesion of pitch particles. Owing to its organophilic character, talc is an effective medium for pitch control. A significant advantage of talc compared with other pitch control agents is that the pitch is incorporated into the end product and leaves the process (Huuskonen and Ahonen 1984).

Plastics: The use of talc as a filler in plastics, both thermoplastics and thermosets, improves its chemical and heat resistance, impact strength, dimensional stability, stiffness, hardness, tensile strength, thermal conductivity, creep resistance, and electrical insulation (Roe and Olson 1983). There is a growing use of polyolefins, which require talc usage, and there is also a growing use of talc filler in polypropylene, especially in Japan and Europe (O'Driscoll 1992). The development of silane coupling agents for talc and other mineral fillers has greatly improved the filler's performance in plastics (Ranney et al. 1972).

Roofing: Talc is used both as a filler and as a surfacing material in the manufacture of roofing. Its function is to act as a stabilizer for the melted asphalt when it is used as a filler, and as a nonsticking agent for asphalt shingles or roll roofing when it is used as a surface material. For both filler and surface usage, talc improves fireproof capability and weather resistance. A low-grade off-color and impure talc is acceptable.

Others: Many synthetic rubbers use ground talc as fillers in their compounding formulations. Talc is used as an insecticide diluent and as a carrier because of its high absorptive capacity, satisfactory bulk density, and low abrasive characteristics. For cosmetics, talc must be free of grit, of fine particle size, and of excellent color.

REFERENCES

Aglietti, E. F. 1994. The effect of dry grinding on the structure of talc. *Appl. Clay Sci.* 9:139–147.

Andrews, P. R. A. 1986. Processing talc in Canada. *Ind. Minerals* (June):63–69.

Berg, R. B. 1979. Talc and chlorite deposits in Montana. *Montana Bur. Mines & Geol. Mem.,* 45, 69p.

Brown, C. E. 1973. Talc. In United States mineral resources. Eds. D. A. Brobst and W. P. Pratt. *U.S.G.S. Prof. Paper.* 820:619–626.

Brown, C. E., and A. E. J. Engel. 1956. Revision of Grenville stratigraphy and structure in the Balmat-Edwards district, northwest Adirondacks, New York. *Bull. Geol. Soc. Amer.* 67:1599–1622.

Burger, J. 1991. Talc—USA and Europe competitive markets. *Ind. Minerals* (January):17–27.

Chidester, A. H., M. P. Billings, and W. M. Cady. 1951. Talc investigations in Vermont: Preliminary Rept., *U.S.G.S. Circ.* 95:1–33.

Chidester, A. H., A. E. J. Engel, and C. A. Wright. 1964. Talc resources of the United States. *U.S.G.S. Bull.* 1167:1–61.

Deer, W. A., R. A. Howie, and J. Zussman. 1992. *Introduction to rock-forming minerals.* Essex: Longman. England, 696p.

Engel, A. E. J., and L. A. Wright. 1960. Talc and soapstone. In *Industrial minerals and rocks.* 3rd ed. Ed. J. L. Gillson, 835–850. New York: Soc. Mining Engineers, AIME.

Evans, B. W., and S. Guggenheim. 1988. Talc, pyrophyllite and related minerals. In Vol. 19 of *Reviews in mineralogy*. Ed. P. B. Ribbe, 225–293. Washington, D.C.: Mineral. Soc. Amer.

Grexa, R. W. 1987. North American talc—competition in every direction. *Ind. Minerals* (June):52–55.

Harben, P. W., and R. L. Bates. 1990. *Industrial minerals, geology and world deposits.* London: Industrial Minerals Div., Metal Bulletin Plc., 312p.

Heystek, H., and E. Planz. 1984. Mineralogy and ceramic properties of some California talcs. *Bull. Amer. Ceramic Soc.* 43:555–561.

Huang, Z. 1983. The geology of industrial rocks and minerals in China, Proc. 9th Forum on the Geology of Industrial Minerals. Ed. S. F. Yundt. *Ontario Geol. Suvr. Misc. Paper* 114:175–179.

Huuakonen, J., and P. Ahonen. 1984. Talc as raw material in the paper industry. *IM Pulp and Paper Survey, 1984.* Eds. P. Harben and T. Dickson, 65–69.

Isokangas, P. 1978. Finland. In Vol. 1 of *Mineral deposits of Europe, Northwest Europe* Eds. S. H. U. Bowie, A. Kvalheim, and H. W. Haslam, 84–85. London: Inst. Min. Metal. Mineral. Soc.

King, P. B., and P. T. Flawn. 1953. Geology and mineral deposits of Precambrian rocks of the Van Horn area, Texas. *Univ. Texas Publ.* 5301:170–172.

Kuzvart, M. 1984. *Industrial minerals and rocks.* Amsterdam: Elsevier, 454p.

O'Driscoll, M. 1992. Talc review—consolidation and competition. *Ind. Minerals* (March):23–37.

Page, B. M. 1951. Talc deposits of steatite grade, Inyo County, California. *California Div. Mines Special Rept.,* 8, 35p.

Piniazkiewicz, R. J., E. F. McCarthy, and N. A. Genco. 1994. Talc. In *Industrial minerals and rocks.* 6th ed. Ed. D. D. Carr, 1049–1069. Littleton, Colorado: Soc. Mining, Metal. & Exploration.

Ranney, M. W., et al. 1972. Silane coupling agents in particulate minerals-filled composites. *27th Reinforced Plastics Technical and Management Conference, SPI, Washington, D.C.,* Feb. 8–11, 1–29.

Roe, L. A., and R. H. Olson. 1983. Talc. In *Industrial minerals and rocks.* 5th ed. Ed. S. J. Lefond, 1275–1301. New York: Soc. Mining Engineers, AIME.

Russell, A. 1988. Industrial minerals of France—current production and developments. *Ind. Minerals* (June):19–34.

———. 1991. India's industrial minerals—completing the picture. *Ind. Minerals* (September):45–55.

Spiegel, R. M. 1974. Medical aspects of talc: Proc. Sym. on Talc, Washington, D.C., May 8, 1973. *U.S.B.M. Circ.* 8639:97–98, compiled by A. Goodwin.

Tatarinov, P. M., ed. 1969. *Textbook of non-metallic mineral deposits.* Moscow: Nedra, 672p.

Taylor, J. R., and A.C. Bull. 1986. *Ceramics glaze technology.* Oxford: Pergamon Press, 263p.

Trauffer, W. E. 1964. New Vermont talc plant makes high-grade flotation product for special uses. *Pit and Quarry* 57:72–76.

Wright, L. A. 1952. Geology of the Superior talc area, Death Valley, California. *California Div. Mines, Special Rept.,* 20, 72p.

———. 1954. Geology of the Silver Lake talc deposits, San Bernardino County, California. *California Div. Mines, Special Rept.,* 38, 36p.

Titanium Minerals

Rutile and ilmenite are the two titanium minerals of economic importance. Their major industrial uses are distinctly twofold: to produce titanium, which is a metal of high strength and low weight, and to produce TiO_2 pigment, which has great whiteness and strong light scattering ability.

Australia leads the world in the production of titanium minerals (25%). Titanium production from Canada, Norway, South Africa, and the United States accounted for 70% of the estimated annual total or 4.0 to 4.5 M tons between 1991 and 1995 (*U.S.B.M. Annual Review* 1995). Other producing nations are India, Sierra Leone, the former USSR, and Malaysia. Canada and South Africa produce almost all of the world's titaniferous slag.

MINERALOGICAL PROPERTIES

1. Rutile

Rutile is essentially TiO_2, but some rutile contains considerable amounts of Fe^{3+}, Fe^{2+}, Nb^{5+}, and Ta^{5+}. Lesser amounts of Cr^{3+}, V^{5+}, or Sn^{4+} may also appear in rutile. A tantalo-rutile with 66.28% TiO_2, 8.64% Nb_2O_5, 15.44% Ta_2O_5, and 8.00% FeO was reported from Clobe Hill, Australia (Deer et al. 1996). In the structure of rutile, Ti^{4+} ions are octahedrally coordinated by oxygen ions, whereas each oxygen is surrounded by three Ti ions that lie in a plane at the corners of an equilateral triangle. By sharing edges, (TiO_6)-chains are formed and extend along the c-axis. The chains are linked by corner-sharing to produce a tetragonal unit cell with a = 4.593Å and c = 2.959Å. The unit cell dimensions increase with increasing temperature. They are a = 4.603Å, 4.616Å, 4.623Å, and c = 2.966Å, 2.977Å, 2.986Å, respectively, at 300°, 600°, and 900°C (Meagher and Lager 1979). Rutile has a hardness of 6 to 6 1/2 and a specific gravity of 4.2 to 5.5. The color of rutile is characteristically reddish brown, but may also be black, violet, yellow, or green. Synthetic rutile is white, but an oxygen-deficient rutile such as $TiO_{1.995}$ is blue. Indices of refraction are ω = 2.605 to 2.613 and ε = 2.899 to 2.901. Both refractive indices and birefringence show little variation with chemical composition (Deer et al. 1996). Rutile has a high dielectric constant which amounts to $\varepsilon_\parallel$ = 173, along the tetragonal c-axis and to $\varepsilon_\perp$ = 89, perpendicular to it with a mean of ε_m = 117 for a random crystal orientation in a ceramic body. TCε is approximately -1000×10^{-6}/K, which is a high negative temperature coefficient of the dielectric constant, an important property for rutile that is to be used in the production of ceramic capacitors (Schüller 1984). Rutile is a semiconductor whose specific conductivity rises rapidly with increasing temperature and is very sensitive to oxygen deficiency. It is an insulator at 20°C, but at 420°C the conductivity increases 10^7 times. Stoichiometric titanium dioxide has a specific conductivity of $<10^{-10}$ Ωcm^{-1}, whereas $TiO_{1.9995}$ yields a value of 10^{-1} Ωcm^{-1}. Its magnetic susceptibility is $(0.078$ to $0.089)\times10^{-6}$ (Whitehead 1983).

TiO_2 is polymorphic. The three naturally occurring forms are rutile, anatase, and brookite. Their stability relationships have not been established. Rutile is thermally stable at all temperatures. Brookite has been converted to rutile at approximately 720°C and one atmospheric pressure experimentally, but the reverse reaction has not been produced (Dachille et al. 1968). In all three polymorphic forms, the Ti^{4+} ion is octahedrally coordinated, but differs in the way that the TiO_6 octahedra are linked. Two edges of each octahedron are shared and form a chain in rutile, whereas three edges are shared in brookite and four in anatase (Meagher and Lager 1979). TiO_2 has a melting point of 1,825°C.

2. Ilmenite

Ilmenite, $FeTiO_3$, is opaque and has an iron-black color. Reflectance in air is in the range of 20 to 23%; in oil it is much less (7 to 8%). The hardness and specific gravity of ilmenite is 5 to 6 and 4.70 to 4.79, respectively. It is weakly magnetic. Ilmenite is isostructural with hematite, Fe_2O_3, with Fe^{2+} and Ti^{4+} sharing the Fe^{3+} octahedral positions in an arrangement in which oxygen ions are closely packed hexagonally. This results in hexagonal symmetry with a = 5.09Å and c = 14.08Å, as compared with a = 5.03Å and c = 13.772Å of hematite. There is a complete series of solid solutions between $FeTiO_3$ and Fe_2O_3 at high temperatures, and the solid solubility decreases with decreasing temperature, which results in a miscibility gap below 525°C. The solid solution series undergoes an order-disorder transition in the ilmenite ($FeTiO_3$)-rich end and Néel transitions in the hematite (Fe_2O_3)-rich end at temperatures above immiscibility (Burton 1984). In the natural occurrence, the amount of Fe_2O_3 is commonly within the limit of 5%. Ilmenite also forms extensive ranges of solid solutions with two other hematite-type minerals. Ilmenite can accommodate 70% $MgTiO_3$ (geikielite) in the system $FeTiO_3$-$MgTiO_3$ and 64% of $MnTiO_3$ (pyrophanite) in the system $FeTiO_3$-$MnTiO_3$. Although naturally occurring solid solutions have been found up to the limits, the amounts of MgO and MnO in ilmenite are normally very minor.

Ilmenite is susceptible to weathering attack, and ilmenite in sand deposits frequently exhibits varying degrees of alteration. Alteration usually takes places in stages, leading to the formation of amorphous Fe-Ti oxides and leucoxene. The latter is a fine-grained mineraloid with a yellowish, brownish, or grayish color, and has a TiO_2 content in the range of 65 to 70% as compared with the 52% TiO_2 for ilmenite. Either amorphous Fe-Ti oxides or leucoxene may contain one or more titanium oxides. Rutile is the most common, but leucoxene may contain anatase or brookite, or a mixture of the dioxides (Temple 1966; Shvetsova 1970). Experimental studies done at high temperatures show that oxidation of ilmenite produces a mixture of hematite, pseudobrookite, and rutile in an approximate molar ratio of 1:5:7, which is similar to naturally occurring "brown leucoxene." The oxidation of ilmenite takes place in nature at temperatures as low as 100°C (Karkhanavala and Momin 1959).

GEOLOGICAL OCCURRENCE AND DISTRIBUTION OF DEPOSITS

Titanium minerals are mined in crystalline rocks as well as from unconsolidated sediments (Force 1991). Nearly all the established economic deposits of titanium minerals that occur in crystalline rocks are associated with anorthositic rocks that range in composition from anorthosite to gabbro. Although these deposits seem to have been formed by magmatic processes, they characteristically occur in high-grade metamorphic terranes (Anderson and Morin 1969; Ashwal 1982). There is evidence that shows that the partitioning of titanium between silicates and oxides follows an equilibrium relationship similar to that for metamorphic rocks (Force 1991). The titanium minerals occur either as massive ores or are disseminated in the rocks, and are of three main types: (1) ilmenite-magnetite, (2) ilmenite-hematite, and (3) ilmenite-rutile.

In unconsolidated sediments, titanium minerals, along with other heavy minerals, become enriched by gravity segregation in beach, dune, and stream sands. The essential elements required to form such a sand deposit are (1) source rocks that are rich in titanium minerals, (2) intense weathering, (3) conduits for adequate transport, and (4) locations for deposition (Lynd and Lefond 1983; Stanaway 1992). Titanium minerals also form concentrations in a large number of fluvial environments (Slingerland and Smith 1986).

At the present time, unconsolidated sand deposits are responsible for more than 50% of the world's supply of titanium minerals. Other sources of titanium are magmatic ilmenite and fluvial rutile (Force 1991).

Australia leads the world in the production of titanium minerals. The other four of the top five producers are Canada, South Africa, Norway, and the United States. Malaysia, the former USSR, Sierra Leone, India, and China are also important producers.

Australia: Australia has the world's most important heavy mineral sand deposits. The east coast, from near Sydney, New South Wales, northward to Fraser Island, Queensland, is a continuous sand beach 1,200 km long and 3 to 12 km wide. As shown in Fig. 95a (Force 1991), the beach along the coast consists of both Holocene and Pleistocene deposits, and of both beach and eolian sands. The beach is plastered against a continental margin of large sedimentary basins and Paleozoic foldbelts. Similar mineralogy occurs in both Pleistocene and Holocene deposits. Rutile is a major component of the heavy mineral assemblages in the southern region and usually amounts to more than 50%. Ilmenite is also present, gradually increasing in quantity to the north, where it becomes the predominant member of the heavy mineral assemblages. In comparison, Pleistocene deposits have ilmenite fractions that are more uniformly altered than those of Holocene deposits (Hails 1969). Other minerals in the heavy mineral assemblages are tourmaline, monazite, zircon, chromite, magnetite, cassiterite, spinel, and garnet. Quartz is the predominant light mineral, whereas carbonates are rare. Feldspar is absent. The immediate sources for the beach and eolian deposits are older beach sand deposits and offshore sands, while the intermediate sources are the Mesozoic sandstones of the Sydney, Clarence, and Moreton basins. The ultimate sources are the Paleozoic foldbelts and their associated granites (Force 1991). The heavy mineral content in the deposits is estimated to be low, and grades of about 1% constitute an ore (McKellar 1975).

In western Australia, heavy mineral deposits occur in two districts: Geographe Bay and Eneabba (Baxter 1977). Three separate strand lines ranging in age from early Pleistocene to Holocene are the sites of heavy-mineral deposits in the Geographe Bay district (Collins et al. 1986; Collins and Hamilton 1986). They are the present beach, the Caper line, and the Yoganup line. The Yoganup line, which is 16 km inland, is the oldest strand line (Fig. 95b) (Force 1991). All the heavy-mineral deposits in the district are dominated by ilmenite, which ranges from 56 to 95% of the heavy minerals, zircon, which ranges from 2 to 18%, and rutile, which ranges from 0.5 to 2%. Ilmenite grains are rounded and poorly sorted. Other heavy minerals in the Geographe Bay district in-

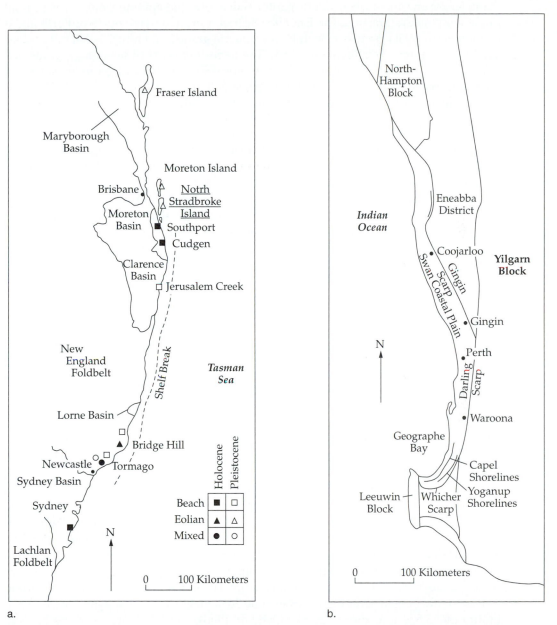

a.

b.

Figure 95 Type and location of sand deposits along (a) eastern Australia coast and (b) western Australia coast, after Force (1991). Underlined are productive deposits.

clude leucoxene, monazite, kyanite, garnet, magnetite, and epidote. Feldspar is present in the light mineral fraction. The Eneabba district is located to the north of Perth and is distinguished by its great size, high heavy-mineral grades, and a rutile-rich mineral assemblage (Baxter 1977; Shepherd 1986). The deposit consists of seven high-grade layers, each one or more meters thick. They are locally cemented by iron hydroxides and a clay matrix, and interbedded with low-grade eolian deposits (Force 1991). The platforms, which consist of these layers, form steps from 130 meters down to 29 meters. Layers in the upper terrace, from 130 to 100 meters, contain 28 to 46% ilmenite, 5 to 8% rutile, and 36 to 61% zircon, whereas layers in the lower terrace contain 53 to 68% ilmenite, 8 to 11% rutile and 15 to 23% zircon. The Eneabba deposits have been intensely weathered, and pseudorutile is the component of the altered ilmenite (Frost et al. 1983).

Canada: An anorthosite-gabbro rock near Allard Lake, Quebec, contains the largest titaniferous magnetite-ilmenite deposit in the world. It is known as the Lac Tio deposit and is a massive, coarse-grained, flat-lying, tabular igneous sheet in anorthosite that has estimated dimensions of 1,100 meters long, 1,000 meters wide, and more than 60 meters thick. It consists of 75% ilmenite and 25% hematite with minor amounts of magnetite, apatite, zircon, spinel, and sulfides (Hammond 1952; Garnar and Stanaway 1994). It appears that the ore is igneous in origin and is related to the ferrodiorite parents. The Lac Tio ore is used to produce high-quality pig iron and titaniferous slag, which has a 75 to 80% TiO_2 content.

Norway: Norway is Europe's main producer of titanium minerals. There are two ilmenite deposits, both in anorthosite. The Storgangen ore deposit, which is located near Hauge, is an elongated intrusion in anorthosite of about 1,600 meters long, 50 meters wide, and 50 meters thick. It is strongly layered, with oxide minerals concentrated in layers near the base of the deposit. The ore contains about 40% ilmenite and 9% magnetite with a gangue that consists mainly of hypersthene and plagioclase with biotite, pyrite, chalcopyrite, and spinel. The ilmenite has magnetite in it as exsolution lamellae, whereas the magnetite is essentially pure (Krause and Pape 1975).

The Tellnes deposit, which is about 40 km east of the Storgangen deposit, consists mainly of an ilmenite-rich norite intruded into anorthosite. The deposit is 2,700 meters long and as much as 400 meters wide with a thickness estimated to be about 200 to 250 meters (Gierth and Krause 1973). The Tellnes ore is much finer-grained than the Storgangen ore, the intergrowth of ilmenite and magnetite is more intimate, and the content of magnetite is 2% versus 9% in Storgangen ore. The ilmenite content is about the same as that in Storgangen ore. Other ores include 36% plagioclase, 15% hypersthene, 3.5% biotite, and 3.5% accessory minerals (Dybdahl 1960).

South Africa: South Africa became one of the world's major producers of titanium minerals when the smelting process for low-TiO_2 ore became successful at the Richards Bay operation. High-grade titaniferous slag and low-manganese iron are the major products. The deposit is in a strip of heavy mineral sands along the modern coast facing the Indian Ocean. It is about 15 km long and 3 km wide. The sand dunes have an average height of 20 meters, but some of them reach a height of 100 meters. The dunes are mostly made of eolian sands of Holocene age (Fockema 1986). Ilmenite is the dominant heavy mineral. Other minerals are rutile, zircon, leucoxene, and monazite.

The United States: The titanium mineral deposits in the Roseland district of Virginia consist of ilmenite and rutile. They are associated with anorthosite and are of two types: disseminated in the border facies of the anorthosite and gneiss, and in the form of dikelike masses in nelsonite (Fish 1962; Fish and Swanson 1964). Coarse-grained rutile is generally developed, and its concentrations are typically about 2%, which matches the TiO_2 content of the metavolcanic country rock. The ilmenite is nearly stoichiometric in composition and commonly free of intergrowth (Force 1991).

The major producing deposits in the United States are in the Jacksonville district, as shown in Fig. 96. The Trail Ridge deposit is part of a 200 km long sand ridge that extends from northern Florida into southeastern Georgia (Force and Rich 1989). The bulk ore sand has a medium grain size of about 0.3 mm and is well sorted. The heavy mineral assemblage consists of 50% altered ilmenite and leucoxene, 15% zircon, 15% staurolite, 5% sillimanite, 5% tourmaline, 3% rutile, and 3% kyanite, with minor amounts of spinel, corundum, and monazite. The Green Cave Springs deposit, located east of Trail Ridge, is a prograduational beach ridge complex about 15 km long and 1 km wide. Its mineralogy is similar to that of the Trail Ridge deposit (Pirkle et al. 1974; Pirkle et al. 1993).

Sierra Leone: The Gbangbama district in the coastal region of Sierra Leone has the most important rutile producing deposits in the world (Lang 1970; Stanaway 1992). Rocks of the Gbangbama Hills consist of garnet amphibolites and garnet granulites, both deeply weathered and both containing rutile. The rutile deposits are radially formed around the Gbangbama Hills in the flat alluvial plains. The detrital mineral assemblage is dominated by rutile with garnet and ilmenite. Common heavy minerals in minor amounts are all present in the deposits.

INDUSTRIAL PROCESSES AND USES

1. Milling and Processes

Milling and beneficiation of ore from crystalline rocks, as adopted in the United States, generally consists of three stages: crushing, grinding, and flotation (Peterson 1966). The primary crushing to −14 mm and the secondary crushing to −64 mm are made in jaw and cone crushers. The magnetite fraction of the ore is removed by vibrating screens and magnetic cobbers. The remaining ore is further reduced in size by grinding in rod and ball mills and is then sent to a flotation plant to produce an ilmenite concentrate. Flotation procedures vary from plant to plant. Generally, after conditioning the pulp with sulfuric acid and oleic acid, a rougher float at pH 6 to 6.5 is made, and then cleaned at a lower pH of 5 to 5.5 (Woditsch and Westerhaus 1993).

Because the ilmenite-hematite ore from Lac Tio, Canada, has intimate intergrowths of the two minerals, a clean ilmenite concentrate cannot be obtained by conventional ore dressing procedures. A smelting process is used to produce pig iron and a titanium-rich slag (Garnar and Stanaway 1994). In this process, the ilmenite-hematite ore is reduced by low-ash coal with or without lime, at 1,500 to 1,700°C, in

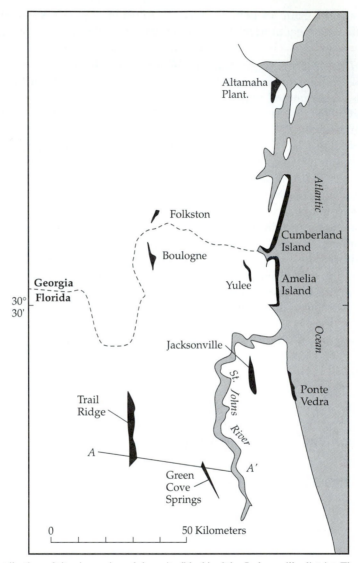

Figure 96 Distribution of titanium mineral deposits (black) of the Jacksonville district, Florida and Georgia, after Force (1991).

graphite electrode furnaces. The lighter titanium slag floats to the top and the iron slag sinks to the bottom and is drawn off from the furnace. The titanium slag is cooled, crushed, and screened prior to packing for shipment. Richards Bay, South Africa, has a similar slagging operation. Slag containing 50 to 55% TiO_2 can be made from low grade ore. Slags of 65 to 70% TiO_2 and 75 to 80% TiO_2 can be produced from medium and high grade ores.

At Narngulu, western Australia, "synthetic rutile" containing 90 to 95% TiO_2 is produced by pyrometallurgical upgrading of an ilmenite feedstock (Cassidy et al. 1986). The iron in the ilmenite feedstock can be converted to the metallic state by using a sub-bituminous coal, both as the source of fuel and as a reducing agent. The reaction is carried out at about 1,200°C in a rotary kiln. The reduced ilmenite is agitated in an aerated 1% ammonium chloride solution at 80°C, and the metallic iron is removed. A sulfur-bearing compound used at the reduction stage, coupled with a weak sulfuric acid wash used after the aeration stage of the process, act to reduce the manganese content of synthetic rutile. The rutile produced has a typical TiO_2 content of 92.5%.

The treatment of beach heavy mineral sands is well developed in Australia, especially with respect to the use of processing equipment for electrical separation and gravity concentration (Blendull et al. 1986; Reaveley and Ritchie 1986; Wright et al. 1986). The dredged beach sands are first treated in a high capacity gravity concentrator by using a Reichert cone, a combined Reichert cone and spiral separator, a Reichert spiral, and a Wright impact tray, or a pinched sluice concentrator. This wet concentration process rejects tailing and produces both a middling that may be recirculated and a rougher concentrate. This rougher concentrate is upgraded in cleaner stages to produce a bulk heavy mineral concentrate. A secondary concentration process is incorporated to remove minerals of light and intermediate densities such as quartz, tourmaline, and other minerals.

The concentrate, which contains about 85% heavy minerals, is first passed through a low-intensity wet drum magnetic separator to remove magnetite. The nonmagnetics from this stage are treated with high-gradient magnetic separators to remove feebly magnetic material. The resulting concentrate is dried and fed to the dry mill where high-tension roll separators and induced magnetic roll separators are used to produce ilmenite and other heavy minerals. A flow sheet illustrating some of the general steps is shown in Fig. 97 (Lynd and Lefond 1983).

2. Industrial Uses

Titanium minerals have two major industrial uses: the manufacture of white pigments and the production of titanium. Ilmenite and titaniferous slag are almost entirely used for pigments, whereas rutile is used for both. Other uses are in electronics and ceramics.

TiO$_2$ Pigments: The scattering power of a pigment depends upon its refractive indices, particle size, and wavelength. The refractive indices of both rutile and anatase are very high, and even with binder they lie in the range between 1.33 (water) and 1.73 (polyester fibers). The scattering power of TiO_2 is at its maximum at a particle size of 0.2 μm. At such a size, the TiO_2 particles scatter light of shorter wavelengths more strongly and, therefore, show a slight blue tint, whereas larger particles have a yellow tone. Lightness and hue are the most important properties for white pigment. For TiO_2, the pigments produced from the chloride process, in which the $TiCl_4$ is purified before combustion, have a higher color purity and very high lightness values (Woditsch and Westerhaus 1993).

There are two basic processes for the production of titanium dioxide pigments: the sulfate process by hydrolysis of titanyl sulfate solutions, and the chloride process by hydrolysis of gaseous titanium tetrachloride. The choice of which of the two to use is partly dictated by the available raw materials. The sulfate process uses ilmenite and titaniferous slag, whereas ilmenite, leucoxene, natural and synthetic rutile, titaniferous slag, and anatase are all used in the chloride process. The sulfate process is a long-established

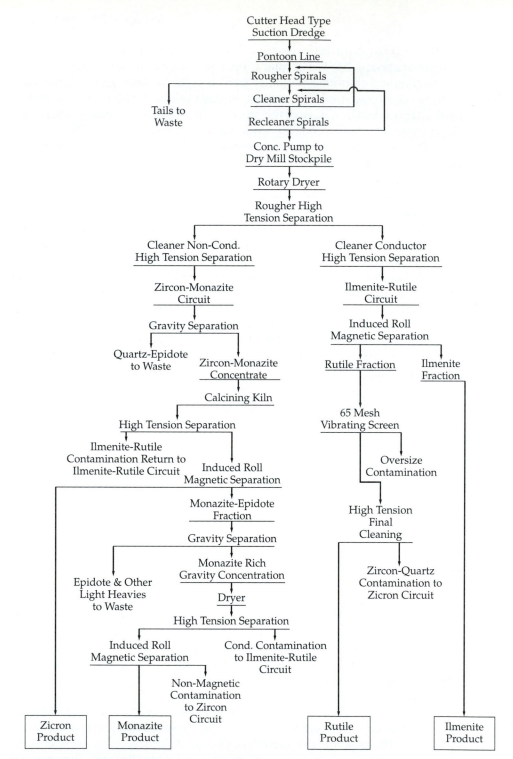

Figure 97 Flow sheet for processing beach sands, after Lynd and Lefond (1983).

process, whereas the chloride process has received increasing industrial usage (Kampfer 1973; Darby and Leighton 1977; Woditsch and Westerhaus 1993).

In the sulfate process, finely ground ilmenite or high-TiO_2 slag is digested in hot, concentrated sulfuric acid to form a solution of ferrous and titanyl sulfates. Any ferric ions present are reduced to a ferrous state by adding scrap iron. After reduction, the solution is clarified by setting and filtration to remove insoluble residues, and cooled in order to remove the bulk of the iron, which is in the form of crystalline, hydrated ferrous sulfate. The solution is then concentrated, and the titania is precipitated by hydrolysis. The hydrous titania is filtered, washed, and calcined at 900° to 1,000°C to produce TiO_2 pigments. Hydrolysis and precipitation of hydrous titania can be hastened by the addition of rutile or anatase seed.

In the chloride process, titanium tetrachloride is prepared by reacting titaniferous minerals and petroleum coke with chlorine and oxygen in a fluidized-bed reactor at 800° to 1,200°C. The raw materials must be as dry as possible to avoid the formation of HCl. The reaction gases are cooled down to a temperature of $<300°C$, a temperature at which the chlorides of other components can be satisfactorily separated from the $TiCl_4$ by condensation or sublimation. The gas is then mainly $TiCl_4$, which is cooled down to 0°C. After purification, titanium tetrachloride is combusted with oxygen at 900° to 1,400°C to form TiO_2 pigment. The resulting fine grained TiO_2 is calcined at about 500° to 600°C to remove any residual chlorine and hydrochloric acid that may have been formed in the reaction. Aluminum chloride is added to the titanium tetrachloride to ensure that all of the titanium is oxidized to rutile.

The estimated tons of raw materials required to produce one ton of TiO_2 are as follows (Lynd and Lefond 1983):

Sulfate Process		Chloride Process	
Ilmenite of slag	1.5 to 2.8	Rutile	1.1 to 1.2
Sulfuric acid	3.0 to 4.0	Chlorine	0.1 to 0.2
Iron scrap	0.1 to 0.2	Petroleum coke	0.1 to 0.2
		Oxygen	0.4 to 0.5
		Aluminum chloride	0.03

Paints and coatings account for the largest consumption of TiO_2 pigment (Loughbrough 1992). The presence of the pigment enables the protective potential of the coating material to be fully exploited, and coatings only a few micrometers thick fully cover the substrate. To improve the weather resistance and lightfastness of the pigment, several types of treatments are used to precipitate colorless inorganic or organic compounds onto the surface. One of the procedures is to apply a dense amorphous silica coating to the titanium pigment. Such coatings are extremely compact and form dense protective skins around the particles of TiO_2. Pigments prepared in this way are extremely durable and have outstanding ease of dispersibility in paint systems (Woditsch and Westerhaus 1993).

TiO_2 pigments are virtually suitable for all printing paper except newsprint, because of their brightness, opacity, printability, and light fastness retention. Although TiO_2 pigments still dominate the industry, alternatives such as kaolin, talc, calcined alumina, and precipitated calcium carbonate have made inroads into the market. However, for white paper that has to be very opaque even when very thin, there are no replacements for TiO_2 pigments. The TiO_2 can be incorporated into the body of the paper or applied as a coating to give it superior quality.

In plastics, TiO_2 pigments are valued because of their high refractive indices, chemical inertness, and whiteness. They can be used to color plastic packaging films < 100 μm thick to protect the packaged goods from view and to allow the film to be printed. In addition, TiO_2 pigments absorb ultraviolet radiation with a wavelength < 4150 Å and thus protect packaged goods from these harmful rays. The TiO_2 pigments are widely used in polystyrene, polyethylene, polyvinylchloride, and polylefin. An organic surface treatment is commonly given to the TiO_2 pigments for their incorporation into plastics. This is usually carried out during the final milling stage. A variety of organic compounds have been used, the most effective being dimethyl siloxane, pentaerythritol, and triethanolamine (Loughbrough 1992; Woditsch and Westerhaus 1993).

TiO_2 pigments are also used in synthetic fibers, enamel, white cement, rubber, and linoleum.

Titanium Metal: Titanium is mainly used in the aerospace industry as a structural metal. Its outstanding properties are high strength to weight ratio and excellent corrosion resistance. It is made by the Kroll process (Barksdale 1966; Gilchrist 1989). The titanium tetrachloride used to manufacture titanium metal is produced by chlorination of rutile, rutile + leucoxene, or other high-TiO_2 concentrates, using fluidized bed chlorinators at temperatures between 850° and 950°C, and with coke as the reductant. The $TiCl_4$ liquid produced is further purified by distillation.

The reduction of $TiCl_4$ is made by magnesium under a helium atmosphere at a temperature above the melting point of magnesium chloride, but below the melting point of titanium. Titanium forms from this process as a spongelike mass that is crushed and leached with acid to remove residual magnesium and magnesium chloride. Titanium sponge is converted to ingot by compacting and double or triple vacuum arcmelting (Lynd and Lefond 1983).

Nonpigmentary TiO_2: TiO_2 has many nonpigmentary uses, including (1) enamels; (2) glass, ceramics, and glass-ceramics; (3) electroceramics; (4) catalysts and catalysts supports; and (5) welding fluxes.

REFERENCES

Anderson, A. T., Jr., and M. Morin. 1969. Two types of massif anorthosites and their implications regarding the thermal history of the crust. In Origin of anorthosite and related rocks. Ed. Y. W. Isachsen. *New York State Museum and Science Service Memoir* 18:57–69.

Ashwal, L. D. 1982. Mineralogy of mafic and Fe-Ti oxide-rich differentiates of Marcy anorthosite massif, Adirondacks, New York. *Amer. Mineral.* 67:14–27.

Barksdale, J. 1966. *Titanium: Its occurrence, chemistry, and technology.* New York: Ronald Press, 696p.

Baxter, J. 1977. Heavy mineral sand deposits of western Australia. *Geol. Surv. Western Australia Mineral. Res. Bull.,* 10, 148p.

Blendull, G., N. Dawson, and D. James. 1986. Utilisation of Readings wet high intensity magnetic separation at Richards Bay Minerals. In Australia: A world source of ilmenite, rutile, monazite and zircon, 81–86. *Sym. Series, Australasian Inst. Mining Metall., Sept.–Oct.,1986, Perth, Australia.*

Burton, B. P. 1984. Thermodynamic analysis of the system Fe_2O_3-$FeTiO_3$. *Phys. Chem. Minerals* 11:132–139.

Cassidy, P. W., D. L. Clements, B. A. Ellis, and P. E. Rolfe. 1986. The AMC Narnguku synthetic rutile plant. In Australia: A world source of ilmenite, rutile, monazite and zircon, 123–129. *Sym. Series, Australasian Inst. Mining Metall., Sept.–Oct., 1986, Perth, Australia.*

Collins, L. B., and N. T. M. Hamilton. 1986. Stratigraphic evolution and heavy-mineral accumulation in the Minninup shoreline, southwest Australia. In Australia: A world source of ilmenite, rutile, monazite, and zircon, 17–22. *Sym. Series, Australasian Inst. Mining Metall., Sept.–Oct., 1986, Perth, Australia.*

Collins, L. B., B. Hochwimmer, and J. L. Baxter. 1986. Depositional facies and mineral deposits of the Yoganup shoreline, southern Perth basin. In Australia: A world source of ilmenite, rutile, monazite, and zircon, 9–16. *Sym. Series, Australasian Inst. Mining Metall., Sept.–Oct.,1986, Perth, Australia.*

Dachille, F., P. Y. Simons, and R. Roy. 1968. Pressure-temperature studies of anatase, brookite, rutile and TiO_2-II. *Amer. Mineral.* 53:1929–1939.

Darby, R. S., and J. Leighton. 1977. Titanium dioxide pigments. In Modern inorganic chemical industry. Ed. R. Thompson. *Special Publ. 31, Chemical Soc., London,* 466p.

Deer, W. A., R. A. Howie, and J. Zussman. 1996. *An introduction to rock-forming minerals.* Essex, England: Longman, 696p.

Dybdahl, I. 1960. Ilmenite deposits of the Egersund anorthosite complex. *XXI Intern. Geol. Congr., Copenhagen, Guidebook to Excursion No. C-10,* 49–53.

Fish, G. E. 1962. Titanium resources of Nelson and Amherst Counties, Virginia 1. Saprolite ores. *U.S.B.M. Rept. Invest.* 6094, 44p.

Fish, G. E., and V. F. Swanson. 1964. Titanium resources of Nelson and Amherst Counties, Virginia 2. Nelsonite. *U.S.B.M. Rept. Invest.* 6429, 25p.

Fockema, P. D. 1986. The heavy mineral deposits north of Richards Bay. *In Mineral deposits of Southern Africa.* Ed. C. R. Anhaenssr, 2301–2307. Johannesburg: Geol. Soc. South Africa.

Force, E. R. 1991. Geology of titanium mineral deposits. *U.S.G.S. Spec. Paper* 259, 112p.

Force, E. R., and F. J. Rich. 1989. Geologic evolution of Trail Ridge eolian heavy mineral sand and underlying peat, northern Florida. *U.S.G.S. Prof. Paper* 1499, 16p.

Frost, M. T., I. E. Grey, I. R. Harrowfield, and K. Mason. 1983. The dependence of alumina and silica contents on the extent of alteration of weathered ilmenite from Western Australia. *Mineral. Mag.* 47:201–208.

Garnar, T. E., and K. J. Stanaway. 1994. Titanium Minerals. In *Industrial minerals and rocks.* 6th ed. Ed. D. D. Carr, 1071–1089. Littleton, Colorado: Soc. Mining, Metal. & Exploration.

Gierth, E., and H. Krause. 1973. Die Ilmenitlagerstâtte Tellnes (süd-Norwegen). *Norsk Geologiske Tidsskrift* 53:359–402.

Gilchrist, J. D. 1989. *Extractive metallurgy.* 3rd. ed. Oxford: Pergamon Press, 431p.

Hails, J. R. 1969. The nature and occurrence of heavy minerals in Pleistocene and Holocene sediments in three coastal areas of New South Wales. *Jour. Royal Soc. of New South Wales* 102:21–39.

Hammond, P. 1952. Allard Lake ilmenite deposits. *Econ. Geol.* 47:634–649.

Kampfer, W. A. 1973. Titanium Dioxide. In Vol. 1 of *Pigment Handbook,* 1–36. Ed. T. C. Patten. New York: Wiley-Interscience.

Karkhanavala, M. D., and A. C. Momin. 1959. The alteration of ilmenite. *Econ. Geol.* 54:1095–1102.

Krause, H., and H. Pape. 1975. Mikroskopische untersuchngen der mineralvergeselischafttung in erz und nebengestein der ilmenitagerstâtte Storgangen (süd-Norwegen). *Norsk Geologisk Tidsskrift* 57:263–284.

Lang, H. D. 1970. Sekundare rutil-lagerstatten in Sierra Leone (West-Africa). *Erzmetall* 23:179–183.

Loughbrough, R. 1992. TiO$_2$ pigment—Overcapacity hits again. *Ind. Minerals.* (June):37–53.

Lynd, L. E., and S. L. Lefond. 1983. Titanium minerals. In *Industrial minerals and rocks.* 5th ed. Ed. S. J. Lefond, 1303–1362. New York: Soc. Mining Engineers, AIME.

McKellar, J. B. 1975. The eastern Australia rutile province. In Economic geology of Australia and Papua New Guinea. Ed. C. L. Knight. *Austral. Inst. Mining Metall. Monograph* 5:1055–1061.

Meagher, E. P., and G. A. Lager. 1979 Polyhedral thermal expansion in the TiO$_2$ polymorphs: refinement of the crystal structures of rutile and brookite at high temperature. *Canadian Mineral.* 17:77–85.

Peterson, E. C. 1966. Titanium resources of the United States. *U.S.B.M. Inform. Circ.*, 8290, 65p.

Pirkle, E. C., W. A. Pirkle, and W. H. Yoho. 1974. The Green Cove Springs and Boulougne heavy-mineral sand deposits of Florida. *Econ. Geol.* 67:1129–1137.

Pirkle, F. L., E. C. Pirkle, J. G. Reynolds, M. A. Pirkle, J. A. Henry, and W. J. Rice. 1993. The Folkston West and Amelia heavy mineral deposits of Trail Ridge, southeastern Florida. *Econ. Geol.* 88:961–971.

Reaveley, B. J., and I. C. Ritchie. 1986. The development of high efficiency spiral separators. In Australia: A world source of ilmenite, rutile, monazite and zircon, 87–98. *Sym. Series, Australasian Inst. Mining Metall., Sept.–Oct., 1986, Perth, Australia.*

Schüller, K. H. 1984. Oxides ceramics. In *Process mineralogy of ceramic materials.* Ed. W. Baumgart, A. C. Dunham, and G. C. Amstutz, 104–124. New York: Elsevier.

Shepherd, M. S. 1986. Australian heavy mineral reserves and world trends. In Australia: A world source of ilmenite, rutile, monazite, and zircon, 61–68. *Sym. Series, Australasian Inst. Mining Metall., Sept.–Oct., 1986, Perth, Australia.*

Shvetsova, I. V. 1970. Mixed rutile-anatase leucoxene. *Dokl. Acad. Sci. USSR, Earth Sci. Sect.* 194:130–133.

Slingerland, R., and N. D. Smith. 1986. Occurrence and formation of water-land placers. *Annual Rev. Earth and Planetary Sci.* 14:113–147.

Stanaway, K. J. 1992. Heavy mineral placers. *Mining Eng.* 44:352–358.

Temple, A. K. 1966. Alteration of ilmenite. *Econ. Geol.* 61:695–714.

Whitehead, J. 1983. Titanium compounds. In Vol. 23 of *Kirk-Othmer's encyclopedia of chemical technology.* 3rd ed. 131–176. New York: John Wiley.

Woditsch, P., and A. Westerhaus. 1993. Titanium dioxide. In *Industrial inorganic pigments.* Ed. G. Buxbaum, 43–71. Weinheim: VCH Verlagsgesellschaft mbH.

Wright, D., R. G. Richards, and M. S. Cross. 1986. The development of mineralsands separation technology—solving today's problem, anticipating tomorrow's. In Australia: A world source of ilmenite, rutile, monazite and zircon, p. 102–123. *Sym. Series, Australasian Inst. Mining Metall., Sept.–Oct., 1986, Perth, Australia.*

Vermiculite

Vermiculite is a clay mineral with a mica-type structure and magnesium-rich composition. It is characterized by its ability to expand when heated rapidly to elevated temperatures. This method produces what is known as expanded or exfoliated vermiculite, which is the most common form of vermiculite used in the construction and agriculture industries.

The United States and South Africa dominate world production of vermiculite. Their combined output amounts to more than 90% of the world's total production, which was estimated to be 500,000 tons/year between 1990 and 1995.

MINERALOGICAL PROPERTIES

Vermiculite belongs to the sheet silicates, and its crystal structure is similar to that of phlogopite, the trioctahedral mica (Shirozu and Bailey 1960; de la Calle and Suquet 1988). The interlayer cation in vermiculite is magnesium, in contrast with potassium in phlogopite. The most characteristic feature that distinguishes vermiculite from phlogopite is the presence of water in the interlayer. The H_2O is linked by a hydrogen bond to an oxygen in the silicate sheet and forms a layer with a hexagonal arrangement. In naturally occurring vermiculite two of these hexagonal layers exist in each interlayer and are held together by the interlayer magnesium. Not all available interlayers are occupied by H_2O molecules under normal atmospheric conditions, and many of the H_2O molecules do not enclose magnesium. A magnesium vermiculite has a basal spacing of $d_{(001)}$ close to 14.3Å, which is reduced to 11.7Å with the removal of one layer of H_2O. The $d_{(001)}$ remains at 14.3Å when treated with glycerol, but collapses to 9Å when both layers of H_2O are removed (de la Calle and Suquet 1988; Deer et al. 1992).

The structural formula for standard vermiculite may be written as $Mg_x(H_2O)_n(Mg,Al,Fe)_3(Si,Al)_4O_{10}(OH)_2$, with "x" being the layer charge per formula unit. The limits of substitution in most natural vermiculite are as shown below:

$$(Mg,Ca)_{0.6-0.9}(Mg)_{4.5-5.0}(Fe^{3+},Al)_{1.5-1.0}$$

$$(Al_{2.2-2.6}Si_{5.8-5.4})O_{20}(OH)_4(H_2O)_{7.0-9.0}.$$

Vermiculite is monoclinic with a = 5.3Å, b = 9.25Å, c = 14.3Å, and β = 97°. Specific gravity, hardness, and refractive indices are 2.3 to 2.75, 1.5 to 2, and 1.520 to 1.583, respectively.

Two endothermic peaks appear in the DTA curves of vermiculite at 140° and 240°C that correspond to the transformation from the two-layer to the one-layer hydrate form, and from the one-layer hydrate to the anhydrous state. Both reactions are reversible up to the limit of 550°C, above which the structure breaks down by endothermic loss of (OH) and by exothermic reactions at higher temperatures (de la Calle and Suquet 1988). During dehydration, the interlayer cations remain in the interlayer positions, therefore the rehydration is strongly affected by the hydration energies of the interlayer cations (Kawano and Tomita 1991). Apparent sintering and melting temperatures were assigned to vermiculite at 1,260°C and 1,315°C, respectively.

When heated rapidly to about 300°C or higher, vermiculite expands by exfoliating at right angles to the cleavage planes. This is due to the rapid conversion of contained water to steam that cannot escape without buckling and separating the structural layers. Vermiculite may also be expanded by treatment with hydrogen peroxide. This appears to relate to the liberation of oxygen by reaction with the interlayer magnesium. This characteristic expansion reduces the bulk density from approximately 650 kg/m^3 to 65 kg/m^3. Individual flakes may expand up to thirty times, but vermiculite as mined generally expands from eight to twelve times. The expansion results in large pores being formed between the platelets, thus increasing the void volume. Color changes in the expansion are dependent on the environment of treatment. An oxidizing atmosphere produces a dull gray or tan color, whereas a reducing atmosphere can produce a bronze or gold color (Deer et al. 1992). Vermiculite that is mixed with mica or with vermiculite-mica mixed layers produces larger expansibility as compared with vermiculite without the additional mica. It has been suggested that the presence of relics of altered mica and the loss of OH groups affect expansion (Justo et al. 1989).

GEOLOGICAL OCCURRENCE AND DISTRIBUTION OF DEPOSITS

Economic deposits of vermiculite are found in ultramafic and mafic igneous and metamorphic rocks that have been intruded by silicic and alkalic igneous rocks. In these rocks, vermiculite is formed from phlogopite, biotite, chlorite, and pyroxene by weathering processes, hydrothermal alterations, or groundwater percolations (Bush 1976).

Occurrences of vermiculite have been reported from more than twenty countries around the world. The United States and South Africa have the largest resources and contribute more than 90% of the world's annual production, which was estimated to have been 500,000 ton/year between 1990 and 1995 (*U.S.B.M. Annual Review 1995*).

The United States: The world's largest vermiculite deposit occurs in the Rainy Creek complex near Libby at the northwest corner of Montana (Boettcher 1967; Bush 1976). The complex represents a composite of successive intrusions of igneous rocks emplaced into quartzite, argillite, and limestone of the Precambrian Belt Supergroup, probably during the middle Cretaceous period. The succession began with the emplacement of a body of coarse-grained biotite pyroxenite that contained a core of coarse-grained biotite. Following this emplacement, magnetite pyroxenite intruded a zone of weakness between the biotite pyroxenite and the Belt Series, forming a ring-shaped dike surrounding the inner pyroxenite body. A younger mass of syenite cut both the magnetite and biotite pyroxenite, and alkalic syenite dikes cut the biotite core (Fig. 98). Vermiculite is formed from surficial alteration by circulating groundwater of biotite in the biotite pyroxenite, and to a smaller extent in the magnetite pyroxenite. Hydrobiotite is formed from hydrothermal alteration. The vermiculite formation is reported to have penetrated to a depth of 300 meters. Some large lenses of vermiculite are 3 to 30 meters thick and 300 meters long, with a vermiculite content of up to 95%.

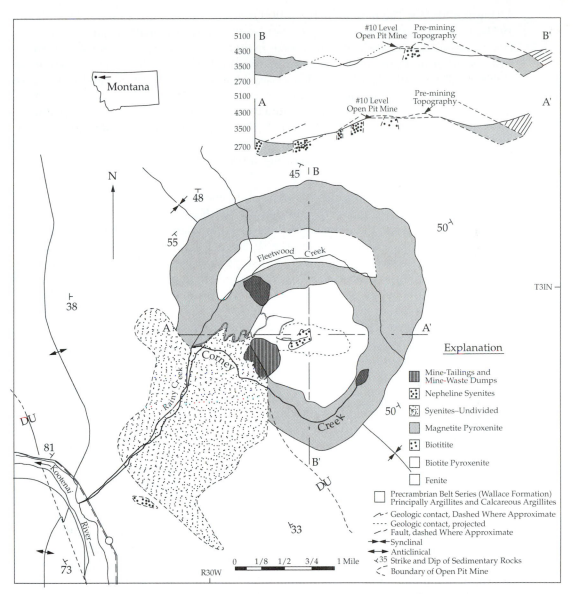

Figure 98   Geological map of the Rainy Creek complex, Libby, Montana, after Boettcher (1967).

The biotite in the core is virtually unaltered (Boettcher 1967). A representative composition of Libby vermiculite is 40.16% SiO_2, 20.63% MgO, 12.01% Al_2O_3, 13.00% Fe_2O_3, 5.93% K_2O, 1.54% CaO, 1.44% TiO_2, and 5.29% H_2O (Strand and Stewart 1983).

In the Blue Ridge province of North Carolina, vermiculite occurs in lenses along the serpentinized point of contact between the dunite and pyroxenite bodies and the metamorphic country rocks, within the crosscutting pegmatites, and along the fractures within the bodies. The lenses are typically 1.5 meters thick and 15 meters long (Bush 1973). In the Enoree district, some 50 km southeast of Greenville, South Carolina, a large number of vermiculite deposits occur in biotite schist and gneiss and were formed from weathering of biotite in enclosed pyroxenite lenses. All the deposits are cut by stringers of pegmatite. Vermiculite lenses are generally 45 meters wide and 140 meters long, with the weathering zone extending down to about 25 meters (Hunter 1950; Maybin and Carpenter 1990). A representative composition for Enoree vermiculite is 39.77% SiO_2, 18.32% MgO, 13.88% Al_2O_3, 12.84% Fe_2O_3, 5.11% K_2O, 1.02% CaO, 2.07% TiO_2, and 6.99% H_2O (Strand and Stewart 1983). The composition of Enoree vermiculite matches the composition of the vermiculite from Libby.

South Africa: A large vermiculite deposit is located near Kruger Park in the pyroxenite rim of the Palabora carbonatite of the northeastern Transvaal. This complex extends 6.5 km in a north-northwesterly direction with a width attaining 2.5 km. Fine mica is most abundant near the surface of the inner rim of the pyroxenite, which is injected by bodies and stringers of serpentine-apatite-magnetite. Vermiculite occurs in the ultrabasics, especially close to serpentinized patches, and is disseminated with apatite. Two types of vermiculite can be distinguished. The honey brown vermiculite (2V = 0) of the north contains 5% H_2O, 12% Al_2O_3, and 6.5% Fe-oxides, and has a twenty-six-fold expansion. The southern black variety (2V = 15°) has a composition of 2.5% H_2O, 7% Al_2O_3, and 8.5% Fe-oxides, with an eleven-fold expansion (de Kun 1965; Clarke 1981).

INDUSTRIAL PROCESSES AND USES

1. Industrial Processes

Milling and processing of vermiculite are generally done in three stages: (1) the vermiculite is separated from the gangue minerals, (2) the vermiculite is delaminated into flakes, and (3) the flakes are classified into industrial size fractions.

In the first stage, the ore is usually broken down into < 20 mm or < 16 mm granules, and the adhering fine material is removed by washing and screening. At Palabora, the vermiculite ore contains 6 to 7% H_2O when it reaches the milling plant, but about 50% of this moisture is eliminated in the drying process. The vermiculite flakes are separated from the accompanying rock by multistage screening in horizontal streams of air by using various suspension procedures (Holz 1983). Wet beneficiation of vermiculite generally consists of two processes: froth flotation for the fine sizes and beneficiation by shape for the coarse sizes. The products from both the United States and South Africa are generally separated into five size grades. They are (Tyler Standard) −3 +10, −6 +14, −8 +28, −20 +65, −50 mesh (Strand and Stewart 1983).

The vermiculite flakes must be expanded or exfoliated for industrial use. During this process, the flakes are expanded five- to eight-fold at 250° to 1,500°C. The time-temperature relationship is critical because an incorrect profile can result in the breakdown of the flakes into powder. Usually a shorter retention time and higher temperature will produce better expansion. At the expanding plant, the vermiculite flakes are fed at a constant rate to vertical, oil, or gas fired furnaces, which expand vermiculite very rapidly at a temperature of about 1,000° C. The vermiculite is then cooled to about 400°C and passed over a separator that removes any remaining impurities. Specifications for expanded products for different industrial uses have been established by the ASTM (Strand and Stewart 1983).

2. Industrial Uses

Vermiculite, in its expanded or exfoliated form, is mainly used in construction and agriculture. Its use extensively overlaps that of perlite, because both have low density and excellent thermal and acoustic insulation characteristics (Power 1986; Loughbrough 1991).

Construction: Loose fill represents the most basic form of exfoliated vermiculite insulation. It is poured between ceiling joists, packed into gaps between brick linings and casings, or used as a simple cavity filler. For use in masonry walls, vermiculite requires a treatment with bitumen to repel water erosion. The formed products of vermiculite are used as insulating blocks, pipe insulation, and refractory bricks (Dickson 1986).

Vermiculite is mixed in various ratios with Portland cement to produce a lightweight concrete. The physical properties of lightweight aggregate concretes for both structural and insulation purposes are given in Table 17 (p. 276). In comparison with perlite, vermiculite has a lighter weight ratio and weaker strength. Exfoliated vermiculite, expanded perlite, and expanded polystyrene are classified as ultra lightweight.

Vermiculite is used to produce insulating plaster by bonding with gypsum or cement. This plaster possesses several characteristics: it is lightweight, has good spreadability, and has good fire retardant, acoustic insulation, and thermal insulation properties. The properties of the plaster depend upon the ratio of vermiculite to gypsum/cement. The greater the proportion of vermiculite, the lighter the weight and the better the spreadability will be, but this results in a relative decrease in the cementitious property of the plaster (Power 1986).

Agriculture: Vermiculite is used in agriculture because of its superior air and water retention properties, and as an inert filler ingredient in cattle feed. Vermiculite, like perlite, has a wide range of applications, which have been categorized by Lin (1989) and discussed in perlite (p. 275). Vermiculite is used in a variety of livestock applications. It is used (1) as a carrier for the absorption of veterinary preparations or of nutrients; (2) as an absorbent for pesticides, fertilizers, oils, or pharmaceuticals; (3) as a digestive additive to the feed mixtures for cattle, poultry, and hogs; (4) as an additive to feed mixes to improve their flow and pourability; (5) as an anticaking agent added to feed mixes; and (6) as a filtering aid to remove toxic concentrations of ammonia from aquacultural systems.

For crop farming applications, vermiculite is used (1) as a substrate in hydroponic farming, (2) as a fertilizer-carrying matrix for controlled slow release, (3) as a medium for physical uptake of seeds in a dry mixture, (4) as a soil conditioner for improving water retention and ventilation, and (5) as an additive for effluent treatment.

Other: Vermiculite, like perlite, is also used in castable refractories up to 1,100 to 1,200°C, especially to cover molten metal for thermal insulation during intermediate storage prior to casting (Dickson 1986). Vermiculite is used in package technology to cushion impacts, for thermal insulation, and to absorb liquids in case of breakage of the container. Experimental data have shown that acid-treated vermiculite appears to be a promising cracking matrix for the conversion of heavy fuels (Suquet et al. 1994).

REFERENCES

Boettcher, A. L. 1967. The Rainy Creek alkaline-ultramafic igneous complex near Libby, Montana. *J. Geol.* 75:526–533.

Bush, A. L. 1973. Lightweight aggregates. In United States Mineral Resources. Eds. D. A. Brobst and W. P. Pratt. *U.S.G.S. Prof. Paper* 820:333–355.

——— 1976. Vermiculite in the United States, Proc. 11th Forum on the Geology of Industrial Minerals. Eds. L. F. Rooney and R. B. Berg. *Montana Bur. Mines & Geol., Special Publ.* 74:145–155.

De La Calle, C., and H. Suquet. 1988. Vermiculite. In Vol. 19 of *Reviews in mineralogy.* Ed. S. W. Bailey, 455–496. Washington, D.C.: Mineral. Soc. Amer.

Clarke, G. M. 1981. The Palabora Complex—triumph over low grade ores. *Ind. Minerals* (October): 45–62.

Deer, W. A., R. A. Howie, and J. Zussman. 1992. *An introduction to rock-forming minerals.* 2nd. ed. Essex, England: Longman, 696p.

Dickson, T. 1986. Insulating refractories, containing the heat. In Raw materials for the refractories industry. Ed. E. M. Dickson, 178–183. *An IM Consumer Survey.*

Holz, P. 1983. Vermiculite expands—in more way than one. *Ind. Minerals* (September): 45–49.

Hunter, C. E. 1950. Vermiculite of the southeastern states. In *Sym. on Mineral Resources of the Southeastern United States.* Ed. F. G. Snyder, 120–127. Knoxville: Univ. Tennessee Press.

Justo, A., C. Maqueda, J. L. Perez-Rodriquez, and E. Morillo. 1989. Expansibility of some vermiculites. *Appl. Clay Sci.* 4:509–519.

Kawano, M., and K. Tomita. 1991. Dehydration and rehydration of saponite and vermiculite. *Clays and Clay Minerals* 39:174–183.

de Kun, N. 1965. *The mineral resources of Africa.* Amsterdam: Elsevier, 740p.

Lin, I. J. 1989. Vermiculite and perlite—For animal feedstuff and crop farming. *Ind. Minerals* (July): 43–49.

Loughbrough, R. 1991. Minerals in lightweight insulation. *Ind. Minerals* (October): 21–35.

Maybin, A. H., and R. H. Carpenter. 1990. Geochemistry: A new approach for vermiculite exploration in South Carolina, Proc. 24th Forum on the Geology of Industrial Minerals. Eds. A. J. M. Zupan and A. H. Maybin. *South Carolina Geol. Surv.,* 57–69.

Power, T. 1986 Perlite and vermiculite, the market overlap. *Ind. Minerals* (November): 39–40.

Shirozu, H., and S. W. Bailey. 1960. Crystal structure of a two-layer vermiculite. *Amer. Mineral.* 51:1124–1143.

Strand, P. R., and O. F. Stewart. 1983 Vermiculite. In *Industrial minerals and rocks.* 5th ed. Ed. S. J. Lefond, 1375–1381. New York: Soc. Mining Engineers, AIME.

Suquet, H., R. Franck, D. F. Lambert, F. Elasse, C. Marcilly, and S. Chevalier. 1994. Catalytic properties of two pre-cracking matrices: A leached vermiculite and an Al-pillard saponite. *Appl. Clay Sci.* 8:349–363.

Wollastonite

Wollastonite is a calcium silicate with a chain-type structure that results in a distinctly acicular crystal habit. This acicularity, which is a property of considerable industrial importance for wollastonite, enables it to be used in ceramics, paint, and plastics.

Wollastonite is one of the minerals that has a short history of industrial use. Its full potential was not recognized until 1950. Worldwide production of wollastonite reached an estimated 325,000 tons in 1995. The United States is the leading producer, followed by China, Finland, India, and Mexico to complete the top five.

MINERALOGICAL PROPERTIES

$CaSiO_3$ is polymorphic. There are three stable forms under atmospheric pressure. Pseudowollastonite, which is stable above 1,126°C, is a cyclosilicate with three-membered rings, $(SiO_3)_3$, linked together laterally by Ca^{2+}. It is triclinic with a = 6.90Å, b = 11,78Å, c = 19.65Å, $\alpha = 90°$, $\beta = 90.30°$, and $\gamma = 90°$. It melts congruently at 1,544°C (Trojer 1968). Both parawollastonite (wollastonite-2M) and wollastonite (wollastonite-Tc) are stable below 1,126°C, and both have an infinite single chain-type structure. Wollastonite single chains differ from pyroxene single chains, as illustrated in Fig. 99 (Liebau, 1985). The repeat unit in the wollastonite chain consists of a pair of tetrahedra joined apex-to-apex as in the (Si_2O_7) group, alternating with a single tetrahedron that has an edge parallel to the chain direction. The three-period chain has a unit length of 7.3Å. Pyroxene has a two-period chain and a chain period length of 5.2Å (Trojer 1968). The arrangements of the $[CaO_6]$ octahedra in both pyroxene and wollastonite are also shown in Figure 99. Parawollastonite (wollastonite-2M) is monoclinic, with a = 15.43Å, b = 7.32Å, c = 7.07Å, and $\beta = 95.40°$, whereas wollastonite (wollastonite-Tc) is triclinic, with a = 7.94Å, b = 7.32Å, c = 7.07Å, $\alpha = 90.03°$, $\beta = 95.37°$, and $\gamma = 103.43°$. The two polymorphic forms are related by a simple packing modification by doubling the "a" dimension of the wollastonite-Tc (Deer et al. 1992).

There is a high-pressure form of $CaSiO_3$ above 3 GPa, which is known as wolstromite, that also has a ring-type structure (Trojer 1969). As pressure increases, wolstromite decomposes to a two-phase assemblage of Ca_2SiO_4 and $CaSi_2O_5$, which in turn combines to form a pervoskite-type $CaSiO_3$ at higher pressures. The limits of stability of the two-phase assemblage are P(GPa) = 7.9 + 0.00147T (°C) and P(GPa) = 9.0 + 0.0021T (°C) (Gasparik et al. 1994).

Natural wollastonite usually approaches its ideal composition, $CaSiO_3$, although at high temperatures extensive solid solutions exist that contain $FeSiO_3$ and $MnSiO_3$. Ferroan and manganoan wollastonites will often form in contact metamorphic rocks (Deer et al. 1992). Wollastonite is generally white, but sometimes colorless, grey, or pale grayish green. Refractive indices for both low-temperature forms are $\alpha = 1.618$, $\beta = 1.630$, and $\gamma = 1.632$, and those for pseudowollastonite are $\alpha = 1.610$, $\beta = 1.611$, and $\gamma = 1.654$. The electrical conductivity at 1,300°C is $0.18\Omega \cdot cm^{-1}$, and the dielectric constant at 20°C is 6.17. Wollastonite and pseudowollastonite are resistant to most industrial

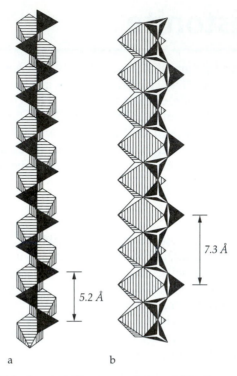

a   b

Figure 99   A comparison of (a) a two-period, pyroxene-chain and (b) a three-period, wollastonite chain, after Liebau (1985). The arrangement of (MO_6) in both chain-type structures is also shown.

gases as well as alkaline solutions, but react with aqueous mineral acids to produce gelatinous SiO_2. Wollastonite is commonly columnar, bladed to fibrous, and its crystals tend to be elongated on the b-direction and flattened on {100} or {001}.

Wollastonite or Wollastonite-Tc is the most abundant of all forms of $CaSiO_3$, and others are rare.

GEOLOGICAL OCCURRENCES AND DISTRIBUTION OF DEPOSITS

Wollastonite is a typical contact metamorphic mineral, occurring within the siliceous limestone near intrusive bodies (Deer et al. 1992). In the metamorphism of limestone, the intrusive introduces Si, Al, Fe, and Mn, to form skarn minerals marked by zones that are distinctively monomineralic or bimineralic. Skarns are the most coarse-grained of all the metamorphic rocks, and the grains are completely unoriented and commonly inter-locking. Common skarn minerals in association with wollastonite are garnet (grossular-ite-andradite series), diopside, epidote-zoisite, tremolite-actinolite, scapolite, and calcite. In these occurrences, wollastonite is the result of the reaction $CaCO_3 + SiO_2 = CaSiO_3 + CO_2$ above approximately 400°C (Harker and Tuttle 1956; Deer et al. 1992). The equilibrium boundary may be displaced to a lower temperature if the fluid phase in the reaction is not pure CO_2, but contains some water (Greenwood 1962). Wollastonite also occurs in some alkaline igneous rocks and regional metamorphic rocks.

The United States has the most extensive wollastonite deposits and is the world's leading producer of wollastonite (125,000 tons). The other four of the top five producing nations are China (85,000), India (60,000), Finland (25,000), and Mexico (15,000). The figures in parenthesis are the production estimates for 1995, as given by Harben and Kuzvart (1996).

The United States: The world's largest deposit of wollastonite is located on the western side of Lake Champlain near Willsboro, New York (Elevatorski and Roe 1983). Wollastonite occurs in a belt of contact metamorphosed limestones and metasomatized sediments that is 9.5 km long and 0.5 km wide. The three main deposits are at Willsboro, Lewis, and Deerhead. The Grenville limestone (Precambrian), on the intrusion of gabbro-anorthosite, was recrystallized and partly replaced by skarn bands that vary in composition from pure wollastonite to pure garnet (grossularite-andradite), and have an average mineralogical composition of 75% wollastonite, 15% garnet, and 19% diopside.

Wollastonite deposits are also mined from the Little and Big Maria Mountains, in California, about 30 km northwest of Blythe; in Arizona from the Mineral Hill area of the Sierrita Mountains and near Rosemont in Pima County; and in Nevada near Yerington in Lyon County (Elevatorski and Roe 1983).

China: Wollastonite deposits occur in at least six provinces: three in the northeastern region: Jilin, Heilongjiang, and Liaoning, and three in the central region along the Yangtze River: Anhui, Jiangxi, and Hubei (Fountain 1986). Of these, the most developed is the Dading Hills mining area in Lishu County, Jilin, where the ore is mined from an open pit, from eleven major and thirty-six minor veins in hosted marble and crystalline limestone. These veins are 2 to 15 meters thick, run from 80 to 300 meters long, and occur as stratified layers to depths of 70 to 90 meters. The ore is white and contains wollastonite (60 to 90%), calcite, quartz, diopside, and garnet (O'Driscoll 1990).

Finland: The main European source of wollastonite is the Lappeenranta deposit in southeastern Finland, close to the border with Russia. Wollastonite occurs as a zone 600 meters long and 45 meters thick in limestone that was metamorphosed by the intrusion of the Rapakivi granite, and which covers an outcrop area of 10,000 km^2. The productive layers are intersected by veins of amphibolites and pegmatites, and are interbedded with calcite, dolomite, and quartz. The wollastonite content ranges from 18 to 20%, with a maximum of up to 60% (Harben and Bates 1990).

India: Wollastonite deposits exist in the states of Madhya Pradesh, Tamil Nadu, and Gujarat, but the major producing deposit is the Belkapahar deposit, near Khila in the Jodhpur division of Rajasthan (Power 1986). Wollastonite occurs in zones and is interbedded with pyroxene and garnet schist.

Mexico: The Santa Fe deposit in Chiapas and the La Blanca District of Zacatecas are the major wollastonite sources in Mexico. The Santa Fe deposit is in an elliptical and dome shape, with a thickness of at least 90 meters and an areal extent that is estimated to be 425 meters long and 120 meters wide. The wollastonite is extremely pure, with a total content of associated minerals of less than 1%, and free of both garnet and calcite. It is brilliant white and acicular in habit. La Blanca wollastonite is also of high quality, but gray wollastonite-marble bands with layers of garnet are abundant at the contact zones where the limestone meets the granite deposits (Elevatorski and Roe 1983).

INDUSTRIAL PROCESSES AND USES

1. Milling and Beneficiation

At Willsboro, the lump ore is crushed to −16 mesh in three stages, and the garnet fraction is removed using high intensity magnetic separators. Four product sizes are then produced in pebble mills, and a high aspect ratio product is produced in an attrition mill. The dry beneficiation process is illustrated in Fig. 100 (Elevatorski and Roe 1983). At Lappeenranta, Finland, the ore is crushed and sorted into limestone, wollastonite, and waste. The beneficiation process involves flotation of calcite and treatment of the wollastonite concentrate in a wet magnetic separator. An optical sorter is also used. At Rajasthan, India, hand-sorting is first used to remove calcite, garnet, diopside, and quartz, and then four grades of wollastonite are produced by crushing and grinding. At plant sites in China and Mexico, beneficiation processes are more or less similar to those already described, including manual sorting, optical and magnetic separation, or flotation (Smith 1981). Industrial products in general fall into two broad classes: milled grade, which has four sizes (in mesh): −200, −325, −400, and 10μm; and one attrition grade, which has a L:D ratio of 20:1.

2. Synthetic Wollastonite

Synthetic wollastonite has been used for many years in industry, mostly in Europe, because of the demand for this mineral, which is of especially high purity. The presence of carbonates, which are the most common minerals found in association with wollastonite in contact metamorphic rocks, causes appreciable loss on ignition. Pure wollastonite has an LOI of 2% (Kienow et al. 1988). The production procedures have been examined extensively (Kurczyk and Wührer 1971; Fekeldjiev and Andreeva 1983; Kotsis and Balogh 1989; Gal'perina et al. 1982; Andreeva and Fekeldjiev 1986; Ibanez et al. 1990). Synthetic wollastonite is of the high-temperature, pseudowollastonite form, which cannot be used as a substitute for the naturally occurring low-temperature wollastonite-2M and wollastonite-Tc forms when fibrous habit is of major importance.

In the synthetic process, the raw materials limestone (or other CaO-bearing materials such as calcite, aragonite sand, chalk, and quicklime) and silica sand (or other SiO_2-bearing materials such as quartz, silica gel, tripoli, diatomite, and glass) are ultimately mixed. The premix must be fine grained and put through a homogenizing and agglomerating process in a silo. Sintering is carried out in a rotary kiln, and the temperature of treatment is up to 1,450°C or higher as required by the reaction that is illustrated in the system $CaO-MgO-SiO_2$ (Osborn and Muan 1964). The sintering

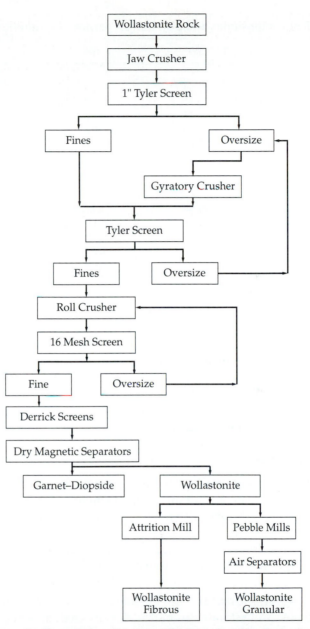

Figure 100 Flow sheet of wollastonite beneficiation at Willsboro plant, New York, after Elevatorski and Roe (1983).

Table 24 Chemical compositions and physical properties of synthetic wollastonite.

Class	SW	SM	SE	SG
Chemical Compositions				
SiO_2	53.59	52.6	52.36	52.58
CaO	45.31	45.7	45.44	45.66
MgO	0.47	0.6	0.61	0.55
Al_2O_3	0.20	0.5	0.38	0.47
Fe_2O_3	0.075	0.2	0.19	0.22
S	—	—	0.0034	0.008
P	—	—	0.0038	0.008
LOI (110° to 1000°C)	0.34	0.4	0.35	0.36
Physical Properties				
brightness% (D 65)	91–93	83	—	—
bulk density (gm/cm^3)	1.1	1.1	1.0	1.0
oil absorption (gm/100gm)	20	15	18	18
softening point	1345°C	1335°C	1335°C	1335°C
melting point	1425°C	1410°C	1410°C	1410°C

process, in order to produce a complete conversion of raw materials to wollastonite, can only be accomplished by means of a significant liquid phase. The treated product is cooled, crushed to −60 mm, and stored. Synthetic wollastonites are classified into two classes depending on their iron content: the extremely low-iron wollastonite (SW), and the low-iron wollastonites (SM, SE, and SG). SE and SG are characterized by extremely low sulfur and phosphorus contents, respectively. Chemical compositions and physical properties selected from Kienow et al. (1988) are tabulated in Table 24. In general, SW and SM are used for ceramics, while SE and SG are for metallurgical applications.

3. Industrial Uses

Wollastonite is mainly used in ceramic formulations and as a filler in paints and plastics. In other applications, wollastonite acts as a natural frit in the production of porcelain enamels, mixes with other components in fluxing formulations used for steel casting, and is used in specialized ultrahigh frequency electronic devices. A limited amount of wollastonite is also used as an asbestos replacement for boards and panels. (Polar 1959; Smith 1981; Power 1986; Sainamthip and Reed 1987).

Ceramic Applications: The production of wall tile consumes the largest amount of wollastonite. Wollastonite promotes dimensional uniformity, low shrinkage, good strength, low warpage, low moisture expansion, and fast firing, with decreased gas evolution. In semivitreous ceramic bodies, the substitution of 1 to 3% wollastonite for flint increases strength and decreases shrinkage, and in vitreous bodies containing 2 to 5% feldspar and quartz, wollastonite lowers the vitrification temperature. In glaze formulations, wollastonite used in place of ground limestone or whiting as the source of CaO can reduce the incidence of faults such as pinholing and eggshell finish, as a result of the consequent reduction in the amount of gas evolved. In combination with feldspars, wollastonite forms a eutectic at 710°C and can result in ready dissolution of the other glaze ingredients in compositions containing these minerals (Taylor and Bull 1986).

Power (1986) listed some important beneficial effects of wollastonite in ceramic bodies and glaze formulations. They are:

1. Acicular crystal habit: contributes mechanical reinforcement, imparts high impact strength, imparts acoustical properties, improves drying rates, reduces shrinkage, and improves pressing quality and green strength.
2. Low thermal expansion: reduces shrinkage, cracking, dunting, crazing, and glaze defects.
3. Loss on ignition: gives minimal gas evolution during firing and produces a smooth surface with diminished pinholing.
4. Chemical purity: maintains a high whiteness value on firing and facilitates a fast firing process for its low sintering temperature and fusing compatibility with alumina and silica.

Paint Applications: Wollastonite is used as a filler in paints because of its favorable properties such as brightness, whiteness, low oil absorption, and particle shape. Brightness and whiteness are important in the production of white and clear color paints. They also promote tint retention. The acicular habit makes wollastonite a good flatting agent, which produces a solid dry film of uniform thickness after application. The low oil absorption (20 to 25 gm/100gm) reduces the binder required in paint (Smith 1981).

Wollastonite has to compete with kaolinite, calcium carbonate, and talc as a filler in paint applications for economic reasons.

Plastic Applications: Wollastonite also has to compete as a filler in plastics for economic reasons with mica, talc, kaolinite, silica, and glass fibers. Its beneficial effects on plastics are based on the following properties, as listed by Power (1986):

1. The acicular habit increases impact, tensile, and flexural strengths.
2. The moderate hardness adds to wear resistance, and high dimensional tolerance reduces shrinkage.
3. High brightness and excellent whiteness reduce pigmentation.
4. Low water absorption promotes surface perfection and stain resistance.
5. Low impurity levels impart good electrical insulation properties.

Polyethylene is the most widely used of all the major thermoplastics. Its general properties are improved by the addition of wollastonite and other mineral fillers. Mineral fillers improve the flexibility and rigidity of polyvinyl chloride, the second most commonly used plastic material after polyethylene. Wollastonite, talc, and mica are commonly used as fillers in polypropylene to increase its stiffness and creep resistance at high temperatures. Wollastonite is mostly used as a filler that improves the impact and tensile strength of polystyrene (Schwartz and Goodman 1982). Thermosetting plastics are used predominantly with mineral fillers and glass fibers to form reinforced plastics.

Wollastonite can be treated chemically with a coupling agent either in a mix or as a coating that promotes a stronger bond at the interface with the plastics. The use of wollastonite that is coated with an amino-type modifying silane in nylon resin imparts improved heat resistance and increased filler levels at lower cost. Titanate coupling agents react with free protons at the mineral interface, resulting in the formation of organic monomolecular layers on the mineral surface. The titanate-treated mineral fillers are hydrophobic, organophilic, and organofunctional, and therefore, exhibit enhanced dispersability and bonding with the polymer. When used in filler polymers, titanates improve impact strength and exhibit melt viscosity (Schwartz and Goodman 1982).

Biomedical Applications: Preliminary experiments have demonstrated that wollastonite is potentially a highly bioactive material. It forms a hydroxylapatite surface layer on exposure to a simulated body fluid that has an ion concentration, pH, and temperature virtually identical with that of human blood plasma, and hence it may be used as a bioactive bone substitute (De Aza et al. 1998). In order to be suitable for implantation into the human body, the CaO/SiO_2 ratio and trace element content must be strictly controlled. Synthetic pseudowollastonite was the form of wollastonite used in experimentation.

REFERENCES

Andreeva, V., and G. Fekeldjiev. 1986. Use of non-conventional raw materials in ceramic production. *Ceramic Int.* 12:229–235.

De Aza, P. N., F. Guitián, S. De Aza, and F. J. Valle. 1998. Analytical control of wollastonite for biomedical applications by use of atomic absorption spectrometry and inductively coupled plasma atomic emission spectrometry. *Analyst* 123:681–685.

Deer, W. A., R. A. Howie, and J. Zussman. 1992. *An introduction to rock-forming minerals.* Essex, England: Longman. 696p.

Elevatorski, E. A., and L. A. Roe. 1983. Wollastonite. In *Industrial minerals and rocks.* 5th ed. Ed. S. J. Lefond, 1383–1390. New York: Soc. Mining Engineers, AIME.

Fekeldjiev, G., and V. Andreeva. 1983. Synthesis of wollastonite at low temperature. Possibilities for its use in ceramic bodies. In *Ceramic powders,* Ed. P. Vincencini, 89–96. Amsterdam: Elsevier.

Fountain, K. 1986. Chinese wollastonite, industry and commerce: Proc., 7th Industrial Minerals International Congress, Eds. G. M. Clarke and J. B. Griffiths, v. 1, pp. 117–125, Monte Carlo, Monaco, April 1–4, 1986, *Metal Bull. PLC,* London.

Gal'perina, M. K., O. S. Grum-Grzhimailo, V. S. Mitrokhin, and N. P. Tarantul. 1982. Synthesizing wollastonite from Tripoli. *Steklo Keram.,* 16–17.

Gasparik, T., K. Wolf, and C. M. Smith. 1994. Experimental determination of phase relations in the $CaSiO_3$ system from 8 to 15 GPa. *Amer. Mineral.* 79:1219–1222.

Greenwood, H. J. 1962. Metamorphic reactions involving two volatile components. *Ann. Rept., Geophys. Lab.* 61:82–85.

Harben, P. W., and R. L. Bates. 1990. *Industrial minerals, geology and world deposits.* London: Industrial Minerals, Div., *Metal Bulletin PLC.,* 312p.

Harben, P. W., and M. Kuzvart. 1996. Industrial Minerals—A global geology. Ind. Minerals Inf. Ltd., *Metal Bulletin PLC,* London, 462p.

Harker, R. I., and O. F. Tuttle. 1956. Experimental data on the $P_{(CO_2)} - T$ curve for the reaction: calcite + quartz = wollastonite + carbon dioxide. *Amer. J. Sci.* 254:239–256.

Ibanez, A., J. M. Gonzalez Pena, and F. Sandoval. 1990. Solid-state reaction for producing β-wollastonite. *Ceramic Bull.* 69:374–378.

Kienow, E., A. Roeder, and J. Stradtmann. 1988. Synthetic wollastonite, diopside, and mayenite and their roles as industrial minerals. In 8th Industrial Minerals International Congress, April 14–17, 1988, Boston. Ed. G. M. Clarke, 45–58. London: *Metal Bulletin, PLC.*

Kotsis, I., and A. Balogh. 1989. Synthesis of wollastonite. *Ceramic Int.* 15:79–85.

Kurczyk, H. G., and J. Wührer. 1971. Synthetic wollastonite and its use in ceramic bodies. *Int. Ceram.* 20:119–125.

Liebau, F. 1985. *Structural chemistry of silicates.* Berlin: SpringerVerlag., 347p.

O'Driscoll, M. 1990. Wollastonite production, tempo rises as markets grow. *Ind. Minerals.* (December):15–23.

Osborn, E. F., and A. Muan. 1964. System CaO-MgO-SiO_2. In *Phase diagrams for ceramists.* Eds. E. M. Levin, C. R. Robbins and H. F. Murdie. Columbus, Ohio: Amer. Ceramic Soc.

Polar, A. 1959. How to use wollastonite in wall tile. *Ceramic Ind.* 72:78–81.

Power, T. 1986. Wollastonite, performance filler potential. *Ind. Minerals.* (January):19–34.

Sainamthip, P., and J. S. Reed. 1987. Fast-fired wall tile bodies containing wollastonite. *Bull. Amer. Ceramic Soc.* 66:1726–1731.

Schwartz, S. S., and S. H. Goodman. 1982. *Plastics materials and processes.* New York: Van Nostrand Reinhold, 965p.

Smith, M. 1981. Wollastonite, production and consumption continue to climb. *Ind. Minerals.* (August):25–33.

Taylor, J. R., and A. C. Bull. 1986. *Ceramics glaze technology.* Oxford: Pergamon Press, 263p.

Trojer, F. J. 1968. The crystal structure of parawollastonite. *Zeit. Krist.* 127:291–308.

——— 1969. The crystal structure of a high-pressure polymorph of $CaSiO_3$. *Zeit. Krist.* 130:185–206.

Zeolites

Zeolites are hydrated aluminosilicates with a three-dimensional framework-type structures that have cavities and channels of molecular dimensions. This open structure allows the movement of large ions during cation exchange and of water molecules during reversible dehydration, and makes zeolites chemically active in absorption, ion exchange, and catalysis.

More than thirty well-characterized species of zeolites occur in nature, and there are a large number of zeolites produced synthetically. Natural zeolites are currently used in dimension stones, water treatment, radioactive waste treatment, and in animal feed and agriculture, whereas synthetic zeolites have major applications in absorption and catalysis.

The important producers of zeolites are the United States, Japan, South Africa, Hungary, Italy, and Cuba.

MINERALOGICAL PROPERTIES

Zeolites are hydrated aluminosilicates of alkalis and alkaline earths, and have infinite, three-dimensional framework-type structures made of SiO_4 and AlO_4 tetrahedra. In comparison with other framework-type structures such as feldspars and quartz, the zeolite structures, with their cavities and channels of molecular dimension, have a much greater degree of openness. These cavities and channels are occupied by H_2O molecules as well as the alkalis and alkaline earths introduced into the structure to achieve the charge balance brought about by $Al^{3+} = Si^{4+}$ substitution. Both the alkalis and alkaline earths and the water molecules have considerable freedom of movement. This allows reversible ion exchange and reversible dehydration to occur (Flanigen 1981; Michiels and de Herdt 1987; van der Waal 1998). Properties related to this distinct structure, as categorized by Breck (1974), include:

1. high degree of hydration,
2. low density and large void volume when dehydrated,
3. large cation exchange capacity,
4. uniform molecular-sized channels in the dehydrated crystals,
5. stability of crystal structures when dehydrated,
6. ionic electrical conductivity,
7. adsorption of gases, and
8. catalytic reaction.

Based upon these properties, zeolites are well suited for molecular sieving, which is defined as the selective adsorption of cations within a sorbent, based on physical dimensions and charge distribution (Barrer 1978).

The chemical composition of zeolites can be represented by the formula:

$$m_{2/n}O \cdot Al_2O_3 \cdot xSiO_2 \cdot yH_2O$$

where "m" is an alkali or alkaline earth cation, "n" is the valence of that cation, "x" is an integer from two to ten, and "y" is an integer from two to seven. There are three re-

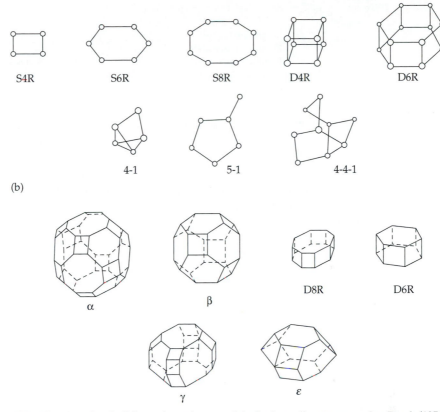

Figure 101   The secondary building units and some polyhedra in zeolite structure, after Breck (1974). Only the positions of SiO_4 and AlO_4 are shown in the secondary building units. The polyhedra shown are α (26-hedron or truncated cuboctahedron), β (14-hedron or truncated octahedron), D8R (δ or double 8-ring), D6R (double 6-ring or hexagonal prism), γ (18-hedron), and ε (11-hedron).

strictions for this formula including (1) the base to alumina ratio is always equal to unity, (2) the (Al+Si):O ratio is always 1:2, and (3) the ratio of SiO_2/Al_2O_3 is always equal to or greater than 2:1 (Flanigen 1981; Barrer 1984a).

The primary units of zeolite structures are the SiO_4 and AlO_4 tetrahedra, the specific array of which forms the subunit of the structure, such as rings of four tetrahedra and rings of six tetrahedra. This subunit was designated a secondary building unit (SBU) by Meier (1968). In addition, the zeolite framework can be considered in terms of polyhedral units such as the truncated octahedron and 11-hedron. These secondary building units and polyhedral units are not unique to zeolites. They are also found in many other framework aluminosilicates. In fact, the truncated octahedron is commonly known as the sodalite unit, and the 11-hedron as the cancrinite unit. Fig. 101 shows the secondary building units and polyhedral units in zeolite structures (Breck 1974).

Based upon a combination of framework topology and secondary building units (Meier 1968), Breck (1974) classified zeolites into seven groups, as listed on next page:

Group	Secondary Building Unit (SUB)
1	Single 4–ring, S4R
2	Single 6–ring, S6R
3	Double 4–ring, D4R
4	Double 6–ring, D6R
5	Complex 4–1, T_5O_{10} unit
6	Complex 5–1, T_8O_{16} unit
7	Complex 4–4–1, $T_{10}O_{20}$ unit

Analcime and clinoptilolite are by far the most abundant zeolites, followed by chabazite, heulandite, erionite, natrolite, laumontite, mordenite, and phillipsite. These are the predominant constituents in zeolite deposits (Hay 1981; Clifton 1987). Of the synthetic zeolites, A, X, Y, and mordenite are in commercial production and are used in many industrial processes (Mumpton and Sheppard 1974; Michiels and de Herdt 1987). Each of Breck's seven groups is represented by one or more of the common natural and synthetic zeolites:

Group 1 (S4R)	—	analcime, laumontite, phillipsite
Group 2 (S6R)	—	erionite
Group 3 (D4R)	—	zeolite A
Group 4 (D6R)	—	chabazite, faujasite, zeolite X and Y
Group 5 (T_5O_{10})	—	natrolite
Group 6 (T_8O_{16})	—	mordenite, synthetic mordenite
Group 7 ($T_{10}O_{20}$)	—	heulandite, clinoptilolite

A brief description of six natural zeolites, each representing a Group, and four synthetic zeolites (A, X, Y, modenite) is given below. This section is based primarily on Breck's comprehensive treatise on zeolite molecular sieves (Breck 1974).

GROUP 1: Analcime, $Na_{16}[(AlO_2)_{16}(SiO_2)_{32}] \cdot 16H_2O$, has a framework structure of the leucite-type that encompasses sixteen large and twenty-four small cavities. Continuous channels that run parallel to the three-fold rotation axes are formed from the large cavities and are occupied by water molecules. The small cavities are situated adjacent to the channels and are occupied by Na^{1+}. Analcime is cubic, with a=13.7Å, and has a Si:Al ratio varying between 1.8 and 2.8. In its structure, there are 4-rings that are perpendicular to the four-fold rotation axes and that vary in UDUD linkage (U: the apical oxygen of (SiO_4) tetrahedron pointing up, and D: down). These rings are connected to form regular 6-rings, which lie parallel to (111), and highly distorted 8-rings, which lie parallel to (100). The water content of analcime varies linearly with silica content. As the silica content increases, the Na^{1+} content required to balance the charge decreases, and there is a concurrent increase in the number of water molecules. On the other hand, if K^{1+} or Cs^{1+} replaces Na^{1+} in analcime, the large cation prefers H_2O sites. The water content thus varies with the degree of ion exchange. The framework density is 1.85 gm/cc, the void fraction, which is determined from water content in the hydrated state, is 0.18, the type of channels is one, and the free aperture of the main channels is 2.6Å.

GROUP 2: The structure of erionite consists of 6-rings arranged in an AABAAC sequence, which is used in the same sense as those used in close-packed spheres in simple structures. It is hexagonal with a=13.26Å and c=15.12Å. The framework can also be considered in terms of the ε-cage linked by D6R units in the c-direction in the sequence of ε-D6R-ε-D6R (Figs. 102a). These columns are crosslinked by S6R units that

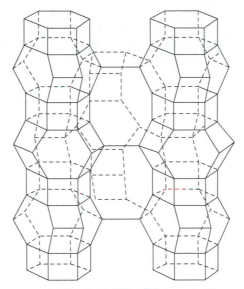

Figure 102a   The erionite structure showing the AABAAC 6-rings stacking sequence, after Newsan (1992).

are perpendicular to the c-axis. The aperture between these cages in the c-direction has a diameter of 2.5Å and is too small to permit the diffusion of most molecules. However, single 8-rings form apertures into any single cage and these have free dimensions of 3.6 by 5.2 Å. In order to diffuse from one cage to another, molecules may have to pass through an 8-ring aperture from one cage to an adjacent cage and then through another 8-ring aperture into an adjacent cage in the original column. Dehydrated erionite has a very stable framework. It shows no change in adsorption capacity after a prolonged period of exposure to H_2O vapor at 375°C. Erionite has a complex composition, with K^{1+} and Ca^{2+} as the main cations. Its formula may be written as $(Ca,Mg,Na_2,K_2)_{4.5}\,[(AlO_2)_9\,(SiO_2)_{27}] \cdot 27H_2O$ with a Si:Al ratio of 3:3.5. The potassium ions show considerable resistance to ion exchange and may have positions in the ϵ-cage. Erionite has a framework density of 1.51 gm/cm^3 and a void fraction of 0.35.

GROUP 3: The structural unit of zeolite A is the double 4-ring (D4R). Its framework is produced by placing the cubic D4R units in the centers of the edges of a cube having an edge length of 12.3Å. This generates truncated octahedral units that are centered at the corners of the cube (Figure 102b). Each corner of the cube is occupied by a truncated octahedron (β-cage) that encloses a cavity with a free diameter of 6.6Å. The center of the unit cell is a large cavity (α-cage) that has a free diameter of 11.4Å. There are two interconnecting, 3-dimensional channel systems. One consists of connected α-cages, 11.4Å in diameter, which are separated by 4.2Å circular apertures and run in the direction of the cubic axis. The other consists of alternating β-cages and α-cages, which are separated by 2.2Å apertures and run in the direction of the 3-fold rotation axis. Zeolite A has a framework density of 1.27 gm/cm^3 and a void fraction of 0.47.

Zeolite A is cubic, with a=24.64Å, and has a composition of $Na_{12}[(AlO_2)_{12}\,(SiO_2)_{12}] \cdot 27H_2O$, with a Si:Al ratio of close to one. In the hydrated state, eight Na^{1+}

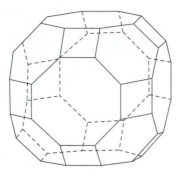

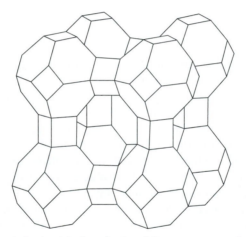

Figure 102b A truncated cuboctahedron (top), and the arrangement of truncated cuboctahedra in the framework of zeolite A (bottom), after Bhatia (1990).

are located near the center of the 6-rings inside the α-cage, and 4 Na^{1+} with H_2O in the 8-rings.

GROUP 4: Zeolites X and Y, and chabazite are members of this group, which has a double 6-ring (D6R) as the secondary building unit. Zeolites X and Y are related to naturally occurring faujasite (Fig. 102c). They are cubic, with a large cell containing 192 $(Si,Al)O_4$ tetrahedra, and a cell dimension of nearly 25Å. Their framework structure is stable and rigid, and has the largest void space of any known zeolite. The chemical compositions for zeolites X and Y are $Na_{86}[(AlO_2)_{86}(SiO_2)_{106}] \cdot 264H_2O$, with a Si:Al ratio of 1.0:1.5, and $Na_{56}[(AlO_2)_{56}(SiO_2)_{136}] \cdot 250H_2O$, with a Si:Al ratio of 1.5:3.0, respectively. Faujasite has a correlated composition of $(Na_2,K_2,Ca,Mg)_{29.5} [(AlO_2)_{59}(SiO_2)_{133}] \cdot 235H_2O$. The aluminosilicate framework consists of truncated octahedra β-cages, which are linked tetrahedrally through double 6-rings that are arranged in the same

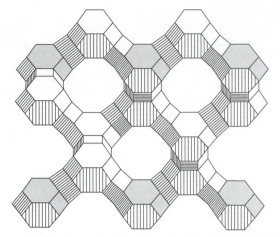

Figure 102c   The structure of zeolite X, zeolite Y, and faujasite viewed along [110]. The corners of each polyhedron represent the centers of SiO_4 or AlO_4 tetrahedra, after Bergerhoff et al. (1958).

fashion as the carbon atoms in the structure of diamond. Each unit cell has eight cavities with diameters of ~13Å. A three-dimensional channel system runs parallel to [110]. The free aperture of the main channel is 7.4Å in both hydrated and dehydrated states. Zeolite X has a framework density of 1.31 gm/cc and a void fraction of 0.50. The corresponding values for zeolite Y are 1.25 gm/cm^3 and 0.48.

Chabazite is rhombohedral with a=9.42Å and α=94°28′, and has a composition of $Ca_2[(AlO_2)_4(SiO_2)_8] \cdot 13H_2O$, with a Si:Al ratio of 1.6:3.0. The framework structure of chabazite consists of double 6-rings arranged in layers in the sequence ABCABC (Fig. 102d). The layers are linked together by tilted 4-rings, and the resulting framework contains large, ellipsoidal cavities, each of which is entered by six apertures formed by 8-rings. When dehydrated, chabazite has a three-dimensional channel system consisting of cavities joined through the 8-rings with a free aperture of 3.1×4.4Å. Chabazite has a framework density of 1.45 gm/cm^3 and a void fraction of 0.47.

GROUP 5: Zeolites of this group are known as the fibrous zeolites. This morphology is a result of the prominent chains in the framework structure. An individual chain has five tetrahedra (4-1 unit) and extends along the c-axis. Three distinct framework structures are produced in this group by three modes that are defined by the location of the linking oxygen atoms: (1) all on the reflection planes (edingtonite), (2) half on the reflection planes and half on the rotation axes (thomsonite), and (3) all on the rotation axes (natrolite). Two types of channels run through the structure that produce small cavities at the intersections. The channel running perpendicular to the c-axis between the adjacent chains has a free diameter of 2.60Å, whereas the diameter of the channel running parallel to the c-axis is 2.08Å. Both exchangeable cations and water molecules are located in the cavities.

Natrolite is orthorhombic, with a=18.30Å, b=18.63Å, and c=6.60Å, and has a composition of $Na_{16}[(AlO_2)_{16}(SiO_2)_{24}] \cdot 16H_2O$ with a Si:Al ratio of 1.44:1.58. The framework cations have an apparent ordered arrangement in the structure. Natrolite has a framework density of 1.76 gm/cm^3 and a void fraction of 0.23 (Breck 1974).

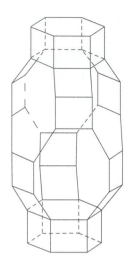

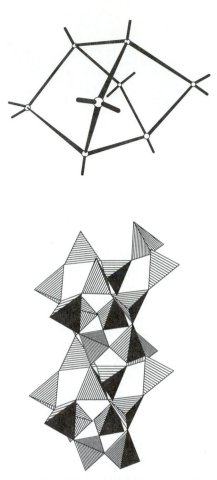

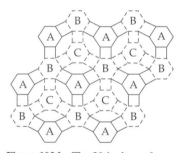

Figure 102d The 20-hedron of chabazite, capped by D6R (top), and the ABCABC... arrangement in chabazite (bottom), after Breck (1974).

Figure 102e The 5-1 unit (secondary building unit) of Group 6 zeolites (top), and the arrangement of the 5-1 units in mordenite (bottom), after Breck (1974).

GROUP 6: Mordenite, $Na_8[(AlO_2)_8(SiO_2)_{40}] \cdot 24H_2O$, is the most siliceous zeolite. Its nearly constant Si:Al ratio of 5 indicates an ordered Si and Al distribution in the structure. Mordenite is orthorhombic with a=18.13Å, b=20.49Å, and c=7.52Å. As shown in Fig. 102e, the framework structure of mordenite consists of chains made of 5-rings that are crosslinked by 4-rings. The SBU is a complex 5-1, T_8O_{16} unit. For the diffusion of large molecules, the dehydrated mordenite has a one-dimensional channel system parallel to the c-axis and a free diameter of 2.8Å, but for the diffusion of smaller molecules the channel system is two-dimensional, with an added channel system parallel to the b-axis. Mordenite has a framework density of 1.70 gm/cm³ and a void fraction of 0.28.

A synthetic type of mordenite, known as "large port" mordenite, has a two-dimensional channel system parallel to the b- and c-axes. The latter one exhibits a free diameter of 6.7Å.

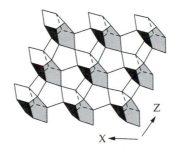

Figure 102f   The 6-1 unit (secondary building unit) of Group 7 zeolites (top), and the arrangement of the 6-1 units in heulandite (bottom), after Breck (1974).

GROUP 7: Clinoptilolite, which is one of the most common zeolites, belongs to this group. The basic unit in the framework structure is the special arrangement of tetrahedra in the form of 4-rings and 5-rings, as illustrated in Fig. 102f. These special units are arranged in sheetlike arrays parallel to (010). Like other sheetlike structures, the low bond density between the sheets is obvious. This one direction of weak bond strength causes structural change during dehydration.

Clinoptilolite is monoclinic with a=7.41Å, b=17.89Å, c=15.85Å, and β=91°29′, and has a composition of $Na_6[(AlO_2)_6 (SiO_2)_{30}] \cdot 24H_2O$ with a Si:Al ratio of 4.25:5.25. It has a framework density of 1.71 gm/cm³ and a void fraction of 0.34. Preliminary structural analysis of clinoptilolite (Chen et al. 1978) shows that it is isostructural with heulandite (Fig. 102f), another zeolite of Group 7, which has a two-dimensional channel system consisting of three channels, parallel to the a-axis, parallel to the c-axis, and at 50° to the a-axis. Free aperture, which is defined by 8-rings and is perpendicular to the a-axis, is 4.0×5.5Å, and the free aperture, which is defined by 10-rings and perpendicular to the c-axis, is 4.0×7.2Å. The pores in clinoptilolite are defined by 10-rings instead of differentiating into 8-rings and 10-rings as in heulandite. The structure is stable up to 700°C in dehydration.

GEOLOGICAL OCCURRENCE AND DISTRIBUTION OF DEPOSITS

Zeolites are commonly formed by the reaction of volcanic glass or tuffaceous materials with pore water. The type of zeolite that will crystallize in a given environment depends on the temperature, pressure, and activities of various ions; and zeolites formed at an earlier stage may react with pore water to produce different types of zeolites. Commonly, for example, phillipsite, mordenite, and clinoptilolite can be replaced by analcime, which is in turn replaced by laumontite (Miyashiro and Shido 1970). The occurrence of zeolite is diversified, and at least six types of geological environments have been recognized. They are saline, alkaline lakes sediments; saline, alkaline soils; marine sediments; sediments open to the percolation of water; hydrothermal-altered sediments; and burial-diagenetic sediments (Hay 1981).

(1) Saline, alkaline lakes sediments: Nearly all modern, saline, alkaline lakes are found in one of two tectonic settings in arid and semiarid regions: block-faulted terrains and trough valleys associated with rifting. These settings establish a closed basin environment that limits the amount of clastic sediments that are able to reach the lake (Surdam and Sheppard 1978). The brine of the restricted lakes in which zeolites form generally has a pH value greater than 9, which probably accounts for the relatively rapid solution of the volcanic glass and precipitation of zeolites. The most distinguishing fea-

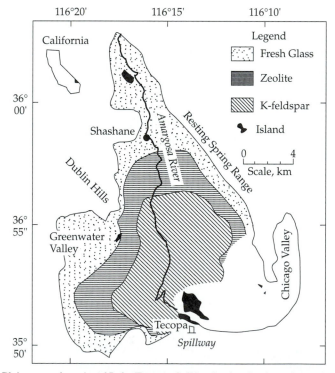

Figure 103 The Pleistocene deposits of Lake Tecopa, California, showing lateral zonation of minerals, after Clifton (1987).

ture of the saline, alkaline lake deposits is the lateral zonation of the minerals. The Pleistocene deposits of Lake Tecopa, California, are characteristic of the closed basin environment (Sheppard and Gude 1968). As shown in Fig. 103 (Clifton 1987), volcanic glass is unaltered in tuff deposited in freshwater along the shore and at inlets of the ancient lake. The glass is succeeded inwardly by a zone of zeolites, and in the central part of the lake basin, by potassium feldspar. In most cases the tuff beds have been altered completely to zeolites or potassium feldspar. Zeolitic tuffs at Lake Tecopa consist of phillipsite, clinoptilolite, and erionite.

(2) Saline, alkaline soils: Zeolites are formed readily from suitable materials at the surface in arid and semiarid regions where pH is high as a result of the concentration of sodium carbonate-bicarbonate by evaporation. The water table is the probable limit of the alteration processes that produce zeolite. Analcime was found in alkaline, saline soils of the eastern San Joaquin Valley, California. The formation of zeolite occurs to a depth of 1.2 meters, and the amount decreases with depth (Balder and Whitting 1968). Analcime, chabazite, natrolite, and phillipsite were reported from alkaline, saline soil profiles at Olduvai Gorge, Tanzania (Hay 1963a).

(3) Marine Deposits: Zeolites form in marine sediments at relatively low temperatures and moderate pH (7 to 8). Phillipsite and clinoptilolite are the dominant zeolites. Others present are analcime, erionite, natrolite, and mordenite (Hay 1978). In the Pacific and Indian Oceans, phillipsite, along with several other zeolites in minor amounts, occurs in post-Miocene brown clays, vitric siliceous and calcareous oozes, and basaltic volcanic sediments. In the calcareous sediments and terrigenous clays of Paleogene and Cretaceous ages in the Atlantic Ocean, clinoptilolite is the dominant zeolite. In sea-floor sediments, zeolites form by the reaction of glasses with pore water, which is similar to sea water in composition, except for the silica content. Phillipsite is commonly associated with basaltic tephra, which is low in silica (less than 20 ppm), and clinoptilolite is associated with siliceous tephra, which is high in silica (about 20 to 40 ppm) (Hay 1978).

(4) Sediments in open hydrological systems: The most voluminous and valuable zeolite deposits belong to this type. Zeolites form from the percolation of groundwater through tuffaceous sediments. The pH and dissolved solids of the groundwater increase as it reacts with the tuffaceous sediments, until zeolites are precipitated. Movement of the groundwater downward through the sediments results in a vertical zonation of water composition and authigenic minerals, including zeolites (Hay 1978; Iijima 1980). The original pyroclastic material was deposited in marine or fluviatile environments or carried by air and deposited on the land surface. Commonly the zeolite deposits are several hundred meters thick and can be traced laterally for tens of kilometers.

The John Day Formation in central Oregon consists of 600 to 900 meters of land-laid silicic tuffs and tuffaceous claystones (Hay 1963b). Zeolites form in the lower part of this formation (300 to 450 meters) in an area of about 5,700 km^2. An upper zone, 300 to 450 meters in thickness, contains fresh glass or glass altered to montmorillonite and opal. Hay (1963b) proposed hydrolysis and solution of silicic glass by subsurface water to account for the development of clinoptilolite (Fig. 104). The Koko Crater, Hawaii, illustrates the alteration of low-silica, mafic tephra in an open system environment. The cone consists of tuffs and lapilli tuffs, and is distinctly divided into two zones. The upper zone, generally 20 meters thick, is characterized by the presence of fresh glass, opal, and montmorillonite, and the lower zone of zeolitic tuff (Hay and Sheppard 1981). Zeolites

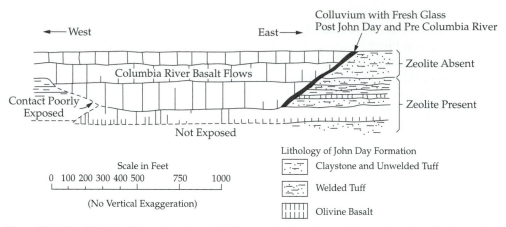

Figure 104 The John Day Formation in central Oregon showing the vertical zonation of zeolite formation in sediments in open hydrological systems, after Hay and Sheppard (1981).

in the upper portion of the zeolitic tuff are phillipsite and chabazite, and phillipsite and analcime are the dominant zeolites in the lower portion of the zeolitic tuff.

(5) Hydrothermal-altered deposits: Zeolites form in sediments that were altered by both alkaline and acidic hydrothermal fluids. Clinoptilolite and mordenite are generally present in the shallow and low-temperature zones, and analcime, heulandite, or laumontite in the deep and high-temperature zones (Iijima 1980). Examples of hydrothermal zeolite deposits are found in Yellowstone Park, Wyoming (Honda and Muffker 1970); Wairakei, New Zealand (Steiner 1953); and Onikobi, Japan (Hay 1981).

(6) Burial-diagenetic deposits: Zeolites associated with burial diagenesis were formed over wide areas in thick volcanoclastic sediments (Boles 1981). Burial diagenetic sequences are typified by vertical zonation that is related to increasing burial depth and temperature. Such vertical zonation can be found in the Triassic sedimentary rocks of Southland, New Zealand. A fresh glass zone is on the surface, and successively lower zones contain mineral assemblages of clinoptilolite-mordenite, heulandite-analcime, laumontite-albite, and prehnite-pumpellyite, which represents a transition into the greenschist facies of regional metamorphism.

In the Niigata oil field, five depth-related zones with characteristic mineral assemblages are displayed. Most of them are formed as replacements of volcanic glass, and some are as replacements of plagioclase (Iijima and Utada 1972). The sequence of zones is as follows: zone I, fresh glass; zone II, clinoptilolite; zone III, clinoptilolite + mordenite; zone IV, analcime+ heulandite; and zone V, albite. In the Tanzawa Mountain, central Japan (Seki et al. 1969), Miocene and younger pyroclastic sedimentary rocks were intruded by granitic or dioritic rocks. There are five mineral zonations that are arranged more or less concentric to the intrusive igneous rocks. Zone I consists of clinoptilolite and stilbite (a Ca-rich Group 7 zeolite); zone II of laumontite; zone III of chlorite, pumpellyite, prehnite, and epidote; zone IV of epidote, biotite, and hornblende; and zone V of biotite, hornblende, diopside, and Ca-garnet. Plagioclase and quartz are present in all zones.

Burial diagenetic zeolites are widely distributed in Cretaceous and Triassic rocks in western North America. In British Columbia, zeolitization occurs in the Triassic Karmutsen Group, Vancouver Island, a 5.5 km thick marine sequence of lava and breccia, tuff, amygdaloid flows, and interlava limestone. Laumontite replaces calcic plagioclase in the volcanic rocks and flows, and the increased albitization of plagioclase and breakdown of analcime to albite occur with increasing depths (Surdam 1973). Zeolitization also occurs (1) to sandstones in the Upper Cretaceous Nanaimo Group, Vancouver Island, and Gulf Islands (Steward and Page 1974), (2) to marine sediments of andesitic origin in the Andrich Mountain area of central Oregon (Dickinson 1962), and (3) to sandstones of the Great Valley sequence in Cache Creek, California (Dickinson et al. 1969). In all cases, laumontite and heulandite are the major zeolites.

The major world producers of zeolites are the United States and Japan (clinoptilolite, analcime, mordenite, heulandite, and laumontite). Other producers are South Africa (clinoptilolite and mordenite), Hungary (clinoptilolite and mordenite), Italy (phillipsite), and Cuba (clinoptilolite, mordenite, analcime, and heulandite). In the United States, prominent zeolite deposits are distributed in California, Oregon, Nevada, Arizona, Wyoming, Utah, and Colorado (Olson 1983).

INDUSTRIAL PROCESSES AND SYNTHESIS

1. Treatment and Characterization

Zeolites are selectively mined after test drilling, and the products are crushed, dried, and milled to a variety of sizes. Roller mills and rotary kilns are generally used. Micronised zeolite grades, zeolite grades, and zeolite granules are produced for various industrial uses (Griffiths 1987). Zeolites of high purity, which are to be used for adsorption processing, must be formed into aggregates of high physical strength and attrition resistance. This is done by using an inorganic binder, generally kaolin clay. The mix is extruded into cylindrical pellets or formed into beads, which are subsequently calcined. Pure zeolite pellets may be produced by hot-pressing without a binder.

Many methods may be used for zeolite characterization. The four prime methods are (1) x-ray powder diffraction, (2) infrared spectroscopy, (3) adsorption capacity measurement, and (4) ionic exchange capacity determination.

2. Synthesis and Production of Zeolites

Synthesis of zeolites is in most cases made under nonequilibrium conditions, and the zeolites synthesized are considered, in a thermodynamic sense, as metastable phases (Breck 1974). In the system $Na_2O-Al_2O_3-SiO_2-H_2O$ for example, at 1,000 atmospheres and excess water, the phases present are albite, analcime, mordenite, hydroxycancrinite, natrolite, nepheline hydrate, hydrosodalite, and montmorillonite, in the temperature range of from 290° to 700°C. Different zeolite phases may be formed metastably at temperatures below 200°C. The nature of starting materials, factors affecting nucleation, and length of reaction time are important in determining the metastable phase assemblages.

The general conditions for zeolite synthesis are as follows (Breck 1974): (1) reactive starting materials such as freshly coprecipitated gels or amorphous solids, (2) relatively high pH introduced in the form of an alkali metal hydroxide or other base, (3) low-temperature hydrothermal conditions with concurrent low autogenous pressure at saturated water vapor pressure, and (4) a high degree of supersaturation of the components of the gel.

Processes for the production of zeolites and commercial molecular sieve materials that rely on basic principles of synthesis may be classified into three types: (1) the preparation of zeolites from reactive aluminosilicate gels or hydrogels, (2) the conversion of clay minerals such as kaolin into zeolite, and (3) the preparation of zeolites from heterogeneous reactant mixtures (Breck 1974). In addition, organic cations are also used in zeolite synthesis (Gilson 1991).

In the aluminosilicate gel process, the gel preparation and crystallization is represented schematically using the Na_2O-Al_2O_3-SiO_2-H_2O system as an example:

$$NaOH(aq) + NaAl(OH)_4(aq) + Na_2SiO_3(aq)$$

$$\downarrow 20°C$$

$$[Na_a(AlO_2)_b(SiO_2)_c \cdot NaOH \cdot H_2O] \text{ gel}$$

$$\downarrow 20° \text{ to } 175°C$$

$$Na_x[(AlO_2)_x(SiO_2)_y] \cdot mH_2O + \text{solution}$$

A flow sheet of hydrogel process is shown in Fig. 105. The raw materials are metered into the makeup tanks in the proper ratios. The crystallization is accomplished in

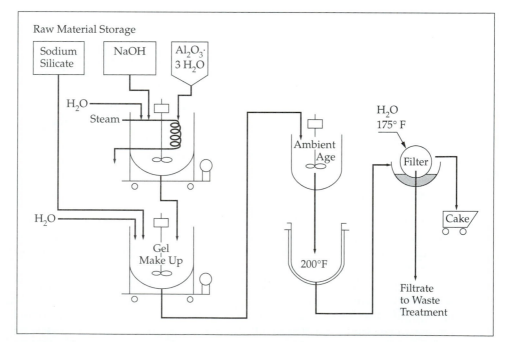

Figure 105   Flow sheet of hydrogel process for the production of synthetic zeolite, after Breck (1974).

a separate crystallizer, after an intermediate aging stop is carried out at an ambient temperature. The processes for making zeolites from hydrogels are summarized in Table 25 (Breck 1974).

In the clay conversion process, kaolin is first dehydroxylated by calcination at 550° to 600°C to metakaolin:

$$2Al_2Si_2O_5 \cdot (OH)_4 \rightarrow 2Al_2SiO_7 + 4H_2O \text{ (reaction 1)}$$

A defect spinel and mullite are produced by further calcination at 925°C and at 1,050° to 1,100°C, according to following equations:

$$2Al_2Si_2O_7 \rightarrow Al_4Si_3O_{12} + SiO_2 \text{ (reaction 2)}$$

$$3Al_4Si_3O_{12} \rightarrow 2Al_6Si_2O_{13} + 5SiO_2 \text{ (reaction 3)}$$

The metakaolin produced by reaction 1 has the correct ratio of SiO_2/Al_2O_3 for zeolite A. SiO_2 must be added to the reactant for conversion to zeolites of higher silica content. In reaction 2, the spinel produced is a defect aluminum-silicon spinel, and the silica produced is very reactive. Mullite is formed in reaction 3, and additional SiO_2 is eliminated to produce cristobalite. The processes used to produce zeolites from kaolin are summarized in Table 25, and a flow sheet for the manufacture of zeolite A is illustrated in Fig. 106 (Breck 1974). The reactants are caustic soda, calcined kaolin, and water that are mixed in the gel make-up tank. An ambient aging step is also used prior to the crystallization step.

Other processes for the manufacture of molecular sieve zeolites are based upon heterogeneous reactant mixtures. For example, the manufacture of large port mordenite (Zeolon) is based on hydrothermal treatment of a reactant mixture of sodium aluminate and various types of amorphous silica (diatomaceous earth or volcanic glass) at 175°C for twenty-four hours.

Zeolites of the Zeolite Scony Mobil (ZSM) type are synthesized in mixed systems containing alkali-metal cations in the presence of organic cations or complexes (Gilson 1991). The generic name, pentasil zeolite, has been proposed to encompass all members of this family of high silica zeolites. The end members are the so-called ZSM-5 and ZSM-11 zeolites. Both have framework structures consisting of a tetrahedral configuration linked through four-, five-, and six-membered rings. The framework encloses a two-dimensional system of intersecting channels with ten-membered rings controlling the size of the pores, which have openings around 6Å. Both sets of channels in ZSM-11 are straight, whereas in ZSM-5 one set is straight and the other is sinusoidal (Fig. 107). During synthesis, aerosil, as a silica source, is added to the aqueous solutions of tetrapropyl and tetrabutyl ammonium (TPA) hydroxide. Sodium aluminate is dissolved in a concentrated NaOH solution and added to the first solution under vigorous stirring. The gel formed this way has the following molar composition:

$$\text{ZSM-5: } (R_2O)_{24}(Na_2O)_{0.3}(Al_2O_3)_{1.0}(SiO_2)_{60}(H_2O)_{1550}$$

$$\text{ZSM-11: } (R_2O)_{3.0}(Na_2O)_{7.0}(Al_2O_3)_{1.0}(SiO_2)_{60}(H_2O)_{900}$$

where R represents the respective tetra-alkyl ammonium molecules. The gel is transferred to an autoclave and stirred at 150°C for 3 hours in the case of ZSM-11, and 6 hours

Table 25 Products, reactants, and processes used in (a) hydrogel and (b) clay mineral processes in the production of zeolites modified, after Breck (1974).

Process	Reactants	Product
Homogeneous gel	Sodium silicate	Type A—powder
	Sodium aluminate	Type X—powder
	Caustic	Type Y—powder
	$Al_2O_3 \cdot 3\,H_2O$	
Heterogeneous gel	$Al_2O_3 \cdot H_2O$	Zeolon—mordenite
	Sodium aluminate	powder
	Silica sol	Type X—powder
	Amorphous solid silica	Type Y—powder
	Sodium silicate	
	Caustic	
Gel preform	Sodium aluminate	Type A—preformed
	Amorphous silica	spheres
	Sodium silicate	Type Y—preformed
		spheres
		Mordenite preform
		Type Y—in a gel
		matrix

(a)

Process	Reactants	Product
Slurry—high purity powder	Metakaolin	Type X
	Sodium silicate	Type Y
	Allophane (36)	
In situ crystallization of preform to yield high purity, binderless pellet	Metakaolin	Type A
	Caustic	Type X
	Sodium silicate	Type Y
	Diatomaceous earth	Zeolon-mordenite
Partial, *in situ* conversion of preformed particle to yield zeolite in clay-derived matrix	Caustic	Type X
	Metakaolin	Type Y
	Calcined kaolin	
	Sodium silicate	
	Raw kaolin	

(b)

in the case of ZSM-5. The products are washed with distilled water, air-dried, and calcined under a flow of dry nitrogen at 500°C. The product is then saturated with dry NH_3 gas at room temperature and exchanged under reflux conditions with 0.5M solutions of NH_4NO_3 at a liquid to solid ratio of 50. This procedure is repeated five times. The products are air-dried and calcined at 500°C in flowing oxygen. The compositions of the end products are $H_{2.9}Na_{0.1}Al_3Si_{93}O_{192}$ for H-ZSM-5 and $H_{2.5}Na_{0.5}Al_5Si_{93}O_{192}$ for H-ZSM-11.

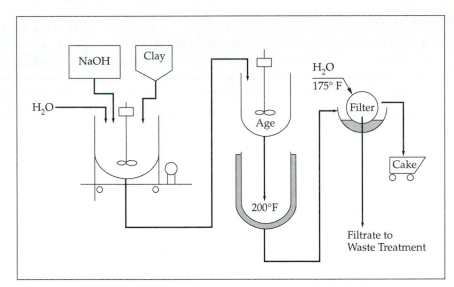

Figure 106   Flow sheet of clay mineral process for the production of zeolites, after Breck (1974).

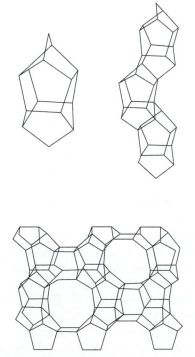

Figure 107   The building unit of ZSM-5 structure (top), and the arrangement of the unit in the form of eight-membered rings on (100) (bottom), after Bhatia (1989).

INDUSTRIAL USES

All industrial applications for zeolites make use of one or more of the physical or chemical properties that characterize both the natural and synthetic materials. These include dehydration and rehydration, structural hydroxyl groups, molecular sieving, diffusion and pore volume, a siliceous composition, and light-color, light weight, and porous structure. For the most part there is little overlap in the industrial applications for natural and synthetic zeolites. The natural zeolites are used in radioactive waste and waste water treatment, coal gasification and natural gas purification, agricultural applications, and construction, whereas the synthetic zeolites are used in adsorption and catalysis (Mumpton 1981; Griffiths 1987).

1. Industrial Uses of Natural Zeolites

Waste Water and Radioactive Waste Treatment: The ion exchange capacity of zeolite is used in the treatment of radioactive waste (Grant et al. 1987, 1988). At the Hanford Nuclear Facility, Washington, solutions containing radioactive Sr^{90} and Cs^{137} were introduced into ion exchangers packed with clinoptilolite from Hector, California. The radioactive ions were extracted and stored in the zeolite, which is either buried as solid waste or purified chemically. Millions of gallons of low-level Cs^{137} waste have been processed by this method at Hanford (Hawkins and Short 1965). A similar process was developed for the treatment of high-level effluent using a chabazite-rich zeolite from Bowie, Arizona (Nelson and Mercer 1963; Mercer et al. 1970a). Chabazite-erionite products, also from Bowie, Arizona, were used in the clean-up operation at the Three Mile Island nuclear power plant site (Griffiths 1987). Similar processes have been developed in several other countries, including Canada, Great Britain, France, Bulgaria, Hungary, Japan, Germany, and the former USSR.

Clinoptilolite ion exchange processes are used to remove ammonium ions from sewage and agricultural effluent (Mercer 1969; Mercer et al. 1970b). These ions are known to be toxic to many forms of aquatic life and contribute greatly to the growth of algae. Removal of up to 99% of the contained ammonium ions can be achieved by clinoptilolite exchange. Treatment plants of various capacities are in use across the United States, from the 0.6 Mgd (million gallons per day) capacity plant at Rosemont, Minnesota, to the 54 Mgd capacity plant in Alexandria, Virginia.

Coal Gasification and Natural Gas Purification: Natural mordenite and clinoptilolite are capable of selectively adsorbing SO_2. This allows high-sulfur coal to be used in the production of electric power (Raux et al. 1973; Anurov et al. 1974). Under static conditions, each gram of zeolite can adsorb 200 mg of SO_2, whereas 40 mg of SO_2 can be adsorbed under dynamic conditions. Certain zeolites are capable of adsorbing nitrogen selectively from air to produce an oxygen-rich gas. On-site zeolite adsorption units will eliminate the problem of producing large amounts of nitrogenous oxides and hydrocarbons if the air is used directly and will also reduce production costs if pure oxygen is used directly in the gasification process. About 25% of coal reserves in the United States are deep-seated and require underground gasification.

Natural zeolites are used in the purification of natural gas by selective adsorption of CO_2, H_2S, and H_2O, of up to 25% from well gas. They are also used in menthane-recovery and purification. In the methane purification process used at the Palos Verde landfill, Los Angeles, California, raw gas containing about 50% CH_4 and 40% CO_2 is routed through two pretreatment vessels to remove moisture and H_2S. The dry gas is then fed into three adsorption columns that are packed with pellets of chabazite and erionite (Bowie, Arizona, deposit) that adsorb CO_2 and other impurities. (Mumpton 1978).

Agricultural Applications: Considerable research and development has been made, especially in Japan, into the use of zeolites as animal feed supplements. Up to 10% clinopilolite has been added to the rations fed to pigs, chickens, and ruminants, resulting in significant increases in feed-conversion values and in the general health of the animals (Mumpton and Fishman 1977). The growth rate of calves was improved by adding zeolites to the feed, because zeolites stimulated their appetites and improved their digestion (Kondo et al. 1969). It has been established that the use of zeolite in calf rearing improves the compactness of the excrement, the atmosphere in the stable, and the daily weight gain of the calf (Griffiths 1987). The use of clinoptilolite (5%) in rations for young pigs has shown it to be effective in reducing the incidence of scours. It appears that the ammonium selectivity of clinoptilolite may act as a nitrogen reservoir in the digestive system of the animal, which allows a slower release of ammonium ions produced by the breakdown of ingested rations in the development of animal protein (England 1975).

The high ion-exchange and adsorption capacities of many natural zeolites make them effective carriers of herbicides, fungicides, and pesticides (Mumpton and Fishman 1977). Clinoptilolite was found to be an excellent carrier of benzyl phosphorothioate to control stem blasting in rice, and to be more than twice as effective as other commercial products as a carrier of the herbicide benthiocarb in eliminating weeds in paddy fields (Mori et al. 1974). Natural zeolites are also used as soil amendments (Minato 1968).

Zeolites have been found to be effective in reducing the transfer of fertilizer and added heavy metals, such as Cu, Cd, Pb, and Zn, from soil to plants (Fujimori and Moriya 1973; Sato 1975; Semmens and Seyfarth 1978). The addition of clinoptilolite to soils has been shown to be effective in reducing the uptake of strontium by plants from soils contaminated with Sr^{90} (Nishita and Haug 1972).

Construction: Devitrified volcanic ash and altered tuff have been used for more than 2,000 years as dimension stones. They are lightweight and easily cut and sawed into building blocks (Mumpton 1981). Many buildings associated with the Mayan centers at Mitla and Monte Alban in southern Mexico were constructed of blocks of massive green tuff consisting of 85 to 90% mordenite and clinoptilolite (Mumpton 1973). The extensive Green Tuff deposit in Tochigi Prefecture produces dimension stone that consists of 80 to 85% clinoptilolite with small amounts of montmorillonite, celadonite, and glass (Iijima and Utada 1972). The Neopolitan yellow tuff in central Italy is used primarily as a common dimension stone. Chabazite and phillipsite are the main components of Neopolitan yellow tuff (Norin 1955). Early ranch houses in western America were built with blocks of erionite.

Natural zeolites are used in the manufacture of pozzolantic cements and concrete for highways, aqueducts, and public buildings (Drury 1954; Norin 1955; Mielenz et al. 1951; Alexiev 1968). Because of its high silica content, the zeolite in pozzolans serves as an agent that neutralizes the excess lime produced by setting concrete. Zeolitic pozzolans have important applications in hydraulic cements where the concrete must withstand continuous underwater corrosion (Mumpton 1983).

2. Industrial Uses of Synthetic Zeolites

In comparison with natural zeolites, synthetic zeolites have higher purity and larger pore size. The two major uses of synthetic zeolites are in adsorption, with a capacity of up to 50% by volume, and catalysis, with wide applicability (Derouane 1984; Dyer 1988).

In adsorption, the sodium form of zeolite A (NaA) readily adsorbs materials such as water, carbon dioxide, sulfur dioxide, and all hydrocarbons that contain one or two carbon atoms. When the sodium ions are exchanged by calcium, the effective pore size increases from 4Å to 5Å (CaA). This increase enables the calcium zeolites to adsorb compounds such as straight-chain paraffins, olefins, alcohols, and cyclopropane. The exchange of NaA with potassium reduces the pore size to 3Å (KA), which adsorbs no hydrocarbon. For the sodium zeolite X (NaX), all molecules with diameters up to 8Å are adsorbed. The exchange of sodium by calcium in this zeolite reduces the effective pore size by about 1Å (CaX). If more than one type of molecule is able to enter the micropore system, the zeolite adsorbs one preferentially to the others, on the basis of polarity or other interaction effects between the molecules and zeolite. Water is a strongly polar molecule, and it is very selectively adsorbed (Breck 1983). For gases such as oxygen and nitrogen, nitrogen is preferentially adsorbed because of the interaction of the quadrupole of the nitrogen molecule with the cations of the zeolite. The diameter of the cross section of the dumbbell-shaped nitrogen is 3.0Å as compared with 2.8Å for oxygen (Barrer 1984b).

KA is the preferred adsorbant for drying cracked gas. With its pore size of 3Å, it excludes all hydrocarbons, including ethylene and other olefins. NaA is used to dry refrigerants because the refrigerant molecules are not adsorbed by zeolite. NaA is also used for the prepurification of natural gas because it removes impurities such as H_2O and CO_2, and excludes hydrocarbons the size of propane and larger. CaA is used to separate and recover normal paraffin hydrocarbons from various hydrocarbon feedstocks such as natural gasoline and kerosene (Dyer 1988).

The best known use for zeolite in catalysis is in hydrocarbon reactions, especially for cracking (Rabo and Poutsma 1971). The rate of acid-catalytic reactions depends on the rate of carbonium ion formation. A carbonium ion can be formed in a number of ways, including abstraction by an olefin of a proton from a Bronsted acid. The presence of polyvalent cations, in particular the rare earths, enhances the acidity of the hydrogen in the hydroxyl group by an inductive effect (Breck 1983). In the industrial catalytic process, the acidic Y zeolite-based catalysts show certain features not found with earlier used acid catalysts. The activity of zeolite Y catalysts is much higher relative to silica alumina gel, and zeolite catalysts produce substantially larger gasoline yields. The composition of the gasoline produced by Y zeolites contains many fewer olefins and substantially more aromatics relative to gasoline with silica alumina (Bhatia 1989). Molecular sieves catalyze organic reactions such as dehydration and polymerization. An alcohol may be dehydrated with a zeolite catalyst to form an olefin. The activity of a zeolite catalyst depends on the cation size and charge. Higher activity is found in zeolite with a higher charge and, therefore, a higher electrostatic field strength. Polymerization is an important reaction since an olefin may polymerize within the pore structure and plug the pores (Venuto and Landis 1968; Dyer 1988).

The ZSM-5 catalyst is considered to be the most versatile of all the shape-selective catalysts (Meisel 1984). It has a catalytic activity that may be tailored to a given

Table 26 Applications and potential applications of molecular sieve catalysts after Breck (1983).

Process	Feed/Products	Competitive Features
Catalytic reforming	Naphtha/gasoline	Activators unnecessary; gasoline with reduced sensitivity; feed pre-treatment minimized.
Polymerization	Low molecular weight olefins/gasoline	Noncorrosive.
Alkylation	Aromatic and low-value olefinic streams/valuable alkylated aromatics	Noncorrosive; feed pre-treatment minimized.
Hydrodealkylation	Toluene/benzene	High activity; improved selectivity.
Hydrogenation	Benzene/cyclohexane	Improved resistance to sulfur poisoning.
Hydrogenation of fats, oils	Unsaturated oils/saturated oils	High selectivity; low isomerization.
Selective hydrogenation	Straight and branched olefins/ n-alkanes and branched olefins	Separation problems minimized.
Methanation	Synthesis gas/methane	High yields; resistant to poisons.
Dehydrogenation	Ethylbenzene/styrene	Improved selectivity.
Dehydration	Alcohols + acids/esters	Improved rates and yields.
Dehydrohalogenation	Alkylhalides/olefins	Molecular size; selectivity.

reaction, exceptionally long life, and a shape-selective pore system that has channels that closely match the size of molecules common to numerous petrochemical processes.

A summary of the applications and potential applications of molecular sieve catalysts in terms of process, feed-product, and competitive features is tabulated in Table 26 (Breck 1983).

REFERENCES

Alexiev, B. 1968. Clinoptilolite des Rhodopes du Nord-Est. *Comp. Rendus, Bulgare Akad. Nauk.* 21:1093–1095.

Anurov, S. A., N. V. Keltsev, V. I. Smola, and N. S. Torocheshnikov. 1974. Adsorption of SO_2 by natural zeolites. *Zh. Fiz. Khim.* 48:2124–2125.

Balder, N. A., and L. D. Whitting. 1968. Occurrence and synthesis of soil zeolites. *Amer. Soil. Sci. Soc. Proc.* 32: 235–238.

Barrer, R. M. 1978. *Zeolites and clay minerals as sorbents and molecular sieves.* New York: Academic Press, 499p.

——— 1984a. Zeolite structure. In *Zeolites: Science and technology.* Eds. F. R. Ribeiro, A. E. Rodrigues, L. B. Rollmann, and C. Naccache, *NATO ASI Series E: Applied Sciences, No. 80, part I,* 35–82. Hague Martinus: Nijhoff Publ.

———— 1984b. Sorption by zeolites. In *Zeolites: Science and technology.* Eds. F. R. Ribeiro, A. E. Rodrigues, L. B. Rollmann, and C. Naccache, *NATO ASI Series E: Applied Sciences, No. 80, part II,* 227–290. Hague: Martinus Nijhoff Publ.

Bergerhoff, G., W. H. Baur, and W. Nowacki. 1958. Über die kristallstruktur des faujasite. *Neues Jahr. Mineral. Monatsheft,* 193–198.

Bhatia, S. 1989. *Zeolite catalysis: Principles and applications.* Boca Raton, Florida: CRC Press, 291p.

Boles, J. R. 1981. Zeolites in low-grade metamorphic rocks. In Vol. 4 of *Reviews in mineralogy.* Ed. P. H. Ribbe, 103–136. Washington: Mineral. Soc. Amer.

Breck, D. W. 1974. *Zeolite molecular sieves.* New York: Wiley-Interscience, 771p.

———— 1983. Synthetic zeolites: properties and applications. In *Industrial minerals and rocks.* 5th ed. Ed. S. J. Lefond, 1399–1413. New York: Soc. Mining Engineers, AIME.

Chen, N. Y., W. J. Reagan, G. T. Kokotailo, and L. P. Childs. 1978. A survey of catalytic properties of North American clinoptilolites. In *Natural zeolites; Occurrence, properties, use.* Eds. L. B. Sand and F. A. Mumpton, 411–420. New York: Pergamon Press.

Clifton, R. A. 1987. Natural and synthetic zeolites. *U.S.B.M. Inform. Circ.,* 9140, 21p.

Derouane, E. G. 1984. Molecular shape-selective catalysis by zeolites. In *Zeolites: Science and technology.* Eds. F. R. Ribeiro, A. E. Rodrigues, L. D. Rollmann and C. Naccache, 347–372. *NATO ASI Series E, Applied Sciences, No. 80, part III.* Hague: Martinus Nijhoff Publ.

Dickinson, W. R. 1962. Petrology and diagenesis of Jurassic andesitic strata in central Oregon. *Amer. Jour. Sci.* 260:481–500.

Dickinson, W. R., R. W. Ojakangas, and R. J. Stewart. 1969. Burial metamorphism of the late Mesozoic Great Valley sequence, Cache Creek, California. *Bull. Geol. Soc. Amer.* 80:519–526.

Drury, F. W. 1954. Pozzolans in California. *Mineral Information Serv.* 7:1–6.

Dyer, A. 1988. *An introduction to zeolite molecular sieves.* Chichester: John Wiley & Sons, 149p.

England, D. C. 1975. Effect of zeolite on incidence and severity of scouring and level of performance of pigs during sucking and early postweaning. *17th Swine Day, Spe. Rept.* 447:30–33. Oregon State University: Agr. Exp. Station.

Flanigen, E. M. 1981. Crystal structure and chemistry of natural zeolites. In Vol. 4 of *Reviews in mineralogy.* Ed. F. A. Mumpton, 19–52. Washington, D.C.: Mineral. Soc. Amer.

Fujimori, K., and Y. Moriya. 1973. Removal and treatment of heavy metals in industrial wastewater. I. Neutralizing method and solidification by zeolites. *Asahi Garasukogyo Gijutsu Shoreikai Kenkyu, Hokoku* 23:243–246.

Gilson, J. P. 1991. Organic and inorganic agents in the synthesis of molecular sieves. In *Zeolite microporous solids: Synthesis, structure, and reactivity.* Eds. E. G. Derouane, F. Lemos, C. Naccache, and F. R. Ribeiro. *NATO ASI Series C: Mathematical and Physical Sciences,* v.352, 19–48. Dordrecht Holland: Kluwer Academic Publ.

Grant, D. C., M. C. Skriba, and A. K. Saha. 1987. Removal of radioactive contaminants from West Valley Waste Streams using natural zeolites. *Environmental Progress* 6:104–109.

Grant, D. C., et al. 1988. Design and optimization of a zeolite ion exchange system for treatment of radioactive wastes. *Amer. Inst. Chem. Engineers, Sym. Series* 264:13–22.

Griffiths, J. 1987. Zeolites cleaning up, from the laundry to Three Mile Island. *Ind. Minerals* (January):19–33.

Hawkins, D. B., and H. L. Short. 1965. Equation for the sorption of cesium and strontium on soil and clinoptilolite. *U.S. Atomic Energy Comm.* IDO-12046, 33p.

Hay, R. L. 1963a. Zeolitic weathering in Olduvai Gorge, Tanganika. *Bull. Geol. Soc. Amer.* 74:1281–1286.

———— 1963b. Stratigraphy and zeolitic diagenesis of the John Day Formation of Oregon. *Univ. California Publs. Geol. Sci.* 42:199–262.

———— 1978) Geologic occurrence of zeolites. In *Natural zeolites: Occurrence, properties, use.* Eds. L. B. Sand and F. A. Mumpton, 135–143. Elmsford, New York: Pergamon.

———— 1981. Geology of zeolites in sedimentary rocks. In Vol. 4 of *Reviews in mineralogy.* Ed. F. A. Mumpton, 53–64. Washington, D.C.: Mineral. Soc. Amer.

Hay, R. L., and R. A. Sheppard. 1981. Zeolites in open hydrologic systems. In Vol. 4 of *Reviews in mineralogy.* Ed. F. A. Mumpton, 93–102. Washington, D.C.: Min. Soc. Amer.

Honda, S., and L. J. P. Muffler. 1970. Hydrothermal alteration on core from research drill hole Y-1, Upper Geyser Basin, Yellowstone National Park, Wyoming. *Amer. Mineral.* 55:1714–1737.

Iijima, A. 1980. Geology of natural zeolites and zeolitic rocks. Ed. L. V. C. Rees, 103–118. *Proc. 5th Intern. Conference on Zeolites, Naples, Italy, June 2–6, 1980.*

Iijima, A., and M. Utada. 1972. A critical review on the occurrence of zeolites in sedimentary rocks in Japan. *Japan Jour. Geol. Geogr.* 42:61–83.

Kondo, K., S. Fujishiro, F. Suzuki, T. Taga, H. Morinaga, B. Wagai, and T. Kondo. 1969. Effect of zeolites on calf growth. *Chikusan No Kenikyum* 23:987–988.

Meier, W. M. 1968. Zeolite structures. In *Molecular sieves.* 10–27 London: Soc. Chem. Ind.

Meisel, S. 1984. Multifarious uses of synthetic zeolites. *Eur. Chem. News* 42:22–27.

Mercer, B. W. 1969. Clinoptilolite in water-pollution control. *Ore Bin* 31:209–213.

Mercer, B. W., L. L. Ames, and P. W. Smith 1970a. Cesium purification by zeolite ion exchange. *Nucl. Appl. & Techn.* 8:62–69.

Mercer, B. W., L. L. Ames, C. J. Touhill, W. J. Van Slyke, and R. B. Dean. 1970b. Ammonia removal from secondary effluents by selective ion exchange. *Jour. Water Poll. Contr. Fed.* 42:R95–R107.

Michiels, P., and D. C. E. de Herdt. 1987. *Molecular sieve catalysts.* Oxford: Pergamon Press, 391p.

Mielenz, R. C., K. T. Greene, and N. C. Schielta. 1951. Natural pozzolans for concrete. *Econ. Geol.* 46:311–328.

Minato, H. 1968. Characteristics and uses of natural zeolites. *Koatsugasu* 5:536–547.

Miyashiro, A., and F. Shido. 1970. Progressive metamorphism in zeolite assemblages. *Litho* 3:251–260.

Mori, Y., Y. Edo, H. Toryu, and T. Ito. 1974 Effect of the particle size and application rate of carrier on the herbicidal effect and the growth rate of dry-seeded rice. *Zasso Ken Kyu* 18:21–26.

Mumpton, F. A. 1973. First reported occurrence of zeolites in sedimentary rocks of Mexico. *Amer. Mineral.* 58:287–290.

———— 1978. Natural zeolites: A new industrial mineral commodity. In *Natural zeolites: Occurrence, properties, use.* Eds. L. B. Sand and F. A. Mumpton. Elmford, New York: Pergamon.

———— 1981. Natural zeolites. In Vol. 4 of *Reviews in mineralogy.* Ed. F. A. Mumpton, 1–18. Washington, D.C.: Mineral. Soc. Amer.

———— 1983. Zeolites: Commercial utilization of natural zeolites. In *Industrial minerals and rocks.* 5th ed. Ed. S. J. Lefond, 1418–1431. New York: SME-AIME.

Mumpton, F. A., and P. H. Fishman. 1977 The application of natural zeolites in animal science and aquaculture. *Jour. Animal Sci.* 4:1188–1293.

Mumpton, F. A., and R. A. Sheppard. 1974. Natural zeolites: Their properties, occurrences and uses. *Minerals Sci. Eng.* 6:19–34.

Nelson, J. L., and B. W. Mercer. 1963. Ion exchange separation of cesium from alkaline waste supernatant solutions. *U. S. Atomic Energy Comm. DOC-HY-76449.*

Newsan, J. M. 1992. Diffraction studies of zeolite. In *Zeolite microporous solids: Synthesis, structure and reactivity.* Eds. E. G. Derouane, I. Lemos, C. Naccache, and F. R. Ribeiro. *NATO ASI Series C: Mathematical and Physical Sci.,* 352: 167–192. Dordrecht: Kluwer Academic Publ.

Nishita, H., and R. M. Haug. 1972. Influence of clinoptilolite on Sr^{90} and Cs^{137} uptakes by plants. *Soil Sci.* 114:149–157.

Norin, E. 1955. The mineral composition of the Neopolitan yellow tuff. *Geologische Rundschau* 43:526–534.

Olson, R. H. 1983. Zeolites. In *Industrial minerals and rocks.* 5th ed. Ed. S. J. Lefond, 1391–1431. New York: Soc. Mining Engineers, AIME.

Rabo, J. A., and M. L. Poutsma. 1971. Structural aspects of catalysis with zeolites: cracking of Cumene and Hexane. In *Molecular sieve zeolites II. Adv. in Chemistry Series* 132:284–314. Amer. Chem. Soc.

Raux, A., A. A. Huang, Y. H. Ma, and I. Swiebel. 1973. Sulfur dioxide adsorption on mordenite. *Amer. Inst. Chem. Engr. Sym. Ser., Gas Purification by Adsorption,* 46–53.

Sato, I. 1975. Adsorption of heavy metal ions on zeolite tuff from the Dirik area, Imagane-cho, Oshima Province, Hakkaido-Fundamental Experiment. *Chika Shigen Chosajo Hokoku* 47:63–66.

Seki, Y., Y. Oki, T. Matsuda, E. Mikami, and K. Okumura. 1969. Metamorphism in the Tanzawa Mountain, central Japan. *Jour. Japan. Assoc. Mineral. Petrol. Econ. Geol.* 61:1–75.

Semmons, M. J., and M. Seyfarth. 1978. The selectivity of clinoptilolite for certain heavy metals. In *Natural zeolites: Occurrence, properties, use.* Eds. L. B. Sand and F. A. Mumpton, 517–526. Elmsford, New York: Pergamon.

Sheppard, R. A., and A. J. Gude, 3rd. 1968. Distribution and genesis of authigenic silicate minerals in tuffs of pleistocene Lake Tecopa, Inyo County, California. *U.S.G.S. Prof. Paper* 597, 38p.

Steiner, A. 1953. Hydrothermal rock alteration at Wairakei, New Zealand. *Econ. Geol.* 48:1–13.

Stewart, R. J., and R. J. Page. 1974. Zeolite facies metamorphism of the late Cretaceous Nanaimo Group, Vancouver Island and Gulf Islands, British Columbia. *Can. Jour. Earth Sci.* 11:280–284.

Surdam, R. C. 1973. Low-grade metamorphism of tuffaceous rocks in the Karmutsen Group, Vancouver Island, British Columbia. *Bull. Geol, Soc. Amer.* 84:1911–1922.

Surdam, R. C., and R. A. Sheppard. 1978. Zeolites in saline, alkaline lake deposits. In *Natural zeolites: Occurrence, properties, use.* Eds. L. B. Sand and F. A. Mumpton. New York: Pergamon.

Van der Waal, J. C. 1998. *Synthesis, characterization and catalystic application of zeolite titanium beta.* Delft, Netherland: Delft University Press, 286p.

Venuto, P. B., and P. S. Landis. 1968. Organic catalysis over crystalline aluminosilicates. *Advances in Catalysis,* v.18, 259–371.

Zircon

As an industrial mineral, zircon is characterized by its physical and chemical properties. Physically, with its high melting point and low thermal expansion, zircon is widely used in refractories and as foundry sands. Chemically, the zirconia, which is produced from zircon by decomposition, is used in a wide range of industrial applications.

The world market for zirconium mineral concentrates has grown in recent years. In 1995, the market was estimated to have reached a total of 900,000 tons. Australia contributed more than 50% of this total. South Africa, the United States, India, and the former USSR contributed the remaining 50%.

MINERALOGICAL PROPERTIES

Zircon, $ZrSiO_4$, always contains a certain amount of hafnium, with the Hf:Zr ratio of 0.02:0.04. A complete series of solutions exist between $ZrSiO_4$ and $HfSiO_4$ (Speer 1982). Phosphorus may be present in zircon, probably replacing silicon, and the structure maintains electrostatic neutrality by accepting rare earths, especially yttrium. Many zircons contain appreciable amounts of uranium and thorium (Deer et al. 1992).

Zircon is tetragonal with a = 6.61Å and c = 5.99Å. In the structure, chains of alternating edge-sharing SiO_4 tetrahedra and ZrO_8 triangular dodecahedra extend parallel to the c-axis and are joined laterally by edge-sharing dodecahedra. These chains are the dominant feature of the zircon structure and are responsible for its prismatic habit and extreme birefringence. Similar chains exist in garnet, but they are cross-linked by octahedra as well as by dodecahedra. In zircon, octahedral voids are present, but contain no cations (Robinson et al. 1971). Under high pressure, the zircon-type $ZrSiO_4$ transforms to the scheelite-type structure with a = 4.73Å and c = 10.48Å, and density increases by 11% (Reid and Ringwood 1969).

Zircon undergoes metamictization by the radioactive decay of thorium and uranium that are contained in the structure (Crawford 1965; Kresten 1970). The intensity of radiation necessary to convert crystalline zircon into an optically amorphous zircon is 10^{15}-10^{16} α/mg (Woodhead et al. 1991). Holland and Gottfried (1955) calculated that zircon appears to be amorphous to X-rays when 20 to 30% of the atoms are displaced. Metamictization causes changes in the physical properties of zircon.

Zircon is normally colorless, but may also be green, blue, red, orange, yellow, brown, and grey. The metamict zircon is typically leaf-green to olive or brownish green in color. The hardness of zircon is 7½ and that of metamict zircons is less. The density of zircon is 4.66 gm/cm^3 and decreases to below 4.0 gm/cm^3 with metamictization. Zircon is classified into two types based on this decrease in density: low zircon (low density) or metamict zircon, and high zircon (high density) or nonmetamict zircon. The refractive indices of zircon are 1.920 to 1.960 (ω) and 1.967 to 2.015 (ε). The refractive index of metamict zircon approaches 1.81 as a limiting value (Holland and Gottfried 1955).

The thermal expansion of zircon, as measured by Subbarao and Gokhale (1968) and Worlton et al. (1972), gave $\alpha_c > \alpha_a$ and a small bulk thermal expansion coefficient

459

of $4 \times 10^{-6}/°C - 5 \times 10^{-6}/°C$ between 25° and 1,300°C. This outstanding property characterizes zircon as ideal for refractories and foundry sands. The thermal conductivity of zircon is 120 ± 10 cal/°C·cm·sec, and of metamict zircon is 23 cal/°C·cm·sec (Crawford 1965). Zircon has the lowest compressibility coefficients of any measured compounds with tetrahedrally coordinated Si (Hazen and Finger 1979). The elastic constants of nonmetamict zircon are ($\times 10^{12}$ dyn/cm^2) $c_{11} = 4.237$, $c_{33} = 4.900$, $c_{14} = 1.136$, $c_{66} = 0.485$, $c_{12} = 0.703$, and $c_{13} = 1.495$. These constants decrease with increasing metamictization (Ôzkan 1976).

ZrO_2 is obtained from zircon by thermal decomposition. Although ZrO_2 occurs in nature as baddeleyite, zircon is its main source. In the system ZrO_2-SiO_2 (Roth et al. 1987), $ZrSiO_4$ decomposes at 1,676°C, slightly below the eutectic temperature between ZrO_2 and SiO_2 at 1,683°C. ZrO_2 has a melting point of 2,690°C and is within the range of 2,050° and 3,050°C, which is considered to be the range for super-refractory oxides (Baumgart 1984). ZrO_2 exists in four polymorphic forms: (1) monoclinic, (2) tetragonal, (3) orthorhombic, and (4) cubic (Whitney 1965; Liu 1980). The monoclinic form transforms to the tetragonal form above 1,100°C, and the transition temperature exhibits a large hysteresis. The tetragonal form transforms at 2,330°C to the cubic form, which is not quenchable. The cubic form, however, can be stabilized at lower temperatures by the introduction of vacancies in the structure. Vacancies are also generated by adding alkaline earth oxides or rare earth oxides to form stable solid solutions. Fully stabilized cubic magnesia-zirconia solid solutions do not withstand thermal cycling, whereas CaO- and Y_2O_3-stabilized ZrO_2 exhibit higher thermal stability (Ruh and Garrett 1967). Partially stabilized zirconia is formed by adding insufficient yttria, calcia, or magnesia to completely stabilize the zirconia, and by slow cooling to produce precipitation of tetragonal zirconia within the grains of stabilized zirconia. Such zirconia has high strength and thermal shock resistance. The orthorhombic ZrO_2 is stable at high pressures (Whitney 1965; Liu 1980).

ZrO_2 has a specific gravity of 5.4 to 5.7 and a hardness of 6.5. Its refractive index, 2.19, is higher than that of zircon. ZrO_2 is inert to most reagents but dissolves slowly in hot concentrated sulfuric acid and concentrated hydrofluoric acid. ZrO_2 reacts with many metal oxides at high temperatures.

GEOLOGICAL OCCURRENCE AND DISTRIBUTION OF DEPOSITS

Zircon is a common accessory mineral of igneous rocks, particularly in siliceous and alkaline plutonic rocks, including granite, granodiorite, syenite, monzonite, and pegmatite. Economic deposits of zircon, however, are of the placer type, which formed as a result of long-term weathering of the igneous rocks and is distributed in river beds and ocean beaches. As a member of the heavy mineral sands, zircon is always found in association with rutile and ilmenite.

Major zircon deposits are distributed on the east and west coasts of Australia, on the Trail Ridge in Florida (the United States), at Richard Bay in South Africa, and in Kerala State, India. Zircon is obtained as a coproduct of rutile and ilmenite from all deposits. Australia leads the world in zircon production, with 50% of the estimated annual output of 900,000 tons for 1995. Other major producers include the United States, South Africa, India, Sri Lanka, Brazil, and the former USSR.

The east Australian heavy mineral deposits are of both Holocene and Pleistocene age, and are composed of both beach and eolian sands. As shown in Fig. 95a (p. 411), the east coast from near Sydney, New South Wales, northward to Fraser Island, Queensland, is a continuous sand beach 1,200 km long and 3 to 12 km wide. It is plastered against a continental margin of large sedimentary basins and Paleozoic foldbelts. Similar mineralogy occurs in both Holocene and Pleistocene deposits. Rutile is a major component of the heavy mineral assemblages in the southern region, whereas ilmenite is predominant in the heavy mineral assemblages in the northern region. Zircon occurs in the deposits of both regions (Connah 1961). In western Australia, heavy mineral deposits occur in two districts: the Geographe Bay and the Eneabba (Baxter 1977). Three separate strand lines, ranging in age from the early Pleistocene to the Holocene, are the sites of heavy mineral deposits in the Geographe Bay district (Collins et al. 1986; Collins and Hamilton 1986). These strand lines are the present beach, the Caper line, and the Yoganup line, with the Yoganup line, 16 km inland, being the oldest (Fig. 95b) (p. 411). All the heavy mineral deposits in the district ranging from 56 to 95% of the heavy minerals, 2 to 16% of the zircon, and 0.5 to 2% of the rutile are dominated by ilmenite. The Eneabba deposits consist of seven high-grade layers, each one or more meters thick. The platforms that contain these layers form steps from 130 meters down to 29 meters. Layers in the upper terrace from 130 meters to 100 meters consist of 28 to 46% ilmenite, 5 to 8% rutile, and 36 to 61% zircon, whereas layers in the lower terrace have 53 to 68% ilmenite, 8 to 11% rutile, and 15 to 23% zircon (Force 1991).

The major producing deposits of zircon in the United States are in the Jacksonville district, as shown in Fig. 96 (p. 414). The Trail Ridge deposit is part of a 200 km long sand ridge extending from northern Florida into southeastern Georgia (Force and Rich 1986). The heavy mineral assemblage consists of 50% altered ilmenite and leucoxene, 15% zircon, 15% staurolite, 5% sillimanite, 5% tourmaline, 3% rutile, and 3% kyanite.

In South Africa, the deposit is in a strip of heavy mineral sands along the modern coast facing the Indian Ocean. It is about 15 km long and 3 km wide. The sand dunes have an average height of 20 meters, but some of them reach a height of 100 meters. The dunes are mostly made of Holocene age eolian sands. Zircon occurs in association with ilmenite, rutile, leucoxene, and monazite (Fockema 1986).

A summary of the chemical data for zircon sands is given in Table 27 (Nielsen et al. 1984).

INDUSTRIAL PROCESSES AND USES

1. Milling and Beneficiation of Zircon

The treatment of heavy mineral sands was discussed in the section on titanium minerals (p. 410–412), and a flow sheet for processing was given in Fig. 97 (p. 416). The dredged beach sands are first treated in a high-capacity gravity concentrator. This wet concentration process produces a tailing for rejection, a middling for recirculation, and a rougher concentrate. This

Table 27 Chemical data of zircon sands (in wt%), after Nielsen et al. (1984).

Assay	Sri Lanka	India	Nigeria	Republic of South Africa	United States, Florida	Australia East	Australia West
constituents							
$(Zr+Hf)O_2$	64.9	64.4	58.2	65.0	65.3	65.7	64.6
Fe_2O_3	0.16	0.23	0.90	0.18	0.07	0.06	0.28
Al_2O_3	0.20	0.60	0.65	0.15	0.17	0.14	0.25
TiO_2	0.71	0.16	0.05	0.08	0.10	0.17	0.32
P_2O_5	0.18	0.13	0.39	0.13	0.05	0.07	0.08
U_3O_8	0.03	0.04	0.11	0.03	0.03	0.03	0.03
Nb_2O_5	<0.01	<0.01	2.0	<0.01	<0.01	<0.01	<0.01
Hf/Hf+Zr	2.2	2.3	6.8	2.3	2.2	2.2	2.2
size, μm (mesh)							
>149 (+100)	0.1	41.8	97.2	5.6	0.4	83.8	20.2
<74–149 (−100–200)	40.7	48.6	2.7	92.7	91.7	14.2	77.9
<74 (−200)	59.2	9.6	0.1	1.7	7.9	2.0	1.9

rougher concentrate is upgraded in cleaner stages to produce a bulk heavy mineral concentrate. This bulk concentrate is passed through a low-intensity wet drum magnetic separator to remove magnetite. The nonmagnetics from this stage are treated with high-gradient magnetic separator to remove feebly magnetic material. The resulting concentrate is dried and fed to the dry mill where high-tension roll separators and induced magnetic roll separators are used to produce zircon and other heavy minerals.

2. Industrial Grades of Zircon

The industrial zircons are generally classified by chemical guidelines into premium and standard grades, although an intermediate grade has been adopted by many users (Clarke 1986). The premium grades from the Australian east coast are exemplified by the products from RZ Mines at Tomago, near Newcastle, New South Wales. A comparison of (1) premium grade, (2) standard grade, (3) grades from Australian west coast, and (4) typical RZM premium grade is given below:

	1	2	3	4
ZrO_2+HfO_2 min%	66	65	65 to 66	66.6
TiO_2 max%	0.1	0.25	0.1 to 0.25	0.08
Fe_2O_3 max%	0.05	0.15	0.05 to 0.15	0.025

Specifications of zircon from the United States and South Africa are given below:

	United States Premium	United States Standard	South Africa Premium	South Africa Standard
ZrO_2+HfO_2 min%	65 to 66	65	65	65
TiO_2 max%	0.15	0.35	0.18	0.3
Fe_2O_3 max%	0.05	0.05	0.1	0.3

Zircons from Brazil, India, Malaysia, and Sri Lanka have a minimum ZrO_2+HfO_2 content, in the range of 65.5 to 64.5, maximum TiO_2 content of 0.1 to 0.5, and maximum Fe_2O_3 content of 0.05 to 0.3.

3. Industrial Uses of Zircon

Zircon has a variety of industrial uses, including foundry applications, investment casting, basic steel production, refractories, abrasives, and porcelain glazes.

Foundry Applications: Among the common foundry sands, zircon is superior to silica, chromite, and olivine, because of the following properties:

1. low thermal expansion and good resistance to thermal shock,
2. high thermal conductivity, high bulk density, and a uniform rate of expansion that gives good chilling properties,
3. unwetted by molten metal, which resists to metal and slag attack,
4. chemically nonreactive with metals,
5. compatible with all binder systems and uses only low binder levels to achieve high bond strength,
6. clean rounded grains from many deposits that readily accept any binder,
7. superior dimensional and thermal stability at elevated temperatures, and
8. neutral or slightly acidic pH (Taylor et al. 1959; Garnar 1983; Clarke 1986; Griffiths 1990).

Zircon sand is used in foundries for shell cores and mold linings. It is also used to make filters for strainer cores at the bottom of the downsprue. Chemical impurities in zircon sand are not critical for foundry applications, as long as the impurities are refractory oxides such as TiO_2 or Al_2O_3. The standard grades satisfy performance requirements for foundry applications.

Investment Casting: Zircon sand is used extensively in casting titanium and superalloys for aircraft parts. In a sand mold, the rate at which a casting freezes is determined primarily by the ability of the mold to accept heat from the cast. The freezing rate has an important bearing on the ability to eliminate solidification shrinkage in a casting. The thermal properties of zircon are suitable for these functions (Taylor et al. 1959). Zircon flour and colloidal silica usually make up the primary coats of investment casting moulds, and zircon sand is often used as a stucco grain. For this usage, stringent specifications are placed upon zircon sand and flour for its contents of trace elements such as Zn, Pb, Bi, and Sn. Their maximum contents must be below 100 ppm range (Clarke 1986).

Basic Steel Production: Zircon sand is used in basic steel production in ladle brick because it minimizes erosion, which extends idle lining life. It is also used in coatings, mortars, and as ladle nozzle fill. For this application, zircon of an intermediate particle size of 110 to 150 microns is the most desirable, and chemical composition is less critical (Garnar 1983).

Refractories: Zircon sand is used to make refractory bricks for lining kilns and glass melting furnaces. It is also used to manufacture ramming mixes and refractory cements (Garnar 1986).

Abrasives: Zircon sand is excellent for blasting abrasives, but its cost has prevented its wide use. Zircon sand is treated at high temperatures to produce ZrO_2 and, in combination with Al_2O_3 and other compounds, to make abrasive products (Garnar 1986).

Porcelain Glazes: Zircon is an efficient and economic opacifier. The opacity may be achieved by incorporating zircon flour either as a mill addition, as a frit component, or by a combination of both. Zircon has a higher solubility in glazes than tin oxide and therefore greater amounts of zircon opacifier are required to match the opacity of a tin oxide glaze. This is offset by an advantageous price structure (Taylor and Bull 1986).

4. Decomposition of Zircon

There are two processes for the production of zirconia. In the plasma arc process, zircon sand is injected into the plasma arc where it is melted and dissociated into zirconia and silica. On leaving the plasma flame the molten particles assume a spherical shape and are cooled rapidly as they pass down the furnace chamber. Solidification occurs with the formation of extremely small zirconia crystallites in an amorphous silica matrix. The reaction product is then treated with boiling aqueous caustic soda to remove the silica, as sodium metasilicate, leaving partly porous spheres of zirconia that contain low levels of silica (Evans and Williamson 1977).

The alternative is to treat zircon with alkali or alkaline earth oxides. Sodium hydroxide reacts with zircon at around 600°C to produce a mixture of sodium zirconate (Na_2ZrO_3), sodium silicate (Na_4SiO_4), and sodium zirconium silicate (Na_2ZrSiO_5). A complete conversion to sodium zirconate and sodium silicate can be accomplished with careful control of the zircon:NaOH ratio. Sodium carbonate requires a higher temperature (1,000°C) to decompose zircon. At mole ratios of $Na_2CO_3:ZrSiO_4 = 1:1$ sodium zirconium silicate is produced, whereas at higher ratios sodium zirconate and sodium silicate are produced (Chukhlantsev et al. 1968):

$$ZrO_2 \cdot SiO_2 + Na_2CO_3 \rightarrow Na_2ZrSiO_5 + CO_2$$

$$Na_2ZrSiO_5 + Na_2CO_3 \rightarrow Na_2ZrO_3 + Na_2SiO_3 + CO_2$$

In these processes the silica is leached from the decomposition products by washing it with water or by treating it with acids. Sodium zirconate is hydrolyzed by water to a hydrous zirconia that is dissolved in HCl or H_2SO_4 for further purification.

Lime or dolomite can be reacted with zircon to produce calcium zirconium silicate, calcium silicate, and calcium zirconate, or mixtures of zirconium dioxide and calcium or magnesium silicate. The end products depend upon the ratio of calcium oxide to zircon and the temperature:

$$CaO + ZrO_2 \cdot SiO_2 \rightarrow CaZrSiO_5 \text{ at } 1,100°C$$

$$2CaO + ZrO_2 \cdot SiO_2 \rightarrow ZrO_2 + Ca_2SiO_4 \text{ at } 1,600°C$$

The second process is generally used because it leads directly to the production of ZrO_2. The decomposition product is leached with HCl to remove the calcium silicate, and the remaining crystalline zirconia is dried and stored (Baldwin 1970).

5. Industrial Uses of Zirconia

Zirconia is used in pigments, refractories, electrical ceramics, solid electrolytes, catalysts, abrasives, glass, and gemstones.

Pigments: Ceramic colors are produced by treating ZrO_2 + SiO_2 with V_2O_5, Pr_2O_3, or Fe_2O_3 in the presence of alkali halides at 850° to 1,000°C under oxidizing conditions (Eppler 1977; Wildblood 1977; Bell 1978). Zircon is formed that has the coloring metal ion incorporated into the structure, thus producing vanadium blue, praseodymium yellow, or iron pink stains. The halides serve as transport agents in the promotion of the reaction.

Refractories: In refractories, neither the monoclinic nor the stabilized cubic form has practical applications. The former has a low thermal conductivity and a volume change on heating to 1,000° to 1,100°C, whereas the latter possesses a high coefficient of thermal expansion that militates against high thermal shock resistance. Partially stabilized zirconia is the form that is used as dies for the hot extrusion of metals, as metal cutting tools, and as tappet facings in high-power diesel engines. In order to be used for industrial applications, zirconia must have high purity, typically 99.9% ZrO_2 + HfO_2, <0.1% SiO_2 and <0.02% Fe_2O_3 (Baldwin 1970).

Adding zirconia to alumina- and mullite-based refractories improves their fracture toughness. This addition is achieved by developing microcracks in the ceramic body, which result from the volume increase associated with the tetragonal-monoclinic phase transition that occurs as the ceramic body is cooled from the sintering temperature (Emiliano and Segadaes 1989; Lucchini and Maschio 1989).

Electrical Ceramics: Zirconia is widely used to produce a piezoelectric effect. This effect results in the development of an electric charge when zirconia is stressed and is due to the displacement of ions within the structure (Baldwin 1970). The most important piezoelectric ceramics are the perovskite-type lead zirconate-titanate $Pb(Zr_xTi_{1-x})O_3$. They are used for sonar devices, strain gauges, ultrasonic cleaners, and spark ignitors. They are prepared by treating a mixture of PbO, ZrO_2, and TiO_2 by solid-state diffusion.

Solid Electrolytes: Cations such as Ca^{2+}, Mg^{2+}, and Y^{3+} are introduced into the monoclinic structure in order to stabilize the cubic form of ZrO_2. This results in the formation of anion vacancies, which are necessary to maintain charge neutrality and account for oxygen conduction at about 1,000°C. The defect structures of solid solutions containing divalent cations have a general formula of $Zr_{1-x}M_x^{2+}O_{2-x}$, and those containing trivalent cations have a formula of $Zr_{1-2x}M_{2x}^{3+}O_{2-x}$ (Hagenmuller and Von Gool 1978). Applications are as oxygen sensors in catalytic convertors, probe analyzers for oxygen content in liquid metals, and solid electrolytes in high temperature fuel cells (Fleming 1977).

Other uses of zirconia are in catalysis, as an abrasive, and in the production of gemstones.

REFERENCES

Baldwin, W. J. 1970. Zircon and zirconates. In part II *High temperature oxides.* Ed. A. M. Eppler, 162–184. New York: Academic Press.

Baumgart, W. 1984. Refractories. In *Process mineralogy of ceramic materials.* Eds. W. Baumgart, A. C. Dunham, and G. C. Amstutz, 81–102. New York: Elsevier.

Baxter, J. 1977. Heavy mineral sand deposits of western Australia. *Geol. Surv. Western Australia Mineral. Res. Bull.,* 10, 148p.

Bell, B. T. 1978. The development of colorants for ceramics. *Rev. Prog. Color. Relat. Top.* 9:48–57.

Chukhlantsev, V. C., Yu. M. Pelezhaev, and K. V. Alyamovskaya. 1968. Reaction of zirconium silicate with sodium carbonate. *Izv. Akad. Nauk SSSR, Neorg. Mater.* 4:745–750.

Clarke, G. 1986. Zircon, in demand as availability squeezed. In *Raw Materials for the Refractories Industry. IM Refractories Survey,* Ed. E. M. Dickson, p. 116–125, Metal Bull. Journals, London.

Collins, L. B., and N. T. M. Hamilton. 1986. Stratigraphic evolution and heavy-mineral accumulation in the Minninup shoreline, southwest Australia, in "Australia, A world source of ilmenite, rutile, monazite, and zircon," p. 17–22, Sym. Series, Australasian Inst. Mining Metall., Sept.–Oct., 1986, Perth, Australia.

Collins, L. B., B. Hochwimmer, and J. L. Baxter. 1986. Depositional facies and mineral deposits of the Ypganup shoreline, southern Perth Basin, in "Australia, A world source of ilmenite, rutile, monazite, and zircon," p. 9–16., Sym. Series, Australasian Inst. Mining Metall., Sept.–Oct., 1986, Perth, Australia.

Connah, T. H. 1961. Beach sand heavy mineral deposits of Queensland. *Geol. Surv. Queenslands Publ.,* 302, 31p.

Crawford, J. H. 1965. Radiation damage in solids: A survey. *Bull. Amer. Ceramic Soc.* 44:963–970.

Deer, W. A., R. A. Howie, and J. Zussman. 1992. *An introduction to the rock-forming silicates.* Essex, England: Longman, 696p.

Emiliano, J. V., and A. M. Segadaes. 1989. Reaction—sintered mullite-zirconia composites—mechanism and properties. In *Zirconia '88, advances in zirconia science and technology.* Eds. S. Meriani and C. Palmonari, 51–66. London: Elsevier.

Eppler, R. A. 1977. Ceramic colorants. In Vol. A5 *Ullmann's encyclopedia of industrial chemistry,* 545–556. Weinheim: VCH Verlagesgellschaft mbH.

Evans, A. M., and J. P. H. Williamson. 1977. Composition and microstructure of dissociated zircon produced by a plasma furnace. *Jour. Mater. Sci.* 12:779–790.

Fleming, W. J. 1977. Physical principles governing nonideal behavior of the zirconium-oxygen sensor. *Jour. Electrochem. Soc.* 124:21–28.

Fockema, P. D. 1986. The heavy mineral deposits north of Richards Bay. In *Mineral deposits of South Africa.* Ed. C. R. Anhaensser, 2301–2307. Johannesburg: Geol. Soc. South Africa.

Force, E. R. 1991. Geology of titanium mineral deposits. *U.S.G.S. Spec. Paper,* 259, 112p.

Force, E. R., and F. J. Rich. 1986. Geologic evolution of Trail Ridge eolian heavy mineral sand and underlying peat, northern Florida. *U.S.G.S. Prof. Paper,* 1499, 16p.

Garnar, T., Jr. 1983. Zirconium and hafnium minerals. *Industrial minerals and rocks.* 5th ed. Ed. S. J. Lefond, 1433–1446. New York: Soc. Mining Engineers, AIME.

Garnar, T., Jr. 1986. Zircon sands of the world and their changing markets, 57–67. *7th Industrial Minerals International Congress, Monaco.* London: Metal Bull.

Griffiths, J. 1990. Minerals in foundry casting. *Ind. Minerals* (May): 39–51.

Hagenmuller, P., and W. Van Gool. 1978. *Solid electrolytes, general principles, characteristics, materials, applications.* London: Academic Press, 345p.

Hazen, R. M., and L. W. Finger. 1979. Crystal structure and compressibility of zircon at high pressure. *Amer. Mineral.* 64:196–201.

Holland, H. D., and D. Gottfried. 1955. The effect of nuclear radiation on the structure of zircon. *Acta Cryst.* 8:291–300.

Kresten, P. 1970. Metamictization of zircon. *Geol. Foeren. Stockhoim Foerh.* 92:110–113.

Liu, L. G. 1980. High pressure phase transformations in baddeleyite and zircon, with geophysical implications. *Earth & Planetary Sci. Letters* 44: 390–396.

Lucchini, E., and S. Maschio. 1989. Alumina-zirconia ceramics-Preparation and properties. In *Zirconia '88, advances in zirconia science and technology.* Eds. S. Meriani and C. Palmonari, 161–170. London: Elsevier.

Nielsen, R. H., J. H. Schuwitz, and N. Nielsen. 1984. Zirconium and hafnium compounds. In Vol. 24 of *Kirk-Othmer encyclopedia of chemical technology,* 3rd. ed., 863–899. New York: John Wiley.

Ôzkan, H. 1976. Effect of nuclear radiation on the elastic moduli of zircon. *Jour. Appl. Phys.* 47:4772–4779.

Reid, A. F., and A. E. Ringwood. 1969. Newly observed high pressure transformations in Mn_3O_4, $CaAl_2O_4$ and $ZrSiO_4$. *Earth & Planet. Sci. Lett.* 6:205–208.

Robinson, K., G. V. Gibbs, and P. H. Ribbe. 1971. The structure of zircon: A comparison with garnet. *Amer. Mineral.* 56:782–790.

Roth, R. S., J. R. Dennis, and H. F. McMurdie, Eds. 1987. *Phase diagrams for ceramics*, v. VI. Westerville, Ohio: Amer. Ceramic Soc.

Ruh, R., and H. Garrett. 1967. Nonstoichiometry of ZrO_2 and its relation to tetragonal-cubic inversion in ZrO_2. *Jour. Amer. Ceramic Soc.* 50: 257–261.

Speer, J. A. 1982. Zircon. In Vol. 5 of *Reviews in mineralogy.* Ed. P. H. Ribbe, 47–112. Washington: Mineral. Soc. Amer.

Subbarao, E. C., and K. V. G. K. Gokhale. 1968. Thermal expansion of zircon. *Jap. Jour. Appl. Phys.* 7:1126–1128.

Taylor, J. R., and A. C. Bull. 1986. *Ceramics glaze technology.* Oxford: Pergamon Press, 263p.

Taylor, H. F., M. C. Flemings, and J. Wulf. 1959. *Foundry engineering.* New York: Wiley, 467p.

Whitney, E. D. 1965. Electrical resistivity and diffusionless phase transformations of zircon at high temperatures and ultrahigh pressures. *Jour. Electrochem. Soc.* 112:91–94.

Wildblood, N. C. 1977. Stable colorants for high temperatures. In High temperature chemistry of inorganic and ceramic materials. Eds. E. P. Glasser and P. E. Potter. *Chem. Soc. Spec. Publ.* 30:12–20.

Worlton, T. G., L. Cartz, A. Niravath, and H. Ôzkan. 1972. Anisotropic thermal expansion and compressibility of zirconium silicate. *High Tem.—High Pres.* 4:463–469.

Woodhead, J. A., G. R. Rossman, and L. T. Silver. 1991. The meta-metamictization of zircon: Radiation dose-dependent structural characteristics. *Amer. Mineral.* 76:74–82.

Index